AF323158

SOYBEAN - BIO-ACTIVE COMPOUNDS

Edited by **Hany A. El-Shemy**

Soybean - Bio-Active Compounds

http://dx.doi.org/10.5772/45866
Edited by Hany A. El-Shemy

Contributors

Mauricio Fonseca, Georgiana Cruz, Luciano Batista, Viviane Silva, Erica Pissurno, Thais Soares, Monique Jesus, Ednilton Tavares De Andrade, Gloria Davila-Ortiz, Alberto Cabrera Orozco, Cristian Jiménez-Martínez, Kenjiro Koga, Oluyemisi Elizabeth Adelakun, Eugene Borodin, Igor Pamirsky, Mikhail Shtarberg, Vladimir Dorovskikh, Alexander Korotkikh, Chie Tarumizu, Kiyoharu Takamatsu, Shigeru Yamamoto, John Bradley Morris, Stefano Tavoletti, Ivan Palomo, Eduardo Fuentes, Luis Guzman, Gilda Carrasco, Elba Leiva, Rodrigo Moore-Carrasco, Neusa Fatima Seibel, Masakazu Naya, Masanao Imai, Sherif M Hassan, Fred Kummerow, Samuel N. Nahashon, Agnes Kilonzo-Nthenge, Katrine Pontoppidan, Dan Pettersson, Anna Lucia Villavicencio, Xiaomeng Li, Ying Xu, Ichiro Tsuji, Enrique Vigueras Santiago, Susana Hernández López, Valeria Abreu, Pilar Teresa Garcia, Joanna McFarlane, Abdolmohamad Rostami, Farinaz Safavi, Praveen V Vadlani, Liyan Chen, Ronald Madl, Weiqun Wang, Li Li, Maria Luisa Brandi

CBS, Edition **2017**

Published by InTech

Janeza Trdine 9, 51000 Rijeka, Croatia

The moral rights of the editor(s) and the author(s) have been asserted.
All rights to the book as a whole are reserved by InTech. The book as a whole (compilation) cannot be reproduced, distributed or used for commercial or non-commercial purposes without InTech's written permission.
Enquiries concerning the use of the book should be directed to InTech's rights and permissions department (permissions@intechopen.com).
Violations are liable to prosecution under the governing Copyright Law.
© The Editor(s) and the Author(s) 2013

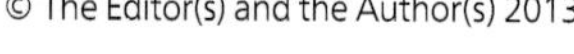

Individual chapters of this publication are distributed under the terms of the Creative Commons Attribution 3.0 Unported License which permits commercial use, distribution and reproduction of the individual chapters, provided the original author(s) and source publication are appropriately acknowledged. More details and guidelines concerning content reuse and adaptation can be found at http://www.intechopen.com/copyright-policy.html.

Notice

Statements and opinions expressed in the chapters are these of the individual contributors and not necessarily those of the editors or publisher. No responsibility is accepted for the accuracy of information contained in the published chapters. The publisher assumes no responsibility for any damage or injury to persons or property arising out of the use of any materials, instructions, methods or ideas contained in the book.

Publishing Process Manager Ana Pantar

Technical Editor InTech DTP team

Cover Designer InTech Design Team

Additional hard copies can be obtained from orders@intechopen.com

Soybean - Bio-Active Compounds, Edited by Hany A. El-Shemy
p. cm.
ISBN 978-953-51-0977-8

Contents

Contents

Contents

Preface

This book provides an overview of the importance of soybean all over the world. The authors contributed with 24 chapters dealing with various topics.

Soybean consumption benefits, especially in several chronic diseases, have been related to its important protein content, high levels of essential fatty acids, vitamins and minerals. Consequently, Chapter 1 provides ideas on Critical Evaluation of Soybean Role in Animal Production Chains Based on the Valorisation of Locally Produced Feedstuff.

We can also learn that new marketing strategies are necessary to make consumers aware of the importance of the overall characteristics of local production chains in defining the quality of a final product and to ensure at the same time a profitable price for the producers.

Chapter 3 brings an overview of the types of equipment used for industrial processing of soybeans for obtaining vegetable oil in small plants and Chapter 4, for example, highlights the international literature which suggests that phytoestrogens have a potentially high clinical impact and the expansion of knowledge on soy, soy foods, and soy products which will lead to novel future developments in the field of cancer treatment.

Other chapters aim at the comprehensive characterization of the antioxidant and antiplatelet activities of bioactive compounds, of soybean and its derivatives, and the extent to which soybean is a health-promoting food.

This book will be useful for soybean researchers and other academic staff and will provide its readers with valuable insight into the last developments in the field.

Hany A. El-Shemy, Professor
Cairo University, EGYPT

Critical Evaluation of Soybean Role in Animal Production Chains Based on the Valorization of Locally Produced Feedstuff

Stefano Tavoletti

Additional information is available at the end of the chapter

http://dx.doi.org/10.5772/52476

1. Introduction

Commodities such as soybean and maize are respectively protein and energy concentrates that represent the basic raw materials used by the animal feeding industry and their prices influence the overall market of agricultural products. In animal feeding, soybean is mostly used as soybean meal which is a by-product of oil seed extraction industry and the availability of this raw material on the international market has led to its worldwide diffusion as the main source of protein for animal feed formulation. Soybean can also be used as raw seed due to its high fat content that makes this grain legume a valid feed to increase both protein and energy concentration of animal diets [1]. The close relationship between oil extraction industry and feed industry together with its high nutritional value for human consumption, as reported by several articles included in the present book, made soybean a perfect crop to be treated as a commodity in the world trade of raw materials.

However, despite all these positive characteristics, the diffusion of soybean and its by-products, together with the overall increased importance of the commodities trade, has triggered in agriculture several downstream effects that had deep consequences on the evolution of agricultural practices. After World War II agriculture initiated a course of progressive structural changes toward the implementation of more intense production processes due to both the need of increasing world food supply and to the progressive reduction of people employed in agriculture. The diffusion of improved varieties, fertilizers, pesticides, advanced agricultural machineries, intensive systems of animal rearing, efficient systems for the storage and transformation of agricultural products led to the abandonment of traditional cropping and animal farming systems [2]. Therefore, important agronomic practices such as crop

© 2013 Tavoletti; licensee InTech. This is a paper distributed under the terms of the Creative Commons Attribution License (http://creativecommons.org/licenses/by/3.0), which permits unrestricted use, distribution, and reproduction in any medium, provided the original work is properly cited.

rotations including a cereals and legumes, cultivation of forage crops for animal feeding, a close link between animal farming and field productions useful to ensure an adequate content of organic matter in the soil were almost completely abandoned due to the diffusion of monocultures [3]. As a consequence, agricultural soil fertility decrease dramatically, as indicated by the dangerous low levels of organic matter content that at present are generally recorded in most countries that have been characterized by such an intensification of agricultural practices, and the use of chemical fertilizers became an indispensable necessity to reach economically valuable productions [4-7]. At the same time animal farming was based on the use of by-products that were available on the market to reduce the costs of production and simplify the animal production system.

At present, globalization together with the rapid economic development actually under way in eastern countries and the world economic crisis are jeopardizing the economic feasibility of many agricultural activities, mainly in the European Union. Those farms that restructured their production processes in order to simply satisfy the demand for raw materials by the food and feed industries and by multiple retailers that manage marketing and commercialization have recently experienced the negative effects of increased costs of production followed by the low prices of agricultural products paid to the farmers. In particular, the increased costs of commodities such as soybean seed and meal that recently happened several times together with the low price of animal products paid to the farmers, mainly in the beef, pork and dairy production chains, made unprofitable the economic activities of animal farms, especially small or medium size farms.

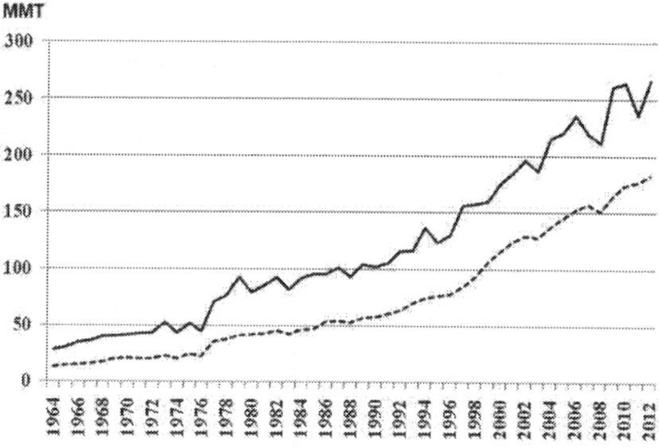

Figure 1. World soybean seed (continuous line) and meal (dashed line) production.

Recently new strategies for agricultural development have emerged due to the interest of consumers toward high quality products and production chains, the increased request of non-standardized food and the attention given to the impact of agricultural activities on the environment, human and animal health [8-10]. Therefore, an increasing number of animal farms adopted more sustainable instead of intensive production systems that were closely linked to the area of production by using locally produced animal feed and reducing their dependence from commodities. Moreover, these farms developed direct commercialization systems trying to make their business more profitable with an emphasis on the quality of both the final products and the production system.

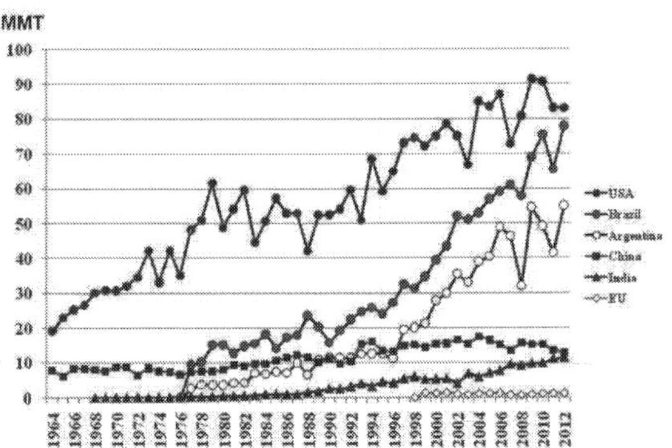

Figure 2. Soybean seed production of USA, Brazil, Argentina, China, India and European Union (EU).

Aim of the present article is to critically evaluate the effects of the large diffusion of soybean on the international market concerning aspects related to the evolution of soybean trade, the effects on agricultural systems where soybean cannot be cultivated, the present dependence from soybean of intensive animal farming systems and the consequences on small or medium farms applying not-intensive animal production chains. An experience under way in the Marche Region (Central Italy) will also be illustrated as an attempt to make agricultural activity both profitable and integrated within local soil and climatic characteristics.

2. Soybean seed and meal trade

Data on soybean seed and meal (oil was not considered) production, import, export and prices were obtained at the Index Mundi website [11] (data source:USDA) where information on all commodities trade is available. Soybean data were usually available from 1964 to 2011 together with estimates referred to the current year 2012, even though for some countries and also for the European Union data availability covered a shorter period of time. Data were organized in an excel data sheet and elaborated to obtain information concerning single countries involved in the soybean international trade. The total worldwide value of seed and meal production, import and export, expressed as Million Metric Tons (MMT), were then calculated by summing the data available for each country for each year. Results were summarized by graphics concerning the overall soybean trade, the characteristics of single countries significantly involved in the international market of soybean and the comparisons among different countries.

Soybean seed production progressively increased from 1964 (28,3 MMT) to 2010 (264,7 MMT) and, although followed by a slight decrement in 2011 (236,4 MMT), soybean seed USDA 2012 estimated production (referred to june 2012) is 266,8 MMT confirming the positive trend for this commodity (Figure 1).

USDA data identified 42 countries characterized by an estimated soybean seed production of at least 0.001 MMT. However since 1964 more than 90% of total soybean seed production was concentrated in 5 countries (USA, Brazil, Argentina, China and India) and USA, Brazil and Argentina covered about 80% of worldwide soybean production (Figure 2).

European Union (Figure 2) produced between 0.6 MMT (year 2008) and 1,4 MMT (year 1999) of soybean seeds and in 2011 EU27 ranked twelfth with a production of 1,1 MMT (0.47% of world production). These data confirm the almost complete dependence of Europe from non-EU and mainly American countries to satisfy the needs of protein concentrates of European animal production chains.

Figure 1 also shows that the production of soybean meal had almost the same trend of world soybean seed production, with a steady increase from 1964 (13,5 MMT) to 2011 (177,4 MMT). However, the relative contribution of each country is different than what has been described for seed production. USA was the highest soybean meal producer until 2009 when it was exceeded by China that, based on the 2012 estimate, at present seems to be the world leader in soybean meal production. China showed a relatively low meal production until 1997 when it started a progressive increase in the production of this by-product of oil extraction (Figure 3).

It is interesting to compare soybean seed and meal amounts produced over time in China; soybean seed production in this country was always lower than 20 MMT, ranging from 6.14 MMT in 1995 to 17.4 MMT in 2004 (Figure 2), whereas soybean meal production was lower than 10 MMT until 1997 but increased from 10 MMT in 1998 to 46.9 MMT in 2011 with an estimated 50.2 MMT for the year 2012 (Figure 3). Starting from 1997 also Brasil and Argentina began to increase their soybean meal production reaching about 28 MMT in 2011, whereas USA, after a constant increase from 1964 to 1996, was characterized by and almost constant meal production of about 35 MMT/year in the 1997-2011 time period (Figure 3).

India, which was the fifth producer of soybean seeds, also showed a constant increase in soybean meal production from 1987 to 2011 although remaining below the 10 MMT level of meal production (Figure 3). On the contrary in the 2001-2011 time period the European Union was characterized by a negative trend of soybean meal production that decreased from 14 MMT in 2001 to less than 10 MMT after 2008 (Figure 3).

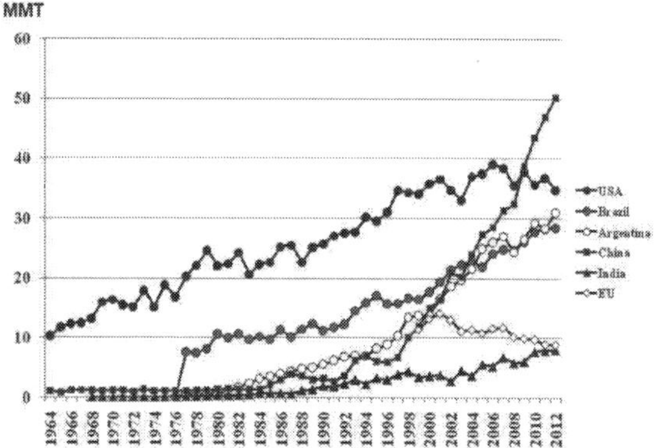

Figure 3. Soybean meal production of USA, Brazil, Argentina, China, India and European Union (EU).

Moreover, in the 1994-2011 time period the relative contribution to the worldwide soybean meal production of Japan decreased from 8% (1994) to 0.82% (2011) and Canada was also characterized by the same negative trend (from 3.12% in 1994 to 0.62% in 2011); this behaviour could be attributed to the low increase in meal production over time that characterized these two countries (from 1.07 to 1.45 MMT for Japan and from 0.42 to 1.10 MMT for Canada) compared to the progressive overall world increase of soybean meal production.

Therefore, the international scenario concerning soybean clearly shows that 3 countries (USA, Brasil and Argentina) handle almost all the world production of soybean seed, whereas China must be added to USA, Brasil and Argentina concerning the production of soybean meal. Conversely European Union has a negligible level of soybean seed production and a low level of soybean meal production as a by-product of oil extraction from imported soybean seeds. This determines that European Union animal production chains rely completely on imported soybean and this situation generates an almost total dependence of European farms from the international trade of these commodities. Data on the import/export of soybean seeds and meal also confirm that EU animal farming is suffering the effects of globalization of the markets, mostly because the dynamics of the international market of soybean is changing as a consequence of the new scenario due to the increased interest toward this commodity by several new countries.

Figure 4 and Figure 5 summarize the world scenario of soybean seed and meal trade, respectively. World seed production increase was also followed by an increase in the amount, expressed as percentage of total world production value, of overall exported seed that was about 25% until 1995, then raised to about 30% between 1996 and 2005 and finally reached the value of 35 % in 2010 with an estimated amount of 36% for year 2012 (Figure 4).

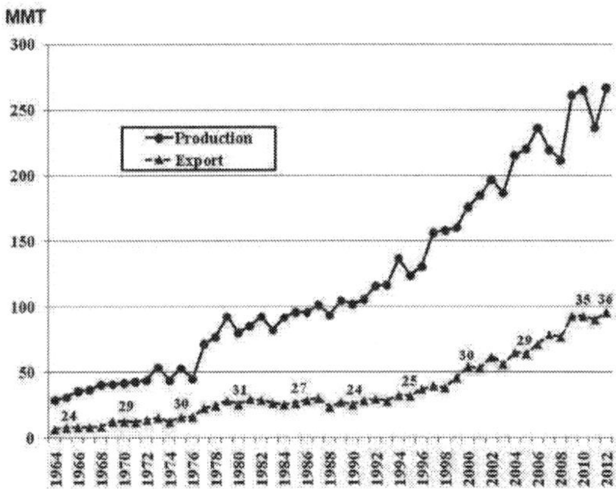

Figure 4. Soybean seed: comparison between world production and export; every five year export values expressed as percentage of total world production are shown (2012 data are estimated values).

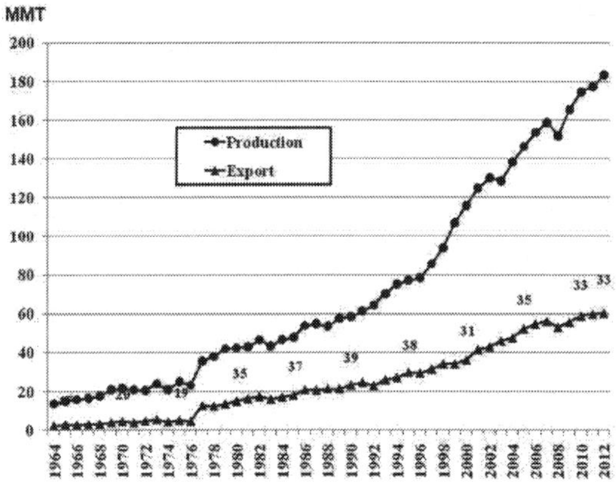

Figure 5. Soybean meal: comparison between world production and export; every five year export values expressed as percentage of total world production are shown (2012 data are estimated values).

A similar trend was shown by global soybean meal production. An increase from 17% to 35% in the percentage of exported meal, expressed as percentage the total world meal production, was registered in the 1965-1980 time period, a further slight increase until 38% was

http://dx.doi.org/10.5772/52476

observed between 1980 and 1995 and subsequently the percentage of exported meal went back to the value of 33% for year 2010, the same value estimated for year 2012 (Figure 5). Therefore, about one third of total world soybean seed and meal production is exported to countries that show a deficit in the internal production of these commodities.

The relative importance of the import/export of soybean seed and meal compared with the internal production is also shown separately for each of the 4 most important countries (USA, China, Brazil and Argentina) together with the situation characterizing the European Union (Figures 6-10).

At present, USA is the highest producer of soybean seed in the world and in 2010 45% of total USA internal production was exported, this level being maintained in 2012 estimate. About 20-25% of internal USA soybean meal has always been exported (Figure 6).

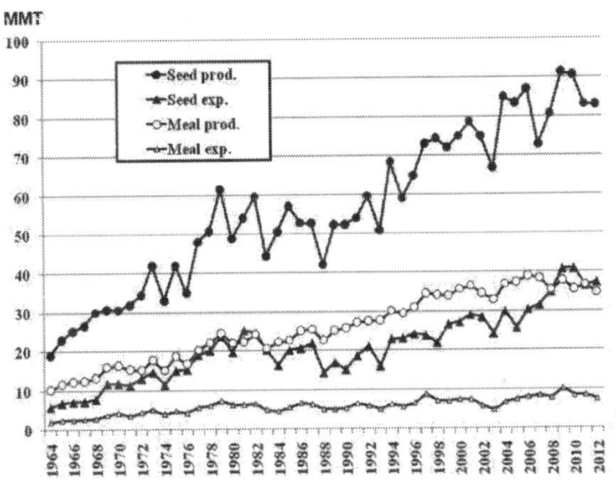

Figure 6. USA: Soybean seed and meal production and export; imported amounts are negligible and therefore are not shown.

Together with USA, also Brazil and Argentina (data available from 1978 to 2012) are characterized by a high amount of internal production which is exported followed by a negligible import of soybean products. In particular, Brazil is getting close to USA levels of production for both soybean seed and meal (Figures 2 and 3) and seed export, that ranged between 12 and 20% in the 1980-1995 time period, steadily increased from 30% in 1996 to 55% in 2011 with an estimate of 45% for year 2012 (Figure 7). A different trend characterized soybean meal Brazilian export that was about 76% from 1978 to 1990 and then decreased to 50% in 2011 with an estimated level of 48% for year 2012. On the whole, about 50% of both seed and meal internal Brazilian productions is exported (Figure 7).

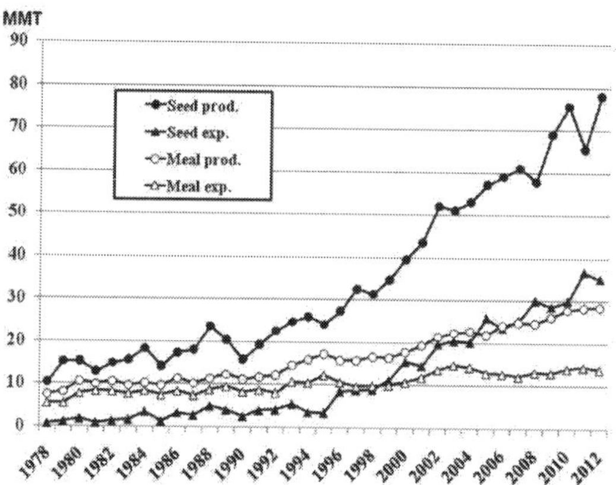

Figure 7. Brazil: Soybean seed and meal production and export; imported amounts are negligible and therefore are not shown.

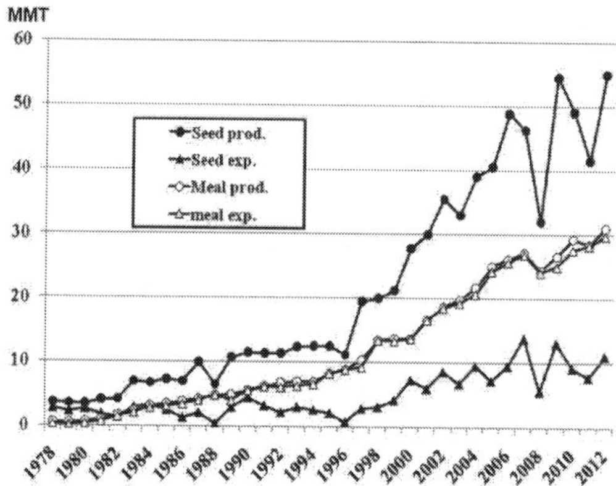

Figure 8. Argentina: Soybean seed and meal production and export; imported amounts are negligible and therefore are not shown.

A large amount of Argentina's seed production in 1978-1980 period was exported, but since 1981 the amount of exported seed dropped drastically to 33-55% until 1985 and thereafter it decreased even more reaching 19% in 2011 with an estimate of 20% in 2012 (Figure 8). On the other end the amount of soybean meal internally produced increased approximately four times and almost 100% of the meal was exported, as clearly shown by Figure 9. Therefore these data suggest that Argentina is essentially growing soybean to export soybean meal.

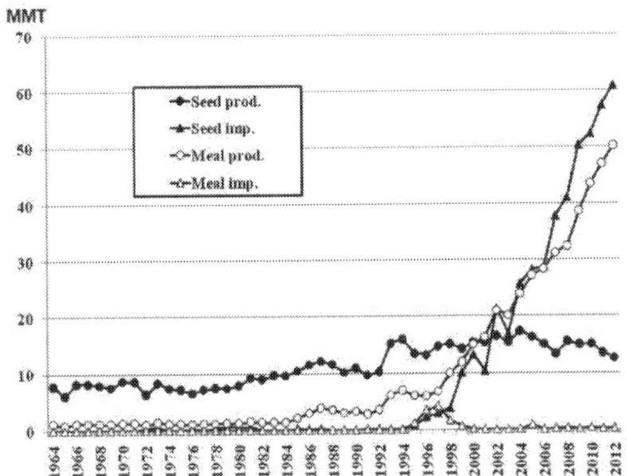

Figure 9. China: soybean seed and meal production and import; exported amounts are negligible and therefore are not shown.

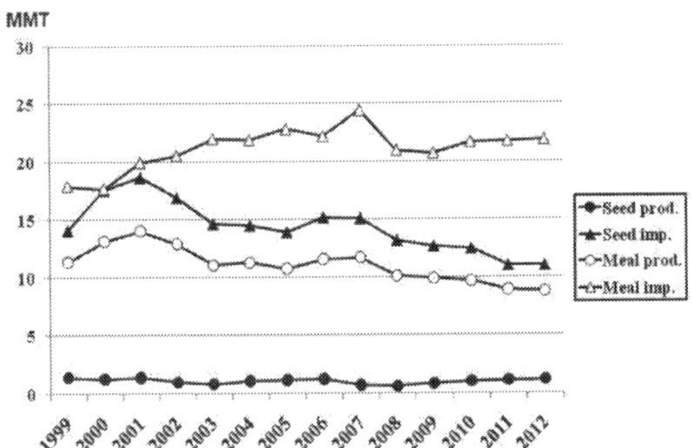

Figure 10. European Union: soybean seed and meal production and import; exported amounts are negligible and therefore are not shown.

On the other hand, an important soybean importer is China (Figure 9) whose inner soybean seed production only slightly increased over time whereas this country has become the largest importer of soybean seed (61 Mt estimated for 2012). All the internal production and import of soybean seed is used for oil extraction and meal by-product production. The large amount of soybean meal produced is almost completely used within the country to support internal animal productions, since China export of meal is negligible. Due to this large volume of import China is getting a predominant role influencing the worldwide exchange of soybean products, sometimes competing with other strong importers such as the European Union (Figure 10). As a matter of fact estimated levels of 2012 EU import of soybean seed and meal are 11 and 21.9 MMT respectively, against an internal production of 1.2 and 8.8 MMT of seed and meal, respectively.

Therefore, costs of production of European animal farms strictly depended on soybean prices that are set by the international market. As shown in Figure 11 monthly price of soybean seeds and meal in the 1983-2011 time period showed a marked change in part due to the trend of world soybean production. In particular, after 2007 seed and meal prices showed a clear average increase compared to the previous years. The unpredictable variation in the price of the basic protein concentrate for animal feeding strongly influenced the incomes of European farmers since it was also related to a general increase in the prices of other commodities and production factors (fertilizers, pesticides, seed, fuels etc.) whereas the prices of animal products to the farmers did not follow the same positive trend.

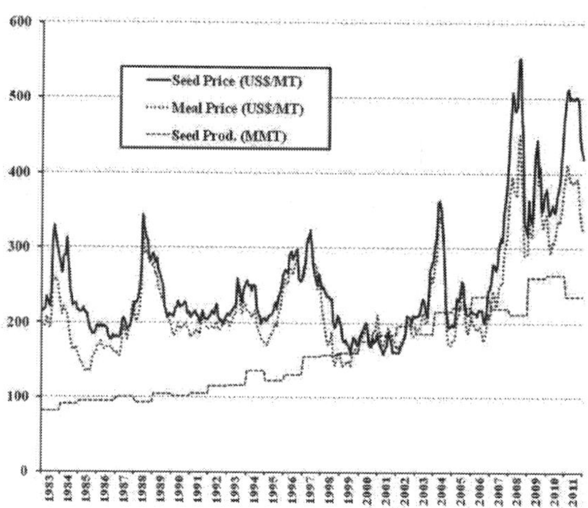

Figure 11. Trend of soybean seed and meal prices (US$ per Metric Ton) compared to the world total seed production (MMT) of each year.

Furthermore, the worldwide diffusion of soybean as protein feed component has almost completely replaced any other protein source for animal feeding and this happened mainly in highly intensive farms trying to maximize animal growth rate, to simplify the management of feedstuff and to standardize meat or milk production systems. The same trend occurred for maize which is currently the most important cereal for animal production chains. Moreover, most of the soybean is produced by using new varieties obtained by genetic engineering [12] whose acceptance by European consumers is strongly debated.

On the whole, globalization of the market of animal products favoured the development of animal farms rearing large numbers of animals determining a high concentration of animals in a reduced number of farms which is typical of intensive farming systems, followed by a progressive reduction in the number of small farms that were not able to compete in such a situation of high costs of production coupled with low prices of products for the farmers.

3. Strategies for partial or total soybean replacement.

Therefore, even though both soybean and maize have extremely interesting nutritional characteristics as feed intended for almost all farm animals, an alternative strategy was necessary to reduce this complete dependence of European animal production chains from the market of the commodities. However, environmental and climatic conditions in most European countries hinder the cultivation of soybeans and maize because they are warm-season crops growing during the spring-summer season making irrigation a fundamental need for the success of these crops. Therefore the limited amount of lands suitable for cultivation of these crops, the high water request making irrigation a fundamental component of the production cost, the low levels of soybean grain production per hectare suggested that attention should have been addressed toward other sources of proteins to reduce dependence on imported soybean.

This is particularly true for European countries characterized by a Mediterranean climate, where rainfall is mainly concentrated during winter until late spring whereas summer is usually characterized by high temperatures coupled with very low and often irregular rainfall. Moreover, the use of water for irrigation is often very expensive and therefore, where it is available, it is devoted to other crops such as orchards and vegetables. As a consequence, in these areas agriculture is essentially based on cool-season crops that could be sown in autumn-end of winter and harvested at the end of spring or during summer. Therefore, these agricultural systems are typically based on rotations between winter cereals and grain legumes such as faba bean (*Vicia faba* var. *minor* L.), field pea (*Pisum sativum* L.), chick pea (*Cicer arietinum* L.) and white lupin (*Lupinus albus* L.). For animal feeding the most commonly used grain legume was faba bean. Recently, genetic selection of field pea new varieties stimulated the diffusion of this crop as protein source for animal feeding. Also chickpea has been proposed as a possible alternative to soybean together with sweet lupin, which is characterized by the highest protein content among the grain legumes alternative to soybean. However, since chickpea is mainly cultivated for human consumption and lupin can be cultivated only in locations with specific soil pH conditions, the main grain legume crops on which to focus attention as possible replacements of soybean are field pea and faba bean.

As a consequence, scientific research was directed to increase knowledge on cultivation, plant breeding and utilization as animal feed of these grain legumes [13-23]. Results showed that in ruminant, monogastric and avian animals at least a partial replacement of soybeans is feasible in intensive animal farming systems, whereas in low input or organic farming systems soybean can be completely replaced by grain legumes that can be grown on farm where soybean cultivation is not feasible [24]. However, despite these encouraging results, grain legume cultivation has suffered a clear reduction in Europe. This trend was related to the development of feed industry that, due to the large quantity of raw materials handled, rely mostly on commodities available on the international market and on the use of industrial protein reach by-products of oil extraction such as rapeseed meal, cottonseed meal, sunflower meal and others.

Therefore, the identification of grain legumes such as faba bean, field pea, chickpea and lupin that could at least partially replace soybean in animal feeding systems and their introduction of in crop rotation systems targeted at supplying animal production chains with locally produced protein concentrates could have several positive effects on European agricultural systems. This set of crops could guarantee, together with forage crops and pastures, the development of animal production chains that were fully integrated with local environmental characteristics. Moreover, the close link between animal farming and field productions supported the maintenance of good soil fertility and organic matter content. Finally, new strategies for the commercialization of final animal products must be undertaken for the full valorisation of the whole production chain.

4. Consequences of simplified cropping systems

The negative trend shown by grain legume diffusion was therefore a consequence of agriculture evolution toward highly specialized intensive production systems that determined the progressive gap between animal productions and field agriculture. This trend led to well-known agricultural and economical drawbacks such as lost of soil fertility, dramatic decrease of soil organic matter content, increased need of inputs (fertilizers and pesticides) to reach the highest agronomic performances, search for high productions to counterbalance the lowering prices of raw materials on both global and local markets. Moreover, a clear separation between farmers that progressively became simple producers of raw materials and industry that managed commercialization and transformation of agricultural products, over time led farmers to lose any possibility of market control. These aspects determined a deep crisis in the agricultural sectors of countries where agriculture was characterized by small or medium sized farms that lost their ability to compete on the market since they were confined to the role of simple low value raw material producers. The diffusion of monoculture, the reduction of forage crops due to the intensive feeding systems mainly based of protein and energy concentrates, the trend toward part-time agriculture, the massive use of chemicals to maximize crop production characterized agriculture for several decades after World War II. As a consequence, market of agricultural products was invaded by standardized products that replaced most of the typical and local productions that previously character-ized agricultural systems strictly integrated with the areas of production.

Recently, the need for a more environmental friendly agriculture together with the increas-ing request by the consumers of high quality products, stimulated farmers to recapture the market of agricultural products [25]. These farms progressively abandoned standardized and intensive agriculture and dedicated to animal and crop productions following the voca-tion of their own local region.

As far as animal production chains are concerned, the reconstitution of local production sys-tems led to the valorisation of animal feeding systems based on locally produced raw mate-rials, both forages and protein (grain legumes) or energy (cereals) concentrates. This allowed these farms to reintroduce rational crop rotations, that were abandoned due to the diffusion of monocultures, by alternately cultivating those cereals and legumes that could also be in-tended for animal feeding. Most of these farms started to commercialize by their own the products of their farms by an action aimed at informing the consumers about all aspects of their production system, receiving economical and professional satisfactions.

This approach led to partly or completely replace soybean in animal diets, to reintroduce forage crops both in field crop rotations and in animal feeding systems, to develop less in-tensive animal farming systems, to stimulate creation of local networks among farmers which could represent a further stimulus for local farms to reintroduce grain legumes and forage crops for animal feeding by making agreements with local animal farms that would withdraw their legume products.

5. Experiences on soybean replacement carried out in the Marche Region (Central Italy)

On the feasibility of local animal production chains a research, funded by the Marche Re-gion (Central Italy), has been carried out by our research group since year 2000 to evaluate the technical and economical possibilities of soybean replacement with grain legumes such as *faba* bean and field pea. At the same time research evaluated the possibility of a total or partial replacement of maize with barley or sorghum grain, since both these crops are valua-ble energy crops for areas characterized by high temperatures and low rainfall during summer, since barley is a cool season cereal whereas sorghum is a worm season but drought resistant cereal crop.

The first objective of the project was to test the agronomic feasibility of both faba bean and field pea in different areas of the Marche Region [26-27]. Very favourable lands, where these crops have been evaluated in optimal agronomic conditions, and more marginal fields were included in field trials. Moreover, several farms including both conventional and or-ganic farms were involved as partners of the project to carry out "on farm" field trials based on large plots that were managed by the farmers themselves. The "on farm" ap-proach allowed the evaluation of the real potential of these crops in the areas under examination, that were mainly located in the inner part of the Marche region were irrigation is not feasible (Figure 12).

Figure 12. Field experimental trial including faba bean and field pea carried out in typical agricultural landscape of inner areas of the Marche Region (Province of Ancona).

Results showed that both faba beans and field pea could be effectively reintroduced in crop rotations with winter cereals such as wheat and barley, reconstituting a correct alternation between nitrogen fixing legumes and cereals. However, the grain productions obviously varied based on the environmental and soil conditions and the seasonal climatic conditions that varied from year to year. This experience allowed the creation of a useful data set indicating that faba beans showed, on the average, a range between 1.0 and 3.5 tons/hectare of grain, the lowest productions being obtained when very dry growing seasons occurred with extremely low values of rainfall during end of winter and spring. Farmers know very well these characteristics of faba beans since this crop has been traditionally used across the whole region mainly as protein grain for beef cattle of the Marchigiana breed. However, plant breeding efforts are requested to stabilize grain production in the variable environmental conditions characterizing the inner areas of Central Italy.

Field pea was characterized by a higher average seed production than faba beans, showing a range between 1.5 and 4.5 tons/hectare. For this crop low production can be due to environmental adverse conditions but also grain loss due to seed shattering during harvesting is a primary cause of production losses mainly when the crop is grown on soils with an irregular surface that makes threshing difficult.

To compare faba bean and field pea with soybean, few farms where irrigation was feasible were asked to try cultivation of soybean. The results showed an average seed production between 2.5 and 3 tons/hectare, similar to a good faba bean or field pea harvest, but costs of irrigation made this crop unprofitable. Moreover, chemical weed control was necessary in order to obtain acceptable seed production because watering also favoured the development of weeds. Herbicides are used also to protect faba beans and field pea against weed competition. However, our field trials carried out in organic farms showed that multi-year rotations including forage crops such as alfalfa, highly competitive cereals against weeds such as barley or wheat, and a higher crop density (number of plants/m^2) than conventional cultivation can avoid the use of herbicides on crops such as faba beans and field pea.

Results of the field trials showed that these grain legumes can be effectively produced in inner areas of the Marche Region where soybean cannot be cultivated and animal farming is traditionally an economic source of income for local farmers. After gathering information on the agronomic feasibility, research has been addressed to verify the possibility of total or partial substitution of soybean with faba beans and/or field pea in beef cattle, dairy cattle and swine feeding systems. Therefore, feeding trials were conducted in one large dairy farm, located in the Province of Ancona, 4 organic farms rearing beef cattle of the Marchigiana breed and located in the Provinces of Macerata and Fermo, one conventional farm located in the Province of Pesaro-Urbino, and one conventional small familiar swine farm, located in the Province of Fermo, rearing pigs using a non-intensive farming system.

Concerning the dairy sector, our experimental trials were carried out while farmers were experiencing the continuous fluctuation of the prices of raw materials, mainly commodities such as soybean meal and corn grain, coupled with the crisis of the dairy sector across all Europe that determined low milk prices despite the increasing costs of production. Therefore, we were able to evaluate both the potential and the limits of soybean and corn partial replacement in a very critical agricultural sector such as dairy farming. Due to the peculiarities of dairy production, based on the animal physiology and on the nutritional characteristics of the raw materials under examination, soybean meal cannot be totally replaced by faba bean and field pea. Based on the results of the "on farm" feeding trial carried out at this dairy farm about 50% of soybean meal present in the ration was replaced by faba bean and field pea. To understand the potential effect of this partial substitution on the local agricultural system it can be considered that the farm was initially using 3kg/cow of soybean meal. Therefore, 1.5 kg of soybean meal were substituted by about 2 kg/cow of faba bean/field pea mix, considering that also part of the corn grain was also partially replaced by these grain legumes due to their starch content. Having the farm an average of 450 lactating cows per day, daily feeding requested about 9 tons/day of faba bean and field pea, which was about 330 tons/year. Based on the results of the field trials, assuming an average field production of 2.5 tons/hectare (2 tons/ha for faba beans and 3 tons/ha for field pea) it can be estimated that this farm could support about 130 hectares cultivated with grain legumes for animal feeding. Assuming a multi-rotation such as wheat-grain legume-barley- 3 years of alfalfa we can roughly estimate that partial substitution of soybean in this case could support an overall agricultural system covering about 790 hectares. Moreover, the presence of a 3 years forage crop such as alfalfa would reduce drastically the use of pesticides and fertilizers and also would increase nitrogen fixation and soil organic matter content, the introduction of organic farming practices to manage cereal and grain legume crops would avoid or at least reduce the use of chemicals and in particular of herbicides also in conventional farms, the milk production would be closely linked to the area of production. However, to make this system working it is necessary to make it economically profitable for the farm. When soybean meal prices increased the use of grain legumes produced by local farms helped the dairy farm to compensate for the increased costs of production. Despite these encouraging results the crisis of the dairy sector is hampering the implementation of this integrated production system that is based on close relationships between the dairy farm and farmers producing raw materials used as animal feed. The identification of different commercialization systems able to fully valorise the quality of the overall production chains is becoming a fundamental step in order to counteract the continuous decrease of milk price on the national and international market.

Beef cattle field trials were carried out at four organic farms located in the inner areas of Provinces of Macerata and Fermo. Differently from the experience previously described in the dairy sector, the small size of these farms, the high quality of both final products and production system and a different approach to the product valorisation based on the direct sales of meat by the producers themselves, allowed research results to be transferred to the final step of the production chain: the product marketing and commercialization.

Feeding trials on organic beef cattle showed that for not intensive production systems soybean can be totally replaced by faba beans and field pea. Moreover, these farms are characterized by self-producing almost all the forage and a high amount of protein (grain legumes) and energy (mainly barley) needed for animal feeding. It is relatively simple for these farms to make arrangements with neighbouring farmers to secure themselves the supply of the raw materials they are not able to self produce. This production system is extremely interesting because, as shown in Figure 13, it is based on rational agronomic crop rotations maintaining both a high level of crop diversification together with a low if not positive environmental impact due to the organic farming practices. The lower animal daily growth rate characterizing organic or non-intensive animal farming (1-1.2 kg/day), compared to growth rate of intensive systems (higher than 1.6 kg/day), is also an aspect of the production system which is valorised in the final product.

Figure 13. Organic animal farm located at Monte San Martino (Province of Macerata) where both experimental field and feeding trials were carried out.

Aim of the research project was also to verify the technical and economic feasibility of GM*free* production chains, that is production systems that do not include genetically modified (GM) feed in the feeding system. Among the commodities, soybean show the highest amount of worldwide production obtained from GM varieties, mainly cultivated in USA or South America [12]. Therefore, since almost soybean seed or meal used in the European Union is imported by American countries, the risk of GM soybean contamination is very high. This is confirmed by the introduction of a threshold of 0.9% technically unavoidable contamination also in organic feedstuff. Therefore, our results demonstrated that also organic farming systems could avoid GM contamination whenever a strict control of the raw material production or origin is made directly from the farmers. Feed composition therefore becomes an index of the raw materials used in the production chains and can be used as further information for the consumers to valorise the value of the final product. For these reason a DNA method has been developed as further result of the project aimed at the identification of the presence of faba beans and/or field pea within feed samples [28]. This could be a simple and not expensive approach to certify the use of local raw materials as feedstuff together with the absence of soybean from the ration. Moreover, an attempt to increase consumers' information about the characteristics of GM *free* organic production has been started as part of dissemination of the project activities and this increased the number of consumers interested in purchasing GM *free* products. On the whole, results on organic beef cattle showed that GM *free* production chains based on feeding systems that rely on locally produced raw materials can be an efficient alternative to intensive production chains. This approach could also be useful to maintain or increase economically effective agricultural systems in inner areas of Central Italy by reducing the dependence from the international market of commodities.

Encouraging results have also been obtained concerning the swine production chain. The farm where feeding trials were carried out had the possibility of rearing pigs both indoor and outdoor (open air). Therefore a feeding trial was conducted to compare one conventional feed (Control) based on the use of soybean meal and corn with an experimental feed where soybean meal was replaced completely by faba beans and field pea and corn was partially replaced with barley. Both feeds were formulated respecting the differences requested between the growth and the finishing phases. No differences in animal growth rate (600 g/day) were detected between the two feeds (Control vs Experimental). At the same time a group of pigs was reared outdoor and fed with the experimental feed. Average daily growth rate was slightly lower than observed in the indoor trial. The same experimental feed was subsequently tested in one organic and one conventional farm and results confirmed that regular growth rates can be obtained when soybean is not included in the feed, with slightly higher average daily gains obtained in the conventional farm (750 g/day). Therefore, non intensive swine production chains could represent another animal farming system that could stimulate the development of production systems linked to the production area, the networking among local farms concerning the exchange of raw materials for animal feeding, the reintroduction of rational not intensive agricultural systems. Commercialization of the final products is again fundamental to guarantee profitability for all the actors of the production chain and for this purpose direct selling is showing to be an effective marketing strategy to reach this objective.

6. Conclusions

The main objective of this paper was to stimulate a critical evaluation of soybean impact on agricultural systems where soybean cannot be cultivated. Notwithstanding soybean positive nutritional characteristics, this commodity may not be the only solution for animal production chains for those countries that may suffer from a complete dependence on import of raw materials.

Field and feeding trials carried "on farm" furnished information on the feasibility of local animal production chains using feeding systems based on locally produced raw materials that could partially or completely replace soybean. This possibility would be a stimulus to recreate networks among local farmers in order to develop local production chains that could restore an economically feasible agricultural systems to farms that are unable to compete on the global market. This also represents an attempt to reintroduce sustainable agricultural systems characterized by a reduced use of chemicals and pesticides due to the cultivation of low input cereals and legumes in rationale crop rotations.

Moreover, the interest toward research results on feedstuff that do not include soybean is also related to the risk of GM contamination due to this commodity. GM free animal production chains are strongly exposed to GM contamination when soybean is included as the main protein source in the feed. The decision to include the 0,9% threshold also for organic farming production chains underlines the real risk of GM contamination and the difficulties to create GM free production chains when the feed is based on the use of the same raw materials characterizing GM animal products.

Our results however showed that for large animal farms, that carryout intensive production systems, it is more difficult than for small farms, characterized by not intensive production chains, to manage soybean replacement. This aspect confirms the almost complete dependence from imported commodities that has been reached in the time by agricultural production sectors aimed at the mass production of large amounts of standardized products. The large volume of raw materials requested followed by the low internal availability of feedstuff that can be used as an alternative to soybean exposes these farms to the risks of international market variations both in the availability and in the price of this commodity.

On the other hand, the implementation of animal production chains based on the use of locally produced feedstuff is a valid approach for small farms producing high quality products using not intensive animal farming systems. These farms can in this way gain a market space for products that can be an alternative to standardized products and at the same time activate agricultural systems well integrated with local environmental features, that make less use of intensive production techniques, reduce the use of fertilizers and pesticides, restore soil fertility. However, new marketing strategies are necessary to make consumers aware of the importance of the overall characteristics of local production chains in defining the quality of a final product and to ensure at the same time a profitable price for the producers.

http://dx.doi.org/10.5772/52476

Author details

Stefano Tavoletti *

Address all correspondence to: s.tavoletti@univpm.it

Dipartimento di Scienze Agrarie, Alimentari ed Ambientali – Università Politecnica delle Marche - Via Brecce Bianche, 60131 ANCONA, Italy

References

[1] Masuda, T., & Goldsmith, P. D. (2009). World Soybean Production: Area Harvested, Yield, and Long-Term Projections. *International Food and Agribusiness Management Association Review*, 12(4), 143-162.

[2] Borsari, B. (2011). Agroecology as a Needed Paradigm Shift to Insure Food Security in a Global Market Economy. *International Journal of Agricultural Resources, Governance and Ecology*, 9(1/2) 1-14.

[3] Altieri MA. (1995). Agroecology: the Science of Sustainable Agriculture. *Westview Press Inc.*

[4] Reicosky D.C, Kemper W.D, Langdale C.W, Douglas C.LJ., 1995. Rasmussen PE. Soil organic matter changes resulting from tillage and biomass production. *Journal of Soil and Water Conservation* 50(3) 253-261.

[5] Tilman, D., Cassman, K. G., Matson, P. A., Naylor, R., & Polasky, S. (2002). Agricultural sustainability and intensive production practices. *Nature*, 418(6898), 671-677.

[6] Montgomery, DR. (2007). Soil erosion and agricultural sustainability. *Proceedings of the National Academy of Sciences of the United States*, 104(33), 13268-13272.

[7] Tilman, D., Balzer, C., Hill, J., & Befort, B. L. (2011). Global food demand and sustainable intensification of agriculture. Proceedings of National Academy of Science; , 108(50), 20260-20264.

[8] Sleper D.A., Barker T.C., Bramel-Cox P.J. editors. 1991. Plant Breeding and Sustainable Agriculture: Considerations for Objectives and Methods. *CSSA Special Publication N.18.*

[9] Jackson, D.L., & Jackson, L.L. (2002). The Farm as Natural Habitat. Reconnecting Food Systems with Ecosystems. *Island Press, Washington DC.*

[10] Jordan, N., Boody, G., Broussard, W., Glover, Keeney. D., Mc Cown, B. H., Mc Isaac, G., Muller, M., Murray, H., Neal, J., Pansing, C., Turner, R. E., Warner, K., & Wyse, D. (2007). Sustainable Development of the Agricultural Bio-Economy. *Science*, 316(5831), 1570-1571.

[11]   Index Mundi.

[12]   International Service fort he Acquisition og Agri-Biotech Applications. *ISAAA: executive summary.*

[13]   Yu, P., Goelema, J. O., Leury, B. J., Tamminga, S., & Egan, A. R. (2002). An analysis of the nutritive value of heat processed legume seeds for animal production using the DVE/OEB model: a review. *Animal Feed Science and Technology*, 99(1), 141-176.

[14]   Froidmont, E., & Bartiaux-Thill, N. (2004). Suitability of lupin and pea seeds as a substitute for soybean meal in high-producing dairy cow feed. *Animal Research*, 53(6), 475-487.

[15]   Prandini, A., Morlacchini, M., Moschini, M., Fusconi, G., Masoero, F., & Piva, G. (2005). Raw and extruded pea (Pisum sativum) and lupin (Lupinus albus var. Multitalia) seeds as protein sources in weaned piglets' diets: effect on growth rate and blood parameters. *Italian Journal of Animal Scince*, 4(4), 385-394.

[16]   Stein, H. H., Everts, A. K. R., Sweeter, K. K., Peters, D. N., Maddock, R. J., Wulf, D. M., & Pedersen, C. (2006). The influence of dietary field peas (Pisum sativum L.) on pig performance, carcass quality, and the palatability of pork. *Journal of Animal Science*, 84(11), 3110-3117.

[17]   Gilbery T.C., Lardy G.P., Soto-Navarro S.A., Bauer M.L. and Anderson V.L. 2007 Effect of field peas, chickpeas, and lentils on rumen fermentation, digestion, microbial protein synthesis and feedlot performance in receiving diets for beef cattle. *J Anim Sci* 85(11) 3045-3053.

[18]   Perella, F., Mugnai, C., Dal, Bosco. A., Sirri, F., Cestola, E., & Castellini, C. (2009). Faba bean (Vicia faba var. minor) as a protein source for organic chickens: performance and carcass characteristics. *Italian Journal of Animal Science*, 8(4), 575-584.

[19]   Volpelli, L. A., Comellini, M., Masoero, F., Moschini, M., Lo, Fiego. D. P., Scipioni, R., & Pea, . (2009). (Pisum sativum) in dairy cow diet: effect on milk production and quality. *Italian Journal of Animal Science*, 8(2), 245-257.

[20]   Crépon, K., Marget, P., Peyronnet, C., Carrouée, B., Arese, P., & Duc, G. (2010). Nutritional value of faba bean (Vicia faba L.) seeds for feed and food. *Field Crops Research*, 115(3), 329-339.

[21]   Jensen ES, Peoples MB,. (2010). Hauggaard-Nielsen H. Faba bean in cropping systems. *Field Crops Research*, 115(3), 203-216.

[22]   Jezierny, D., Mosenthin, R., & Bauer, E. (2010). The use of grain legumes as a protein source in pig nutrition: A review. *Animal Feed Science and Technology*, 157(3), 111-128.

[23]   Masey, O'Neill. H. V., Rademacher, M., Mueller-Harvey, I., Stringano, E., Kightley, S., & Wiseman, J. (2012). Standardised ileal digestibility of crude protein and amino acids of UK-grown peas and faba beans by broilers. *Animal Feed Science and Technology*, 175(3), 158-167.

http://dx.doi.org/10.5772/52476

[24] Blair, R. (2011). Nutrition and Feeding of organic cattle. *CABI, UK*, 13978184593758.

[25] Jarosz, L. (2012). Growing inequality: agricultural revolutions and the political ecology of rural development. *International Journal of Agricultural Sustainability*, 10(2), 192-199.

[26] Tavoletti, S., Mattii, S., Pasquini, M., & Trombetta, M. F. (2004). La reintroduzione delle colture proteiche nelle filiere agrozootecniche marchigiane. *L'Informatore Agrario*, 21-31.

[27] Migliorini, P., Tavoletti, S., Moschini, V., & Iommarini, L. (2008). Performance of grain legume crops in organic farms of central Italy. *In: Neuhoff D, Halberg N, Alföldi T, Lockeretz W, Thommen A, Rasmussen IA, Hermansen J, Vaarst M, Lueck L, Caporali F, Jensen HH, Migliorini P, Willer H (eds.) Cultivating the future based on science. Proc. Of 16th IFOAM Organic World Congress*, 16-20, 978-3-03736-023-1.

[28] Tavoletti, S., Iommarini, L., & Pasquini, M. (2009). A DNA method for qualitative identiWcationof plant raw materials in feedstuff. *European Food Research Technology*, 229(3), 475-484.

Chapter 2

Advanced Techniques in Soybean Biodiesel

Mauricio G. Fonseca, Luciano N. Batista,
Viviane F. Silva, Erica C. G. Pissurno, Thais C. Soares,
Monique R. Jesus and Georgiana F. Cruz

Additional information is available at the end of the chapter

http://dx.doi.org/10.5772/52990

1. Introduction

The planet where we inhabit, live, has experienced a great transformation period in the most different fields. The evolution followed by a great technological development, on the other side has caused an imbalance both in society itself, as in the material environment in which we live. The planet earth has visibly demonstrated how has been affected by this imbalance and how it has naturally reacted. In recent years, all this has being reported through the literature, studied by different scientific research groups, as also observed by what is reported in the media in general, even in the form of documentaries and films, as the documentary performed by former USA vice president Al Gore, an inconvenient truth (an inconvenient truth, 2006). All learning takes a certain time to begin to be assimilated and been put in practice effectively, so humanity has learned, been advised by the latest natural disasters of this century, as in the case of Japan's earthquake, tsunami, the strong hurricanes that plague the northern hemisphere summer, as in fact glaciers melting that were called eternal, the poles of this planet, strong climate change experienced over the past years and the major pollution in large urban cities where population are forced to live in many different fields, has signaling how much real acts, changes are necessary to continue to be possible living an inhabited planet.In this century, XXI, the world main problems, which it has experienced, are related to the scarcity of natural resources such as water, which had been mismanaged, contaminated by urban and industrial solid waste disposal, and in relation to generation and use of energy the most diverse shapes.

These energy sources can be broadly classified into three categories: fossil fuels (coal, oil and natural gas), renewable (hydroelectric, wind, solar and biomass) and nuclear sources. Among those can be highlighted Biomass, where all organic matter that is produced by

© 2013 Fonseca et al.; licensee InTech. This is a paper distributed under the terms of the Creative Commons Attribution License (http://creativecommons.org/licenses/by/3.0), which permits unrestricted use, distribution, and reproduction in any medium, provided the original work is properly cited.

this process is called biomass. This has a great advantage over fossil fuels, it's less polluting, because its processes do not add carbon dioxide to the atmosphere, the environment. The biomass process reduces the carbon dioxide amount in atmosphere through the photosynthesis, performed by increasing the planted green areas, to cultivate the seeds crops. Research and development departments have been engaged in fuels discovery that do not cause much environment damage and that can replace fossil fuels, reducing the toxic emissions level, replacing the rare fossil fuel used to date. In the midst of these researches has been observed that the use of vegetable oils has shown great ability to make this one a possible alternative renewable energy (Agarwal & Das, 2000).A related problem in the replacement of diesel for oil plant was related to physical and chemical factors such as high viscosity, low volatility which results in incomplete combustion, leading to formation of carbon deposits in the engine and a high unsaturations degree (Meher et al, 2008), factor that reduces the power of the fuel at the lowest level of cetane,but also favors oxidation. Studies have shown that vegetable oils characteristics can be modified through four ways (Shrivastava & Presad, 2002): By pyrolysis, microemulsification, dilution and transesterification process. The latter originates the alkyl esters that constitute what is called biodiesel.

2. Biodiesel

The use of vegetable oils as an alternative fuel for diesel engine was discovered more than100 years ago, in the Paris world exhibition in 1900, when Rudolph Diesel used peanut oil in an engine ignition (Shay, 1993).This predicted saying, "The use of vegetable oils as fuel engine may be negligible in the present moment, but in the future may become so important as oil and coal as energy sources. The biodiesel term is a subject still under discussion. Some definitions consider biodiesel as a mixture of any vegetable oils with fuel, diesel and fossil derivative others consider the alkyl esters mixture from vegetable oils or animal fats with fossil fuels. Under the chemical aspect biodiesel, an alternative fuel can be defined as alkyl esters derived from fatty acids obtained from oils, vegetables or animal fats, which suffering a chemical reaction, transesterification with short chain alcohols such as methanol and ethanol (Pinto et al, 2005). Transesterification: Chemical reaction between an ester (RCOOR ') and an alcohol (R"COH) resulting in a new ester (R"COOR') and an alcohol (RCOH).

This reaction type, used in biodiesel production is the reaction between the triglycerides, main components of vegetable oils and fats that react with short chain alcohols, methanol and ethanol, resulting in two products, methyl esters derived from fatty acids, and the second product glycerol formation. Transesterification reaction rates can be affected by some aspects: The catalyst type (acid or alkaline), purity of reactants (mainly water content), free fatty acid content and alcohol/vegetable oil molar ratio (Helwani et al, 2009). The biodiesel reaction can be optimized specially by three factors: The first is an increase in the temperature. An increase in temperature increases reaction rate in exponential, allowing the reactants to be more miscible, obtaining a higher reaction rate to take place.

This parameter is limited by the solvent, reactant boiling point. The second factor to improve reaction yield, vigorous mixing, possibilities a higher collision rate between the reactants, been obtained a reaction mixture plus homogenized, yielding a higher rate of methyl esters obtained. In general alcohols and triglyceride sources are immiscible, vigorous mixing possibilities the obtaining of alcohol dispersed as fine droplets, increasing the contact surface between the two immiscible reactants (Stamenkovic et al, 2008). The use of a secondary solvent, a co-solvent as THF, possibilities a higher miscibility of the alcohol in the triglyceride phase, obtaining a better mixing of the two phases and hence a more reactions to take place, improving the biodiesel yield. The following Figure 1 illustrates a biodiesel type of reaction.

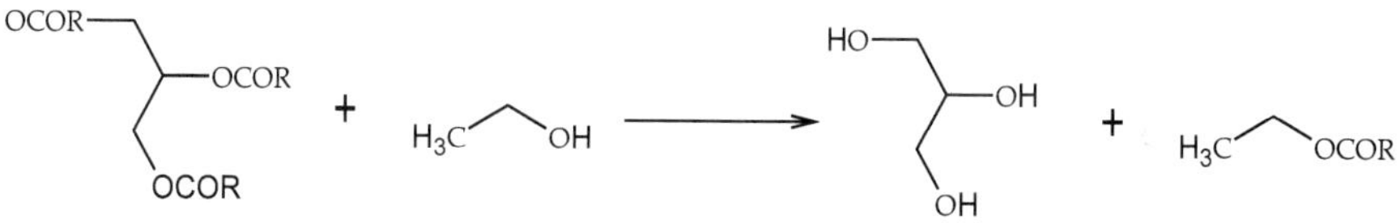

Figure 1. General transesterification reaction to produce biodiesel

In general terms, these reactions take place under homogeneous catalysts, acid or base catalysts, enzyme or through the use of heterogeneous catalysts. The selection of appropriate catalyst depends on the amount of free fatty acids in the oil. Heterogeneous catalyst provides high activity; high selectivity, high water tolerance properties and these properties depend on the amount and strengths of active acid or basic sites. Basic catalyst can be subdivided based on the type of metal oxides and their derivatives. Similarly, acidic catalyst can be subdivided depending upon their active acidic sites (Singh & Sarma, 2011). Generally, a basic catalyst gives better yields than the acids catalysts in both homogeneous and heterogeneous catalysts. The better results of homogeneous catalysts are related to the fact that base catalysts are kinetically much faster than heterogeneously catalyzed transesterification and are economically viable. There are many factors which govern the path of transesterification reactions, between these can be stand out the following parameters: the nature o raw material, the optimum experimental conditions, as the ratio oil/methanol, the temperature and the catalyst concentration, for example. Comparing heterogeneous catalyst with homogeneous catalysts can be observed that the use of solid heterogeneous use more extreme reaction conditions, higher pressure and temperature due the fact of the difficulty in the limited mass transfer between the three phase system solid-liquid-liquid immiscible (catalyst, oil, methanol). The main advantages in the use of solid catalysts are related to the easy work up when compared with homogeneous catalysts. Solid catalysts are separated just by filtration and centrifugation and are environmentally friendly, because they are reusable and reduce the amount of wasted, treated water used. Among the heterogeneous catalysts, we can highlight the use of zeolites (Suppes et al, 2004), clays, ion exchange resins and oxides.

3. Catalysts

3.1. Heterogeneous catalysts

The useof heterogeneous catalysts (Wang & Yang, 2007 and Leclercq et al, 2001) has as major advantage the reaction work-up, i.e., post-treatment reaction, separation and purification steps, since these can be easily removed and can be reused. Another interesting factor is the fact that this type of catalysis, there is no formation of by products, such as saponification (Suppes et al, 2001; Tomasevic et al, 2003 and Gryglewicz, 1999). The greatest difficulty encountered in using this reaction type is directly related to problems in relation between the diffusion systems, oil /catalyst /methanol.

3.2. Homogeneous catalysts

3.2.1. Basic catalysts

Basic Catalysis (Zhou et al, 2003) are procedures that use in general alkoxides of sodium and potassium, carbonates and hydroxides of these elements. Among these three groups it is found that alkoxides catalysts are financially unfavorable because they are more expensive but also difficult to handle because they are hygroscopic, and facilitate the achievement of side products such as derivatives of saponification, but have the advantage of carrying out the reactions in milder temperatures, produces high levels of esters derived from fatty acids and do not have corrosive properties as acid catalysts. A solution used to minimize the soap formation when biodiesel has a high free fatty acid content or water is the use of 2 or 3% mol of K_2CO_3 that will form the corresponding bicarbonate salt instead of water.

In the following, table 1, it's possible to find diverse types of heterogeneous catalysts used to obtain biodiesel of soybean, cotton seed, *Jatropha curcas*, palm, rape oil and sunflower.

Among the studies using soybean oil to obtain biodiesel, can be stand out the work developed by Wang et al, using CaO, SrO as a solid catalyst used in a heterogeneous process to obtain biodiesel. Cao, is a typical basic solid catalyst used in the most different ways. This compound has many advantages as a reusable due to its long catalyst lifetime, higher activity and requirement of only mild reaction conditions. At the example of table 1, is observed that in the best conditions to obtain biodiesel in yield of 95% is necessary a temperature of 65ºC, a molar ratio of MeOH/Oil of 5 and even a little reaction time from 0.5 to 3 hours. Even with all these specific positive factors, solid acid catalysts have been very useful at many industrial processes. Acid catalysts contain a large variety of acid sites with different strength of Bronsted, Lewis acidity, which is considered a good advantage at the transesterification process. These catalysts are even very useful, when is necessary to obtain biodiesel from oils rich in FFA, free fatty acids, because they convert the FFA into FAME prior to the biodiesel production, avoiding by this way the problem encountered at base catalysts, the soap formation.

Vegetable oil	Catalysts	Ratio MeOH/Oil	Reaction time (h)	Temperature (ºC)	Conversion (%)	References
Soybean	Calcined LDH (Li-Al)	15	1-6	65	71.9	Li
Soybean	La/zeolite beta	14.5	4	160	48.9	Furata
Soybean	MgOMgAl$_2$O$_4$	3	10	65	57	Schumaker
Soybean	MgO, ZnO, Al$_2$O$_3$	55	7	70-130	82	Trakarnpruk
Soybean	Cu and Co	5	3	70		Shu
Soybean	CaO, SrO	12	0.5-3	65	95	Wang
Soybean	ETS-10	6	24	120	94.6	Arzamendi
Cotton seed	Mg-Al-CO$_3$ HT	6	12	180-210	87	Wang
Jatropha Curcas	CaO	9	2.5	70	93	Albuquerque
Palm	Mg-Al-CO$_3$ (hydrotalcite)	30	6	100	86.6	Huaping
Rape	Mg-Al HT	6	4	65	90.5	Zeng
Sunflower	NaOH/Alumina	6-48	1	50	99	Liu
Sunflower	CaO/SBA-14	12	5	160	95	Suppes
Blended vegetable	Mesoporous silica loaded with MgO	8	5	220	96	Barakos

Table 1. Different heterogeneous catalysts used for transesterification of vegetable oils.

In the following figure 2, is exemplified the mechanism of base catalyzed transesterification. The mechanism can be resumed in the following way. In the first step the methoxide anion attaches to the carbonyl carbon atom of the triglyceride. In the second step, the oxygen picks up an acid H$^+$ from the alcohol. In the last step a rearrangement of the tetrahedral intermediate results in the formation of biodiesel and glycerol.

3.2.2. Acid catalysts

Sulfur and chlorides compounds are the most commonly used acid catalysts. This type of catalysis[(Mohamad & Ali, 2002) has as main advantages the absence of products derived from saponification reactions, higher yields but has some disadvantages such as the fact that the reactions are performed in a highly corrosive and reactive post-treatment, where the medium, the rinse water should be neutralized.

4. Enzymes

Enzymes are a fourth class of compounds used to produce biodiesel (Fukuda et al, 2001). In general its use is complicated by the fact that the enzyme generally are a specific material,

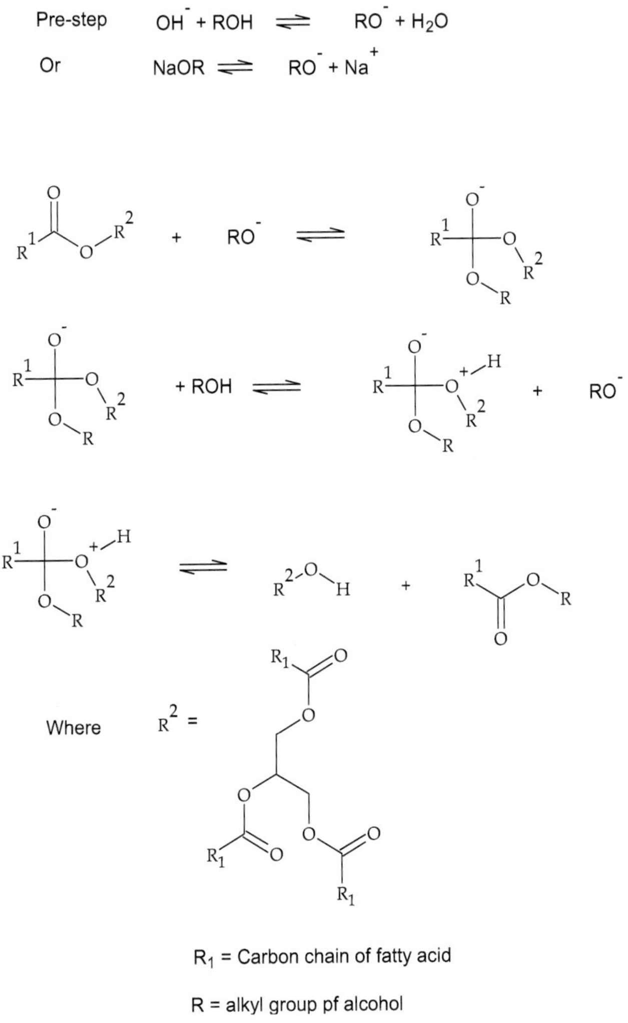

$$\text{Pre-step} \quad OH^- + ROH \; \rightleftharpoons \; RO^- + H_2O$$

$$\text{Or} \quad NaOR \; \rightleftharpoons \; RO^- + Na^+$$

Figure 2. Mechanism of base catalyzed transesterification.

and extremely expensive in relation to this type of reaction, are sensitive to the presence of methanol and ethanol, which causes deactivation of the same (Salis et al, 2008). Literature (Modi et al, 2007) shows that this problem can be circumvented by the water (Kaieda et al, 2001 & Kaieda et al, 1998) use and organic (Raganathan, 2008 & Harding et al, 2008) solvents such as dioxanes and petroleum ether, for example.

R = Alkyl group of the alcohol

R_1 = Carbon chain of fatty acid

R_2 =

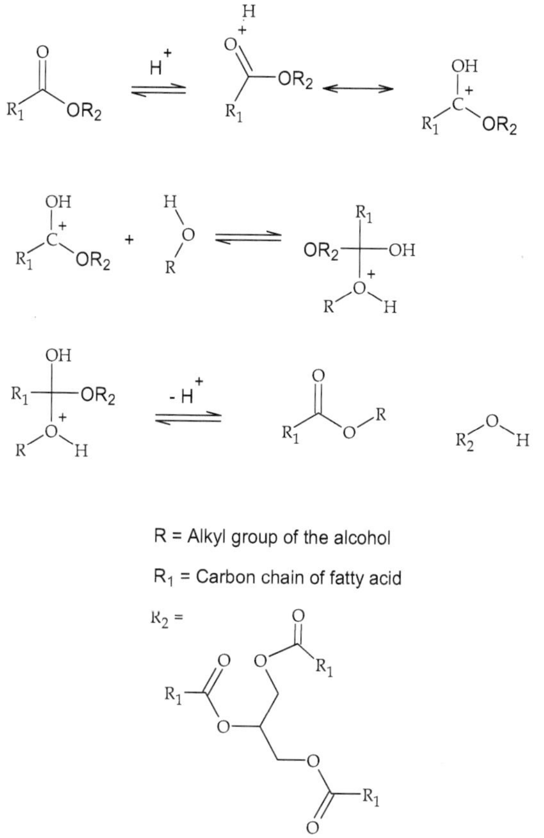

Figure 3. Mechanism of acid catalyzed transesterification

5. Non catalytic fatty acid alkyl ester production

The use of supercritical methanol process (Marulanda, 2012) to obtain biodiesel has been a useful method when the feed stock oil contains high amount of free fatty acids. This methodology has solved the problem encountered at the use of solid catalysts, low reaction rates due to low level of mass transfer, limitation between liquid and solid phase of catalysts and reactants. By this process, the dielectric constant of liquid methanol which tends to decrease in the supercritical state, increase the oil in to methanol solubility, resulting in a single phase oil/methanol system (Lee et al, 2012). In the next page, is exemplified the process of obtaining biodiesel by supercritical method, figure 4.

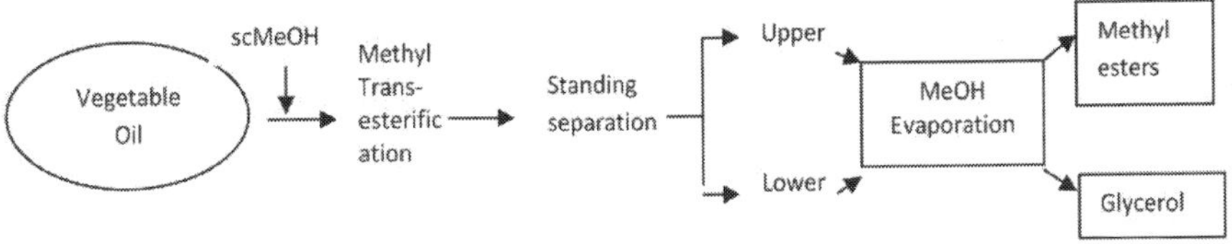

Figure 4. Schematic process of biodiesel production by supercritical method.

The methodology using supercritical methanol to obtain biodiesel by transesterification has reached the possibility to realize the reactions under mild, relatively moderate reactions to avoid the thermal degradation of fatty acid methyl esters (FAME). The reactions has been investigated in a wide range of reaction conditions (T = 200 – 425 ºC, time = 2 – 40 min. and P = 9.6 – 43 MPa). Thermal stability studies of methyl esters has showed that the best reaction conditions by supercritical methanol methodology to obtain biodiesel, consists in the temperature of 270ºC and reaction pressure of 8.09 MPa. The use of a co-solvent, as propane, CO_2 and heptane, diminishes, decrease the reaction temperature and the pressure needed to achieve a high yield of biodiesel obtained. The following table 2 exemplifies the investigated reactions conditions to obtain biodiesel.

Oil (co-solvent)	T (ºC)	P (MPa)	MeOH/Oil ratio	Time (min)	B/C	Yield (%)	Refs.
Soybean	100-320	32	40	25	C	96	He H. et al
Soybean (CH_3H_8/MeOH = 0.05)	280	12.8	24	10	B	98	Cao W. et al
Soybean (CH_3H_8/MeOH = 0.05)	288	9.6	65.8	10	B	99	Hegel P. et al
Soybean (CO_2/MeOH = 0.1)	280	14.3	24	10	B	98	Han H. et al
Soybean (CO_2/MeOH = 0.1)	350-425	10-25	3-6	2-3	C	100	Anitescu G. et al

Table 2. Examples of biodiesel production, experiment data using supercritical methanol.

6. Sources to obtain biodiesel

The sources for biodiesel production are chosen according to availability of the same in each country, region, taking into account the relative low cost of production and favorable economies of scale. For example, the use of refined oil would not be favorable due to high production costs and low production scale, on the other hand the use of seeds, algae and fat have a low production cost and greater availability than refined oils or recycled, which is a favorable factor for the production of biodiesel from these elements. When choosing a source of biodiesel production plants, a relationship is taken into account is how much they produce and the yield of oil per hectare. Following some examples of studied seeds: soybean, Babassu(*Orbiginiasp.*), castor oil, fish oil, microalgae(*Chorellavulgaris*) (Miao & Wu, 2006), tobacco

(Usta, 2005), *JatropaCurcas* (Berchmans & Hirata, 2008), Karanja (*Pongamiaglabra*) (Meher et al, 2006), salmon (El-Mashad et al, 2008), cooking oils, among others. All biodiesel sources are chosen according to the chemical composition of their fatty acids in relation to the size of their chains, unsaturation degree and the presence of other chemical functions, as these factors influence the biodiesel quality.

7. New advanced techniques to obtain biodiesel

In recent years soybean biodiesel has achieved a high level of advanced techniques to improve its production. Has been developed a methodology using microwave assistance to improve the esters conversion rates,, using heterogeneous catalysts, nano Cao, for example which facilitates the interaction between the molecules (Hsiao et al, 2011). Dr Hsiao and his research group has proved that by this methodology, is possible to obtain a higher biodiesel yield in less time. There are two factors that influence this reaction type. The use of nanocompounds facilitates the interaction between the molecules, nanocompounds possibility a high contact surface between the molecules. The microwave methodology reduces even the reaction time due to changing the electrical field activates the smallest variance degree of ions and molecules leading to molecular friction, enabling the initiation of molecule, chemical reaction. This methodology also provides an easier access to susceptible bonds, so, increases the chemical interaction, been obtained products in less time and higher yields. Microwave methodologies has proved to be part of desirable green chemistry, cause it is a safe, comfortable and clean way of working with chemical reactions. Microwave flow system assistance through homogeneous catalysis is another example which has improved the biodiesel production in less time, depending of some factors as reaction residence time, catalyst amount and temperature at the exit point (Encinar et al, 2012). In attempt to improve the microwave assisted methodology to obtain biodiesel another technique was added, used, the ultrasound. Microwave and ultrasound developed methodology has proved to be very efficient when used together. In this process, the first step used is ultrasound, cause ultrasonic field induced an effective emulsification and mass transfer that increases the rate of ester formation due to ultrasonic mixing causes cavitations of bubble near the phase boundary between the methanol and seeds oil, facilitating the thoroughly mixing, interaction between the oil and the reactant, methanol (Hsiao et al, 2010).Another technique has been also developed using ultrasonic irradiation with vibration ultrasonic. Instead of using heterogeneous or homogeneous transesterification catalyst, was used the enzyme methodology, through the application of Novozym 435. This methodology has proved to be efficient in enzymatic reaction to obtain biodiesel. Was observed that the use of ultrasonic added to vibration is a further factor to obtain higher values of biodiesel, cause the movement increase might facilitate the interaction between the substrate and the active site (Yu et al, 2010).Transesterification of soybean oil was achieved using ultrasonic water bath and two different commercial lipases in organic solvents (*n*-hexane) for example (Batistella et al, 2012). Dimethyl carbonate is a useful alternative to obtain biodiesel, this one is nontoxic, cheapness product and the reaction obtained product, glycerol carbonate is a value added substance with various useful applications. This can be obtained through enzymatic transesterification of soybean oil in organic solvents in mild conditions (Seong et al, 2011).The products, biodiesel are obtained in more time, but by other side this methodology has many advantages: It is an easy to use methodology, the enzyme can be reusable and the reaction work up is chemically friendly, cause it's not necessary the treatment of water used to purify biodiesel by the traditional, usual transesterification by homogeneous basic catalyst and is obtained an added value product, the glycerol carbonates. Heterogeneous catalysis using subcritical methanol is an advance in the soybean biodiesel obtaining methodology. By this technique is possible to use less amount of catalyst and have as main advantages the catalyst reusable and the separation, obtaining from reaction medium through centrifugation. The use of small amount of a catalyst, K_3PO_4, 0.1%wt, insoluble in methanol has transformed the reaction in subcritical methanol more available, cause has reduced the temperatures from 350ºC to 160ºC and the methanol molar ratio from 42 to 24, for example. The catalyst can be reusable at least three times. (Yin et al, 2012). KF Modified calcium magnesium oxide catalyst is an example of heterogeneous catalysis to obtain soybean biodiesel and even to recycle the catalysts due to be easily removed from the reaction through centrifugation and the use of a reaction under atmosphere pressure and 65ºC of temperature. This new catalyst has even improved the ester methyl yield from 63.6% (CaO-MgO catalyst) to 97.9% (KF-MgO-CaO) (Fan et al, 2012). Response surface methodology is an applicable technique to improve the results in obtaining soybean biodiesel. This methodology verifies the main parameters to optimize the biodiesel production process (Silva et al, 2011).The use of a process entirely independent from petroleum has been reached by the use of ethanol, obtained from a renewable source, sugar cane and seed oil. Gomes and his research group has developed a methodology to obtain ethyl biodiesel and even has developed a methodology to simplify the work-up process. In order to optimize the separation step of glycerol from biodiesel, many techniques have been studied. The microfiltration through ceramic membrane has demonstrated to be a useful technique to obtain biodiesel. This methodology is environmentally friendly cause reduces the amount of used water to purify the biodiesel. This technique simplify the entire purification process, the biodiesel is obtained by transesterification, after the end of reaction is added acidified water, this process facilitates the separation in two phases, the organic one, rich in oil and the aqueous, which posses the soaps converted in water soluble salts, catalysts, glycerol and other water soluble substances (Gomes et al, 2011). In water, glycerol forms greater droplets that are retained during the microfiltration step, been the biodiesel obtained by this way with glycerol content lower than 0.02% wt, the limit of free glycerol specified by Brazilian regulation. Mesoporous silica catalyst was used in a heterogeneous catalysis of soybean transesterification with methanol. $La_{50}SBA\text{-}15$ is used to obtain an ethyl biodiesel in mild conditions after 6 hours. The main advantage of this technique is the use of a heterogeneous catalysis to obtain ethyl biodiesel in mild conditions, don't use the usual high temperatures 473K and lower amounts of the ratio Oil/ Alcohol, from 36 to 20. This lower amount of alcohol facilitates the phase separation organic/ water, less amount of alcohol causes an easier purification process, because diminishes the possibility of emulsion formation (Quintela et al, 2012). The heterogeneous catalysis of biodiesel has reached an advance with the development of methodologies using membranes.

These membranes can be prepared in a simple way by the use of clays as hydrotalcite and poly (vinyl alcohol). The biodiesel is prepared by transesterification and can be obtained in mild conditions, 60ºC in a volume ratio oil/methanol of 5:60. The methyl biodiesel by this methodology can be obtained in 90%. The catalyst can be reused at least by 7 times (Guerreiro et al, 2010). An alternative method to obtain biodiesel is the enzymatic-catalytic way. In general this methodology has as main advantage the enzymatic selectivity, is a reusable catalytic and facilitates the separation, purification process. Methodology using a immobilized lipase onto a nanostructure has been developed due to the good transesterification activity of lipase and the use of electrospinning method to obtain nanofibrous membranes, which have larger surface area and porous structure that can lower the substrate resistance and facilitate enzyme immobilization, generating a reusable catalyst for biodiesel synthesis. The use of this methodology simplifies the separation process, where after the reaction, glycerol can be removed by centrifugation and the biodiesel obtained in the 90% range (Li et al, 2010).

Variable	Base Catalyst	Acid Catalyst	Lipase Catalyst	Supercritical Alcohol	Heterogeneous Catalyst
Reaction temperature (ºC)	60 - 70	55 - 80	30 - 40	239 - 385	180 - 220
Free fatty acid in raw material	Saponified products	Esters	Methyl esters	Esters	Non sensitive
Water in raw materials	Interfere with reaction	Interfere with reaction	No influence		Non sensitive
Yields of methyl esters	Normal	Normal	Higher	Good	Normal
Recovery of glycerol	Difficult	Difficult	Easy		Easy
Purification of methyl esters	Repeated washing	Repeated washing	None		Easy
Production cost of catalyst	Cheap	Cheap	Relatively expensive	Medium	Potentially cheaper

Table 3. Example of different technologies to produce biodiesel.

8. Non usual methods of soybean biodiesel analysis of cold properties and oxidation state

Several methods have been used to characterize biodiesel, and each methodology analyses some aspects of biodiesel as cold properties and oxidation process. Most legislation assumes a small group of tests to determination of biodiesel quality. Eighteen percents of Brazilian biodiesel production uses soybean as oil sources. Several nations have been establishing standards and legislation about biodiesel. Mainly determinations include iodine value, acid content, specific mass, esters content among others. The aim of this work are the availability of methodologies that wasn`t includes on official methodologies.

8.1. Cold properties

Official methodologies of biodiesel are: Cold Filter Plugging Point (CFPP), Cloud Point and Pour Point. Pour point indicates a moment of initial crystallization, but this methodology has low accuracy. For studies with more complexity other methodologies has better performance.

8.2. Differential scanning calorimetry

One of most usual and versatility methodology is a differential scanning calorimetry, these methodology are based on monitoring the difference in energy provided/released to/ by the sample (reagent system) in relation to a reference system (inert) as a function of temperature when both systems are subjected to a controlled temperature program. The changes in the temperature of the sample are caused by phase rearrangements, dehydration reaction, dissociation or decomposition reactions, oxidation or reduction reaction, gelatinization and other chemical reactions. DSC evaluates absorption or energy liberation to determine the initial of the reaction. A typical curve of biodiesel is presented at figure 1:

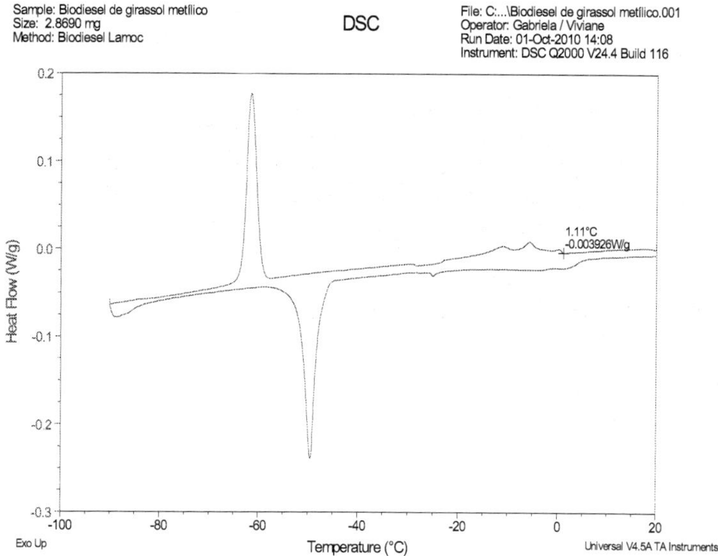

Figure 5. Differential Scanning Calorimetry of biodiesel.

Small signal at temperature of -1.1ºC is due a crystallization of saturated ester notably palmitic and stearic methyl esters, shaped signal of -60ºC is caused by crystallization of unsaturated esters (oleic, linoleic and linolenic methyl esters). These temperatures can change by cooling rate, so is important promoting standardization for independent analysis.

DSC trying to be associates with pour point results, because, a priori, both methods analyze a formation of firsts crystals, Formation of crystals initiate with nucleation of crystals that precedes crystallization is dependent on the formation and growth of aggregates or clusters of molecules. These aggregates must overcome a critical size in order to keep a steady growth and become a crystal of detectable dimensions [Avrami, 1940; Avrami, 1941]. At the stage of crystal growth, molecules of solute adsorb on the crystal surface and the process depends on the diffusion of material from the liquid phase to the solid phase which is being formed. Any of these stages can control crystal growth. The added substance must be capable to interfere with one of these stages: either avoiding or delaying the growing of aggregates to a critical size or reducing crystal growth rate [Mullin,2001].

In this point it's crucial to understand the difference of results obtained by pour point analyses and DSC. Calorimetric method is very sensible and detect energy associate of crystallization phenomena under of critical crystal size. While pour point detection occur only when crystal reached a minimum size.

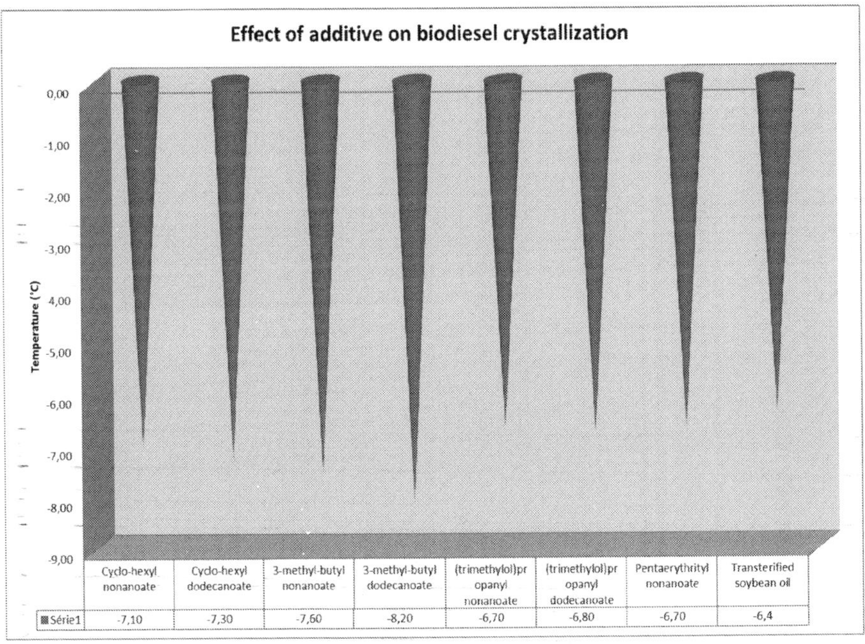

	Cyclo-hexyl nonanoate	Cyclo-hexyl dodecanoate	3-methyl-butyl nonanoate	3-methyl-butyl dodecanoate	(trimethylol)propanyl nonanoate	(trimethylol)propanyl dodecanoate	Pentaerythrityl nonanoate	Transterified soybean oil
Série1	-7,10	-7,30	-7,60	-8,20	-6,70	-6,80	-6,70	-6,4

Figure 6. Impact of additives into onset temperature of biodiesel crystallization based on Soares et al, 2009

Soares et al (2009) present a work in which several esters derived from branched chain, cyclic monohydroxylated alcohols or polyhydroxylated alcohols were added to methyl transesterified soybean oil (biodiesel like) to investigate their effect on the transesterified soybean oil crystallization. In this work were added kinetic studies, were conducted in order to detect differences in crystallization mechanisms due to differences in additive structures.

They obtained a depression of onset crystallization point about 2ºC when used an additive (0,08mol/100g of transesterifed soybean oil)

Still about this work were determined induction times of crystallization. Induction time indicates how many time initiate crystallization at determined temperature. This time tends to reduce while temperature decreases. This information is important because permitted association with limit time to storage before crystallization.

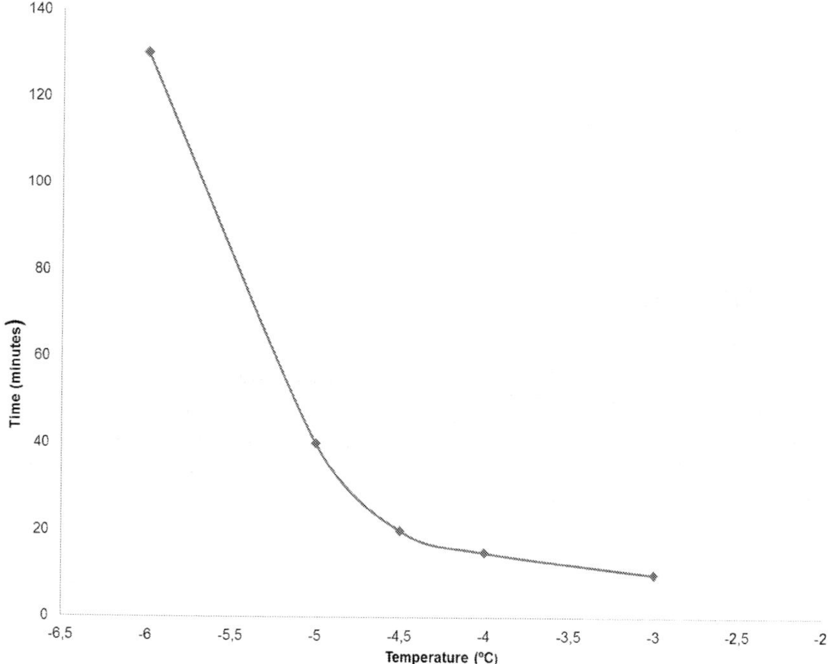

Figure 7. Induction time of crystallization based on Soares et al (2009)

These kinetic models describe how the extension of the phase transformation of a given material occurs as a function of time and temperature. Their equation is based on the suppositions of isothermal condition of crystallization, aleatory homogeneous or heterogeneous nucleation and that the new phase growth rate is temperature dependent. Avrami admitted that a number of tiny nuclei (aggregates of subcritical size) are already present in the phase to be transformed and that these aggregates must grow to a critical size to start a steady growth. By simplifying his statistical treatment presented in the calculation of transformed matter, he came to the generalized expression:

$$\alpha = 1 - \exp(-kt^n) \tag{1}$$

Where α is the volume fraction transformed (crystallized mass); k is dependent on a shape factor, on nucleation probability, on nucleation and growth rates and on the dimensionality of crystal growth while n reflects the mechanism of nucleation and growth and the crystal morphology [Avrami, 1940; Avrami, 1941]. Several works has been using DSC as assessment cold properties (ref), with better accuracy and precision.

8.3. Oxidative properties

Oxidative properties of biodiesel commonly assessment by EN 14112 called rancimat and Iodine value, but several methods have been used by analysis of oxidation state of oil and biodiesel. Oxidation products from these compounds as Petrooxy, differential scanning calorimetry (DSC), Pressure Differential Scanning Calorimetry (PDSC) (Dufaure et al, 1999) iodine value (IV) and mainly Rancimat Method. Each method is based at one step, intermediate compounds or reactants of biodiesel oxidation.

At the PetroOxy, the sample is inducted to oxidation through an intense oxygen flow, manipulating by this way the stability conditions through a specific apparatus. The analysis time is recorded as the required time to the sample absorbs 10% of oxygen pressure. Analysis is based on oxygen consumption (reactant) but not detect products.

The differential scanning calorimetry (DSC) monitors the difference in energy provided/ released between the sample (reagent system) and the reference system (inert) as a function of temperature when both, the system are subjected to a controlled temperature program. Changes in temperature sample are caused by rearrangements of induced phase changes, dehydration reaction, dissociation or decomposition reactions, oxidation or reduction reaction, gelatinization and other chemical reactions. DSC evaluated absorption or energy liberation for determining initial reaction. This process can present some problems because of formation of lipid alkyl radical is an endotermic process and others reactions are exotermic (Santos et al, 2011). The time for secondary product formation from the primary oxidation product, hydroperoxide, varies with different oils. Secondary oxidation products are formed immediately after hydroperoxide formation in olive and rapeseed oils. However, in sunflower and safflower oils, secondary oxidation products are formed when the concentration of hydroperoxides is appreciable (Guillen and Cabo 2002).

At the Rancimat technique, oxidative stability is based at the electric conductivity increase (Hadorn & Zurcher, 1974.). The biodiesel is prematurely aged by the thermal decomposition. The formed products by the decomposition are blown by an air flow (10L/ 110 ºC) into a measuring cell that contains bi-distilled, ionized water. The induction time is determined by the conductivity measure and this can be totally automatized. Rancimat is the most used technique to determine finalized biodiesel stability, under oxidative accelerated conditions, according to standard EN14112. This technique evaluated final products of thermal decomposition.

The differential scanning calorimetry (DSC) monitors the difference in energy provided/released between the sample (reagent system) and the reference system (inert) as a function of temperature when both in the system are subjected to a controlled temperature program. Changes in temperature sample are caused by rearrangements of induced phase changes, dehydration reaction, dissociation or decomposition reactions, oxidation or reduction reaction, gelatinization and other chemical reactions. DSC evaluated consumption or energy liberation for determining initial reaction.

The Pressure Differential Scanning Calorimetry (P-DSC) is a thermo analytical technique that measures the oxidative stability using a differential heat flow between sample and reference thermocouple under variations of temperatures and pressure. This technique differs from the Rancimat for being a fast method and presents one more variable - the pressure, allowing to work at low temperatures and using a small amount of sample (Candeia, 2009). All of these methods evaluated one aspects of oxidation process. But to predict behavior or design an adequate biodiesel its necessary to associate a oxidation process with structural properties or composition.

Iodine Value (IV) has been used for a long time to quantify unsaturated bonds on vegetable oil and, actually, biodiesel. Iodine Value is considering mainly structural method to assessment oxidation stability. Although currently some authors agree that this method is not necessarily the better method to evaluate stability.

Agreement with oxidation mechanism of fatty acids its very common associated presence of unsaturated with tendency of low stability, but Jain and Sharma (2010) presents weak relationship (R^2= 0,4374) between unsaturated esters content and induction period for several biodiesels, its associate this discrepancy with differ technology productions or presence of impurities.

Knothe (2007) bring an important discussion about relevance of Iodine value, its associate iodine value with some others structural index as APE and BAPE and associate to several properties of biodiesel in his work; he gives examples of how different mixtures of methyl esters of the three most common unsaturated FA—oleic, linoleic, and linolenic—can achieve nearly identical IV just slightly below the value of 115. This is an important fact because if oxidative stability depends of how many times one biodiesel can resist to be oxidized, and Linolenic acid reacts more fast then linoleic and oleic acid, this mixtures should present a differing oxidation behavior independently to have same iodine Value.

One of innovative uses of Thermogravimetric analysis is based on gum formation during biodiesel storage. Figure 4 shows a impact of storage at high temperature storage (85°C) of biodiesel. High resistance of biodiesel after four weeks is probably due formation of polymeric species as trimmers or tetramers of unsaturation esters.

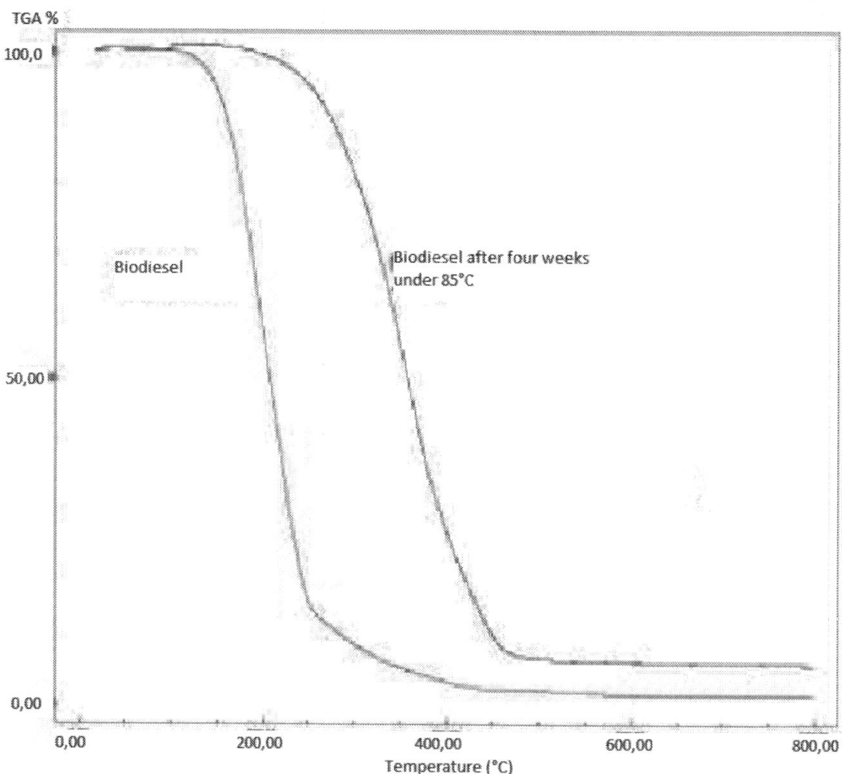

Figure 8. Formation of high molecular level species tends to increase oil viscosity

In presence of oxygen molecule polymer are forming by C-O-C linkages (Jonhson et al,1957; Wexler, 1964; Formo et al, 1979) and C-C linkages, while under an inert atmosphere only polymers with C-C linkages primordially are founded. High contents of polyunsaturated fatty acid chains enhance oxidative polymerization in fatty oils (Korus et al, 1983).Trimers and other fatty acid polymers presents higher thermal stability enhanced Termogravimetric curves of biodiesel, therefore this methodology can be used to determine biodiesel oxidation stage.

9. Conclusion

Many advanced techniques are obtained throughout the world to obtain and characterize biodiesel. To follow all these methodologies is necessary a constant research in many different science fields as chemistry, theoretical chemistry and even physics methodologies. This chapter contains only a simple updated tool to help to verify all the nowadays latest news, but never forget the most important tool to use is your brain, the determination and interest in the studied subject to unravel the science frontiers.

Author details

Mauricio G. Fonseca[1*], Luciano N. Batista[1], Viviane F. Silva[1], Erica C. G. Pissurno[1], Thais C. Soares[1], Monique R. Jesus[1] and Georgiana F. Cruz[2]

1 INMETRO – National Institute of Metrology, Quality and Technology, Metrological Chemistry Division Xerém, Duque de Caxias, Rio de Janeiro, Brasil

2 UENF – North Fluminense University, Engineering and exploitation of petroleum laboratory, Macaé, Rio de Janeiro, Brasil

References

[1] Agarwal, A. K., & Das, L. M. (2000). Biodiesel development and characterization for use as fuel in compression ignition engines. *Am. Soc. Mech. Eng. J., Eng. Gas Turbine Power*, 123, 440-447.

[2] Albuquerque, M. C. G., Jimenez-Urbistondo, I., Santamaria-Gonzales, J., Merida-Robles, J. M., Moreno-Tost, R., Rodriguez-Castellom, E., Jimenez-Lopez, A., Azevedo, D. C. S., Cavalcante Jr, , , C. L., Maireles-Torres, P., et al. (2008). CaO supported on mesoporous sílica as basic catalysts for transesterifications reactions. *Appl. Catal. A: Gen*, 334, 35-43.

[3] Anitescu, G., Deshpande, A., & Tavlarides, L. L. (2008). Integrated technlogy for supercritical biodiesel production and Power cogeneration. *Energy Fuels*, 22, 1391-1399.

[4] Antunes, W. M., Veloso, C. O., & Henriques, C. A. (2008). Transesterification of soybean oil with methanol catalyzed by basic solids. *Catal. Today*, 133-135, 548 EOF-554 EOF.

[5] Avrami, M. (1940). Kinetics of phase change. II: transformation-time relations for random distribution of nuclei. *J Chem Phys.*, 8, 212-224.

[6] Avrami, M. (1941). Granulation, phase change, and microstructure: kinetics of phase change. III. *J Chem Phys.*, 9, 177-184.

[7] Arzamendi, G., Campoa, I., Arguinarena, E., Sanchez, M., Montes, M., & Gandia, L. M. (2007). Synthesis of biodiesel with heterogeneous NaOH/Alumina catalysts: Comparison with homogeneous NaOH. *Chem. Eng. J.*, 134, 123-130.

[8] Barakos, N., Pasias, S., & Papayannakos, N. (2008). Transesterification of triglycerides in high and low quality oil feeds over an HT2 hydrotalcite catalyst. *Bioresource Technol*, 99, 5037-5042.

[9] Batistella, L., Lerin, L. A., Brugnerotto, P., Danielli, A. J., Trentin, C. M., Popiolski, A., Treichel, H., Oliveira, J. V., & Oliveira, D. (2012). Ultrasound-assisted lipase-catalyzed transesterification of soybean oil in organic solvent system. *UltrasonicsSonochemistry*, 19, 452-458.

[10] Candeia, R. A., Silva, M. C. D., Carvalho, Filho. J. R., Brasilino, M., Bicudo, T. C., Santos, I. M. G., & Souza, A. G. (2009). Influence of soybean biodiesel content on basic properties of biodiesel diesel blends. *Fuel*, 88, 738-743.

[11] Dufaure, C., Thamrin, U., & Moulongui, Z. (1999). Comparison of the thermal behaviour of some fatty esters and related ethers by TGA-DTA analysis. *Thermochim. Acta,,* 368, 77-83.

[12] Berchmans, H. J., & Hirata, S. (2008). Biodiesel production from crude Jathropa Curcas L. Seed oil with a high content of free fatty acids. *Bioresour. Technol.,,* 99, 1716-1721.

[13] Cao, W., Han, H., & Zhang, J. (2005). Preparation of biodiesel from soybean oil using supercritical methanol and co-solvent. *Fuel*, 84, 347-351.

[14] El -Mashad, H. M., Zhang, R., & Avena-Bustillos, R. J. (2008). A two steps process for biodiesel production from salmon oil. *Bisosys. Engin.,,* 99, 220-227.

[15] Encinar, J. M., González, J. F., Martínez, G., Sánchez, N., & Pardal, A. (2012). Soybean oil transesterification by the use of a microwave flow system. *Fuel*, 95, 385-393.

[16] Fan, M., Zhang, P., & , Q. (2012). Enhancement of biodiesel synthesis from soybean oil by potassium fluoride modification of a calcium magnesium oxides catalyst. *Bioresource technology*, 104, 447-450.

[17] Formo, M. W., Jungermann, E., Noris, F., & Sonntag, N. O. (1979). *Bailey's Industrial Oil and Fat Products,,* 1(4), Daniel Swern, Editor: John Wiley and Son,, 698-711.

[18] Fukuda, H., Konda, A., & Noda, H. (2001). Bioidesel fuel production by transesterification of oils. *J. Biosc. Bioeng.,,* 92, 405-416.

[19] Furata, S., Matsuhasbi, H., & Arata, K. (2004). Biodiesel fuel production with solid superacid catalysis in fixed bed reactor under atmospheric pressure. *Catal. Commun*, 5, 721-723.

[20] Gomes, M. C. S., Arroyo, P. A., & Pereira, N. C. (1999). Biodiesel production from degummed soybean oil and glycerol removal using ceramic membrane. *Journal of Membrane Science*, 378, 453-461.

[21] Gryglewicz, S. (1999). Rapeseed oil methyl esters preparation using heterogeneous catalysts. *BioresourTechnol*, 70, 249-253.

[22] Guerreiro, L., Pereira, P. M., Fonseca, I. M., Martin-Aranda, R. M., Ramos, A. M., Dias, J. M. L., Oliveira, R., & Vital, J. (2010). PVA embedded hydrotalcite membranes as basic catalysts for biodiesel synthesis by soybean oil methanolysis. *Catalysis Today*, 156, 191-197.

[23] Guillen, M. D., & Cabo, N. (2002). *Food Chemistry,*, 77(4), 503-510.

[24] Hadorn, H., & Zurcher, K. (1974). Zur bestimmung der oxydationsstabilitat von olen und fetten. *Deutsche Lebensmittel Rundschau.*, 70, 57-65.

[25] Han, H., Cao, W., & Zhang, J. (2005). Preparation of biodiesel from soybean oil using supercritical methanol and CO_2 as co-solvent. *Process Biochem,.*, 40, 3148-3151.

[26] Harding, K. G., Dennis, J. S., von, Blottnitz. H., & Harrison, S. T. L. (2008). A life-cycle comparision between inorganic and biological catalyse for the production of biodiesel. *J. Clean. Produc.*, 16(13), 1368-1378.

[27] He, H., Wang, T., & Zhu, S. (2007). Continuous production of biodiesel fuel from vegetable oil using supercritical methanol process. *Fuel,*, 86, 442-447.

[28] Hegel, P., Mabe, G., Pereda, S., & Brignole, E. A. (2007). Phase transitions in a biodiesel reactor using supercritical methanol. *Ind. Eng. Chem. Res.,*, 46, 6360-6365.

[29] Helwani, Z., Othman, M. R., Aziz, N., Fernando, W. J. N., & Kim, J. (2009). Technologies for production of biodiesel focusing on green catalytic techniques: A review. *Fuel Processing Technologies*, 90, 1502-1514.

[30] Hsiao, M., , C., Lin, C., , C., Chang, Y., , H., Chen, L., & , C. (2010). Ultrasonic mixing and closed microwave irradiation-assisted transesterification of soybean oil. *Fuel*, 89, 3618-3622.

[31] Hsiao-C, M., Lin-C, C., & Chang-H, Y. (2011). Microwave irradiation-assisted transesterification of soybean oil to biodiesel catalyzed by nanopowder calcium oxide. *Fuel*, 90, 1963-1967.

[32] Huaping, Z., Zongbin, W., Yuanxiao, C., Ping, Z., Shije, D., Xiaohua, L., & Zongqiang, M. (2006). Preparation of biodiesel catalyzed by solid super base of calcium oxide and its refining process. *Chinese J. Catalysis*, 27, 391-396.

[33] Johnson, O. C., & Kummerow, F. A. (1957). Chemical Changes Which Take Place in an Edible Oil During Thermal Oxidation. *JAOCS*, 34, 407-409.

[34] Kaieda, M., Samukawa, T., Matsumoto, T., Ban, K., Kondo, A., Shimada, Y., Noda, H., Nomoto, F., Obtsuka, K., Izumoto, E., & Fukuda, H. (1999). Biodiesel fuel production from plant oil catalyzed by rhizopus orizae lipase in a water containing system without an organic solvent. *J. Biosc. Bioeng.,*, 88(6), 627-631.

[35] Kaieda, M., Samukawa, T., Kondo, A., & Fukuda, H. (2001). Effect of methanol and water contents on production of biodiesel fuel from plant oil catalyzed by various lipases in a solvent free system. *J. Biosc. Bioeng.,*, 91(1), 12-15.

[36] Knothe, G. H. (2007). Some aspects of biodiesel oxidative stability. *Fuel Processing Technology*, 88, 677-699.

[37] Korus, R. A., Mousetis, T. L., & Lloyd, L. (1982). Polymerization of Vegetable Oils. *American Society of Agricultural Engineering,,* Fargo, ND,, 218-223.

[38] Leclercq, E., Finiels, A., & Moreau, C. (2001). Transesterification of rapessd oil in the presence of basic zeolites and related solid catalysts. *J. Am. Oil Chem. Soc*, 78, 1161-1165.

[39] Lee, S. B., Lee, J. D., & Hong, I. K. (2011). Ultrasonic energy effect on vegetable oil based biodiesel synthetic process. *Journal of Industrial Engineering Chemistry*, 17, 138-143.

[40] Lee, S., Posarac, D., & Ellis, N. (2012). An experimental investigation of biodiesel synthesis from waste canola oil using supercritical methanol. *Fuel*, 91, 229-237.

[41] Li, S., , F., Fan, Y., , H., Hu, R., , F., Wu, W., & , T. (2011). Pseudomonas cepacia lipase immobilized onto electrospun PAN nanofibrous membranes for biodiesel production from soybean oil. *Journal of Molecular Catalysis B: Enzymatic,,* 72, 40-45.

[42] Li, E., & Rudolph, V. (2008). Transesterification of vegetable oil to biodiesel over MgO-functionally mesoporous catalysts,. *Energy and Fuels*, 22, 143-149.

[43] Liu, X., He, H., Wang, Y., & Zhu, S. (2008). Transesterification of soybean oil to biodiesel using CaO as a solid base catalyst. *Fuel,,* 87, 216-221.

[44] Liu, X., He, H., Wang, Y., Zhu, S., & Piao, X. (2007). Transesterification of soybean oil to biodiesel using CaO as a solid base catalyst. *Cat. Commun*, 8, 1107-1111.

[45] Marulanda, V. F. (2012). Biodiesel production by supercritical methanol transesterification : process simulationand potential environmental impact assessment. *Journal of Cleaner Production,,* 33, 109-116.

[46] Meher, L. C., Dharmagada, V. S. S., & Naik, S. N. (2006). Optimization of Alkaly-catalyzed transesterification of Pongamia pinnata oilfor production of biodiesel. *Bioresour. Technol.,,* 97, 1392-1397.

[47] Meher, L. C., Sagar, D. V., & Naik, S. N. (2008). Technical aspects of biodiesel production by transesteriofication- Review. *Renew. Sustain Energy Ver.*, 10, 248-268.

[48] Miao, X., & Wu, Q. (2006). Biodiesel production from heterotrophic microalgal oil. *Bioresour. Technol.*, 97, 841-846.

[49] Modi, M. K., Reddy, J. R. C., Rao, B. V. S. K., & Prasad, R. B. N. (2002). Lipase mediated conversion of vegetables oils in to biodiesel using ethyl acetate as acyl acceptor. *Bioresource Technology*, 98(6), 1260-1264.

[50] Mohamad, I. A. W., & Ali, O. A. (2002). Evaluation of the transesterification of waste palm oil into biodiesel. *BioresourTechnol*, 85, 225-256.

[51] Mullin, J. W. (2001). *Crystallization*, 4, Boston: Butterworth-Hetnemann,, 181.

[52] Pinto, A. C., Guarieiro, L. L. N., Rezende, M. J. C., Ribeiro, N. M., Torres, E. A., Lopes, A. W., Pereira, P. A. P., & Andrade, J. B. (2005). Biodiesel: An overview. *J. Braz. Chem. Soc*, 16(6B), 1313-1330.

[53] Quintella, S. A., Saboya, R. M. A., Salmin, D. C., Novaes, D. S., Araujo, A. S., Albuquerque, M. C. G., Cavalcante Jr, , & , C. L. (2012). Transesterification of soybean oil using ethanol and mesoporous silica catalyst. *Renewable Energy*, 38, 136-140.

[54] Raganathan, S. V., Narasiham, S. L., & Muthukumar, K. (2008). An overview enzymatic production of biodiesel. *Bioresour. Technol*, 99(10), 3975-3981.

[55] Salis, A., Pinna, M., Manduzzi, M., & Solinas, V. (2008). Comparision among immobilised lipases on macroporous polypropilene toward biodiesel synthesis. *J. Molec. Catal. B. Enzim.,*, 54(1), 19-26.

[56] Santos, V. M. L., Silva, J. A. B., Stragevitch, L., & Longo, R. (2011). *Fuel*, 90(2), 811-817.

[57] Schumaker, J. L., Crofcheck, C., Tackett, S. A., Santillan-Jimenez, E., Morgan, T., Crocker, M., Ji, Y., & Toops, T. J. (2008). Biodiesel synthesis using calcined layered double hydroxide catalysts. *App. Cat. B: Env*, 82, 120-130.

[58] Seong-J, P., Jeon, B. W., Lee, M., Cho, D. H., Kim-K, D., Jung, K. S., Kim, S. W., Han, S. O., Kim, Y. H., & Park, C. (1993). Enzymatic coproduction of biodiesel and glycerol carbonate from soybean oil and dimethyl carbonate. *Enzyme and Microbial Technology*, 48, 505-509.

[59] Shay, E. G. (1993). Diesel fuel from vegetable oil: Status and opportunities. *Biomass Bioenergy*, 4(4), 227-242.

[60] Shrivastava, A., & Presad, R. (2002). Triglycerides-based diesel fuels. *Renewable Sustainable Energy Rev*, 4, 111-133.

[61] Shu, Q., Yang, B., Yuan, H., Qing, S., & Zhu, G. (2007). Synthesis of biodiesel from soybean oil methanol catalyzed by zeolite beta modified with La^{3+}. *Catal. Commun.,*, 8, 2159-2165.

[62] Siddharth, J., & Sharma, J. P. (2010). Stability of Biodiesel and its Blends: A Review. *Renewable and Sustainable Energy Reviews*, 14(2), 667-678.

[63] Silva, G. F., Camargo, F. L., & Ferreira, A. L. O. (2011). Application of response surface methodology for optimization of biodiesel production by transesterification of soybean oil with ethanol. *Fuel Processing Technology*, 92, 407-413.

[64] Singh, Couhan. A. P., & Sarma, A. K. (2011). Modern heterogeneous catalysts for biodiesel production: A comprehensive review. *Renewable and Sustainable Energy Reviews*, 15(9), 4378-4399.

[65] Soares, V. L., Nascimento, R. S. V., & Albinante, S. R. (2009). Ester-additives as inhibitors of the gelification of soybean oil methyl esters in biodiesel. *Journal of Thermal Analysis and Calorimetry*, 97, 621-626.

[66] Stamenkovic, O. S., Lazic, Z. B., Todorovic, M. L., Veljkovic, V. B., Skala, D. U., & (2004, . (2004). The effect of agitation intensity on alkali-catalyzed methanolysis of sunflower oil. *Bioresour. Tech.*, 98(14), 2688-2699.

[67] Suppes, G. J., Dasari, M. A., Doskocil, E. J., Mankidy, P. J., & Goff, M. J. (2004). Transesterification of soybean oil with zeolite and metal catalyst. *Appl. Catal. A: Gen*, 257, 213-223.

[68] Suppes, G. J., Bockwinkel, K., Lucas, S., Botts, J. B., Mason, M. H., & Heppert, J. A. (2001). Calcium carbonate catalyzed alcoholysis of fats and oils. *J. AM. Oil Chem. Soc.*, 78(2), 139-146.

[69] Tomasevic, A. V., & Marinkovic, S. S. (2003). Methanolysis of used frying oils. *Fuel Process Technol*, 81, 1-6.

[70] Trakarnpruk, W., & Porntangjitlikit, S. (2008). Palm oil biodiesel synthesized with potassium loaded calcined hydrotalcite and effect of biodiesel blend on elastomers properties. *Renew Energy*, 33, 1558-1563.

[71] Usta, N. (2005). Use of tobacco seed oil methyl esterin a turbocharged indirect injection diesel engine. *Biomass and Bioenergy*, 28, 77-86.

[72] Wang, L., & Yang, J. (2007). Transesterification of soybean oil with nano-MgO or not in supercritical and subcritical methanol. *Fuel*, 86(3), 328-333.

[73] Wang, Y., Zhang, F., Yu, S., Yang, L., Li, D., Evans, D. G., & Duan, X. (2008). Preparation of macrospherical magnesia-rich magnesium aluminate spinel catalysts for methanolysis of soybean oil. *Chem. Eng. Sci*, 63(17), 4306 EOF-4312 EOF.

[74] Wang, Y. D., Al-Shemmeri, T., Eames, P., Mcnullan, J., Hewitt, N., Huang, Y., et al. (2006). An experimental investigation of the performance of gaseous exhaust emissions of a diesel engine using blends of a vegetable oil. *Appl. Thermal Eng.*, 26, 1684-1691.

[75] Wexler, H. (1964). Polymerization of Drying Oils. *Chemical Reviews*, 64, 591-611.

[76] Yin-Z, J., , Z., Shang-Y, Z., Hu-P, D., & Xiu-L, Z. (2012). Biodiesel production from soybean oil transesterification in subcriticalmethanol with K3PO4 as a catalyst. *Fuel*, 93, 284-287.

[77] Yu, D., Tian, L., Wu, H., Wang, S., Wang, Y., , D., & Fang, X. (2010). Ultrasonic irradiation with vibration for biodiesel production from soybean oil by Novozym 435. *Process Biochemistry*, 45, 519-525.

[78] Zeng, H., , Y., Feng, Z., Deng, X., Li, Y., & , Q. (2008). Activation of Mg-Al hydrotalcite catalyst for transesterification of rape oil,. *Fuel*, 87(13-14), 3071-3076.

[79] Zhou, W., Konar, S. K., & Boocock, D. G. V. (2003). Ethyl esters from the single-phase base-catalyzed ethanolysis of vegetable oils. *J. Am. Oil Chem. Soc.*, 80(4), 367-371.

Chapter 3

Facilities for Obtaining Soybean Oil in Small Plants

Ednilton Tavares de Andrade,

Luciana Pinto Teixeira, Ivênio Moreira da Silva,

Roberto Guimarães Pereira,

Oscar Edwin Piamba Tulcan and

Danielle Oliveira de Andrade

Additional information is available at the end of the chapter

http://dx.doi.org/10.5772/52477

1. Introduction

In this chapter we will study the processes, procedures and types of equipment used in industrial processing of soybeans for obtaining vegetable oil and its adequatestorage in small plants.

Soybeans, despite its low oil content (18% to 22%) is the second most important oilseed crop of the world, after palm oil. In 2010, it represented 27.3% of vegetable oil total produced worldwide, compared with 33.7% palm oil (pulp and almond), 15.6% rapeseed oil and 8.7% sunflower oil, together are account for 85.3% of vegetable oil total produced globally [12]. The high protein content (37% to 40%) of soybean is the main feedstock in the manufacture of feed for domestic animals. Almost 70% of meal protein that makes up the animal feed comes from soybean [7].

The demand for vegetable oils will grow, mainly by increased consumption / capita in emerging countries. The average annual consumption of edible oil of a citizen of a developed country is about 50 liters, while the world average is about 20 liters / head / year. Another factor that will contribute to this increase is the use as biofuel (biodiesel and H-Bio), the new lever to consumption of vegetable oil.

The soybean oil currently holds the 2nd position in the world supply of oils and fats, according to Oilworld. In 1990, production of oil stood at around 16.1 million tonnes, followed by palm oil with 10.8 million tons. Other vegetable oils were the significant world production of rapeseed and sunflower, both with approximately 8 million tons, and cotton and peanuts, with

---

© 2013 Tavares de Andrade et al.; licensee InTech. This is a paper distributed under the terms of the Creative Commons Attribution License (http://creativecommons.org/licenses/by/3.0), which permits unrestricted use, distribution, and reproduction in any medium, provided the original work is properly cited.

approximately 4 million tons each. Although interchangeable, each one of these oils has specific characteristics that makes it more or less appropriate depending on its final use [12].

While the supply of vegetable oils is large, each of these oils have specific characteristics that make them more or less suitable for use as a biofuel [12]. The restriction on the use of soy for biodiesel is compared to the low oil content in their grains. The oil yield per hectare of soybean, considering an average oil content of 20% and within the grain yield per area of 400 to 800 kg in a crop that produces 2000 to 4000 kg / ha, respectively [21]. The yield of soybean is around 2.8 to 2.95 t / ha.

According to the USDA (U.S. Department of Agriculture United States) for the 2010/2011 harvest, the estimate of global soybean production was of 256.1 million tons (Table 1), down 1.46% compared to 259.89 million tons produced in 2009/2010. Likewise, the global consumption of 2010/2011 was estimated at 255.284.000 tons, an increase of 7.5% compared to 237.430.000 tons achieved in the previous crop. Still, world ending stocks of the product in 2010/2011 will be at 58.21 million tons, 3.26% below the world ending stocks of previous crop (2009/2010) of 60.17 million tons.

Country	Harvested Area (Million Hectares)	Production (1000 MT)
United States	29.800	83.172
Brazil	25.000	68.500
Argentina	18.600	46.500
China	7.650	13.500
India	10.270	11.000
Paraguay	2.600	5.000
Canada	1.542	4.246
Russia	1.180	1.749
Ukraine	1.100	2.200
Uruguay	1.000	1.700
Bolivia	900	1.580
World Total	103.094	245.065

Table 1. Estimates of global soybean production for 2011/2012 harvest [30]

2. The Product: The Meal and Soybean Oil

The products of the processes are simplified form the crude oil and meal (cake) semi-defatted, which can be used for the preparation of animal feed. It is planned to be used for human consumption, since such use would imply the need for greater care and sophistication in hygiene and microbiological control. There is no waste in the process, which may cause environmental problems.

Compared with the traditional processes, the main disadvantage is the high content of residual oil in the semi-defatted meal (about 8%). Soybean oil is very unsaturated and may lead to rancidity bran under these conditions. In a small scale production, it is expected that the soy meal is consumed as it is produced, ie the time of storage would be small. Furthermore, the oil in the meal will replace the oil which is usually added in the preparation of feeds.

The second biggest problem is the allocation of oil produced. Because crude soybean oil, cannot be consumed as food without the refining process - the taste is very bad. Although oil extracted by extraction with low concentrations of phosphatides, which is equivalent to the degummed oil, its storage is not recommended for long periods. Hardly the big oil refiners will buy this product.

In the case of commodities, the market promotes an intense coordination of the system, controlling the prices of commodities. As soybean production-level family does not benefit from economies of scale, the tendency is to seek new ways to add value to the product, incorporating new features into grains. The differentiation of the grains can open prospects for more efficient production, processing or use, making common grains, marketed as mere commodities, specializing in products with high added value and commercial [11, 10].

3. Production Process

There are two production processes oils and fats. For materials with high oil contents (over 30%), it uses the pressing process. For raw materials with lower levels of oil, it uses the solvent extraction. In the extraction by pressing the residual oil content of the raw material is around 10%, while in this extraction solvent content can be reduced to less than 1%.

In industrial processes, typically raw materials rich in oil is pressed up to a residual oil content of about 20% and the remaining oil is extracted by solvent. Thus, the soybean oil (20% oil) is usually extracted only solvent as sunflower oil (45% oil) is partially removed by pressing and the remaining solvent.

The oil obtained in these processes, known as crude oil, generally undergoes a purification process (refining) before being consumed as food. The only exception is the commercial-scale oil (olive oil) olive oil that is consumed without refining (oil "virgin"), although other oils such as sesame, sunflower, peanut oils can be consumed raw. The soybean oil, cotton and canola are consumed only after refined.

The residues of extraction, pie, if the pressing, bran, in the case of solvent extraction, less than 20% are used for human consumption. They are generally used for the preparation of animal feed.

The oil extraction process can be divided into three phases. The first involves the pre-cleaniness, drying and storage of product to be processed. The second step concerns the preparation of the grains for the oil extraction, by facilitating the extraction processes, such as the loss of grain, conditioning or heating, lamination, and expander. Finally, the third stage involves the extraction itself, which may develop by pressing or solvent. Figure 1 details the phases and steps of the extraction process.

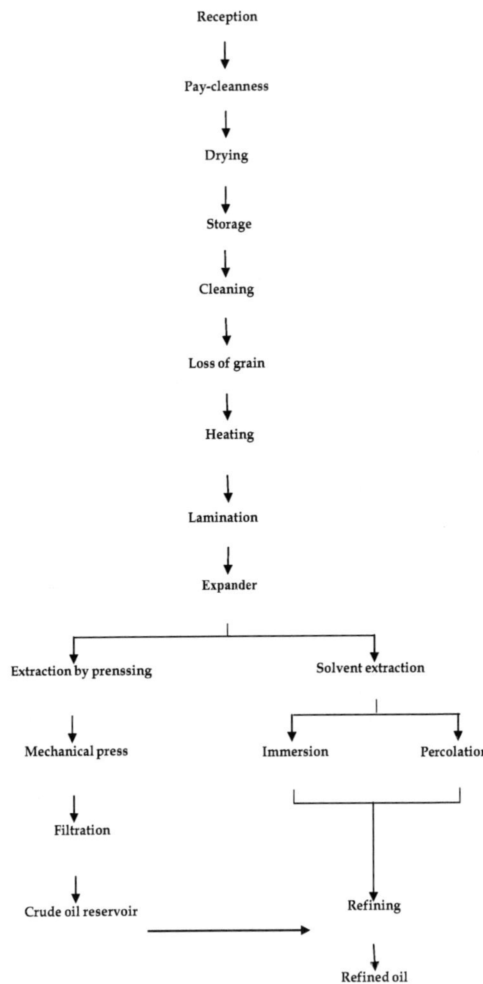

Figure 1. Flowchart for the production of soybean oil

3.1. Storage of raw material

Considering itself it importance of all previous the productive stages the harvest, since the election of the seeds to the cultural treatments, to one adjusted storage, the grains must pass for two important stages: the first one is the daily pay-cleanness, in which all the impurities must be removed, therefore it intervenes directly with the income of the process of oil extraction, useful life of the involved machines, beyond serving as inoculate of plagues and harmful microorganisms. The second stage is the process of drying that has as objective to guarantee the reduction of the moisture content of the product, of form to minimize the deterioration processes during the storage.

The stage of storage is essential for the maintenance of the quality of the raw material to be processed, influencing directly in the final product quality [23]. During the storage of the grains, that must be carried through in silos or specifically projected warehouses, the characteristics of temperature and moisture content of the product they must be monitored. Such characteristics are determinative in the minimization of the losses for deteriorations caused for microorganisms and attack of plagues, beyond being essential to prevent the occurrence of oxidative processes in the interior of the grains.

In what it says respect the quality of the oil during the storage, the variations in the moisture content of the product, as well as of the temperature can provide to enzymatic and oxidative reactions of the present oil in the interior of the grain, providing alterations in the characteristics and disposal of acid the fatty gifts. For the case of the soy grains, the oil if finds deposited in lipid bodies (the Spherosomes) distributed throughout its endosperm. Therefore, the control of such characteristics throughout all the previous stage the extraction of the oil is of basic importance for the guarantee of a quality by-product, as much in artisan scale as in industrial scale. In accordance with [23], low the quality of the crude oil influences in the increase of the losses and expenses with refining, providing a lesser income.

3.2. Preparation of the raw material

The extraction of the oil presents as by-products, beyond the involved lipid fraction with the rude oil, also proteins and carbohydrates that constitute the pie or bran, much used in the food industry, as much for animals as for human beings. Therefore, for the guarantee of the quality and not contamination of by-products, the value aggregation, and the integral exploitation of the processed product, must be carried through the preparation of the grain the form to guarantee a maximum separation between the oil and the bran.

In what the oil production says respect proceeding from the soy grain, that presents about 20% of oil, two methods extraction can be used: the solvent extraction for and the extraction for pressing. Although some authors to characterize the extraction for pressing of the soy as not being economically interesting in reason of its low oil content, indicating the extraction for solvent, in this book will be presented the preparations of the necessary raw material for the two forms of extraction.

The purpose of the stage of preparation of the grain for the extraction of the oil involves to provide the increase of the susceptibility of disruption of lipid organelles, contained in endosperm, through thermal and mechanical treatments. For this, after the withdrawal of the mass of grains of the storing unit, with moisture content of approximately 10%, in wet base (b.w.), is indicated, again the separation of the impurities that still can be contained in the mass of grains.

After this stage, must be carried through the form grain in addition to lead the unfastening and the separation of the rinds, that are abrasive, and to favor the uniformity of the size of particles to be processed. The grain in addition can be carried through by means of a breaker called equipment of coil that consists of the disposal of two corrugated cylinders, parallel made use that turns with different speeds, while the mass of grains is lead between them. During this process, for the withdrawal of the remaining rind a pneumatic aspiration is carried through, jointly with bolters splitting.

In this process of in addition the grains the milling of the grain is not indicated, since it negative intervenes with the separation between the rind and the remaining structure of the grain, the cotyledon; beyond, for the case of the extraction for solvent, to make it difficult the separation of the solvent and the oil of the bran. Another important determinative characteristic in the success of this stage is the maintenance of the grains with moisture content of, approximately, 10%, in b.u., of form to prevent the embouchement of the machine for higher the moisture content cases, or the dust production for very low moisture contents.

After the grain in addition, must be carried through the baking of the processed mass of grains. This stage has as objective to provide to a fast increase of the moisture content of the mass of grains, producing a bigger plasticity to the mass, and minimizing the dust production. This process is presented as a facilitator to rupture of the Spherosomes cell walls, in order to facilitate the leakage oil.

In accordance with [18], the increasing of the moisture of the flakes, the cell wall disruption and subsequent increase in the permeability of cell membranes, facilitates the exit of the oil, reducing its viscosity and its surface tension, which allow agglomeration of the oil droplets and its subsequent extraction. The baking can be carried through by warm vertical cookers through chambers with warm vapor, or horizontal rotating drums for a warm tubular beam the vapor.

After this preparation of the raw material, mainly for the case of the extraction with solvent, is indicated the accomplishment of plus others two stages: the lamination and the expansion. These two stages have the objective in the distance to provide the minimization of between solvent and the oil, favoring the extraction process.

The lamination if bases on the flake attainment from pieces of grains, with the objective to minimize the internal resistance in favor of the extraction of the oil. For the confection of flakes, the grains broken and with raised temperature more are lead, in the rolling mill, enter a pair of made use smooth cylinders horizontally that 0,3 mm jam pieces of the grains of soy in blades of 0,2, forming flakes [23]. In this process, the flake production with homogeneous thickness directly is related with the efficiency of extraction of the oil and, mainly, with the quality and pureness of the produced bran, since this characteristic influences in the interaction between the oil and the solvent during the extraction.

The following stage is called of expansion. In this stage, the flakes are, again, humidified with warm water vapor and for attrition throughout screws without end that lead the material until a perforated plate. In reason of the difference of pressure before and after the ticket for this plate, the warm and humidified flake suffers the process from expansion. In accordance with [23], this expansion occurs in reason of the starch presence in the grain.

The ticket of the mass of grains for the expander, or expander - extruder, propitiates greater porosity to the mass and permeability to solvent, favoring the contact between solvent and the expanded cell, guaranteeing a more efficient percolating between the oil the solvent. According [23], the use of the expander implies in the increase of about 40% in the capacity of extraction for solvent. After the expansion, the material must be cooled until the temperature of extraction of the oil, and depending on the conditions of the processing, the mass must pass for the form drying to guarantee a bigger efficiency in the extraction process.

3.3. The extraction

The extraction of the oil for being carried through two different methods: the extraction mechanics through the pressing, or by means of the chemical extraction for solvent. In situations special, of form if to get the maximum efficiency of extraction it can be used the two methods sequentially. To follow, the two forms of oil extraction will be presented proceeding from oil seeds, as it is the case of the soy.

3.3.1. Extraction for pressing

The method of extraction for pressing consists of the withdrawal of the oil by means of the application of a external pressure on the mass of grains, through the pressing mechanics. As main involved advantages with the use of the press mechanics for the oil extraction low cost of installation is its, not the use of solvent, and not the necessity of posterior refining of the oil, what it implies in the reduction of the processing cost and, consequently, of the gotten oil, favoring the use of the same one for small producers.

Currently, the press more common mechanics is the continuous press of screw, also call of expeller, that hopper of feeding is composed for one, that leads the material to be pressed by means of a screw without end of step interrupted for steel rings, made use parallel; to the end of the set, they find if a cone tip that regulates the speed of exit of the pie and the chamber pressure on the mass.

The pressing process if develops from the introduction of the mass of grains in hopper that it feeds the screw without end, compressing it against steel rings, providing the elimination of the oil for the orifices. The extraction speed depends directly on the imposed pressure, that initially must understand of 300 the 400 kg cm^{-2}, but throughout the process due to the gradual accumulation of mass in the interior of the press, the pressure can be superior the 1,000 kg cm^{-2} [23].

During the process, the mass of grains equally is pressed, preventing the resorption of the oil for other parcels of the mass. After the ticket for the press, the rude oil must be filtered with the purpose to separate the solid residues proceeding from the remaining pie.

3.3.2. *Extraction for solvent*

The oil extraction for solvent can be used as only method of extraction, or same as comple‐ ment to the extraction mechanics, in the daily pay-extraction form. This method of extrac‐ tion if bases on the absorption of the solvent for the lipid cells, where in its interior it has the dissolution of the oil, that, later, for leaching, he is loaded for the exterior of the cell. This process, in adjusted conditions, approximately removes 99% of the oil contained in the mass of grains [23].

However, for the guarantee of the efficiency of the process, it is essential the adjusted prepa‐ ration of the grain and the choice of to be used extractor, of form to guarantee the maximum contact of the solvent with the cellular wall. Of this form, how much bigger it is the amount of cells breached throughout the preparation of the mass of grains, faster it is the extraction process, since the solvent will only go to dissolve the free oil, not needing to carry for diffu‐ sion the dissolved oil to the external region to the cell.

In accordance with [23], the transport of lipid throughout the cellular membrane occurs in function of the variation of its permeability (initially impermeable to the lipid) in function of the difference of the internal and external osmotic pressures to the cells. The increase of in‐ tracellular pressure in virtue of the action of the solvent it provides the expansion of the membrane and, consequently, the dilatation of the pores of the cellular membrane, allowing the ticket of the solvent oil solution and for the extracellular region, which had to the gradi‐ ent of existing concentration [23, 26].

The extraction process occurs in higher temperatures, seen its influence in the viscosity of the solvent oil mixture and, and in the solubilization of the oil in the solvent. The extraction speeds from beginning to end of the process of solvent extraction progresses differently. Ini‐ tially, when the oil of better quality is extracted, the process if develops of fast form, due to the biggest gradient of concentration, however, throughout the process this speed diminish‐ es, and the extracted oil to the end presents minor quality, in reason, mainly, of the presence of other cellular composites that provide losses throughout the refining.

From this process, the oil extraction for solvent can be developed of two forms, for immer‐ sion, where the mass of grains is kept immersed in the solvent for a definitive period of time, or by percolating, which the mass of grains is made use in layers to guarantee opti‐ mum contact of the solvent that it passes freely between the grains and it is renewed when it has saturation. To the end of the extraction, as much the solvent mixed to the oil, as the sol‐ vent gift next to the remaining bran can be recouped and be reintroduced in the process. For the grains to solvent extraction with low concentrations of oil is indicated by immersing the system [2].

For the extraction of the oil, frequent the commercial hexane is used as solvent. Proceeding from the refining of the oil, the hexane presents determinative characteristics for its good performance as solvent, as: it is a composition to apolar total being miscible in oil at the same time that it is immiscible in water, presents low latent heat of boiling, and it does not react with the constituent material of the equipment used for the extraction. Although its fa‐ vorable characteristics physicist-chemistries, this solvent requires special care with the security in the industrial plant and with its manuscript, seen its high inflammability, explosion and toxicity [23].

4. Refining

The refining if characterizes as the set of operations carried through after the process of extraction for the removal of residues gifts in the crude oil that can affect the color, stability, aroma, and flavor, beyond its physical characteristics. These residues are preceding from drag mechanics, and/or solubilization of other substances in the oil or solvent the occurrences during the extraction process.

The refining takes place in two stages: the first step is the physical removal of substances, while the second involves the refining processes through neutralization, clarification, and deodorization. In accordance with [31], initially the oil passes for a stage of physical separation, where, in a tank, the separation for gravity of insoluble substances is carried through. After this stage, the oil passes for the degumming that consists of the removal of phospholipids, sugars, resins, and breaks up of soluble proteins in water.

In accordance with [23], the presence of phospholipids in the soy oil favors the occurrence of losses, in reason of the formation of depositions with presence of about 35% of oil in the deep one the tanks of deposition, with aspect similar to a gum. In this in case that, the phospholipids can be presented of two forms: hydratable, being liable of withdrawal with addition of water and centrifugation, and not hydratable that needs the addition of acid citric phosphoric or for becomes it hydratable for its posterior withdrawal.

Still of according to [31], after the degumming process, must be proceeded with the neutralization, that the withdrawal of fatty acid involves, pigments, remaining phospholipids of the degumming, and soluble sulfur composites in water. In the neutralization process it has the caustic soda water addition, with fatty acid the purpose to reduce the text of free, to clot the phosphatides and the gums, to degrade part of the dyes gifts, and to load the insoluble substance for the clotted material. Of this process it has the production of you leave organic sodium or soaps, what it results in the necessity of one another stage, the *laundering*.

The laundering is a process that results in the withdrawal of the soap produced during the neutralization. It is based on the *distilled* water addition the temperature of 85-95°C (about 10-20% of the volume of oil) for the elimination of the soda water and the foam of the oil, however case the separation between the oil and the water is difficult indicates it dilution in the aluminum sulfate water. The laundering process can more than proceed a time until the total exemption from soap from the oil. To finish the process laundering, in reason of the water addition, the drying must be proceeded from the oil.

After the drying, the oil is directed to the *clarification (bleaching)* that objective the pigment elimination, residual products of the oxidation, metals weighed, and soaps, of form to guarantee the improvement of the flavor, odor and oxidative stability of the oil. Bleaching occurs from the adsorption of specific adsorbents substances for the elimination of substances that confer coloration, sulfur, soaps and metals to the soy oil. The adsorbents substances can be some types of silicates, diatoms lands, clays acid activated, silica and active coal [23].

In accordance with [14], throughout the refining, the clarification produces an oil with bigger susceptibility to the oxidation, being indicated for the storage the use of absent nitrogen stream bed of oxygen.

After the clarification, the oil follows for its last stage of refining, the deodorization. The deodorization of characterizes for the substance elimination formed during the storage and processing of the grain and other natural substances of the oil that can provide to awkward flavor and odor. For the withdrawal of these substances to use it distillation with chain of vapor to the vacuum, of form to guarantee the protection of the oil to the effect of the atmospheric oxidation, prevention of hydrolysis for the vapor, and the reduction of the necessary amount of vapor for the process.

For some types of oils, as it is the case of the palm oil and peanut, the physical refining is proceeded after the accomplishment from the deodorization. The physical refining consists of the elimination for evaporation of fatty acid the free gifts in the oil, providing bigger income of the fine oil. For the case of the soy oil, the physical refining is not indicated, in reason of its low acidity after to the end of the process and the difficulty in the degumming process.

5. Quality

Every product has a number of characteristic attributes. It's called quality, whose existence will define the success or failure in their marketing. This quality is mainly observed by two fundamental aspects: the first relates to the consumer who seeks desirable characteristics, whether from an economic standpoint, nutritional, aesthetic, etc. The second aspect refers to the legality, where the product goes through a series laboratory analysis and is classified into pre-established standards and its final quality is attested.

According [13], quality can be defined as a set of characteristics that will directly impact on the acceptability of the product. In this context, the subjectivity of the sensory analysis of each person are determinants for to preset of the quality, that can involve the appearance, texture, flavor and aroma.

The color, size, shape, integrity and consistency are factors directly involved in the appearance of the product, already the relateds with the physical senses of touch, and mouth are determined on the according to texture; the flavor factors involve the taste, the aroma, and correlate with the olfactory sensitivity. Besides these, you can also to relate the after taste that occurs according to a secondary analysis of the product [13].

Associated with these organoleptics factors for the acceptance of the product are also the nutritional value, the presence of toxic substances and the final price. All these characteristics are to related to the type of raw material and production method. In the case of vegetable oils, with is the case of the soybean oil, these factors are directly related to the conditions of storage and processing of the beans, as has been seen above, as well as the characteristics, extraction and treating the produced oil, and form of storage and shipping.

Overall, the changes and loss of quality of the agricultural products are related with the growth and activity of microorganisms, the action of enzymes, chemical reactions, attack by insects and rodents, and physical changes caused by mechanical agents. According [17], the major causes of food deterioration are, respectively, the attack of microorganisms and oxidative processes.

The vegetable oils have fewer characteristics in your reactive molecule, unlike proteins and carbohydrates presents in the grains. The reactivity of the oil is concentrated, mostly, in the hydrolysis that produces free fatty acids according to the moisture content present, lipolytic enzymes and temperature, and the oxidation reaction of lipid compound that is a function of the concentration of oxygen in the medium.

The hydrolysis process provides the breakdown of triglycerides and therefore the increase of free fatty acids, that influences the acidity of the vegetable oil. High levels of acidity in vegetable oils leads to high losses during the refining stage [23]. However, in general, for to monitor the deterioration of the vegetable oils is accomplished due to the oxidative reactions occurred in the triglyceride molecule.

Oils	Fatty acids composition (Wt.-%)								
	12:00	14:00	16:00	18:00	18:01	18:02	18:03	22:01	IV
Canola	0	0	4.5	1.5	58.5	25	9	1.5	118
Corn	0	0	11	2	36	50	1		121
Sunflower	0		6	4.5	32	57.5			126
Soybean	0		8	4	28	53	7		130

Table 2. Vegetable oils fatty acid composition

According [24], all vegetable oils consist primarily of triglycerides. The triglycerides have a three-carbon backbone with a long hydrocarbon chain attached to each of the carbons. These chains are attached through an oxygen atom and a carbonyl carbon, which is a carbon atom that is double-bonded to second oxygen. The differences between oils from different sources relate to the length of the fatty acid chains attached to the backbone and the number of carbon–carbon double bonds on the chain. Most fatty acid chains from plant based oils are 18 carbons long with between zero and three double bonds. Fatty acid chains without double bonds are said to be saturated and those with double bonds are unsaturated (Misra and Murthy, 2010).

In general, vegetable oils are made especially for fatty acids with chains between 12 and 24 carbons: Lauric (C12:0), Myristic (C14:0), Palmitic (C16:0); Palmitoleic (C16:1) Stearic (C18:0), Oleic (C18:1), Linoleic (C18:2); Linolenic (C18:3); Arachidic (C20:0); Gadoleic (C20:1); Behenic (C22:0), Erucic (C22:1); Lignoceric (C24:0). The proportions of the fatty acid composition can be determined by gas chromatography method. The weight composition of fatty acids found in soybean oil was: C16:0=11.6%, C16:1=0.1%, C18:0=32%, C18:1=20.4%; C18:2=57.1% and C18:3=5% [24].

Average values of the chemical composition of vegetable oils and values of iodine index are presented in the Table 2 [15] and physical properties of soybean oil in Table 3. The iodine Index or Number is one characteristic of the oil that measures its index of unsaturation, when evaluating the amount of iodine necessary to saturate the double links of the molecule.

PROPERTY	SOYBEAN OIL
Density kg/L (20 °C)	0.92037
Viscosity mm²/s (40 °C)	30.787
Flash Point °C	332
Cloud Point °C	-2
Pour Point °C	-14
Copper strip corrosion	1b

Table 3. Physical properties of soybean oil [24]

Both the hydrolysis reaction as the oxidation reaction can also be influenced by temperature, light, presence of unsaturated fatty acids, moisture content, and product type. As a result of these reactions is the development of different flavors and odors that compromise the quality and acceptability of the product. For these changes organoleptics give the name of rancidity. Addition, for the production of biodiesel, the presence of free fatty acids in vegetable oils affect the process of separation of soaps, and influence on the reactions of esterification, compromising the efficiency of the process.

Thus, in that it involves, specifically, the oxidation of lipids, the main causative factors are the composition of fats, the presence of oxygen, temperature, and luminosity. The composition of fats influence on the presence of unsaturation in the molecule, as the unsaturation increases the susceptibility of oxygen absorption, that the more present in the environment, more available for the reactions is presented. Both the temperature and luminosity influence, proportionally, in the rates of reactions. For the specific case of light, its presence influences on accelerating the development of rancidity in fats.

According [8] and [27], the quality of the oil can be followed depending on the index of acid determined by the presence of free fatty acids; index of saponification which demonstrates the presence of oils or fats high proportion fatty acids, color; foaming, viscosity, density, and index of peroxide that is determined by the presence of iodine. According to [24], the Table 3 shows the properties of the soybean oil.

It is shown in Table 4 that vegetable oils present corrosion within the pattern established for diesel oil (Standard corrosion = 1) according to ANP.

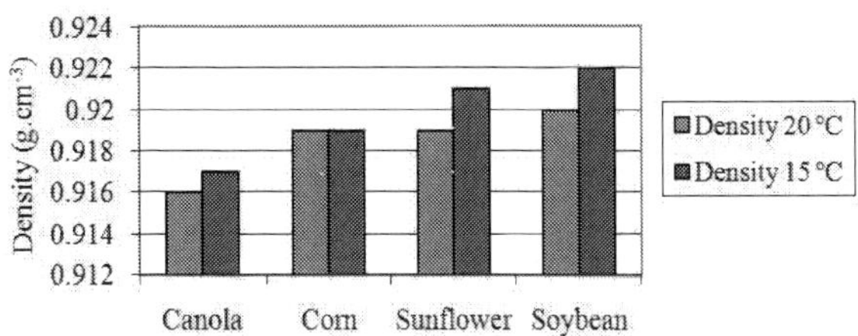

Figure 2. Density in temperatures of 15 and 20 °C

Vegetable Oil	Corrosivity
Canola	1a
Sunflower	1a
Corn	1a
Soybean	1b

Table 4. Corrosivity of vegetable oils at temperature of 100 °C

Looking at Table 5, 6 and 7, canola oil has the lowest density of 0.917 g.cm^{-3} at a temperature of 15 °C and 0.916 g.cm^{-3} at a temperature of 20 °C and soybean oil has the highest density, 0.922 g.cm^{-3} at a temperature of 15 °C and sunflower and corn oil showed a higher density at a temperature of 20 °C, 0.919 g.cm^{-3}. According to the specifications of the ABNT standard, diesel has a density of around 0.865 g.cm^{-3} (Figure 2).

In Table 3 the mean values and standard deviation of the viscosity of vegetable oils are shown.

Vegetable Oil	Kinematic Viscosity (mm².s^{-1})	Standard deviation
Canola	35.5278	0.0081
Sunflower	31.7275	0.0726
Corn	33.7713	0.0409
Soybean	31.6107	0.0093

Table 5. Kinematic viscosity values of vegetable oils obtained experimentally

The test results of kinematic viscosity of vegetable oils are shown in Figure 3 below:

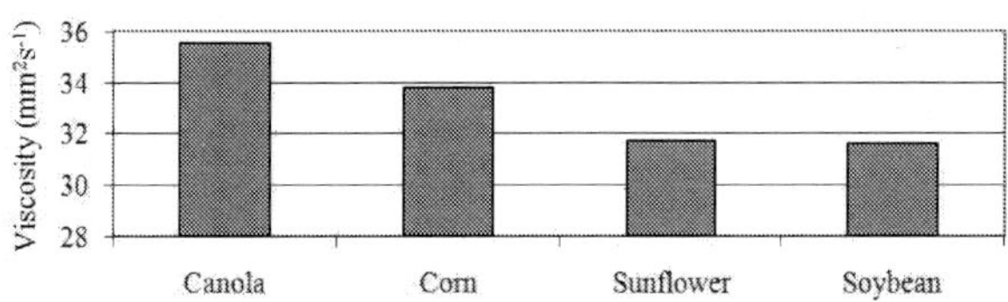

Figure 3. Kinematic viscosity of vegetable oils

Table 6 show the temperatures measured in the tests of cloud.

Vegetable Oil	Temperatures of Cloud Point	Temperatures of Flow Point
Canola	-1 ºC	-20 ºC
Sunflower	1 ºC	-18 ºC
Corn	3 ºC	-12.5 ºC
Soybean	-2 ºC	-18.5 ºC

Table 6. Temperatures of cloud points and flow points of vegetable oils

Vegetable oils have a density slightly larger than that of diesel fuel and below water, they are products of easy handling and processing. Their viscosities are far from being similar to diesel fuel, but this allows their use in cases of thermo-chemical conversion without major problems, and are easy to carry through pipelines without large deformation of tension and energy.

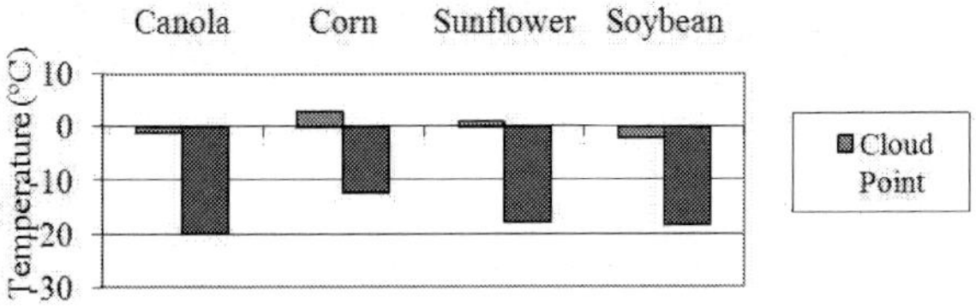

Figure 4. Results obtained experimentally for cloud point and flow point of vegetable oils

6. Control the Production Management of Soybean Oil

The acquisition process of vegetable oils occurs through of relatively simple procedures, starting with the attainment of oilseeds, proceeding to mechanical extraction, filtering and marketing of oil. This understanding contributes to the study of the entire production chain.

Generally the oil extracted is sent to the food industry, or the production of biofuels, and in some cases is used as feedstock in other industrial processes. Soybean oil has its applications in the first two cases cited. [16] agribusiness of the soybean oil has a high processing capacity, with oil production exceeding 50 million liters per year.

Vegetable oil	Density 15°C	Standard deviation	Density 20°C	Standard deviation	Corrosion	Kinematic Viscosity (mm^2.s^{-1})	Standard deviation	Cloud point	Flow point
Canola	0.917	0.0005	0.916	0.0005	1a	35.5278	0.0081	-1 °C	-20 °C
Sunflower	0.921	0.0010	0.919	0.0005	1a	31.7275	0.0726	1 °C	-18 °C
Corn	0.919	0.0005	0.919	0.0005	1a	33.7713	0.0409	3 °C	-12.5 °C
Soybean	0.922	0.0000	0.920	0.0006	1b	31.6107	0.0093	-2 °C	-18.5 °C

Table 7. Physical characterization of vegetable oils [29, 1]

It is known specific attributes which must be met after defining the destination of oil extracted, not all characteristics of a product can be easily evaluated. In the case of agri-food products there is the existence of personal or subjective parameters that hinder your judgment. Thus, the oil-producing units paying attention to product quality, this will help to stay in the market and seeing new opportunities.

[3] reported that the quality of a product is resulting from the joining of items:

a.   Quality of product design: that is the result of the activities that translate the knowledge of market needs and opportunities in information technology for the production.

b.   Quality of design process: is results from the efficient translation of the design specifications of the product in process at various levels, such as process flow diagram, layout, design tools and equipment, project work etc..

c.   Quality of conformation: is the result of the efficiency level of actual production.

d.   Quality of services associated with the product: it is the result of the quality level of the stages of distribution and marketing, beyond quality of after-sales services.

[16] reported that the limit of the concept of quality has advanced in segments related to the management of the productive sector. It is clear that the companies' survival now depends on the quality of products or services it offered. Several agents have sought to monitor quality through product recalls, allowing verification of consignments dispatched by identifying the weaknesses of the production complex, such as inefficiency in logistics and handling conditions and storage of the product.

The production of vegetable oil in small units can be designed more efficiently by understanding the concept of agro-industrial chain, because according to [20] in recent years the so-called industrialization of agriculture has led to increasing dependence on agriculture to industry. Thus small farmers can participate in a competitive market where strategic alliances allow to produce and market their products. Given this perception, competitiveness is to migrate the level of individual firms to competition between agribusiness chains. Thus, because it is an integrated system there is a need to coordinate the supply chain, and the search for quality is now undertaking the various agents.

According [3] the coordination of quality is a set of activities planned and controlled by a coordinator agent, in order to improve quality management in the chain and ensure product quality through a process of transaction information, helping to improve the matching of customers and to reduce costs and losses at all stages of the chain.

This concept of quality coordination, "to plan", "controlling" and "improve" the quality management, "the process of information transaction" and "coordinating agent" are:

- Planning: Planning activities in order to create a process capable of producing products that meet consumers;

- Control: Control processes and activities with the objective of evaluating the performance of real quality and act if there is a diversion;

- Improve: set of activities that aims to improve the quality of processes and products;

- Transition process of information: the acquisition, management and distribution of information throughout the production chain;

- Coordinating agent: a key to the coordination of quality aims to make information related to product quality and quality management are identified, communicated and controlled throughout the chain.

The coordinator agent can be a company, a group of companies, an association of representatives of the segments of the agribusiness chain or a third party contracted to perform the tasks for the agent coordinated.

The coordination process quality in agro-industrial chain of soybean oil, produced in small units, can be strengthened by adopting measures to encourage the involvement of its agents. Usually the resources of small producers to cover production are limited, requiring therefore of a reliable outlook for the sale of its production. At this point, there is the need of productive sector strategies are aligned with unusual strategies for the development of the productive chain.

The literature dedicated to the study of agribusiness systems suggests practices to be implemented from the producer to the customer. Measures such as discounted prices, flexibility in time and remuneration for services, contribute to the strengthening of alliances among its members.

The strategies that comprise the agribusiness chain agents, shall consider agribusiness products are subject to special features such as seasonality of production, the need for special conditions for transport and storage and other care as they are perishable. The following paragraphs deal with aspects related to the quality of agro-industrial products, particularly vegetable oils.

6.1. Methodologies applied to quality management

Methodologies will be presented that can be applied to quality management in the production of vegetable oils in small units.

6.1.1. Application of ISO 9000

The International Organization for Standadization is responsible for standardization in global character, is the ISO 9000 group linked the quality management systems, which was created searching to standardize the rules for products and industries.

In addition to ISO 9000, regarding the standards focused on vocabulary and fundamentals, there is the importance of ISO 9001 that dealing with Quality Management Systems and Guidelines for SGQs auditing.

A process model of ISO 9001 may be seen as an implementation of the PDCA cycle, according [19], in which:

- PLan: It means setting objectives and processes necessary to deliver products or services in accordance with the requirements and policies of the organization;

- Do: It means the implementation of procedures;

- Check: It means to monitor and measure processes and products by comparing them to the policies, objectives and product requirements and report the results;

- Action: It means to take actions to continually improve process performance.

The most common steps for the implementation of ISO 9001 are:

- Conviction of top management

- Choice of coordinating implementation

- Assessment of current situation

- Preparation of schedule

- Training leveling

- Establishment of working groups

- Specific training

- Process Mapping

- Development / deployment documentation

- Training of internal auditors

- Performing internal audits

- Corrective Actions

- Training end

- Pre-certification audit, and

- Setting and maintaining the quality management system ISO 9001

6.2. Hazard Analysis and Critical Control Points (HACCP)

According [4] the HACCP is based in the analysis method of the mode and effects and fault causes or failure, mode and effect analysis (FMEA), developed by Kaoru Ishikawa in the Japanese industry. The HACCP allows to identify and assess hazards associated with different stages of the food chain, and define the necessary means for its control. The HACCP should be considered as a quality system, a rational practice, organized and systematic, indicated to provide the necessary confidence that a agri-food product will meet the health and safety requirements expected by the consumer.

Studies based on the concept of prevention, developed by [5], concluded that the HACCP represents an advance in food safety, when the adoption of preventive measures promoted the effective design of food safety and the processes in which, a priori, was analyzed quality (microbiological, physico-chemical and sensory) of the products already processed.

6.3. Traceability

At first glance, it is believed to be unnecessary to use tools that require high technological standard in small units of oil extraction. However, products offered in different ways may present a promising alternative for small farmers. Traceability aims to identify the origin of the product from the farm to the consumer. This principle is indifferent to processes in which the product has undergone. Thus, it is expected to reach the final with a quality product, and with a known source.

To [9] an efficient screening process should consist of the following elements:

Standards and / or the quality that aim to protect / secure;

Procedures allowed, prohibited, tolerated and required;

Grace periods or transition established as provided in the rules;

Requirement that producers are provided with proof of purchase, sales, everything that allows inspectors to check compliance of standards by the operator (owner of the process);

Periodic visits to the default setting;

Visits "surprise" the establishment.

According to [19] in soybean production, traceability is an essential tool for identification and separation of genetically modified organisms that are restricted in certain countries or markets. Thus, traceability systems are necessary for the management and development of the agribusiness chain, ensuring quality food offered.

6.4. Good Manufacturing Practices

Such standards can be applied in the production of oil intended for human consumption. Procedures are designed to get quality products. The Good Manufacturing Practices (GMP), according to Tomich et al. (2005), are a set of standards used in products, processes, services and buildings, for the promotion and certification of quality and food safety.

According to [6] requires that each establishment has its Manual of Good Manufacturing Practices, which details the conditions of hygiene and sanitary food handling procedures, cleaning equipment, utensils, facilities and buildings of establishments, in addition to establishment of minimum health requirements of buildings, facilities, equipment and tools, control of water supply, health and hygiene of food handlers, the integrated control of pests and vectors, and control and quality assurance of final products.

The following legislation provides for the establishment of Good Manufacturing Practices (GMP), in Brazil, whereas other related items: Standards of Identity and Quality, sanitary inspection, Standard Operating Procedures and Checklist of GMP.

- Ordinance no. MS 1.428/1993

- Ordinance no. 368/1997 of the MAP

- Ordinance no. 326/1997 of the MS / SVS

- Resolution RDC / ANVISA no. 275/2002

6.5. Cleaner Production: CP

The methodology of CP is the application of an economic strategy, environmental and technical, integrated processes and products in order to increase the efficiency of use of raw materials, water and energy, not the generation, waste minimization and recycling with environmental and economic benefits to the productive processes.

To have success in the development of cleaner production, are required to exercise responsible environmental management and technology assessment, the initial effort is fundamental in changing attitudes, including a subjective evaluation of the entire process.

All factors imply aggregate values and services, with the primary concern of consuming less material and generate less pollution.

[25] studied the management of agro-industrial residues. As a result it was proposed a roadmap for Environmental Management System (EMS). Operational actions were suggested as: Mass Balance identifying and quantifying infrastructure resources; Anticipation and monitoring the adoption of measures to prevent accidents or damage to the ecological nature of the life cycle analysis of products, Cleaner Production (CP) and Reverse Logistics, which is to collect, package, transport and dispose of waste that were generated in the activities of obtaining Raw Material and managerial actions suggesting adoption of sustainable strategies and policies, Environmental Management System as a mechanism to monitor the administrative and managerial performance the organization, Environmental Audit and Environmental Education.

7. From Control to Management

Currently, the concern with quality extends beyond the aspects mentioned above, no longer a mere bureaucratic requirement of regulatory agencies and inspection, but a fundamental strategy and essential to ensure competitiveness. The quality is replaced by a much broader approach, involving all levels of the organization and process.

8. Quality attributes

The main attributes that describe the desirable characteristics in grain for the quality of the bran are high in protein concentration, profile and level of amino acids, especially lysine, and high energy. To control the meal must consider factors such as moisture content, oil content, protein content, the urea activity and the rate of protein digestibility. And in the case of crude oil, emphasizes the determination of fatty acids and their state of rancidity.

These attributes directly influence the quality and safety of the product and the forms of controls are given in the text. It should be laid down the acceptable quality level (EQS) according to the intended use of the product. Obviously, the level of residual bran oil obtained by pressing has maximum and minimum other than the one extracted by solvent. The determination of the EQS should consider the characteristics of the process to optimize your control.

This item will be established quality standards obtained for soybean oil for immediate and latent analysis.

9. Process Control

As animal feed, the control of raw materials and products may be carried out with simpler and less frequently. The analyzes have suggested the purpose of controlling the quality of raw materials, products and the process yield. The quality and characteristics of the raw material has great influence on product quality and yield. Soybeans should be free of mold, broken grains, greens and other defects.

During the extrusion process, there is an internal friction of grain against the internal elements of the extruder rising to temperature and pressure with the complete inactivation of the activities: ureatics, antitrypsin and hemoglutinant (anti-nutritional factors). Exposure of the grain at high temperature and pressure for an extremely short time (20-40 seconds) favors the obtaining a quality product without compromising the nutritional quality of soybean. Therefore, factors such as temperature and pressure during the process should be monitored and recorded continuously. In annex presents tables to assist in process control. Below we suggest some analysis should be performed periodically.

10. Installations and Equipments

The following relate to the main points that should be taken into consideration when choosing a site to be deployed to agribusiness:

* the potential for obtaining the raw material in the region should be higher than the projected demand of the plant and enable future expansion in production;

* water supply reliable and good quality (drinking water);

* providing sufficient electrical power without interruption;

* availability of skilled labor, including technical personnel;

* no contaminants of any kind on the outskirts of agribusiness;

* road infrastructure in working condition and easily accessible;

* availability of sufficient area to implement the agricultural industry and its future expansion.

All new property before it began, it must seek approval of their facilities by the regional body of the Labor Ministry or agency responsible, and this after doing a preview, issue a certificate of approval for facilities. This procedure is adopted in order to ensure that the new establishment activities free of accident hazards and / or occupational diseases, which is why the establishment does not meet regulations will be subject to the impediment of its operation until the standard is met.

Figure 5. Seed cooker

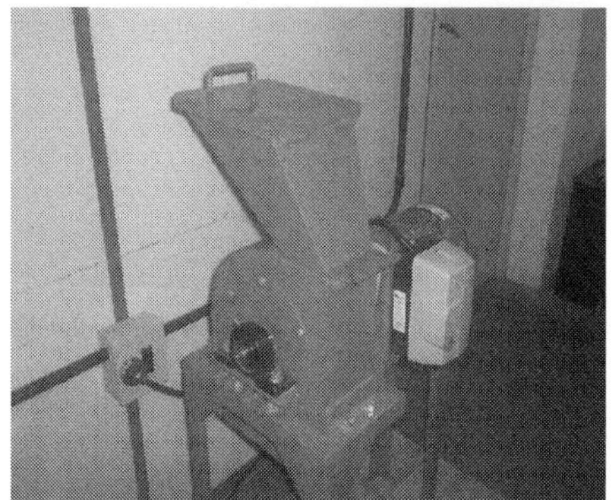

Figure 6. Crusher seeds

Figure 7. Press small

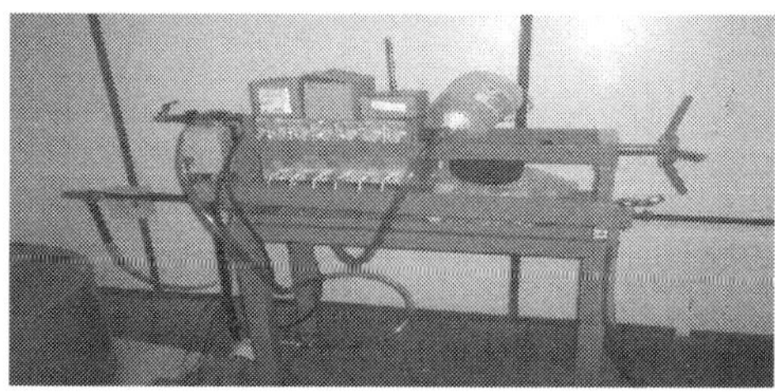

Figure 8. Filter with collect the oil bucket

Figure 9. Small plant for the production of biodiesel

Figure 10. Settling tank to the glycerin

A unit of oil extraction on the farm can be mounted in a simplified form with the elements described below. Figure 5 shows a seed cooker being installed, its use may be waived in work with seeds whose oil extraction is done cold, as in the case of sunflower seeds. The crusher shown in Figure 6 can be used to prepare larger seed. Equipment such as the press shown in Figure 7 can perform pressing in an average yield of 40 kg per hour and is suitable for small applications. The filtration must be done efficiently preventing the oil is lost, thus a device for filtering, with collect bucket oil is of great importance in the process (Figure 8). A small plant, for production of oil and biodiesel, you can compose the oil processing unit (Figure 9). Glycerin resulting product from the transesterification of vegetable oils may be separated by centrifugation or decantation. In the second case, the composition of glycerin biodiesel more tanks can be conducted as shown in Figure 10, which should preferably have conical structure inside facilitating the decanting of the glycerine.

11. Conclusion

In conclusion we have with the expansion of soy agribusiness, there is a need to add value to the product. And in case of small farmers or family farmers in the production of oil will allow your own farm profit by adding value, besides the possibility of using this product on the farm.

The values of cloud points, close to zero degrees Celsius, and flow, near minus fifteen degrees, makes manipulation of these oils possible in tropical climates, where temperatures barely reach those levels, do not show stoppage problems, flow and clogging the lines of the process involved with the use of oil.

Tests for corrosion highlights low levels of attack in copper sheets, and oil is a product that does not generate large losses of material in the areas of the process which they have contact with, on the other hand the formation of clogging points and false surfaces as a result of polymerization of oil when working at higher temperatures can be more dangerous.

The physical and chemical characteristics evaluated in this work present a great relation with the composition of the tested oils, where there is a relation between the density, viscosity and corrosion index with the iodine index of the fluid. On the other hand, the cloud and flow points are related with the presence of saturated and unsaturated fatty acids. In respect to the saturation of the oil, it can be favorable in problems such as corrosion, of low temperatures or in the case where lower viscosity values are needed.

Author details

Ednilton Tavares de Andrade[1], Luciana Pinto Teixeira[3], Ivênio Moreira da Silva[3], Roberto Guimarães Pereira[2], Oscar Edwin Piamba Tulcan[4] and Danielle Oliveira de Andrade[3]

1 Federal Fluminense University/TER/PGEB/PGMEC, Brazil

2 Federal Fluminense University/TEM/PGMEC/PGEB/MSG, Brazil

3 Federal Fluminense University/TER/PGMEC, Brazil

4 National University of Colombia, Bogota, Colombia

References

[1] Andrade, D. O., Tulcan, O. E. P., Andrade, E. T., & Pereira, R. G. (2008). Determination of the physical characterization of vegetable oil. *International Conference of Agricultural Engineering*, 4.

[2] Bockisch, M. (1993). Fats and oils handbook. *Champaign, AOCS*, 345-718.

[3] Borrás, Miguel. Ângelo, & Toledo, José. Carlos. (2006). A Coordenação de Cadeias Agroindustriais: Garantindo a Qualidade e Competitividade no Agronegócio. *Zuin, Luiz Fernando Soares; Queiroz, Timóteo Ramos. Agronegócios: Gestão e Inovação. São Paulo: Saraiva*, 464, cap. 2.

[4] Borrás, Miguel. Ângelo, & Toledo, José. Carlos. (2006). Qualidade dos Produtos Agroalimentares: A Importância da Gestão da Qualidade no Agronegócio. *Zuin, Luiz Fernando Soares; Queiroz, Timóteo Ramos. Agronegócios: Gestão e Inovação. São Paulo: Saraiva*, 464, cap. 7.

[5] Brum, Jaime. Victor Ferreira. (2004). Análise de Perigos e Pontos Críticos de Controle em Indústria de Laticínios de Curitiba- PR.

[6] Carrazza, Luis. Roberto, Maciel, Luis. Gustavo, Borges, Moacir. Chaves, Cordeiro, Augusto. César Rodrigues., & Ávila, João. Carlos Cruz. (2011). Caderno de Normas Fiscais, Sanitárias e Ambientais Regularização de Agroindústrias Comunitárias de Produtos de Uso Sustentável da Biodiversidade. *SPN- Instituto Sociedade, População e Natureza(ISPN).Brasília- DF*, 43.

[7] Dall', Agnol. A., Lazarotto, J. J., & Hirakuri, M. H. (2010). Desenvolvimento, Mercado e Rentabilidade da Soja Brasileira. *Circular Técnico 74. Embrapa Soja*, April, 18.

[8] Dobarganes, M. C. (2000). Frying fats: quality control. In: IUPAC workshop on fats, oils and oilseed analysis. *Rio de Janeiro. EMBRAPA. Book of conferences*, 29-45.

[9] Dulley, Richard. Domingues, & Toledo, Alessandra. A. Gayoso Franco. de. (2003). Rastreabilidade dos Produtos Agrícolas. *Informações econômicas, SP*, 33(3), mar.

[10] Embrapa Soja. (2012b). Uso industrial. http://www.cnpso.embrapa.br/index.php?op_page=27&cod_pai=31, Accessed: 26 March.

[11] Embrapa Soja. (2012a). Uso. http://www.cnpso.embrapa.br/index.php?op_page=25&cod_pai=29, Accessed: 26 March.

[12] FAO (Food and Agricultural Organization of the United Nations). (2012). http://faostat.fao.org/site/339/default.aspx, Accessed: 26 March.

[13] Gava, A. J., Silva, C. A. B., & Frias, J. B. G. (2008). Tecnologia de alimentos: princípios e aplicações. *São Paulo: Nobel.*

[14] Hui, Y. H. (1996). Baley's industrial oil and fats products. *5. Ed. Nova York, Wiley,* 4.

[15] Knothe, G., Dunn, R., & Bagby, M. O. (2003). Biodiesel: The Use of Vegetable Oils and Their Derivatives as Alternative Diesel Fuels. *Fuels and Chemicals from Biomass. Washington, D.C.: American Chemical Society.*

[16] Lima, Suzana. Maria Valle., & Castro, Antonio. Maria Gomes. de. (2010). A Agroindústria de Óleo Vegetal para a Produção de Biodiesel. *Castro, Antonio Maria Gomes de; LIMA, Suzana Maria Valle; SILVA, João Flávio Veloso. Complexo Agroindutrial de Biodiesel no Brasil: Competitividade das Cadeias Produtivas de Matérias-Primas. Brasília, DF: Embrapa Agroenergia,* cap. 6.

[17] Lindley, M. G. (1998). The impact of food processing on antioxidants in vegetable oil, fruits and vegetables. *Trend in Food Science and Technology,* 9.

[18] Mandarino, José. Marcos Gontijo., & Roessing, Antônio. Carlos. (2001). Tecnologia para produção do óleo de soja: descrição das etapas, equipamentos, produtos e subprodutos. *Londrina: Embrapa Soja,* 40.

[19] Martins, Roberto. Antônio. (2009). Gestão da Qualidade Agroindustrial. *BATALHA, Mário Otávio. Gestão Agroindustrial. GEPAI. 3ª Ed. São Paulo: Atlas,* cap. 8.

[20] Mendes, Judas. Tadeu Grassi., & Padilha, Junior. João Batista. (2007). Agronegócio: Uma abordagem Econômica. *São Paulo: Pearson Prentice Hall.*

[21] Moraes, R. M. A. (2007). Potencial da soja na produção de biodiesel. http://www.cisoja.com.br/index.php?p=artigo&idA=1, Accessed: 26 March. 2012.

[22] Moraes, R. M. A. (2007). Potencial da soja na produção de biodiesel.. http://www.cisoja.com.br/index.php?p=artigo&idA=1, Accessed: 26 March 2012.

[23] Oetterer, M., Regitano-D'Arce, M. A. B., & Spoto, M. H. F. (2006). Fundamentos de Ciência e Tecnologia de Alimentos. *Barueri, SP: Manole,* 612.

[24] Pereira, R. G. ., Tulcan, O. P. ., Lameira, V. J. ., Espirito, Santo., Filho, D. M. ., & Andrade, E. T. (2011). Use of Soybean Oil in Energy Generation. *Dora Krezhova. (Org.). Recent Trends for Enhancing the Diversity and Quality of Soybean Products. Recent Trends for Enhancing the Diversity and Quality of Soybean Products. Rijeka: InTech,* 01, 301-320.

[25] Schenini, Pedro. Carlos. (2011). Gerenciamento de Resíduos da Agroindustria. *II Simpósio Internacional sobre Gerenciamento de Resíduos Agropecuários e Agroindustriais- II SIGERA.15 a 17 de março de- Foz do Iguaçu, PR, Volume I- Palestras.*

[26] Schneider, F. H. (1980). Zur extraktiven lipid-freisetzung aus pflanzlichen zellen. *Fette Seifen Antstrichmittel, Hamburgo,* 82(1), 16-23.

[27] Shahidi, F., & Wanasundra, U. N. (1998). Methods for measuring oxidative rancidity in fats and oils. *Akoh, C & Min, D. O. Foods lipids- chemistry, nutricion and biotechnology. Nova York, Marcel Dekker.*

[28] Shahidi, F., & Wanasundara, U. N. (1998). Omega-3 fatty acid concentrates: Nutritional aspects and production technologies. *Food Sci. Technol, 9,* 230-240.

[29] Tulcan, O. E. P., Andrade, D. O., Andrade, E. T., & Pereira, R. G. (2008). Analisys of physical characteristics of vegetable oil. *International Conference of Agricultural Engineering, 4.*

[30] USDA- United States Department of Agriculture. (2012). http://www.fas.usda.gov/psdonline/psdQuery.aspx, Accessed: 26 March.

[31] Young, V. (1980). Processing of oils and fats. *Hamilton, R. J. & Bhati, A. (eds). Fats and oils: chemistry and technology. Londres, Appl. Sci. Publ,* 135-165.

Chapter 4

Phytoestrogens and Colon Cancer

B. Pampaloni, C. Mavilia, E. Bartolini, F. Tonelli,
M.L. Brandi and Federica D'Asta

Additional information is available at the end of the chapter

http://dx.doi.org/10.5772/54065

1. Introduction

Colorectal carcinoma (CRC) represents the most frequent malignancy of the gastrointestinal tract in the Western world in both genders. There is a wide variation of incidence rate for both colonic and rectal cancer among the populations of different countries: up to a 30-40-fold difference is seen between North America (Canada, Los Angeles, San Francisco), New-Zealand (non–Maori), Northern Italy (Trieste), Northern France (Haut- and Bas-Rhin) in which the rate of CRC is around 50/100,000 inhabitants, and India (Madras, Bangalore, Trivadrum, Barshi, Paranda, Bhum, Karunagappally), Algeria (Setif), and Mali (Bamako) in which the rate is around 3/100,000 [1]. It is estimated that approximately 6% of the United States population will eventually develop a CRC, and that 6 million of American citizens who are living will die of CRC [2].

The geographic differences in CRC incidence are due more to environment, life-style, and diet than to racial or ethnic factors. Demonstration of this fact is that migrants from low to high incidence areas have the same incidence as the host country within one generation, having assimilate western lifestyle and diet [3].

Colonoscopy to screen asymptomatic adults older than 50 years allows an estimation of the prevalence of adenomatous polyps or CRC: in North America CRC is found in 2%, and advanced adenoma (more than 1 cm in diameter) in 10% [4[.

Population-based studies have investigated several environmental factors as contributors to the initiation of sporadic colorectal carcinogenesis. High-calorie diet, high red meat consumption, overcooked red meat consumption, high saturated fat consumption, excess alcohol consumption, cigarette smoking, sedentary lifestyle, and obesity are considered to increase the incidence of CRC, while consumption of fiber, fresh fruit and vegetables, and a high-calcium diet could have a protective effect [5]. A recent review [6] provided an over-

© 2013 Pampaloni et al.; licensee InTech. This is a paper distributed under the terms of the Creative Commons Attribution License (http://creativecommons.org/licenses/by/3.0), which permits unrestricted use, distribution, and reproduction in any medium, provided the original work is properly cited.

view of the epidemiological evidence supporting the roles of diet, lifestyle, and medication in reducing the risk of colorectal cancer. Similarly, many studies that implicate effects of dietary agents in various types of cancers are available and suggest that much of the suffering and death from cancer could be prevented by consuming a healthy diet, reducing tobacco use, performing regular physical activity, and maintaining an optimal body weight [7]. Even if several epidemiological and experimental studies support the role of these factors in the genesis of CRC, other well-designed prospective and randomized clinical trials conducted in recent years report conflicting evidence, in particular on the role of the diet component in the etiology of CRC [8, 9].

Meanwhile, the great majority of CRC are sporadic, with 2 to 6% of them related to a hereditary disease due to mutations of highly penetrant autosomic dominant genes. Mutations of *APC* tumor suppressor gene is responsible for familial adenomatous polyposis (FAP), and mutations of the mismatch repair (*MMR*) genes are related to hereditary non polypoid colorectal cancer (HNPCC or Lynch's syndrome). Mutations of *MLH1* and *MSH2* are responsible for more than 90% of the family affected by HNPCC. In these familial events, the onset of CRC is greatly anticipated in comparison to the sporadic counterpart which is usually diagnosed after 50 years of age. However, an increasing incidence rate of CRC not clearly related to the presence of inheritable or predisposing colonic diseases was observed in individuals less than 40 years of age in recent decades [10]. Furthermore, an enhanced risk for CRC and colonic adenomas is present in individuals whose first-degree relatives are affected by CRC, especially if the tumor occurs before the age of 60 [8]. Possible factors of this inherited susceptibility to CRC are polymorphisms of genes deputed to glutathione synthesis such as *GSTP1*, *GSTM1* and *GSTT1* genes [11].

Prognosis of CRC is in relationship to local and distant tumor progression. Deep penetration of carcinogenic cells in the colonic wall, invasion of adjacent organs, diffusion in lymph nodes or peritoneum, and distant metastases must be evaluated for staging of the disease and correct therapeutic planning. One third of all colorectal tumors are located in the rectum: prognosis of distally sited rectal cancer is worse than that of proximally sited rectal cancer or of colonic cancer. Despite great advances in population screening, early diagnosis, surgical interventions, and complementary therapies, long-term survival for CRC remains in the range of 50-60%.

Tumor formation in humans is a multistage process involving a series of events, and generally occurs over an extended period. During this process, several genetic and epigenetic alterations lead to the progressive transformation of a normal cell into a cancer cell. These cells acquire various abilities that transform them into malignant cells: they become resistant to growth inhibition, proliferate without dependence on growth factors, replicate without limit, evade apoptosis, and invade, metastasize, and support angiogenesis. Mechanisms by which cancer cells acquire these capabilities can vary considerably, but most of the physiological changes associated with these mechanisms involve alteration of signal transduction pathways [7].

It is commonly agreed that the first step of colorectal tumorigenesis is the shift of the proliferative zone in the glandular crypts, accompanied by the development of aberrant crypt foci, and followed by the formation of an adenomatous polyp. These pathological features are considered the precursor of the carcinoma in a temporal sequence that also can be completed in several years. However, CRC is not a homogenous disease: several histological types can be distinguished such as tubular or villous, mucinous, serrated, medullary, signet-ring, squamous cell, adenosquamous, small cell, and undifferentiated, and different molecular basis can also be recognized in histologically similar tumors. In recent years, the identification of the genetic mutations of hereditary forms of CRC has clarified two fundamentals types of carcinogenesis. The first is similar to that described for the development of the FAP, and is characterized by a progressive accumulation of genetic changes starting from a biallelic inactivation of *APC*. Additional mutations either of oncogenes *KRAS* and *p53* or of oncosuppressor genes (*DCC* and *DPC4*) are necessary for the neoplastic progression and invasivity [12]. The genetic alterations are responsible for an increased mucosal proliferation and a reduced apoptosis, causing a clonal cellular expansion. The second, similar to the CRC arising in the HNPCC, is due to inactivation of *MLH1* or of other *MMR* genes. Repetitive sequences of DNA, sited in non-encoding microsatellite regions throughout the genome, are specifically found in this type of CRC, hence, the definition of micro satellite instability (MSI). The mechanism responsible for the carcinogenesis is epigenetic due to an extensive DNA methylation. Rarely in this type of CRC both proto-oncogenes (*KRAS, p53*) and oncosuppressor genes (*APC, TGFBRII, IGF2R, BAX*) are mutated or inactivated [13]. The former genetic mechanism explains the most frequent form of sporadic CRC characterized by the sequence adenoma-carcinoma and a long period for the formation of cancer; vice versa, the last mechanism is only present in 15% of sporadic CRC, and can have the character of an accelerated carcinogenesis.

Improved knowledge of the molecular mechanisms of colorectal carcinogenesis allows a rationale chemopreventive use in individuals who have an increased risk of developing colorectal adenomas or cancer. Both natural or synthetic agents have been employed to prevent or suppress the colorectal tumorigenesis. In particular, in experimental animals, cohort and clinical case-control studies have shown inverse association between the use of either anti-inflammatory non steroidal drugs (NSAIDs), estrogens or phytoestrogens, and incidence of both colonic adenomas and CRC. NSAID use appears to prevent the occurrence of carcinogen-induced animal colonic tumors [14] and to decrease the number and size of colo-rectal polyps in FAP (Familial Adenomatous Polyposis) patients [15]. Randomized placebo controller trials showed that aspirin reduced the risk of colorectal adenomas in populations with an intermediate risk of developing adenomas [16]. Furthermore, NSAIDs or selective COX-2 inhibitors reduce the in vitro growth of human colon cancer cell [17]. The effect of NSAIDs is mediated by cell cycle arrest due to inhibition of the Wnt-signaling pathway that favors the phosphorilation of beta-catenin and by induction of apoptosis [18, 19].

The fact that estrogens have an effect in decreasing the risk of colo-rectal cancer is shown by the following data:

1.   Several epidemiologic studies show a smaller incidence of sporadic CRC in the female gender. Also the occurrence of CRC in HNPCC is lower in females than in males;

2.   women who are multipare are a reduced risk of CRC in confront to nullipare;

3.   epidemiologic studies of postmenopausal women show that users of HRT have a significant reduction of CRC development in respect to women who had never used HRT. The risk appears to be halved with 5-10 years of HRT use [20, 21];

4.   use of non-contraceptive hormones for more than 5 years reduces by (OR = 0.47, 95 percent CI = 0.24-0.91) the risk of colon cancer [22].

2. Nutrition and colon cancer

It is now believed that 90–95% of all cancers are attributed to lifestyle, with the remaining 5–10% attributed to faulty genes [7]. Almost 30 years ago epidemiological research suggested that appropriate nutrition could prevent approximately 35% of cancer deaths, and up to 90% of certain cancers could be avoided by dietary enhancement [23, 24].

Colon cancer is a multifactorial disease that results from the interaction of different factors such as aging, family history, and dietary style. Identifying modifiable factors associated with colorectal cancer is of importance, the ultimate goal being primary prevention, and particularly the role of diet in the aetiology, initiation, and progression of colorectal cancer remains an area of important research. Moreover, several components of food can exert a potent activity also in the later stages of cancer. Several studies have indicated that inhibition of metastasis by genistein, one of the most important constituents of soy foods, represents an important mechanism by which it is possible to reduce mortality associated with solid organ cancer.

Many plant-derived dietary agents have multitargeting properties and are therefore called nutraceuticals. A nutraceutical (a term formed by combining the words "nutrition" and "pharmaceutical") is simply any substance considered to be a food or part of a food that provides medical and health benefits. During the past decade, a number of nutraceuticals have been identified from natural sources. Nutraceuticals are chemically diverse and target various steps in tumor cell development [7].

Several epidemiological studies have consistently shown an inverse association between consumption of vegetables and fruits and the risk of human cancers at many sites. Wickia & Hagmannc (2011) recently reported that many case-control and cohort studies are dealing with the effect of fruits and vegetables on cancer incidence [25]. Early data indicated a beneficial effect [26] and, as recently as 2008, Freedman et al. found a reduced occurrence of head and neck cancers with increased fruit and vegetable consumption [27].

The concept that a diet that is high in fiber, especially from fruits and vegetables, lowers risk of colorectal cancer has been in existence for more than 4 decades. The majority of case-control studies have shown an association between higher intake of fiber, vegetables, and possibly fruits, and lower risk of colon cancer [28]. A meta-analysis of six case-control studies found that a high intake of vegetables or fiber was associated with an approximate 40%–50% reduction in risk for colon cancer [29]. Similarly, a pooled analysis of 13 case-control studies reported an approximately 50% lower risk of colon cancer associated with higher intake of fiber [30].

Increasing intake of fruits, vegetables, or fiber is unlikely to prevent a large proportion of colorectal cancers, particularly among the US population, which has a food supply already fortified with folate and other dietary factors that might protect against colorectal neoplasia. There is also little evidence that concentrated sources of one type of fiber are efficacious, although fiber-rich diets have health benefits for other gastrointestinal conditions, such as diverticular disease and constipation, and possibly other chronic diseases [6].

All evidence supporting the decreased risk include results from a few studies of adenomatous polyps (which may progress to colorectal carcinomas). Fruit and grain intake also appears to be inversely related to risk of colorectal cancer and polyps, although less consistently than vegetables. These potentially protective associations may result from the high levels of dietary fibres, antioxidants (e.g., beta-carotene, vitamin C), or other anticarcinogenic constituents (e.g., protease inhibitors, phytoestrogens) in these vegetables, fruits, and grains. However, the association of adenomatous polyps of the large bowel with intake of vegetables, fruits, and grains has not been studied to any great degree, and existing data on these associations are not entirely consistent. Because adenomatous polyps are precursors to colorectal cancer, studying polyps instead of cancer might allow one to measure the diet of relatively asymptomatic subjects closer to the time of the initial neoplastic process. [31].

A recent meta analysis and data review, conducted by Magalhães B. [32], substantiates that the risk of colon cancer was increased with patterns characterized by high intake of red and processed meat, and decreased with those labelled as 'healthy.'

There are many plausible mechanisms by which intake of vegetables, fruits and "healthy foods" may prevent carcinogenesis.

Plant foods contain a wide variety of anticancer phytochemicals with many potential bioactivities that may reduce cancer susceptibility [7,33, 34].

3. Soyfoods and colon cancer

Many epidemiologic studies evidence a lower rate of hormone-related cancers among Asian populations which are characterized by regular consumption of soy based foods. Soy is a major plant source of dietary protein for humans. A review of epidemiologic studies (most of which were case-control studies published before 2000) suggested an inverse association between high soy intake and colon cancer risk in humans [35]. Moreover, migration studies show that Japanese immigrants in the United Status have incidence rates of colorectal cancers very near to the rates among the whites in the country [6]. Thus the protective effect of soy foods and isoflavones is a matter of interest in the etiology of colorectal cancer.

Soy and soy foods contain a wide variety of chemical compounds, biologically active, that may contribute, individually or synergistically, to the health benefits of this plant; in particular, polyphenols are considered to possess chemopreventive and therapeutic properties against cancer.

Among these compounds, certainly, there are isoflavones, the most important and abundant of which is genistein, which also have estrogenic properties. In fact, in recent decades, there have been several studies showing that isoflavones are promising candidates for cancer prevention [36, 37, 38, 39].

Data associating soybean consumption with reduced cancer rates have been used as evidence for a role of isoflavones in cancer prevention. However, soybeans are also a rich source of trypsin inhibitor, other proteins with health benefits, phosphatidyl inositol, saponins, and sphingolipids, all of which have potential health benefits. All of these soybean constituents demonstrate tumor preventive properties in animal models. Research by Birt et al. demonstrated that 20% by weight of dietary soy protein significantly reduced rat intestinal mucosa levels of polyamine, a biomarker of cellular proliferation for colorectal cancer risk [39].

Surely, soy foods are complex foods, and it is difficult to assume that associations which suggest protective properties of soy foods are due only to a single constituent. Because of the association between diets in Japan and China and lower rates of cancers, such as those of the breast, prostate, and colon, than in Europe and the United States, many investigators have assumed that this is due to soy food consumption in Japan and China.

Other factors in the Asian diet may be responsible, and it's important to evaluate the possible confounding dietetic factors in the studies.

Several studies suggest that soy foods, the predominant source of isoflavones, are associated with reductions in cancer rate, but they do not consistently appear to be the primary protective component of the Asian diet.

Wu et al. noted the difficulties in assessing the relationship between the level of intake and protection. Case control and prospective epidemiological investigations that have provided a suggestion of protection against cancer by soy foods have not provided adequate information on the bioactive constituents in the soy foods, the portion size, or other components that may be protective in the diets of people who eat soy foods [40].

Isoflavones and flavonoids may be rapidly and predominately glucuronidated in the GI mucosa, if genistein can be considered a model for all of these phenolic compounds [41]. Further, glucuronidation occurs in the liver. Genistein undergoes biliary excretion, with more than 70% of a dose recovered in bile within 4 hr after dosing in rats. Although genistein may be absorbed well initially, a maximum of 25% of an oral genistein dose would be eliminated in rat urine. About 20–25% of an oral dose of genistein (predominantly as its glucoside from soy foods) is recovered in human urine [42, 43].

The presence of hydroxylated and methylated genistein metabolites correlated positively with inhibition of cancer cell proliferation, but genistein sulfates were not associated with antiproliferative effects of genistein, suggesting that some types of metabolism of the isoflavones may be crucial for their action [44].

Witte, et al, showed that higher consumption of tofu (or soybeans) was inversely associated with polyps. Tofu (or soybeans) contain a number of potentially anticarcinogenic constituents, including isoflavones, saponins, genistein, and phytosterols. They were able to look at tofu (or soybeans) as a single food item (i.e., separate from legumes) because almost 15 percent of our multiethnic study population reported consuming tofu (or soybeans) at least once a week. The strongest association observed was for vegetables—including those high in carotenoids, cruciferae, and broccoli—as well as garlic and tofu (or soybeans), and these associations were found even after adjusting for dietary fiber, folate, beta-carotene, vitamin C, and other commonly measured antioxidants [31].

Men tend to have a slightly higher incidence of colorectal cancer than women of similar age (American Cancer Society, 2007), and oestrogen seems to be implicated for this decreased risk in women. Epidemiological studies and results of a Women's Health Initiative (WHI) clinical trial provide strong evidence that colorectal cancer is hormone sensitive because the cancer risk is reduced by post-menopause hormone therapy [35]

In effect, many epidemiological and experimental studies suggest a protective role of estrogens against colorectal cancer. The decrease in the number of deaths from large bowel carcinoma observed in the United States in the last 40 years was significantly higher in women (30%) as compared to men (7%). A link was observed between oral contraceptive use and a reduction of colorectal cancer, whereas there was a higher than expected frequency of colorectal tumors among non users [45].

Interestingly, as reported by Barone et al., although several experimental studies have confirmed a protective role of estrogens for CRC, few studies have been conducted, and with conflicting results, on the possible protective effect of estrogens against the development of adenomatous polyps in the colon, although it is well known that the development of adenocarcinoma mostly involves polyp formation [46].

Gender differences in the incidence and behavior of colorectal cancer (CRC), as well as epidemiologic data indicating a protective effect of hormone replacement therapy in women, have further supported the concept of hormonal influence on the development of CRC. It has been suggested that the protective effect of estrogens (or phytoestrogens) may be mediated through activation of ERβ, which has been shown to be the predominant subtype of ER in the gastrointestinal tract [47].

ERs are nuclear receptors belonging to the steroid hormone receptor superfamily which have the characteristic of being activated upon binding of the ligand. If the ligand is not present, ERs bind to a shock protein. Otherwise, when the ligand is present, the ERs make a stable dimer and initiate the specific estrogenic response, with transcription of the target genes. Two main types of ER have been identified: alfa (ERα) and beta (ERβ). They are the so-called ligand-activated transcriptional factors through which estrogens exert their effects on various tissues and have a different tissue distribution. ERα is mainly present in the mammary glands and in the utterus; ERβ is mainly present in endothelial cells, the urogenital tract, the central nervous system, and the colonic mucosa. Experimental data have demonstrated that CRC express an elevated number of estrogen receptors (ERs), but while ERα is detected in very low levels either in normal or pathological colonic mucosa (adenoma and carcinoma), ERβ expression is high in the normal colonic mucosa, and progressively decreased in the pathological mucosa in relationship to the cellular differentiation and CRC stage.

The observation that the level of ERβ protein is lower in malignant tumors than in normal tissue of the same organ has fostered the hypothesis that ERβ may function as a tumor suppressor, protecting cells against malignant transformation and uncontrolled proliferation.

ERβ is present in various isoforms: studying different types of colonic tumoral cells, isoform 1 of ERβ is found in the Lo-Vo, HCT8, HCT116, DLD-1 and isoform 2,3,4 and 5 only in the HCT8 and HCT116. It has not been well investigated whether the function of the various isoforms of ERβ, but loss of the expression of isoform 1 of ERβ, is accompanied by undifferentiated proliferation, mucinous histological type, and tumor progression [48]. It is accepted that the binding of estrogens to the ERβ blocks the activity of AP-1 on the genes involved in the cellular proliferation and provokes an activation of p53. Conversely, SERM, such as tamoxifene and raloxifene, induce an antiproliferative effect in human colorectal cell lines by a citostatic or cytotoxic effect [49]. Several observations on the CRC cellular cultures and on the experimental mouse with germinal mutation of APC have clarified the role of the ER and estrogenes for colorectal cancerogenesis: 17β estadiol decreases the proliferation in vitro of the HCT116, Lo-Vo and DLD1 cells, but increases the proliferation of the HCT8 cells. However, the effect on the last type of cells is completely changed by increasing the level of RRb by transfection with ERβ. The overexpression of ERβ can have an inhibitory effect on the proliferation. In the transfected HCT8 cells the levels of CD4 and CP21, which are onco-suppressor genes, are significantly increased, and the level of cyclinE, which have oncogenic activity, significantly decreased, in respect to normal HCT8 [50].

ERβ is lower in the adenomatous polyps of FAP patients and in the intestinal adenomas which develop in APC Min+/- mouse than in the colonic normal mucosa. The restoration of normal levels of ERβ obtained with dietary phytoestrogenes is accompained by regression or disappearance of the polyps in the experimental animal. Patients with sporadic adenomas in the colon show an increase of apoptoic activity, and ERβ expression of the colonic mucosa, if their diet is supplemented by phytoestrogenes [45]. These data strongly support a pivotal role of ERβ in a protective action against the initiation and progression of colorectal carcinogenesis.

Many epidemiologic studies evidence a lower rate of hormone-related cancers among Asian populations which are characterized by regular consumption of soy based foods. Soy is a major plant source of dietary protein for humans. Among other components, soy contains large amounts of phytoestrogens.

As proposed for estrogens, genomic and non-genomic mechanisms have also been suggested for phytoestrogens to explain their biological activities

As reported by several authors in the past, genomic pathways are mediated through the ability of phytoestrogens to interact with enzymes and receptors, and cross the plasma membrane. In this way, they bind ERs and induce the transcription of estrogen-responsive genes, stimulate cell growth in the breast, and modify ER transcription itself. However, some of their effects are not due to interaction with ERs, and are therefore denominated non-genomic effects. For example: inhibition of tyrosine kinase and DNA topoisomerase, suppression of angiogenesis, and antioxidant effects [33, 36, 46].

The bioavailability of phytoestrogens (determined by: absorption, distribution, metabolism (bioconversion in the gut and biotransformation in the liver) and excretion) and their activity is highly variable and changes with respect to several factors, such as administration rules, dosage, metabolism and interaction with other pharmacological substances. Moreover, their biological effect is influenced by the type of target tissue, the number and type of ERs expressed in the tissue, their serum concentration, and sex steroid hormone concentration [51, 52].

Phytoestrogens, present in soy and soy-based food, may act through hormonal mechanisms to reduce cancer risk by binding to estrogen receptors (ER) or interacting with enzymes involved in sex steroid biosynthesis and metabolism [53].

Although cancer incidence in women is much lower than in men in both countries, there is also a difference when the 2 countries are compared. Japanese men as well as women have a lower colorectal cancer incidence than their American counterparts, although mortality is quite similar when related to specific incidence data. In hormone-dependent cancers such as those of the breast and prostate, incidence is exceedingly low in Japan (and was even lower in earlier decades) compared with that in the United States. Mortality, again in proportion to incidence, is rather similar. Numerous reports have suggested that this difference in tumor incidence is probably due to consumption of soy as a staple food in Asian countries in contrast to Western industrialized countries. These substances, through their potential to act as selective estrogen receptor modulators, may affect vitamin D–related inhibition of tumor growth by upregulating extrarenal synthesis of 1,25- D3. Genistein, the most prominent phytoestrogen in soy, is known to regulate other P450 enzymes, such as 5-reductase and 17-hydroxysteroid dehydrogenase, which are essential for metabolism of sex hormones [54].

In vitro studies of DLD1 colon adenocarcinoma cells have linked the effects of soy with estrogen receptor beta. Experiments conducted on this cell line, with or without ER-β gene silencing by RNA interference (RNAi), have shown that soy isoflavones decreased the expression of proliferating cell nuclear antigen (PCNA), extracellular signal-regulated kinase (ERK)-1/2, AKT, and nuclear factor (NF)-κB. Soy isoflavones dose-dependently caused G2/M cell cycle arrest and downregulated the expression of cyclin A. This was associated with inhibition of cyclin dependent kinase (CDK)-4 and upregulation of its inhibitor p21 expressions. ER-β gene silencing lowered soy isoflavone-mediated suppression of cell viability and proliferation. ERK-1/2 and AKT expressions were unaltered and NF-κB was modestly upregulated by soy isoflavones after transient knockdown of ER-β expression.

Soy isoflavone-mediated arrest of cells at G2/M phase and upregulation of p21 expression were not observed when ER-β gene was silenced. These findings suggest that maintaining the expression of ER-β is crucial in mediating the growth-suppressive effects of soy isoflavones against colon tumors. Thus, upregulation of ER-β status by specific foodborne ER-ligands such as soy isoflavones could potentially be a dietary prevention or therapeutic strategy for colon cancer [55].

4. Genistein and isoflavones: Other mechanisms of action

In addition to estrogenic/antiestrogenic activity, some mechanisms of action have been identified for isoflavone/flavone prevention of cancer: antiproliferation, induction of cell cycle arrest and apoptosis, prevention of oxidation, induction of detoxification enzymes, regulation of host immune system, and changes in cellular signaling [39, 56, 57]. It is expected that also combinations of these mechanisms may contribute to cancer prevention.

Gene silencing due to the promoter methylation provides an opportunity for clinical intervention, as gene-re-expression can be induced by a variety of DNA demethylating agents.

Recent studies show that genistein may affect DNA methylation, serves as a natural demethylation agent, and that it is specifically effective on colon cancer cells from early-stage colon cancer [58]. WNT family members are highly conserved, secreted signaling molecules that play important roles in both tumorigenesis and normal development and differentiation. Study of Hibi *et al.* evidences that genistein treatment affected the DNA methylation of *WNT5a*, and that *WNT5a* downregulation is correlated with hypermethylation of its promoter in human colon cancer patients [60, 59].

Moreover, genistein may inhibit cancer progression by inducing apoptosis or inhibiting proliferation, and the mechanisms by which genistein exerts its anti-tumor effects has been the subject of considerable interest [61, 62, 63].

Genistein has been shown to induce epigenetic changes in several cancer cell lines and in the in vivo animal models. [64].

The presence of hydroxylated and methylated genistein metabolites correlated positively with inhibition of cancer cell proliferation, but genistein sulfates were not associated with antiproliferative effects of genistein, suggesting that some types of metabolism of the isoflavones may be crucial for their action.

Genistein is a known inhibitor of protein-tyrosine kinase (PTK), which may attenuate the growth of cancer cells by inhibiting PTK-mediated signaling mechanisms [65]. Sakla et al. (2007) recently reported that genistein inhibits the protooncogene HER-2 protein tyrosine phosphorylation in breast cancer cells as well as delaying tumor onset in transgenic mice that overexpress the HER-2 gene. These data support its potential anti-cancer role in chemotherapy of breast cancer. However, effects independent of this activity have also been demonstrated [66, 67].

Soy isoflavone supplemented diets also prevented the development of adenocarcinomas in the prostate and seminal vesicles in a rat carcinogenesis model [68].

Phytoestrogens, present in soy based food, may act through hormonal mechanisms to reduce cancer risk by binding to estrogen receptors (ER) or interacting with enzymes involved in sex steroid biosynthesis and metabolism [53]. Moreover, genistein may inhibit cancer progression by inducing apoptosis or inhibiting proliferation, and the mechanisms by which genistein exerts its anti-tumor effects have been the subject of considerable interest [61, 62, 63].

Studies demonstrate that ERβ is highly expressed in superficial and crypt epithelium of the normal colon in both genders. ERβ expression was highly correlated among all cell types in both genders, and the strongest correlation was observed between surface and crypt ERβ expression. This finding suggests that there may be an intersubject difference in ERβ expression that is manifested in all cell types. ERβ expression was significantly lower in colon cancer cells compared with normal colonic epithelium, and there was a progressive decline in ERβ expression that paralleled the loss of cancer cell differentiation. The present findings are consonant with previous results reported by Foley and colleagues [69], who also detected a loss of ERβ protein expression in malignant colon tissue by western immunoblotting. Another immunohistochemical study of ERβ in 55 patients with colorectal adenocarcinomas showed that 32% of all tumors in both genders were ERβ-negative; the 10% cut-off threshold was used to distinguish ERβ-positive from negative tumors [70].

Studies conducted with ER subtype-specific ligands and those performed with estrogen receptor b-knockout mice (ERβKOs) have illustrated the involvement of ERβ in cellular anti-inflammatory pathways and tissue homeostasis in the colon. These results suggest that ERβ-specific ligands may be promising targets in the pharmaceutical and therapeutical treatment of inflammatory bowel disease and the prevention of CRC. ERβKOs suggest that ERβ-specific agonists and ERβ-selective phytoestrogens like genistein (GEN) and coumestrol may serve as potential regulators of intestinal tissue homeostasis [71, 72, 73].

Schleipen et al. investigate the influence of ERα and ERβ-specific agonists, and of genestein on cell proliferation and apoptosis of the small intestine and the colon. Recent data indicate that ERβ-specific agonists and GEN inhibit epithelial proliferation of the prostate and mammary gland, and can even impede prostate cancer development [74, 76, 75]. It can therefore be assumed that ERβ-specific agonists may also inhibit the proliferation of the intestinal epithelium. To prove this hypothesis in the study, ovariectomized rats were treated with 17β-Estradiol (E2), the phytoestrogen GEN and ER subtype-specific agonists for ERα and ERβ for 3 weeks.

Genistein has been shown to induce epigenetic changes in several cancer cell lines and in *in vivo* animal models [64]. Recent studies show that genistein may affect DNA methylation, serves as a natural demethylation agent, and is specifically effective on colon cancer cells from early-stage colon cancer [58].

WNT family members are highly conserved, secreted signaling molecules that play important roles in both tumorigenesis and normal development and differentiation. Study of Hibi *et al.* evidence that genistein treatment affected the DNA methylation of *WNT5a*, and that *WNT5a* down-regulation is correlated with hypermethylation of its promoter in human colon cancer patients [59, 60]. Aberrant WNT signaling is considered one of the most correlated factors in over 90% of both benign and malignant colorectal tumors [77].

Many epigenetic silencing and activating events have been discovered in the WNT pathway that are also related to aberrant WNT signaling, including aberrant expression of *sFRP1*, *DKK1*, and *APC* [78, 79]. Therefore, Wang and Chen investigate the effect of genistein on WNT pathway regulation in colon cancer development [58]. This study showed that: genistein treatment selectively induced *WNT5a* expression in specific colon cancer cell lines; *WNT5a* showed the lowest expression compared to other more advanced tumor cell lines; and the novel finding that *WNT5a* mRNA expression was upregulated by genistein in this early-stage colon cancer cell line.

These results support the notion that genistein serves as a natural demethylation agent and that it is specifically effective on colon cancer cells from early-stage colon cancer. Genistein treatment affected the DNA methylation of *WNT5a*. It has been shown that *WNT5a* downregulation is correlated with hypermethylation of its promoter in human colon cancer patients [59, 60].

Wang and Chen studies showed that the time dependent induction of WNT5a by genistein in colon cancer cell line SW 1116 was correlated with decreased methylation of a CpG island within its promoter, as determined by bisulfate sequencing [58].

Demethylation of CpGs inhibition of Dnmt and MBD2 activity, and activation of the histones by acetylation and demethylation at the BTG3 promoter followed by genistein treatment, were observed in renal cancer cells [80]. Using the mouse differential methylation hybridization array, alteration of DNA methylation in specific genes in mice was observed following feeding of a diet containing genistein compared to that in mice fed a control casein diet [81].

Other direct evidence that genistein affected DNA methylation was that maternal exposure to dietary genistein altered the epigenome of offspring in viable yellow agouti (Avy/a) mice. Overall, the potential of genistein as an effective epigenome modifier, which may greatly impact CRC metastasis, highlights the potential ability of dietary genistein to improve CRC prognosis [82].

Downregulation by promoter hypermethylation occurs in cell lines from earlier stages of colon cancer but not in cell lines from later stages.

These findings suggest that maintaining the expression of ER-β is crucial in mediating the growth-suppressive effects of soy isoflavones against colon tumors. Upregulation of ER-β by specific foodborne ER-ligands, such as soy isoflavones, could potentially be a dietary prevention strategy for colon cancer. [55].

Genistein has been shown to inhibit cancer metastasis through its ability to regulate nearly every step of the metastatic cascade, including cell adhesion, migration invasion, and angiogenesis. The effect of genistein on the metastaic cascade involves many metastasis suppressor or related signaling pathways, such as NFKappaB. Genistein can affect both of these processes, as well as modulate key regulatory protein such as Akt and nuclear factor κB (NF-κB). In general, low-to-mid micro molar concentrations of genistein are required for these effects in cell-culture-based models, although, interestingly, effects in animal models have been observed at lower concentrations. Genistein inhibits critical pathways in cancer invasion and can specifically target MEK4. This inhibition results in inactivation of the MEK4 pathway, decreased MMP-2 production, and decreased cell invasion. Genistein also activates Smad1, which is activated by the endoglin signaling pathway, and causes decreased cell invasion. Additionally, genistein inhibits FAK activation, resulting in increased cell adhesion. At this time, it is unclear whether the activation of Smad1 and FAK are due to genistein's inhibition of MEK4 or via a different signaling mechanism [83].

Several reports have demonstrated that genistein can induce cell cycle arrest and that it can therapeutically modulate key regulator cell cycle proteins at concentrations ranging from 5 to 200 μM [84]. It is important to note that these concentrations are greater than the blood levels that are observed with dietary consumption, indicating that this is likely not the primary mechanism by which genistein inhibits metastasis. However, it is theoretically possible to achieve these levels in humans, and various animal studies have also demonstrated that genistein can reduce the primary tumor size in certain contexts.

Studies by Wentao et al. show that genistein inhibits EGF induced loss of FOXO3 activity by targeting the PI3K/ Akt pathway. Downstream, genistein inhibits EGF induced FOXO3 disassociation from p53(mut), which further promotes FOXO3 activity and leads to increased expression of the p27kip1 cell cycle inhibitor, which inhibits proliferation in colon cancer cells. The author demonstrated that one of the anti-proliferative mechanisms of genistein in colon cancer cells is to promote FOXO3 activity by inhibiting EGF-induced FOXO3 phosphorylation (inactivation) via the PI3K/Akt pathway. Active FOXO3 negatively regulates proliferation of colon cancer cells and shows that its inactivation is an essential step in EGF-mediated proliferation [85, 86].

5. Conclusion

Several studies shown that consumption of fiber, fresh fruit and vegetables, a high-calcium diet could have a protective effect on the increased risk of colorectal cancer, and suggest that much of the suffering and death from cancer could be prevented by consuming a healthy diet, reducing tobacco use, performing regular physical activity, and maintaining an optimal body weight [5].

Soy is one of the most consumed foods worldwide. Soy foods contain larger amounts of phytoestrogens of which the isoflavon genistein is surely the biologically most important.

This compound, in recent years, has received much attention in the field of oncology research, as it exerts a wide range of biological effects of direct relevance to cancer.

Phytoestrogens and in particular genistein, have shown to be an important tool for the inhibition of cancer metastasis, exerting effects on both the initial steps of primary tumor growth as well as the later steps of the metastatic cascade.

The international literature suggests that phytoestrogens have potentially a high clinical impact and the expansion of knowledge on soy, soy foods, and soy products will lead to novel future developments in the field of cancer treatment.

Phytoestrogens in soy foods	
Foods	**Total isoflavons (mg/100 g)**
Miso	41.45
Natto	82.29
Roasted soybeans	148.50
Soy beans	154.53
Soy cheese american	17.95
Soy flour (textured)	172.55
Soy milk	10-200
Soy milk curd, dried	83.30
Soy milk fortified or unfortified	10.73
Soy milk skin or film (Foo jook or yuba), cooked	44.67
Soy milk skin or film (Foo jook or yuba), raw	196.05
Soy protein concentrate	94.65
Soy protein drink	81.65
Soy protein isolate	91.05
Soy yogurth	33.17
Tempeh	60.61
Tofu (dried frozen)	83.20
Tofu raw regular with calcium and sulphate	22.73
Tofu yogurt	16.30

Table 1. Isoflavone Content of Selected Soy Foods (USDA Database 2008)

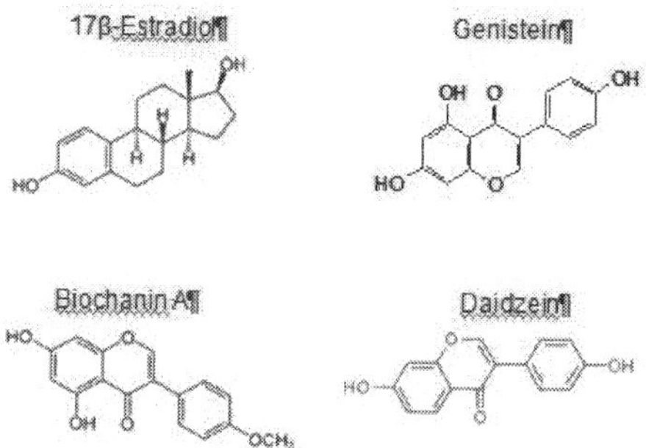

Figure 1. Chemical structures of soy phytoestrogens are similar to17 beta estadiol

Author details

B. Pampaloni[1], C. Mavilia[1], E. Bartolini[1], F. Tonelli[2], M.L. Brandi[1] and Federica D'Asta[3]

1 Department of Internal Medicine, School of Medicine, University of Florence, Florence, Italy

2 Department of Clinical Physiopathology, School of Medicine, University of Florence, Florence, Italy

3 Department of Neurosciences, Psychology, DrugArea and Child Health, School of Medicine, University of Florence, Florence, Italy

References

[1] Parkin DM, Whelan SL, Ferlay L et al Cancer incidence in five continents, vol VII. IARC Scientific Publications No.143. Lyon: IARC,1997

[2] Jemal A, Siegel R, Ward E et al: Cancer statistics, 2006. CA Cancer J Clin 2006;56:107

[3] Whittemore As, Wu-Williams AH, Lee M et al. Diet, physical activity, and colorectal cancer among Chinese in North America and China. J Natl Cancer Inst 1990, 82:915

[4] Lieberman DA, Weiss DG, Bond JH et al. Use of colonoscopy to screen asymptomatic adults for colorectal cancer. Veterans Affairs Cooperative Study Group 380. N Engl J Med 2000;343:162

[5] Libutti SK, Saltz LB, Rustgi AK, Tepper JE. Cancer of the colon in DeVita VT, Hellman S, Rosenberg SA eds Cancer: Principles & Practis of Oncology. 7th Edition, Lippincott &W, Philadelphia, 2005, pag 1062

[6]   Chan Andrew T., Giovannucci Edward L. "Primary Prevention of Colorectal Cancer" Gastroenterology 2010;138:2029–2043

[7]   Gupta Subash C., Ji Hye Kim, Sahdeo Prasad & Bharat B. Aggarwal "Regulation of survival, proliferation, invasion, angiogenesis, and metastasis of tumor cells through modulation of inflammatory pathways by nutraceuticals" Cancer Metastasis Rev (2010) 29:405–434

[8]   Fuchs Cs, Giovannucci EL, Colditz GA et al. A prospective study of family history and the risk of colorectal cancer. New Engl J Med 1994;331:1669

[9]   Alberts DS, Martinz ME, Roe DE et al. Lack of effect of high-fiber cereal supplement on the recurrence of colorectal adenomas. Phoenix Colon Cnacer Prevention Physicians'Network. N Engl J Med 2000;342:1156

[10]   O'Connell JB, Maggard MA, Liu JH et al. Rates of colon and rectal cancers are increasing in young adults. Am Surg 2003; 69: 866-872

[11]   Welfare M, Monesola Adeokun A, Bassendine MF, Daly AK. Cancer Epidemiol Biomarkers Prev 1999;8:289

[12]   Vogelstein B, Fearon ER, Hamilton SR et al. Genetic alterations during colorectal tumor development. N Engl J Med 1988; 319: 525

[13]   Grady WM. Genomic instability and colon cancer. Cancer Metast Rev 2004;23:11

[14]   Jacoby RF, Seibert K, Cole CE et al. The cyclooxygenase-2 inhibitor celecoxib is a potent preventive and therapeutic agent in the Min mouse model of adenomatous polyposis. Cancer Res 2000;60:5040-44

[15]   Giardiello FM, Hamilton SR, Krush AJ et al. Treatment of colonic and rectal adenomas with sulindac in familial adenomatous polyposis. N Engl J Med 1993;328:1313-16

[16]   Baron JA, Cole BF, Sandler RS et al. A randomized trial of aspirin to prevent colorectal adenomas N Engl J Med 2003; 348:891-99

[17]   Sheng H, Shao J, Kirkland SC et al. Inhibition of human colon cancer cell growth by selective inhibition of cyclooxygenase-2 J Clin Invest. 1997 May 1;99(9):2254-9.

[18]   Picariello L, Brandi Ml, Formigli L et al. Apoptosis induced by sulindac sulfide in epithelial and mesenchymal cells frm human abdominal neoplasms. Eur J Pharmacol 1998;360:105-112

[19]   Boon EMJ, Keller JJ, Wormhoudt TAM et al. Sulindac targets nuclear beta-catenin accumulation and Wnt signaling in adenomas of patients with familial adenomatous polyposis and in human colorectal cancer cell lines Br J Cancer. 2004 Jan 12;90(1): 224-9

[20]   Calle EE, Miracle-McMahill HL, Thorn MJ, Heath CW. Estrogen replacement therapy and risk of fatal colon cancer in a prospettive color of post-menopausal women. J Natl cancer Inst 1995;87:517-523

[21] Koo JH, Leong RWL. Sex differences in epidemiologic, clinical and pathological characteristics of colorectal cancer. J Gastroenterol Hepatol 2010; 25:33-42

[22] Jacobs EJ, White E, Weiss NS. Exogenous hormones, reproductive history and colon cancer. Cancer Causes Contr 1994;5:359-366

[23] Doll, R., & Peto, R. (1981). The causes of cancer: Quantitative estimates of avoidable risks of cancer in the United States today. Journal of the National Cancer Institute, 66, 1191–1308. 10.

[24] Hardy, G., Hardy, I., & Ball, P. A. (2003). Nutraceuticals—A pharmaceutical viewpoint: Part II. Current Opinion in Clinical Nutrition and Metabolic Care, 6, 661–671

[25] Wickia A. & Hagmannc J., "Diet and cancer" Swiss Med Wkly. 2011;141:w13250

[26] Block G, Patterson B, Subar A. Fruit, vegetables, and cancer prevention: a review of the epidemiological evidence. Nutr Cancer. 1992;18:1–29.

[27] Freedman ND, Park Y, Subar AF, et al. Fruit and vegetable intake and head and neck cancer risk in a large United States prospective cohort study. Int J Cancer. 2008;122:2330–6.

[28] Levi F, Pasche C, Lucchini F, La Vecchia C. Dietary fibre and the risk of colorectal cancer. Eur J Cancer 2001;37:2091–2096.

[29] Trock B, Lanza E, Greenwald P. Dietary fiber, vegetables, and colon cancer: critical review and meta-analyses of the epidemiologic evidence. J Natl Cancer Inst 1990;82:650–61.

[30] Howe GR, Benito E, Castelleto R, et al. Dietary intake of fiber and decreased risk of cancers of the colon and rectum: evidence from the combined analysis of 13 case-control studies. J Natl Cancer Inst 1992;84:1887–1896.

[31] Witte JS., Matthew PL, Cristy LB, Eric R.L, Harold D.F and Robert W. H. Relation of Vegetable, Fruit, and Grain Consumption to Colorectal Adenomatous Polyps; Am J Epidemiol. 1996 Dec 1;144(11):1015-25.

[32] Magalhães B, Peleteiro B, Lunet N. Dietary patterns and colorectal cancer: systematic review and meta-analysis.Eur J Cancer Prev. 2012 Jan;21(1):15-23.

[33] Adlercreutz, H. (1990). Western diet and Western diseases: some hormonal and biochemical mechanisms and associations. Scand J Clin Lab Invest 50 (Suppl. 201), 3– 23.

[34] Waladkhani, A. R., & Clemens, M. R. (1998). Effect of dietary phytochemicals on cancer development. Int J Med 1, 747– 753.

[35] Spector D, Anthony M, Alexander D, Arab L. "Soy consumption and colorectal cancer". Nutr Cancer 2003;47:1–12

[36] Adlercreutz, H. (1995). Phytoestrogens: epidemiology and a possible role in cancer protection. Environ Health Perspect 103 (suppl. 7), 103– 112.

[37]   Knight, D. C., & Eden, J. A. (1996). A review of the clinical effects of phytoestrogens. Obstet Gynecol 87, 897– 904.

[38]   Knekt P, Järvinen R, Seppänen R, Hellövaara M, Teppo L, Pukkala E, Aromaa A. Dietary flavonoids and the risk of lung cancer and other malignant neoplasms. Am J Epidemiol. 1997 Aug 1;146(3):223-30.

[39]   Birt DF, Hendrich S, Wang W. Dietary agents in cancer prevention: flavonoids and isoflavonoids. Pharmacol Ther. 2001 May-Jun;90(2-3):157-77.

[40]   Wu, A. H., Ziegler, R. G., Nomura, A. M. Y., West, D. W., Kolonel, L. N., Horn-Ross, P. L., Hoover, R. N., & Pike, M. C. (1998). Soy intake and risk of breast cancer in Asians and Asian Americans. Am J Clin Nutr 68, 1437S–1443S.

[41]   Zhang Y, Wang GJ, Song TT, Murphy PA, Hendrich SUrinary disposition of the soybean isoflavones daidzein, genistein and glycitein differs among humans with moderate fecal isoflavone degradation activity. J Nutr. 1999 May;129(5):957-62.

[42]   Watanabe S, Yamaguchi M, Sobue T, Takahashi T, Miura T, Arai Y, Mazur W, Wähälä K, Adlercreutz H. Pharmacokinetics of soybean isoflavones in plasma, urine and feces of men after ingestion of 60 g baked soybean powder (kinako). J Nutr. 1998 Oct; 128(10):1710-5.

[43]   Zhang Y, Song TT, Cunnick JE, Murphy PA, Hendrich S. Daidzein and genistein glucuronides in vitro are weakly estrogenic and activate human natural killer cells at nutritionally relevant concentrations. J Nutr. 1999 Feb;129(2):399-405

[44]   Peterson TG, Ji GP, Kirk M, Coward L, Falany CN, Barnes S. Metabolism of the isoflavones genistein and biochanin A in human breast cancer cell lines. Am J Clin Nutr. 1998 Dec;68(6 Suppl):1505S-1511S.

[45]   Barone M., Katia Lofano, Nicola De Tullio, Raffaele Licino, Francesca Albano, Alfredo Di Leo. "Dietary Endocrine and metabolic factors in the development of colorectal cancer" J Gastrointest Canc 2012, 43: 13-19

[46]   Barone M., Sabina Tanzi, Katia Lofano, Maria Principia Scavo, Raffaella Guido, Lucia Demarinis, Maria Beatrice Principi , Antongiulio Bucci and Alfredo Di Leo Estrogens, phytoestrogens and colorectal neoproliferative lesions, 2008, Genes Nutr (2008) 3:7– 13

[47]   Nüssler NC, Reinbacher K, Shanny N, Schirmeier A, Glanemann M, Neuhaus P, Nussler AK, Kirschner M. Sex-specific differences in the expression levels of estrogen receptor subtypes in colorectal cancer. Gend Med. 2008 Sep;5(3):209-17.

[48]   Fiorelli G, Picariello L, Martineti V et al. Functional estrogen receptor beta in colon cancer cells. Biochem Biophys Res Communic 1999; 261: 521-527

[49]   Picariello L, Fiorelli G, Martineti V et al. Anticancer Res 2003; 23:2419-24

[50] Martineti V, Picariello L, Tognarini I et al ERB is a potent inhibitor of cell proliferation in the HCT8 human colon cancer cell line through regulation of cell cycle components. Endocrine-Related Cancer 2005; 12:455-469

[51] Wiseman H (1999) The bioavailability of non-nutrient plants factors: dietary flavonoids and phytoestrogens. Proc Nutr Soc 58(1):139–146

[52] Xu X, Harris KS, Wang HJ, Murphy PA, Hendrich S (1995) Bioavailability of soybean isoflavones depends upon gut microflora in women. J Nutr 125(9):2307–2315

[53] Cotterchio Michelle, Beatrice A. Boucher, Michael Manno, Steven Gallinger, Allan Okey, and Patricia Harper. "Dietary Phytoestrogen Intake Is Associated with Reduced Colorectal Cancer Risk". J. Nutr. 136: 3046–3053, 2006

[54] Cross Heide S., Eniko° Ka' llay, Daniel Lechner, Waltraud Gerdenitsch, Herman Adlercreutz, H and H. James Armbrecht Phytoestrogens and Vitamin D Metabolism: A New Concept for the Prevention and Therapy of Colorectal, Prostate, and Mammary Carcinomas 2004 American Society for Nutritional Sciences.

[55] Bielecki A, Roberts J, Mehta R, Raju J. "Estrogen receptor-β mediates the inhibition of DLD-1 human colon adenocarcinoma cells by soy isoflavones". Nutr Cancer. 2011;63(1):139-50

[56] Chrzan BG, Bradford PG. (2007). Phytoestrogens activate estrogen receptor β1 and estrogenic responses in human breast and bone cancer cell lines. Mol Nutr Food Res 51: 171–177.

[57] Murata IC, Itoigawa M, Nakao K, Kumagai M, Kaneda N, Furukawa H. (2006). Induction of apoptosis by isofl avonoids from the leaves of Millettia taiwaniana, in human leukemia HL-60 cells. Planta Med 72 (5): 424–429.

[58] Wang Z. and H. Chen, "Genistein Increases Gene Expression by Demethylation of WNT5a Promoter in Colon Cancer Cell Line SW1116" Anticancer Research 30: 4537-4546 (2010)

[59] Ying, J, Li, H, Yu, J, Ng, KM, Poon, FF, Wong, SCC et al. "WNT5A exhibits tumor-suppressive activity through antagonizing the Wnt/β-catenin signaling, and is frequently methylated in colorectal cancer". Clinical Cancer Res 14(1): 55-61, 2008 54,

[60] Hibi K, Mizukami H, Goto T, Kitamura Y, Sakata M, Saito M et al: WNT5A gene is aberrantly methylated from the early stages of colorectal cancers. Hepatogastroenterology 56(93): 1007- 1009, 2009.

[61] Wang HK. "The therapeutic potential of flavonoids". Expert Opinion on Investigational Drugs. 2000;9(9):2103–2119

[62] Barnes S, Peterson TG. "Biochemical targets of the isoflavone genistein in tumor cell lines". Proceedings of the Society for Experimental Biology and Medicine Society for Experimental Biology and Medicine New York, NY. 1995; 208(1):103–108

[63] Sarkar FH, Li Y. "Soy isoflavones and cancer prevention". Cancer Investigation. 2003; 21(5):744–757)

[64] Zhang Yukun and Hong Chen "Genistein attenuates WNT signaling by up-regulating sFRP2 in a human colon cancer cell line" Experimental Biology and Medicine 2011; 236: 714–722).

[65] Szkudelska K., L. Nogowski, Genistein – a dietary compound inducing hormonal and metabolic changes, J. Steroid. Biochem. Mol. Biol. 105 (2007) 37–45.

[66] Sakla M.S., N.S. Shenouda, P.J. Ansell, R.S. Macdonald, D.B. Lubahn, Genistein affects HER2 protein concentration, activation, and promoter regulation in BT-474 human breast cancer cells, Endocrine 32 (2007) 69–78.

[67] Abler A., J.A. Smith, P.A. Randazzo, P.L. Rothenberg, L. Jarett, Genistein differentially inhibits postreceptor effects of insulin in rat adipocytes without inhibiting the insulin receptor kinase, J. Biol. Chem. 267 (1992) 3946–3951

[68] . Onozawa M., T. Kawamori, M. Baba, K. Fukuda, T. Toda, H. Sato, et al., Effects of a soybean isoflavone mixture on carcinogenesis in prostate and seminal vesicles of F344 rats, Jpn. J. Cancer Res. 90 (1999) 393–398.

[69] Foley EF, Jazaeri AA, Shupnik MA, Jazaeri O, Rice LW. Selective loss of estrogen receptor beta in malignant human colon. Cancer Res 2000, 60, 245–248.

[70] Witte D, Chirala M, Younes A, Li Y, Younes M. Estrogen receptor beta is expressed in human colorectal adenocarcinoma. Hum Pathol 2001, 32, 940–944

[71] Harris,H.A. et al. (2003) Evaluation of an estrogen receptor-beta agonist in animal models of human disease. Endocrinology, 144, 4241–4249.

[72] Wada-Hiraike,O. et al. (2006) New developments in oestrogen signalling in colonic epithelium. Biochem. Soc. Trans., 34, 1114–1116. ,Weige,C.C. et al. (2009) Estradiol alters cell growth in nonmalignant colonocytes and reduces the formation of preneoplastic lesions in the colon. Cancer Res., 69, 9118–9124.

[73] Weige,C.C. et al. (2009) Estradiol alters cell growth in nonmalignant colonocytes and reduces the formation of preneoplastic lesions in the colon. Cancer Res., 69, 9118–9124.

[74] Lechner,D. et al. (2005) Phytoestrogens and colorectal cancer prevention. Vitam. Horm., 70, 169–198;

[75] Mak,P. et al. (2010) ERβeta impedes prostate cancer EMT by destabilizing HIF-1alpha and inhibiting VEGF-mediated snail nuclear localization: implications for Gleason grading. Cancer Cell, 17, 319–332

[76] 76. Sarkar,F.H. et al. (2006) The role of genistein and synthetic derivatives of isoflavone in cancer prevention and therapy. Mini Rev. Med. Chem., 6, 401–407

[77] Giles RH, van Es JH and Clevers H: Caught up in a Wnt storm: Wnt signaling in cancer. Biochim Biophys Acta 1653(1): 1-24, 2003.

[78] Aguilera O, Fraga MF, Ballestar E, Paz MF, Herranz M, Espada J et al: Epigenetic inactivation of the Wnt antagonist DICKKOPF-1 (DKK-1) gene in human colorectal cancer. Oncogene 25(29): 4116-4121, 2006.

[79] Suzuki H, Watkins DN, Jair KW, Schuebel KE, Markowitz SD, Chen WD et al: Epigenetic inactivation of SFRP genes allows constitutive WNT signaling in colorectal cancer. Nat Genet 36(4): 417-422, 20?.

[80] Shahana Majid, Altaf A.Dar1, Ardalan E.Ahmad, Hiroshi Hirata, Kazumori Kawakami, Varahram Shahryari, Sharanjot Saini, Yuichiro Tanaka, Angela V.Dahiya, Gaurav Khatri and Rajvir Dahiya BTG3 tumor suppressor gene promoter demethylation, histone modification and cell cycle arrest by genistein in renal cancer. Carcinogenesis vol.30 no.4 pp.662–670, 2009

[81] Kevin J. Day et al Genistein Alters Methylation Patterns in Mice 2002 American Society for Nutritional Sciences]

[82] Qian Li and Hong Chen "Epigenetic modifications of metastasis suppressor genes in colon cancer metastasis" Epigenetics 6:7, 849-852; July 2011

[83] Pavese Janet M. & Rebecca L. Farmer & Raymond C. Bergan: Inhibition of cancer cell invasion and metastasis by genistein. Cancer Metastasis Rev (2010) 29:465–482

[84] Ramos, S. (2007). Effects of dietary flavonoids on apoptotic pathways related to cancer chemoprevention. The Journal of Nutritional Biochemistry, 18(7), 427–442

[85] Wentao Qi1, Christopher R Weber2*, Kaarin Wasland1 and Suzana D Savkovic1: Genistein inhibits proliferation of colon cancer cells by attenuating a negative effect of epidermal growth factor on tumor suppressor FOXO3 activity. BMC Cancer 2011, 11:219

[86] Dijkers PF, Medema RH, Pals C, Banerji L, Thomas NS, Lam EW, Burgering BM, Raaijmakers JA, Lammers JW, Koenderman L, Coffer PJ: Forkhead transcription factor FKHR-L1 modulates cytokine-dependent transcriptional regulation of p27(KIP1). Molecular and Cellular Biology 2000, 20(24):9138-914 36

Bowman-Birk Protease Inhibitor as a Potential Oral Therapy for Multiple Sclerosis

Farinaz Safavi and Abdolmohamad Rostami

Additional information is available at the end of the chapter

http://dx.doi.org/10.5772/54066

1. Introduction

1.1. The Bowman-Birk Protease Inhibitor (BBI)

Legume seeds contain different kinds of proteins and protease inhibitors. Serine proteases are a large sub-group of the protease family [1] and they play a role in various pathological conditions such as cancer and thrombotic and inflammatory diseases [2]. Thus they are excellent targets for treatment of many disorders.

Various plant species and, in particular, legumes contain a great number of serine protease inhibitors. The Bowman-Birk protease inhibitor belongs to a family of serine protease inhibitors that has been widely studied for the past 60 years [3, 4].

The soybean-derived Bowman-Birk protease inhibitor (BBI) is a small protein consisting of 71 amino acids and 7 disulfide bonds [4]. BBI is a double-headed serine protease inhibitor, with two functional active sites at opposite sides of the molecule, which inhibits both trypsin and chymotrypsin-like proteases [1,3] (Figure 1). It is a water-soluble protein that is resistant to acidic conditions and proteolytic enzymes [3]. These characteristics make it a good candidate for use as an oral agent for therapeutic purposes.

Crude soybean contains a small amount of BBI and may have components that counter some of the beneficial effects of BBI. Bowman-Birk Inhibitor Concentrate (BBIC) is a soybean extract enriched in BBI [5]. Researchers prefer to use BBIC in their studies because a smaller amount of BBIC contains the proposed dose of BBI compared to crude soybean.

In rodents, BBI is detectable in the blood, tissue and urine after ingestion [6]. Interestingly, BBI can be detected in the central nervous system (CNS) of animals even when the blood-

© 2013 Safavi and Rostami; licensee InTech. This is a paper distributed under the terms of the Creative Commons Attribution License (http://creativecommons.org/licenses/by/3.0), which permits unrestricted use, distribution, and reproduction in any medium, provided the original work is properly cited.

brain barrier is intact. In human studies, although the BBI level could not be detected in blood after oral BBIC dosing, it could be measured in urine [6].

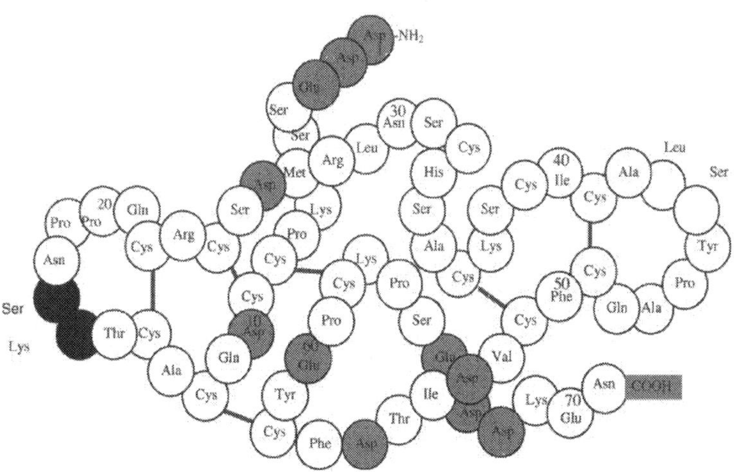

Figure 1. Crystal structure of soybean-derived Bowman-Birk protease inhibitor

The ability of certain serine protease inhibitors to prevent the malignant transformation of cells was shown two decades ago [7, 8]. BBI prevents/suppresses carcinogenesis in a variety of in vitro and in vivo systems [8, 9].

Several human clinical trials to evaluate the effect of BBIC have been completed or are in progress [10-12]. To date, in completed clinical trials, neither toxicity nor neutralizing antibodies against BBIC have been reported in patients receiving BBIC [9].

2. Multiple Sclerosis (MS)

MS is the second cause of disability in young adults and is considered to be a demyelinating disease of the central nervous system (CNS) along with chronic inflammation, demyelination and gliosis [13]. Lesions are characterized by periventricular cuffing and infiltration consisting mainly of T lymphocytes and macrophages, leading to myelin destruction. Recently neuronal degeneration and axonal involvement have also been shown in MS lesions [14]. Current findings therefore raise some doubts about the original assumption that MS is exclusively a white matter disease.

Based on the MS inflammatory phenotype, it has been considered an autoimmune disorder in which peripherally activated myelin-reactive T cells enter the CNS and begin an immunologic cascade that subsequently causes myelin damage.

Activated antigen presenting cells (APCs) and auto-reactive T cells produce pro-inflammatory cytokines, including IL-23, IFN-γ, TNF-α, IL-17, that enhance cell-mediated immunity in the CNS [15-17]. Conversely, other cytokines, such as IL-10, IL-27, IL-4 and TGF-β, play an immunoregulatory role and may be protective in MS [17-20].

Despite extensive research, only a few pharmacotherapeutic agents (e.g., IFN-β, glatiramer acetate, and mitoxantrone) are available, all of which are administered by injection, demonstrate mild to moderate efficacy and have potential side effects [21,22]. A new oral therapeutic agent (Fingolimod) was approved by the FDA and shows potential benefits in MS patients [23].

Recently, in two phase three clinical trials, BG-12 (dimethyl fumarate), a newly proposed oral drug for the treatment of multiple sclerosis, showed a significant reduction in relapse rate and number of MRI lesions in treated patients compared to the placebo group [68, 69].

Development of a new, effective and oral therapy for MS, with fewer side effects, is therefore desirable.

3. Experimental Autoimmune Encephalomyelitis (EAE)

Experimental Autoimmune Encephalomyelitis (EAE) is an autoimmune animal model of MS. Immunization with myelin peptides in different strains of mice induces chronic or relapsing types of the disease, which makes EAE a good tool for studying disease mechanisms and testing therapeutic agents [24] To date, three of the four therapies currently approved for MS were first tested in this animal model [24].

After immunization with myelin protein, APCs present myelin on the surface of MHC II and produce pro-inflammatory cytokines. Dendritic cell-derived IL-12 and IL-23 lead to development of myelin-specific Th1 and Th17 cells, respectively. Th1 and Th17 cells are the two main culprits in pathogenesis of EAE and MS [25]. Auto-reactive T cells enter the CNS and facilitate recruitment of other immune cells such as monocytes and neutrophils. Accumulation of inflammatory cells within the CNS promotes myelin damage, axonal loss and clinical manifestations in affected animals [24].

Recently it has been shown that dendritic cells are also able to produce another cytokine from the IL-12 family called IL-27. Compared with IL-12 and IL-23, IL-27 elicits different immunoregulatory effects. IL-27 inhibits encephalitogenicity of T cells and suppresses EAE disease [26]. In addition, it stimulates IL-10 production in T cells and induces Tr1 cells [17]. IL-10 is a widely studied immunoregulatory cytokine, which virtually all immune cells are able, in different conditions, to release and which suppresses inflammatory response [27]. IL-10 also plays a significant role in suppression of EAE [28-30].

In general, if a therapeutic agent is able to stimulate IL-10 production and Tr1 cells, it could be an excellent candidate for MS therapy.

4. Proteases in inflammation

Several proteases are associated with the pathogenesis of inflammatory disorders [24, 31]. Proteolytic enzymes are involved in activation and migration of immune cells, cytokine and chemokine activation/inactivation and complement function [32].

Various studies demonstrate that neutrophil serine proteases induce proinflammatory activity of both IL-32 and IL-33 cytokines [33, 34]. They are also able to convert inactive forms of IL-1 and IL-18 to the active form of these cytokines [35]. Cytotoxic T cell-derived proteases called granzymes are also involved in inflammation. Granzymes promote T cell entry into the site of inflammation. In addition, they stimulate B cell proliferation [36].

The complement cascade contains different enzymes that activate each other and proteases that play a role in initiation of the cascade, which results in formation of the membrane attack complex [37].

In general, proteases are involved in all aspects of the immune response and play a significant role in inflammation.

5. Proteases in pathogenesis of EAE and MS

Modulators of neuronal and endogenous proteolysis show a different pattern in spinal cords of EAE rats compared to control animals. This finding indicates higher activity of some proteases in EAE than in control groups, which makes specific proteases good potential biomarkers for disease activity or therapeutic targets in the EAE model and MS [38]. Various types of proteases, including lysosomal proteases and matrix metalloproteinases (MMPs), are highly expressed in MS lesions [24, 39-42]. Serine proteases such as plasmin, cathepsin G, chymase and trypsin activate inert MMP proenzymes to their active forms [24, 41, 42].

GelatinaseB (MMP-9] increases the number of leukocytes entering the site of inflammation and promotes myelin breakdown [39, 43]. Plasmin is a serine protease that mainly participates in the coagulation cascade. It has been demonstrated that plasmin directly induces myelin destruction and demyelination [44].

Levels of gelatinase and tissue plasminogen activator (t-PA) are also increased in MS lesions and in the cerebrospinal fluid (CSF) of active MS patients [46, 47]. Reactive astrocytes and infiltrating lymphocytes, macrophages and microglia express MMP-2, MMP-9 and t-PA in early active MS plaques [24, 41, 45, 47].

6. Anti-inflammatory effects of BBI

BBI suppresses the function of several proteases such as leukocyte elastase, trypsin and human cathepsin G released from human inflammatory cells. [48-50]. Mast cell chymase stimulates migration of lymphocytes and purified T cells, and BBI inhibits this enzyme quite efficiently [49]. In addition, BBI significantly suppresses the chemotactic activity of chymase, thus suppressing lymphocyte migration [51].

Stimulated human polymorphonuclear leukocytes produce reactive oxygen species (superoxide and hydrogen peroxide) that may damage cell membranes by reacting with phosopholipids to form peroxides [52]. BBI is able to suppress the production of reactive oxygen species and inhibits their destructive effects [53]. Macrophage-derived proteases and free radicals are also associated with inflammation. BBI down-regulates NO and PGE2 inflammatory pathways in LPS-activated macrophages [54]. Activated macrophages also induce neurotoxicity in the CNS. Anti-inflammatory effects of BBI prevent macrophage-induced neurotoxicity [55].

Serine protease inhibitors can prevent conversion of pro-MMPs to enzymatically active forms [56, 57]. BBI inhibits generation of active MMP-1 and MMP-9 in vitro, and BBIC reduces MMP-2 and -9 activity in supernatants of spleen cells [58].

The aforementioned mechanisms may be particularly relevant in the context of the pathogenesis of multiple sclerosis and myelin destruction in the CNS.

BBI may have significant immunomodulatory effects and can be an excellent potential candidate for treatment of inflammatory and autoimmune diseases.

7. BBI and other protease inhibitors in treatment of inflammatory disease

The role of proteases in inflammation has been reviewed in previous sections. Based on the fact that proteases are actively involved in inflammation, they can be a good therapeutic target in suppression of inflammatory response and treatment of inflammatory diseases.

RWJ-355871 is a synthetic protease inhibitor that effectively suppresses allergic inflammatory diseases of the respiratory system [59]. 4-(2-Aminoethyl) benzenesulfonyl fluoride (AEBSF) is another protease inhibitor that attenuates ovalbumin-induced allergic airway inflammation in its animal model [60].

Several studies have reported that protease inhibitors diminish inflammatory response in inflammatory bowel diseases [1]. Nafamostat is a serine protease inhibitor that suppresses dextran sulfate sodium-induced colitis and diminishes inflammatory infiltration in the colon [61]. BBI is able to suppress gland inflammation in the gastrointestinal tract and shows a strong anti-inflammatory effect in the acute colitis model [62]. In addition, in a completed clinical trial [12], BBI demonstrates anti-inflammatory effects and a degree of amelioration of clinical disease and remission rate in patients with ulcerative colitis. We have also shown that administration of oral BBIC significantly inhibits experimental autoimmune neuritis (EAN) in rats [63, 64].

All of the above findings show the potential immunomodulatory and therapeutic effect of BBI in autoimmune diseases.

8. Immunoregulatory effect of BBI in the EAE model

We have shown that oral treatment of BBIC in MBP-induced EAE in rats, reduces disease severity from clinical score 3 (complete hind limb paralysis) to less than 1 (flaccid tail) compared to control animals. In addition, BBIC treatment significantly diminished demyelination in the peripheral nerve tissue of treated animals [58]. We have also shown that both BBI and BBIC suppress clinical and pathologic manifestations of chronic and relapsing EAE in B6 and SJL mice. In addition, the therapeutic effect of oral BBI is dose-dependent, and oral administration of higher amounts of BBI inhibits EAE more efficiently [65] (Figure 1).

BBI treatment also decreased pathogenicity of myelin-reactive T cells and induced milder disease in the adoptively transferred EAE model (unpublished data).

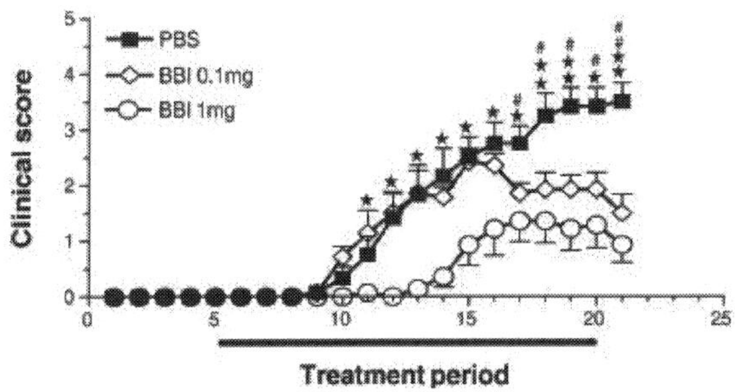

Figure 2. Effect of oral BBI compared to PBS treatment in EAE. Mice that received BBI showed significantly less severe disease compared to control group. The therapeutic effect of BBI is dose-dependent. Elsevier Publications Ltd. has kindly granted us permission to reproduce this figure from Touil et al., 2008.

BBI inhibits invasion of immune cells through the blood-brain barrier (BBB). BBI-treated mice showed dramatically lower numbers of CNS-infiltrating MNCs than control animals [58, 65, 66]. In addition, BBI suppresses generation of active MMP-1 and MMP-9 in vitro, and BBIC reduces MMP-2 and -9 activity in supernatants of spleen cells [58]. Consistent with other findings, BBI decreased migration of splenocytes in Boyden's chamber assay [65].

However, BBI may inhibit release of active MMP-2 and MMP-9 at the blood-brain barrier and prevent immune cell infiltration into the CNS; it might decrease expression of adhesion molecules on immune cells or invasiveness of immune cells, resulting in an altered cytokine pattern of inflammatory cells that hinders their migration from peripheral immune organs to the site of inflammation.

In order to clarify immunoregulatory mechanisms of BBI, the direct effect of BBI on immune cells was evaluated, and it was shown that splenocytes produce a higher amount of IL-10 following BBI treatment [65, 66]. Several reports have demonstrated the immunoregulatory effect of IL-10 in the EAE model of multiple sclerosis. To determine whether the immuno-modulatory effect of BBI depends on IL-10, we have compared the therapeutic effect of oral

BBI in EAE in WT and IL-10 KO mice. Although BBI-treated WT mice showed less severe disease, BBI treatment did not affect clinical disease in BBI-treated IL-10 KO mice compared to the control group [66] (Figure 3).

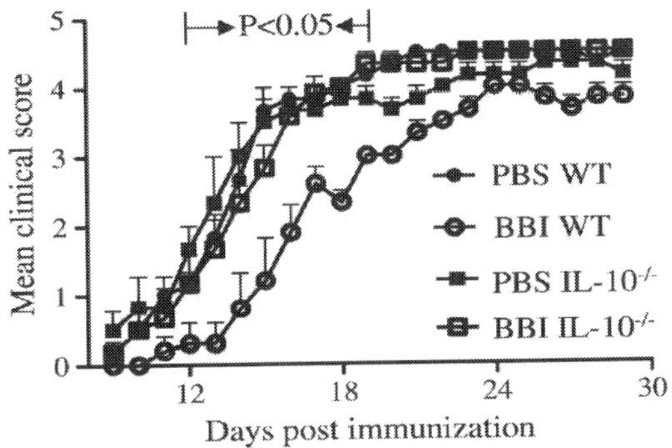

Figure 3. Effect of oral BBI compared to PBS treatment in EAE in WT and IL-10 KO mice. Although BBI-treated WT mice showed significantly less severe disease compared to the control group, there was no significant difference in treated and control IL-10 KO mice. Figure reproduced from Dai et al., 2012 with the kind permission of Elsevier Publications Ltd.

Different types of immune cells can release IL-10 cytokine [27]. However, BBI treatment induces IL-10 mainly in CD4[+] T cells [66]; it increases IL-10 production in CD8[+] T cells (unpublished data), demonstrating that BBI has a strong ability to activate IL-10 producing pathways in T cells. Exploring these underlying mechanisms will be a major focus of our future studies.

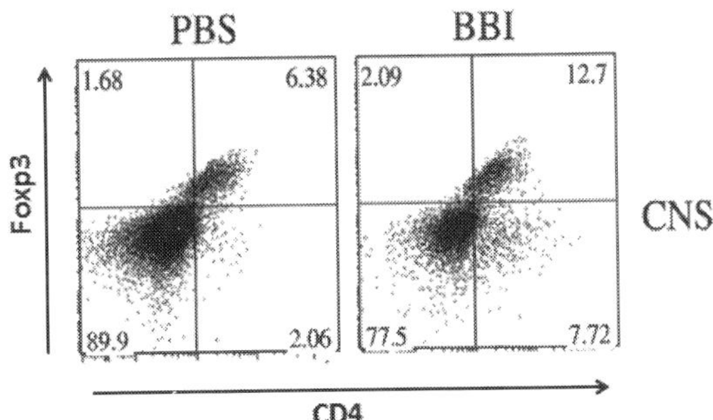

Figure 4. Higher expression of Foxp3[+] Treg cells in CNS infiltrating cells after oral treatment with BBI, Figure reproduced from Dai et al., 2012 with the kind permission of Elsevier Publications Ltd.

Treg cells are a subgroup of CD4[+] T cells that expresses the Foxp3[+] transcription factor. They produce IL-10 in the CNS and can suppress EAE disease [67]. Oral administration of BBI also induces Treg cells in the CNS, which might be one of the underlying mechanisms of the therapeutic effect of BBI in EAE [66] (Figure 4)

BBI also induces IL-10 in other types of effector T cells, and the immunomodulatory effect of BBI might be related to an increase in Tr1 cells. Should this be the case, BBI can be used to induce regulatory T cells and for treatment of autoimmune diseases such as multiple sclerosis.

9. Conclusion

BBI is a soybean-derived serine protease inhibitor. It can be administered orally with several immunomodulatory characteristics and no major side effects. Our observations have shown that BBI dramatically decreases severity of EAE and that its therapeutic effect is mediated through IL-10. In addition, BBI decreases infiltration of inflammatory cells across the BBB and inflammation in the CNS. BBI has potential as a safe and effective oral therapy for multiple sclerosis and other autoimmune diseases.

Acknowledgments

This study was supported by a grant from the National Institutes of Health (NIH 1R01AT005322). We are very grateful to Katherine Regan for editing the chapter.

Author details

Farinaz Safavi and Abdolmohamad Rostami

Department of Neurology, Thomas Jefferson University, Philadelphia, PA, USA

References

[1] Clemente A, Sonnante G, Domoney C. Bowman-Birk inhibitors from legumes and human gastrointestinal health: current status and perspectives. Curr Protein Pept Sci. 2011;12:358-373.

[2] Losso JN. The biochemical and functional food properties of the bowman-birk inhibitor. Crit Rev Food Sci Nutr. 2008;48:94-118. doi: 10.1080/10408390601177589.

[3] Birk Y. The Bowman-Birk inhibitor. Trypsin- and chymotrypsin-inhibitor from soybeans. Int J Pept Protein Res. 1985;25:113-131.

[4] Odani S, Ikenaka T. Studies on soybean trypsin inhibitors. 8. Disulfide bridges in soybean Bowman-Birk proteinase inhibitor. J Biochem. 1973;74:697-715.

[5] Kennedy, A. R. 1993. Overview: anti-carcinogenic activity of protease inhibitors. In Protease inhibitors as cancer chemopreventive agents. W. Troll, Kennedy A. R., ed. Plenum Press, New York. 9-64.

[6] Wan XS, Lu LJ, Anderson KE, Ware JH, Kennedy AR. Urinary excretion of Bowman-Birk inhibitor in humans after soy consumption as determined by a monoclonal antibody-based immunoassay. Cancer Epidemiol Biomarkers Prev. 2000;9:741-747.

[7] Kennedy AR, Little JB. Protease inhibitors suppress radiation-induced malignant transformation in vitro. Nature. 1978;276:825-826.

[8] Yavelow J, Collins M, Birk Y, Troll W, Kennedy AR. Nanomolar concentrations of Bowman-Birk soybean protease inhibitor suppress x-ray-induced transformation in vitro. Proc Natl Acad Sci U S A. 1985;82:5395-5399.

[9] Kennedy AR. Chemopreventive agents: protease inhibitors. Pharmacol Ther. 1998;78:167-209.

[10] Malkowicz SB, McKenna WG, Vaughn DJ, Wan XS, Propert KJ, Rockwell K, et al. Effects of Bowman-Birk inhibitor concentrate (BBIC) in patients with benign prostatic hyperplasia. Prostate. 2001;48:16-28. doi: 10.1002/pros.1077.

[11] Armstrong WB, Kennedy AR, Wan XS, Taylor TH, Nguyen QA, Jensen J, et al. Clinical modulation of oral leukoplakia and protease activity by Bowman-Birk inhibitor concentrate in a phase IIa chemoprevention trial. Clin Cancer Res. 2000;6:4684-4691.

[12] Lichtenstein GR, Deren JJ, Katz S, Lewis JD, Kennedy AR, Ware JH. Bowman-Birk inhibitor concentrate: a novel therapeutic agent for patients with active ulcerative colitis. Dig Dis Sci. 2008;53:175-180. doi: 10.1007/s10620-007-9840-2.

[13] Peterson LK, Fujinami RS. Inflammation, demyelination, neurodegeneration and neuroprotection in the pathogenesis of multiple sclerosis. J Neuroimmunol. 2007;184:37-44. doi: 10.1016/j.jneuroim.2006.11.015.

[14] Geurts JJ, Kooi EJ, Witte ME, van der Valk P. Multiple sclerosis as an "inside-out" disease. Ann Neurol. 2010;68:767-8; author reply 768. doi: 10.1002/ana.22279.

[15] Cua DJ, Sherlock J, Chen Y, Murphy CA, Joyce B, Seymour B, et al. Interleukin-23 rather than interleukin-12 is the critical cytokine for autoimmune inflammation of the brain. Nature. 2003;421:744-748. doi: 10.1038/nature01355.

[16] Langrish CL, Chen Y, Blumenschein WM, Mattson J, Basham B, Sedgwick JD, et al. IL-23 drives a pathogenic T cell population that induces autoimmune inflammation. J Exp Med. 2005;201:233-240. doi: 10.1084/jem.20041257.

[17] Fitzgerald DC, Zhang GX, El-Behi M, Fonseca-Kelly Z, Li H, Yu S, et al. Suppression of autoimmune inflammation of the central nervous system by interleukin 10 secret-

ed by interleukin 27-stimulated T cells. Nat Immunol. 2007;8:1372-1379. doi: 10.1038/ni1540.

[18] Bettelli E, Das MP, Howard ED, Weiner HL, Sobel RA, Kuchroo VK. IL-10 is critical in the regulation of autoimmune encephalomyelitis as demonstrated by studies of IL-10- and IL-4-deficient and transgenic mice. J Immunol. 1998;161:3299-3306.

[19] Lenz DC, Swanborg RH. Suppressor cells in demyelinating disease: a new paradigm for the new millennium. J Neuroimmunol. 1999;100:53-57.

[20] Zhao Z, Yu S, Fitzgerald DC, Elbehi M, Ciric B, Rostami AM, et al. IL-12R beta 2 promotes the development of CD4+CD25+ regulatory T cells. J Immunol. 2008;181:3870-3876.

[21] Hafler DA, Slavik JM, Anderson DE, O'Connor KC, De Jager P, Baecher-Allan C. Multiple sclerosis. Immunol Rev. 2005;204:208-231. doi: 10.1111/j.0105-2896.2005.00240.x.

[22] Boster A, Edan G, Frohman E, Javed A, Stuve O, Tselis A, et al. Intense immunosuppression in patients with rapidly worsening multiple sclerosis: treatment guidelines for the clinician. Lancet Neurol. 2008;7:173-183. doi: 10.1016/S1474-4422[08]70020-6.

[23] Devonshire V, Havrdova E, Radue EW, O'Connor P, Zhang-Auberson L, Agoropoulou C, et al. Relapse and disability outcomes in patients with multiple sclerosis treated with fingolimod: subgroup analyses of the double-blind, randomised, placebo-controlled FREEDOMS study. Lancet Neurol. 2012;11:420-428. doi: 10.1016/S1474-4422[12]70056-X.

[24] Cuzner ML, Opdenakker G. Plasminogen activators and matrix metalloproteases, mediators of extracellular proteolysis in inflammatory demyelination of the central nervous system. J Neuroimmunol. 1999;94:1-14.

[25] Chen SJ, Wang YL, Fan HC, Lo WT, Wang CC, Sytwu HK. Current status of the immunomodulation and immunomediated therapeutic strategies for multiple sclerosis. Clin Dev Immunol. 2012;2012:970789. doi: 10.1155/2012/970789.

[26] Fitzgerald DC, Ciric B, Touil T, Harle H, Grammatikopolou J, Das Sarma J, et al. Suppressive effect of IL-27 on encephalitogenic Th17 cells and the effector phase of experimental autoimmune encephalomyelitis. J Immunol. 2007;179:3268-3275.

[27] Saraiva M, O'Garra A. The regulation of IL-10 production by immune cells. Nat Rev Immunol. 2010;10:170-181. doi: 10.1038/nri2711.

[28] Santambrogio L, Crisi GM, Leu J, Hochwald GM, Ryan T, Thorbecke GJ. Tolerogenic forms of auto-antigens and cytokines in the induction of resistance to experimental allergic encephalomyelitis. J Neuroimmunol. 1995;58:211-222.

[29] Cash E, Minty A, Ferrara P, Caput D, Fradelizi D, Rott O. Macrophage-inactivating IL-13 suppresses experimental autoimmune encephalomyelitis in rats. J Immunol. 1994;153:4258-4267.

[30] Rott O, Fleischer B, Cash E. Interleukin-10 prevents experimental allergic encephalo-myelitis in rats. Eur J Immunol. 1994;24:1434-1440. doi: 10.1002/eji.1830240629.

[31] Vaday GG, Baram D, Salamon P, Drucker I, Hershkoviz R, Mekori YA. Cytokine pro-duction and matrix metalloproteinase (MMP)-9 release by human mast cells follow-ing cell-to-cell contact with activated T cells. Isr Med Assoc J. 2000;2 Suppl:26.

[32] Scarisbrick IA. The multiple sclerosis degradome: enzymatic cascades in develop-ment and progression of central nervous system inflammatory disease. Curr Top Mi-crobiol Immunol. 2008;318:133-175.

[33] Novick D, Rubinstein M, Azam T, Rabinkov A, Dinarello CA, Kim SH. Proteinase 3 is an IL-32 binding protein. Proc Natl Acad Sci U S A. 2006;103:3316-3321. doi: 10.1073/pnas.0511206103.

[34] Lefrancais E, Roga S, Gautier V, Gonzalez-de-Peredo A, Monsarrat B, Girard JP, et al. IL-33 is processed into mature bioactive forms by neutrophil elastase and cathepsin G. Proc Natl Acad Sci U S A. 2012;109:1673-1678. doi: 10.1073/pnas.1115884109.

[35] Guma M, Ronacher L, Liu-Bryan R, Takai S, Karin M, Corr M. Caspase 1-independ-ent activation of interleukin-1beta in neutrophil-predominant inflammation. Arthritis Rheum. 2009;60:3642-3650. doi: 10.1002/art.24959.

[36] Mullbacher A, Waring P, Tha Hla R, Tran T, Chin S, Stehle T, et al. Granzymes are the essential downstream effector molecules for the control of primary virus infec-tions by cytolytic leukocytes. Proc Natl Acad Sci U S A. 1999;96:13950-13955.

[37] Walport MJ. Complement. First of two parts. N Engl J Med. 2001;344:1058-1066. doi: 10.1056/NEJM200104053441406.

[38] Jain MR, Bian S, Liu T, Hu J, Elkabes S, Li H. Altered proteolytic events in experi-mental autoimmune encephalomyelitis discovered by iTRAQ shotgun proteomics analysis of spinal cord. Proteome Sci. 2009;7:25. doi: 10.1186/1477-5956-7-25.

[39] Bever CT,Jr, Rosenberg GA. Matrix metalloproteinases in multiple sclerosis: targets of therapy or markers of injury? Neurology. 1999;53:1380-1381.

[40] Halonen T, Kilpelainen H, Pitkanen A, Riekkinen PJ. Lysosomal hydrolases in cere-brospinal fluid of multiple sclerosis patients. A follow-up study. J Neurol Sci. 1987;79:267-274.

[41] Hartung HP, Kieseier BC. The role of matrix metalloproteinases in autoimmune damage to the central and peripheral nervous system. J Neuroimmunol. 2000;107:140-147.

[42] Kieseier BC, Seifert T, Giovannoni G, Hartung HP. Matrix metalloproteinases in in-flammatory demyelination: targets for treatment. Neurology. 1999;53:20-25.

[43] Leppert D, Lindberg RL, Kappos L, Leib SL. Matrix metalloproteinases: multifunc-tional effectors of inflammation in multiple sclerosis and bacterial meningitis. Brain Res Brain Res Rev. 2001;36:249-257.

[44] Romanic AM, White RF, Arleth AJ, Ohlstein EH, Barone FC. Matrix metalloproteinase expression increases after cerebral focal ischemia in rats: inhibition of matrix metalloproteinase-9 reduces infarct size. Stroke. 1998;29:1020-1030.

[45] Cuzner ML, Gveric D, Strand C, Loughlin AJ, Paemen L, Opdenakker G, et al. The expression of tissue-type plasminogen activator, matrix metalloproteases and endogenous inhibitors in the central nervous system in multiple sclerosis: comparison of stages in lesion evolution. J Neuropathol Exp Neurol. 1996;55:1194-1204.

[46] Gijbels K, Masure S, Carton H, Opdenakker G. Gelatinase in the cerebrospinal fluid of patients with multiple sclerosis and other inflammatory neurological disorders. J Neuroimmunol. 1992;41:29-34.

[47] Maeda A, Sobel RA. Matrix metalloproteinases in the normal human central nervous system, microglial nodules, and multiple sclerosis lesions. J Neuropathol Exp Neurol. 1996;55:300-309.

[48] Larionova NI, Gladysheva IP, Tikhonova TV, Kazanskaia NF. Inhibition of cathepsin G and elastase from human granulocytes by multiple forms of the Bowman-Birk type of soy inhibitor. Biokhimiia. 1993;58:1437-1444.

[49] Gladysheva IP, Larionova NI, Gladyshev DP, Tikhonova TV, Kazanskaia NF. The classical Bowman-Birk soy inhibitor is an effective inhibitor of human granulocyte alpha-chymotrypsin and cathepsin G. Biokhimiia. 1994;59:513-518.

[50] Tikhonova TV, Gladysheva IP, Kazanskaia NF, Larionova NI. Inhibition of elastin hydrolysis, catalyzed by human leukocyte elastase and cathepsin G, by the Bowman-Birk type soy inhibitor. Biokhimiia. 1994;59:1739-1745.

[51] Tani K, Ogushi F, Kido H, Kawano T, Kunori Y, Kamimura T, et al. Chymase is a potent chemoattractant for human monocytes and neutrophils. J Leukoc Biol. 2000;67:585-589.

[52] Halliwell B. Free radicals, reactive oxygen species and human disease: a critical evaluation with special reference to atherosclerosis. Br J Exp Pathol. 1989;70:737-757.

[53] Frenkel K, Chrzan K, Ryan CA, Wiesner R, Troll W. Chymotrypsin-specific protease inhibitors decrease H2O2 formation by activated human polymorphonuclear leukocytes. Carcinogenesis. 1987;8:1207-1212.

[54] Dia VP, Berhow MA, Gonzalez De Mejia E. Bowman-Birk inhibitor and genistein among soy compounds that synergistically inhibit nitric oxide and prostaglandin E2 pathways in lipopolysaccharide-induced macrophages. J Agric Food Chem. 2008;56:11707-11717. doi: 10.1021/jf802475z.

[55] Li J, Ye L, Cook DR, Wang X, Liu J, Kolson DL, et al. Soybean-derived Bowman-Birk inhibitor inhibits neurotoxicity of LPS-activated macrophages. J Neuroinflammation. 2011;8:15. doi: 10.1186/1742-2094-8-15.

[56] Bawadi HA, Antunes TM, Shih F, Losso JN. In vitro inhibition of the activation of Pro-matrix Metalloproteinase 1 (Pro-MMP-1] and Pro-matrix metalloproteinase 9

(Pro-MMP-9] by rice and soybean Bowman-Birk inhibitors. J Agric Food Chem. 2004;52:4730-4736. doi: 10.1021/jf034576u.

[57] Losso JN, Munene CN, Bansode RR, Bawadi HA. Inhibition of matrix metalloprotei-nase-1 activity by the soybean Bowman-Birk inhibitor. Biotechnol Lett. 2004;26:901-905.

[58] Gran B, Tabibzadeh N, Martin A, Ventura ES, Ware JH, Zhang GX, et al. The pro-tease inhibitor, Bowman-Birk Inhibitor, suppresses experimental autoimmune ence-phalomyelitis: a potential oral therapy for multiple sclerosis. Mult Scler. 2006;12:688-697.

[59] Maryanoff BE, de Garavilla L, Greco MN, Haertlein BJ, Wells GI, Andrade-Gordon P, et al. Dual inhibition of cathepsin G and chymase is effective in animal models of pulmonary inflammation. Am J Respir Crit Care Med. 2010;181:247-253. doi: 10.1164/rccm.200904-0627OC.

[60] Saw S, Kale SL, Arora N. Serine protease inhibitor attenuates ovalbumin induced in-flammation in mouse model of allergic airway disease. PLoS One. 2012;7:e41107. doi: 10.1371/journal.pone.0041107.

[61] Cho EY, Choi SC, Lee SH, Ahn JY, Im LR, Kim JH, et al. Nafamostat mesilate attenu-ates colonic inflammation and mast cell infiltration in the experimental colitis. Int Im-munopharmacol. 2011;11:412-417. doi: 10.1016/j.intimp.2010.12.008.

[62] Ware JH, Wan XS, Kennedy AR. Bowman-Birk inhibitor suppresses production of superoxide anion radicals in differentiated HL-60 cells. Nutr Cancer. 1999;33:174-177. doi: 10.1207/S15327914NC330209.

[63] Fujioka T, Purev E, Kremlev SG, Ventura ES, Rostami A. Flow cytometric analysis of infiltrating cells in the peripheral nerves in experimental allergic neuritis. J Neuroim-munol. 2000;108:181-191.

[64] Olee T, Powell HC, Brostoff SW. New minimum length requirement for a T cell epit-ope for experimental allergic neuritis. J Neuroimmunol. 1990;27:187-190.

[65] Touil T, Ciric B, Ventura E, Shindler KS, Gran B, Rostami A. Bowman-Birk inhibitor suppresses autoimmune inflammation and neuronal loss in a mouse model of multi-ple sclerosis. J Neurol Sci. 2008;271:191-202. doi: 10.1016/j.jns.2008.04.030.

[66] Dai H, Ciric B, Zhang GX, Rostami A. Interleukin-10 plays a crucial role in suppres-sion of experimental autoimmune encephalomyelitis by Bowman-Birk inhibitor. J Neuroimmunol. 2012;245:1-7. doi: 10.1016/j.jneuroim.2012.01.005.

[67] Rynda-Apple A, Huarte E, Maddaloni M, Callis G, Skyberg JA, Pascual DW. Active immunization using a single dose immunotherapeutic abates established EAE via IL-10 and regulatory T cells. Eur J Immunol. 2011;41:313-323. doi: 10.1002/eji.201041104; 10.1002/eji.201041104.

[68] Gold R, Kappos L, Arnold DL, Bar-Or A, Giovannoni G, Selmaj K, Tornatore C, Sweetser MT, Yang M, Sheikh SI, Dawson KT; DEFINE Study Investigators.Placebo-

controlled phase 3 study of oral BG-12 for relapsing multiple sclerosis. N Engl J Med. 2012 Sep 20; 367[12]:1098-107.

[69] Fox RJ, Miller DH, Phillips JT, Hutchinson M, Havrdova E, Kita M, Yang M, Raghupathi K, Novas M, Sweetser MT, Viglietta V, Dawson KT; CONFIRM Study Investigators.Placebo-controlled phase 3 study of oral BG-12 or glatiramer in multiple sclerosis. N Engl J Med. 2012 Sep 20;367[12]:1087-97.

Chapter 6

Soybean Oil Derivatives
for Fuel and Chemical Feedstocks

Joanna McFarlane

Additional information is available at the end of the chapter

http://dx.doi.org/10.5772/54067

1. Introduction

Plant-based sources of hydrocarbons are being considered as alternatives to petrochemicals because of the need to conserve petroleum resources for reasons of national security and climate change [1]. Changes in fuel formulations to include ethanol from corn sugar and methyl esters from agricultural products are examples of this policy in the United States and elsewhere as biofuels from efficiently grown and processed biomass are claimed to be carbon neutral. In the United States, the mandate to include biofuels has been implemented as the Renewable Fuels Standards (RFS1 and RFS2) [2] with biobased diesel fuel as one of the categories. The production of biodiesel in the United States has varied considerably over the last few years, but was 241×10^6 gallons in the first quarter of 2012, a high number but one that still only represents 2% of the total volume of diesel fuel produced for heating and vehicles [3]. Most of the biodiesel comes from soybean oil, more than double the contribution of the other major feedstocks combined: canola oil, yellow grease, and tallow.

Replacements for commodity chemicals are also being considered, as this value stream represents much of the profit for the oil industry and one that would be affected by shortages in oil or other fossil fuels. While the discovery of large amounts of natural gas associated with oil shale deposits have reduced this as an immediate concern for instance the estimated recoverable reserves in the Western US have now reached 800×10^9 bbls [4] -research into bio-based feedstock materials continues for the expected long-term benefit. In particular, this chapter reviews a literature on the conversion of bio-based extracts to hydrocarbons for fuels and for building block commodity chemicals, with a focus on soybean derived products.

© 2013 McFarlane; licensee InTech. This is a paper distributed under the terms of the Creative Commons Attribution License (http://creativecommons.org/licenses/by/3.0), which permits unrestricted use, distribution, and reproduction in any medium, provided the original work is properly cited.

2. Fuels

Although commercially produced, more economical conversion of methyl esters from soybean triglycerides is an active area of research to make the product more cost competitive in comparison with standard petrochemical diesel [5]. The processes of esterification and transesterification to produce methyl esters that can be burned directly in compression -gnition engines has been reviewed elsewhere [6, 7].The fatty acid chains on the lipid molecule that constitutes soybean oil, also called a triacylglycerol or TAG, are split from the glycerol backbone and esterified with an alcohol, generally methanol, in the presence of a homogeneous base or acid catalyst, Reaction 1.

Commercial processing of biodiesel through homogeneous catalysis suffers from high feedstock costs and batch processing that requires long residence times to achieve good conversion. Hence, ongoing research continues to explore methods how to best use low-quality feedstocks, and to reduce the reagent requirements, energy usage, processing time, and complexity [8]. Figure 1 shows results from simulation of a continuous process to make biodiesel, varying temperature (a) and methanol content (b) to determine conditions for the optimal production of high quality grade biodiesel. As the process is limited by kinetics and mass transfer, the effect of mixing has also been investigated by considering the available volume fraction of reagents (c) [9], defined as the molar ratio of reagents in the reaction zone versus the overall reagent volume in the vessel. The available volume can be changed by increasing the contact zone between the immiscible reagents where the reactions take place. The interfacial surface area is dependent on the intensity of mixing in the multiphase system. Interfacial area can also be increased by reducing the size of the dispersed phase droplets, such as by bubbling reagent methanol into the oil through a frit. Novel approaches to biodiesel production continue to be explored, particularly for lower grade and waste feedstocks, such as the direct extraction of fatty acid chains through use of a solvent such as an ionic liquid to pretreat esterification to the methyl ester [10]. Other work has examined methanol-based transesterification of waste cooking oil under quite mild conditions (110°C in 2 h) in contact with tungsten oxide solid acid catalysts, giving yields of fatty acid methyl esters (FAME) that are close to the American Society of Testing and Materials (ASTM) standard for biodiesel [11]. The authors of that study, Komintarachat and Chuepeng, reported several advantages of working with a WO_x/Al_2O_3 catalyst. Prior separation of free fatty acids, in their sample of waste cooking oil reported as 15%, was not necessary to achieve high yields in a one step process. In addition, they found the catalyst has desirable properties for scale-up, being low cost, reusable, and less reactive than traditional homogeneous catalysts.

The choice of acid or base homogeneous catalysis depends on the concentration of free fatty acids (FFA) in the triglyceride feedstock. Virgin soybean oil has a low FFA content, <4%, and so can be converted to biodiesel by transesterification without an acid-catalyzed esterification pretreatment. However, oil that has been degraded by heat, such as waste oil, requires a two-step conversion. FFA produced during heating have to be esterified, otherwise they become saponified during transesterification. New processes are being developed to simplify the conversion of waste oil, such as the use of a supported heteropolyacid catalyst that simultaneously promotes both the esterification and transesterification processes [13].

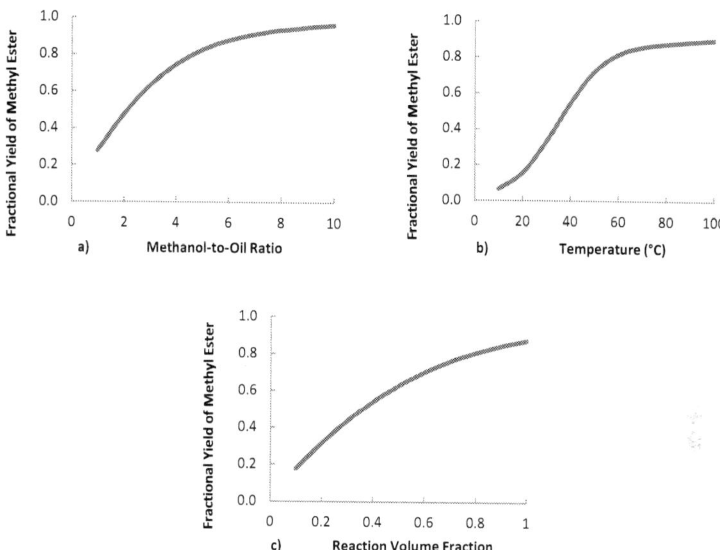

Figure 1. Results of parametric studies on methyl ester production in a continuous reactor showing (a) asymptotic approach to a maximum yield with methanol-to-oil molar ratio and (b) with reactor temperature. The dependence of yield on volume fraction simulated the effect of mixing in the reactor. These calculations were programmed in Mat-Lab® [12], for nominal reaction conditions of 50°C, 7:1 methanol-to-oil molar ratio, and an effective reaction volume of 60% [9].

$$C_3H_5(CO_2R^1)(CO_2R^2)(CO_2R^3) + CH_3OH \rightarrow CH_3O_2R^1 + C_3H_5(OH)(CO_2R^2)(CO_2R^3)$$
$$C_3H_5(OH)(CO_2R^2)(CO_2R^3) + CH_3OH \rightarrow CH_3O_2R^2 + C_3H_5(OH)_2(CO_2R^3) \qquad (1)$$
$$C_3H_5(OH)_2(CO_2R^3) + CH_3OH \rightarrow CH_3O_2R^3 + C_3H_8O_3$$

Triglyceride + 3 Methanoln3 Methyl Esters + Glycerine

The chemical conversion to FAME produces a low viscosity, high-cetane number fuel that can be mixed directly with conventional diesel. The physical properties of diesel and biodiesel, or FAME, are compared in Table 1 [14]. The properties in the table are given for the liquid phase at 25°C and for vapor phase at 527°C, corresponding to pre-ignition conditions in a compression ignition engine. Although similar in carbon chain length and cetane number, biodiesel differs from diesel significantly in its vapor pressure, liquid viscosity, and vapor diffusion coefficient. The properties of the biodiesel depend on the length and unsaturation of the fatty acid chains, Table 1. Where the data are lacking for the soybean-derived methyl esters, the vapor phase properties for biodiesel have been replaced by those of methyl oleate.

Physical Property	Diesel*	Biodiesel*
Density [kg·m⁻³]	762*	884*
vapor pressure [Pa]	2.22*	6.10×10^{-5}-1.18×10^{-3}*
surface tension [J·m⁻²]	2.68×10^{-2}*	2.49×10^{-2}*
liquid viscosity [Pa·s]	3.14×10^{-3}*	9.10×10^{-3}*
liquid thermal conductivity [J·m⁻¹·s⁻¹·K⁻¹]	0.144*	0.100*
latent heat [J·kg⁻¹]	3.60×10^{5}*	3.38×10^{5}***
liquid specific heat [J·kg⁻¹·K⁻¹]	2.27×10^{3}*	2.01×10^{3}*
vapor specific heat [kJ·mol⁻¹·K⁻¹]	0.643#	0.848#
vapor diffusion coefficient [m²·s⁻¹]	8.50×10^{-6}#	9.44×10^{-7}#
vapor viscosity [Pa·s]	1.00×10^{-5}#	1.21×10^{-5}**#
vapor thermal conductivity [J·m⁻¹·s⁻¹·K⁻¹]	4.35×10^{-2}#	2.60×10^{-2}#

* Values at 25°C for liquids

Values at 527°C for vapors

** Values for methyl oleate as a representative biodiesel component.

Table 1. Diesel ($C_{14}H_{30}$) versus biodiesel properties at 25 or 527°C

Although successfully blended up to 20 volume% for commercial and military use [16], methyl ester content in vehicle fuel is limited by a number of factors, including the perform‐ ance in cold weather, the effect of oxygen content on engine components (particularly in the case of older engines), shelf-life and thermal stability [17], and higher NO_x emissions from engines that are not tuned to handle the higher temperature conditions of methyl ester com‐ bustion [18]. Results from simulations presented in Figures 2 and 3 show on a microscopic scale how the combustion of biodiesel can differ from diesel (represented as n-heptane in the engine simulations) in terms of temperature and emissions [15]. The development of en‐ gines that can accommodate biodiesel have focused on the effects of physical properties on spray parameters and droplet formation that will greatly affect the ignition conditions and combustion characteristics, Figures 2 and 3, and enhanced by early oxidation in the low tem‐ perature heat release phase of combustion, Figure 4. Figure 4 shows the progression of the combustion of 20% biodiesel as a function of crank angle position, with 360° corresponding to top-dead-center. The key radicals in the low temperature heat release portion of the cycle include OH• and HO_2•, but OH• dominates after the main ignition event.

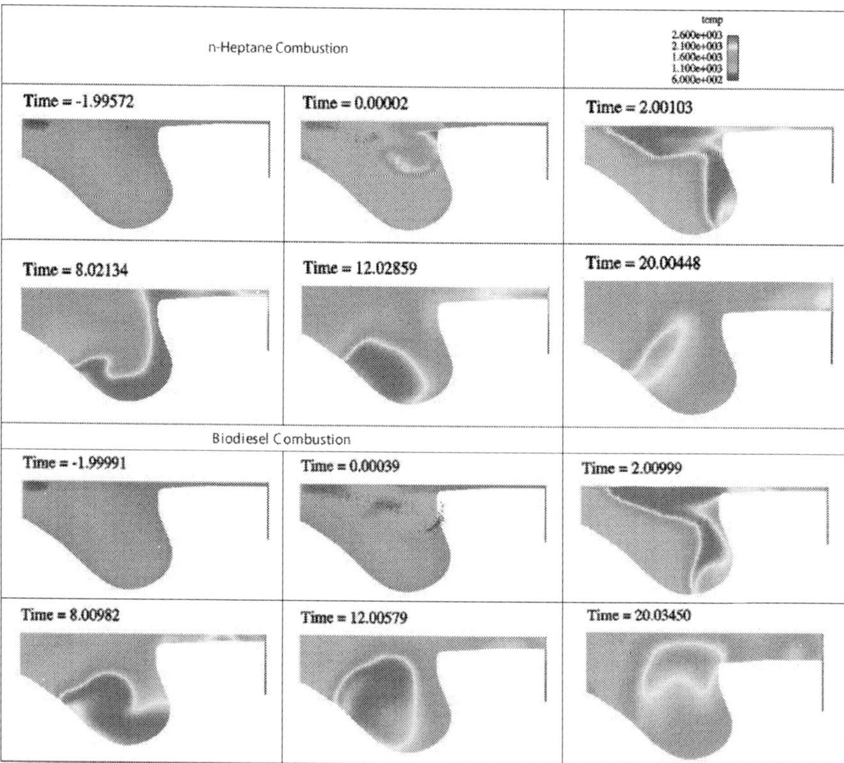

Figure 2. Predicted differences in the temperature in n-heptane and biodiesel combustion after injection into the cylinder. The n-heptane, representing diesel fuel, shows combustion occurring very rapidly after injection, 2×10^{-8}s. The biodiesel, while slower to vaporize and ignite, shows a higher temperature at 12 ms after injection. The simulations were done assuming a cycle of 2000 Rev/min. The temperature key in the upper right is given in degrees K. Reprinted with permission from SAE paper 2008-01-1378 Copyright © 2008 SAE International [15].

These factors have led to interest in synthesizing a hydrocarbon fuel starting with methyl esters, a so-called "green diesel" that will maintain the high cetane number of biodiesel, but will achieve better performance in an automobile: through enhanced mixing, injection, and combustion; reduced downstream issues such as NO_x emissions; and better upstream handling associated with fuel manufacture and distribution. Bunting and colleagues have reviewed the development of fungible and compatible biofuels [20]. That report considers a wide variety of products, from ethanol to pyrolysis oils as well as soy-derived biodiesel. Concerns that arise when developing alternative fuels include refining, blending, and distribution, regulatory barriers, verification of performance, and changes in operating practices throughout the distribution system. In this chapter, we focus on the chemistry of the fuel.

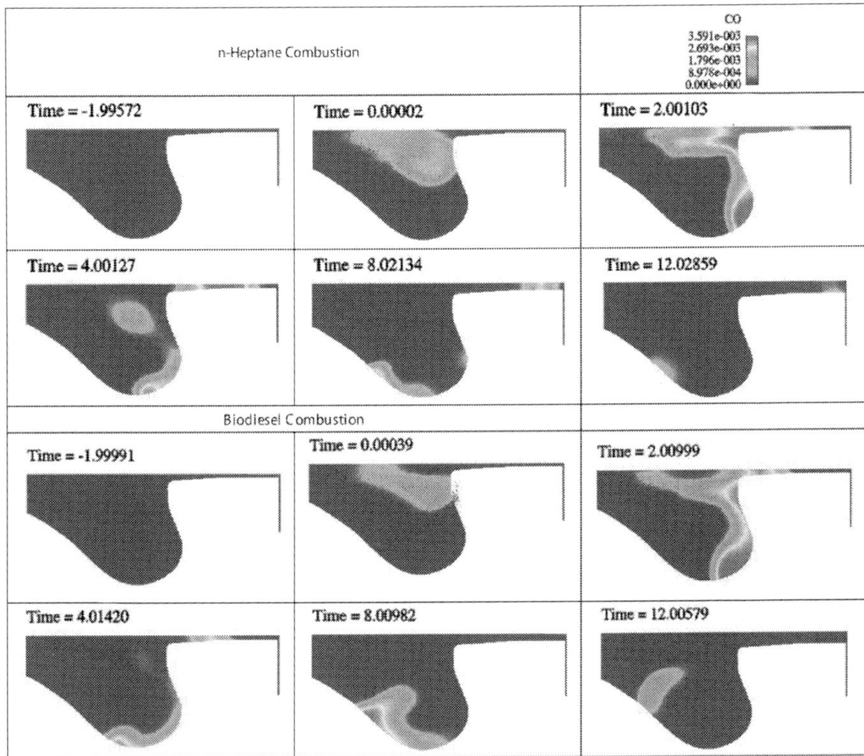

Figure 3. Predicted differences in the CO concentration in n-heptane and biodiesel combustion after injection into the cylinder. As with the temperature profiles, the production of CO from biodiesel lagged that from n-heptane during the event. However, the final concentration of CO was higher for biodiesel than for n-heptane. The CO concentration key in the upper right is given in mole fraction. Reprinted with permission from SAE paper 2008-01-1378 Copyright © 2008 SAE International [15].

Unsaturated methyl esters have more affinity for water and contaminants than does hexadecane, a typical component of diesel fuel. Water affinity is often expressed in the form of the octanol-water distribution coefficient or K_{ow}, with lower values of K_{ow} indicating more hydrophilic compounds, K_{ow}=moles(octanol)/moles(water). K_{ow} values for a few select components of fossil-based diesel [21] and long chain methyl esters typical of those derived from soy oil [22] are presented in Table 2. Water can be problematic in fuel distribution systems which are made of low carbon or low grade stainless steel, but water can be gravity separated when the fuel is held in storage vessels. Separation is less likely to occur with oxygenated fuels, particularly those that have degraded to shorter chain components through autooxidation. Because of the issues with materials compatibility, potential contamination of pipelines by residues, and high viscosity at low temperatures, biodiesel must be added to standard diesel fuel at a terminal loading facility, where the fuel is mixed and then loaded onto trucks for distribution. However, mixing at a distribution terminal affords less quality control than at a refinery, with the latter having the ability for online testing of properties and composition, followed by adjustment to meet ASTM specifications if necessary [23].

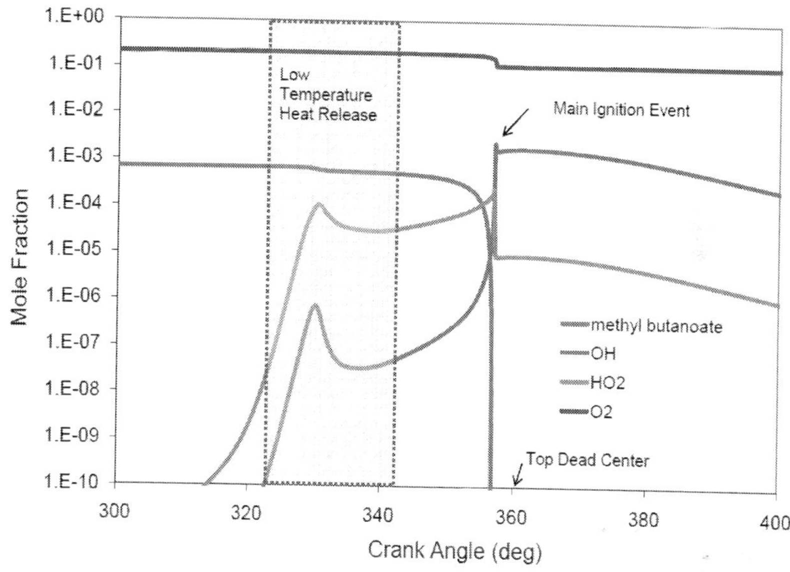

Figure 4. Combustion of a 20% blend of methyl butanoate (simulating biodiesel) and n-heptane (simulating petroleum diesel) showing the importance of reactive species during different times of the cycle [19].

Biodiesel derived from soybean oil comprises long fatty acid chains, C16-C18, with a high cetane number, and a high degree of unsaturation, 84-87%, for better cold flow properties (reduced viscosity) relative to other plant-based methyl esters. However, the unsaturation can also lead to issues with shelf life and thermal stability [28] in comparison with more hydrogenated oils such as palm oil. The process of oxidation and its effects on the properties of biodiesel has been studied using chemical and thermal analysis by Tan and colleagues [29], and reviewed by Mushbrush [17] and Knothe [30]. Oxidation of the double bonds can occur through an autocatalytic mechanism simplistically depicted below, Reaction (2), initiated by hydrogen abstraction from an unsaturated carbon atom. The greater the unsaturation, the more stable the allylic radical, $R_n\bullet$, thus a precursor mono-unsaturated FFA has greater stability that the doubly- and triply-unsaturated chains. Once formed, the radical can combine with O_2, allowing formation of a secondary intermediate in the chain, a reactive hydroperoxide [31, 32], Reaction (3). Light increases the rate of decomposition, because photosensitization allows a direct reaction between the O_2 and the carbon-carbon double bond offering another pathway to oxidation. The hydroperoxides, once formed, can convert to a variety of products, cyclized five and six membered rings, malonaldehyde or $C_3O_2H_4$, hydroxy and epoxy esters, and allylic hydroxyl- and ketone compounds, among other oxygenated derivatives. Cleavage reactions form reactive radicals, continuing the process to produce volatile compounds such as carbonyls, alcohols, esters, and short chain hydrocarbons, Reactions (4a and b), some of which react to form furans, aldehydes, ketones, lactones, alkynes, and aromatics [33]. Because of an associated increase in viscosity and acid number, these oxidation products are generally undesirable in a combustion engine [34].

	Log K_{ow}	water solubility (mg/L)at 25°C
Biodiesel Components		
methylpalmitate, C16:0 (10-12%) (hexadecanoic acid, methyl ester)	7.38	4.00×10^{-3}
methyl stearate, C18:0 (3-4%) (octadecanoic acid, methyl ester)	8.35	3.01×10^{-3}
methyloleate, C18:1 (23-25%) (octadecenoic acid, methyl ester)	7.45	3.68×10^{-3}
methyllinolate, C18:2 (53-56%) (octadecadienoic acid, methyl ester)	6.82	2.10×10^{-2}
methyllinolenate, C18:3 (6-8%) (octadecatrienoic acid, methylester)	6.29	9.18×10^{-2}
Diesel #2 Components		
Monoaromatics and small cyclic compounds (10.0%)	2-5 Hexylbenzene is 5.52	1.02
cycloparaffins (34.0%)	3-5 Cyclohexane is 3.44	55 [27]
Naphthalenes and PAH (14.7%)	3-5 Naphthalene is 3.3	30
n- and i-paraffins (41.3%)	3.3-7.06 for short chain HC [21] 8.2 for $C_{16}H_{34}$	9×10^{-4}

Table 2. Hydrophobicity expressed as octanol-water partition coefficients [24, 25] for organic derivatives of petroleum and biodiesel [26]

$$\begin{aligned}
R_n\text{-H} &\rightarrow R_n\cdot + H\cdot \\
R_n\cdot + O_2 &\rightarrow R_nOO \\
R_nOO\cdot + R_{n'}\text{-H} &\rightarrow R_nOO\text{-H} + R_{n'}
\end{aligned} \tag{2}$$

$$R_n\text{-C=C-}R_{n'} + 2O_2 \rightarrow R_n\text{-C}(OOH)\text{-C=C-}R_{n'-1} + R_{n-1}\text{-C=C-C}(OOH)\text{-}R_{n'} \tag{3}$$

$$\begin{aligned}
R_n\text{-CH=CH-C}(O\cdot)\text{H-}R_{n'} &\rightarrow R_{n'}\text{-CHO} + R_n\text{-CH=CH}\cdot \text{ (a)} \\
R_n\text{-CH=CH-C}(O\cdot)\text{H-}R_{n'} &\rightarrow R_n\text{-CH=CH-CHO} + R_{n'}\cdot \text{ (b)}
\end{aligned} \tag{4}$$

Besides light and heat, oxidative stability is also greatly influenced by the choice of storage tank materials and the presence of minor components or contaminants in the mixture. Hence, deterioration can be slowed by the use of additives in the fuel. Commonly used anti-oxidants include phenol derived compounds such as tert-butyl hydroquinone (TBHQ) [35], butylatedhydroxytoluene (BHT), butylatedhydroxanisole (BHA), and propyl gallate. Effective additive concentrations of 1000 mg kg^{-1} (1000 ppm) do not appear to affect combustion or physical properties [36], but are sufficient to increase the induction period for autooxidation by binding with the peroxy radicals as shown created in Reaction (2) [37]. Phosphorylated antioxidants, including phosphites, phosphonites and phosphines [38], either hinder hydrogen atom extraction or promote the decomposition of hydroperoxides [39], Reaction (5). They are often used in combination with the phenolic antioxidants for additional efficacy. Organosulfites can also be used to stabilize methyl esters, as they react with hydroperoxides to form sulfates [40].These compounds would be less than desirable as fuel additives; however, as the non-radical decomposition is catalyzed by the presence of acid, and ultimate products include SO_x and acids H_2SO_3 and H_2SO_4.

$$R_nOO\cdot + A\text{-H} \leftrightarrow R_nOO\text{-H} + A\cdot \tag{5}$$

3. Chemical conversions

For soybean oil and soybean-derived feedstocks to be used as drop-in replacements for petroleum derived products, deoxygenation processing has to be undertaken. Depending on the desired products, this can involve a number a steps, listed below. Not all of the steps are needed for each product. In general, the desire is to shift increase the carbon-to-oxygen ratio to be closer to that of petroleum, and reduce the carbon to hydrogen ratio, as depicted in Figure 5, plotted with data collected by Choudhary [41]. The various catalysts used to achieve the deoxygenation of triglycerides has been reviewed in a number of publications, for instance by Morgan [42].

i.   Hydrogenation (saturation of double bonds)[43]

ii.   Thermal cracking – heating in an inert atmosphere without addition of H_2

iii.   Acid- or base-catalyzed cracking over metal oxides or zeolites [44]

iv.   Hydrodeoxygenation (HDO removal of O as H_2O) – usually at higher H_2 pressure and lower temperatures than the cracking processes. The catalyst has transition metal + heteroatom (S or N) like $NiMo/Al_2O_3$ or $CoMo/Al_2O_3$. Non-sulfided forms have also been studied to a lesser extent, Ni/Al_2O_3 or Ni/SiO_2.

v.   Decarboxylation (remove O as CO_2) – more H_2 efficient than HDO, unless product CO_2 becomes methanized. Often uses supported platinum catalysts in a batch reaction, for example: Pt/Al_2O_3 or Pd/C, 270-360°C, 17-40 bar H_2.

vi.   Decarbonylation or removal of oxygen as CO – same catalysts as decarboxylation

vii.   Removal of other heteroatoms (S, N, P, metals) – especially from used cooking oil or if sulfur and nitrogen compounds have been added in earlier processes to maintain the catalyst activity.

viii. Various side reactions including: hydrocracking, water-gas shift, methanization, cyclization, and aromatization

ix.   Isomerization [45], often deliberately designed to produce better fuel characteristics, such as cold flow behavior. For example, linear paraffins, n=16-18, freeze at 18-28°C, while iso-paraffins of the same carbon number freeze at -11 to 3°C

x.   Co-processing of soybean oil with diesel fuels in an oil refinery by fluidized catalytic cracking (FCC) – achieves isomerization as well as separations. May get inhibition of deoxygenation because of S groups in the diesel fuel components and vice versa.

4. Oil feedstocksfor hydrocarbon fuels

Deoxygenation drives the overall process for converting soybean oil into hydrocarbons. Processing requirements will be similar for both fuels and chemical feedstocks if the product from deoxygenation can be introduced as a feed in a petrochemical refinery where it will undergo further reactions and separation. If the goal is to only make hydrocarbon fuels; however, separations may not be as important after deoxygenation as they would be for isolating particular building block chemicals. Conversion in a smaller scale independent biorefinery may be feasible, solving the issue of the distributed production of soybeans. The drawback to this scheme is that if the fuels are to be introduced directly into the distribution system, issues related to quality control of the product may become siginificant. In this case, processing will have to account for the variability of bio-based feedstocks, even within a particular crop. Hence, for fuel production at a distributed processing facility, the goals would include achieving sufficient deoxygenation to allow incorporation upstream of the distribution point, enabling pipeline transportation, and providing reliability and quality control.

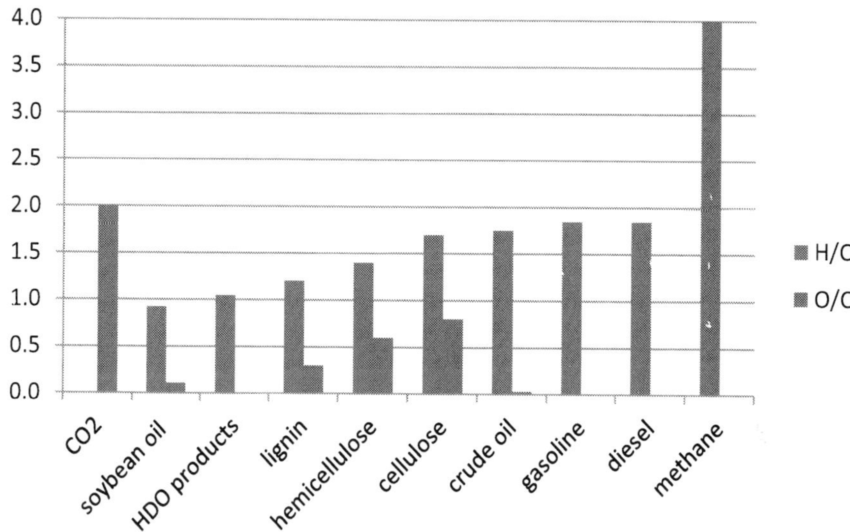

Figure 5. Progression in H/C mole ratio and O/C mole ratio for different sources of organic materials

Thermal cracking has the advantage of not requiring hydrogen for processing. Uncatalyzed thermal cracking has been used to convert soybean oil to hydrocarbons, holding the oil at 430-440°C under vacuum for over one hour [46]. Not including the free fatty acid byproducts, more than half of the products comprised linear alkanes (51%). Other significant hydrocarbon products were identified by gas chromatography – mass spectroscopy as cycloalkanes (11%), alkenes (20%), cycloalkenes (5%), aromatics (8%), and polyaromatic hydrocarbons or PAH (5%). The authors contend that the rings, both saturated and aromatic, came from cyclization of the fatty acid chains rather than from a Diels Alder addition. The latter is usually considered the mechanism for ring formation from olefins, but in the case of the soybean oil conversion, the precursor dienes were not observed. The FFA also became decarboxylated, releasing hydrogen for saturation of double bonds, and producing byproducts CO_2 and CO. The significant fraction of PAH could be problematic for direct combustion of the resulting fuel, as these compounds can survive conditions through the combustion pathway in the engine and be emitted into the atmosphere. However, pretreatment, to reduce the acid number and separation of the FFA, would allow this feedstock to be transported to a refinery for further processing. In thermal cracking, the most important variables governing product distribution include the temperature followed by the residence time [46-48]. Other literature has described investigations of catalysts that have shown promise for thermal cracking, including supported Ni, Pt or Pd on carbon. In particular, Morgan and coworkers have discovered a Ni/C catalyst and demonstrated a 92% conversion of soybean-derived triglyceride at 400°C, with a 70% yield in liquid form [42].

Biogas oil, which is a mixture of normal and isoparaffins having boiling points close to that of diesel, may be made from selective hydrotreating of natural triglycerides. Although some research on thermal cracking has suggested that introduction of hydrogen is not necessary [47], the HDO process allows the conversion to be carried out at lower temperatures, with fewer issues related to byproduct char and gas formation. Hydrodeoxygenation has also received much attention in the literature because this process shows promise to operate with less hydrogen than needed for hydrogenation [41]. HDO, in fact, involves a series of hydogenolysis and hydrogenation steps and is analogous to hydrodesulfurization of petroleum. Catalysts have already been developed for hydroprocessing of heavy oils. HDO investigations have been carried out on a number of seed oils, including soybean oil. For instance, rapeseed oil was deoxygenated at 260-280°C under 3.5 MPa H_2. The rate of reaction ranged over 0.25-4 h^{-1} when tested with a number of different catalysts, in order of performance: Ni-Mo/Al_2O_3> Mo/Al_2O_3>Ni/Al_2O_3 [49]. HDO of sunflower oil was performed in DMSO, with a NiMo/Al_2O_3/F catalyst, in a bench-scale continuous operation at high pressure. The oil hydrocracking entailed combined processes, including olefinic double bond saturation, oxygen removal, and isomerization [50], to give a conversion of 90%. HDO of sunflower oil (310-360°C, 2.0MPa) on a Pd/SAPO-31 catalyst gave excellent conversion to C17 and C18 straight chain and branched alkanes. However, the catalyst became fouled after a few hours [51]. Mixtures of sunflower oil and gas oil have been hydroprocessed over a sulfided catalyst, NiO(3%)–MoO_3(12%)–γ-Al_2O_3 incorporating 0, 15 or 30 wt.% zeolite beta (BEA). The reaction took place at conditions of 330°C, 60 bar, at a weight hourly space velocity (WHSV) of 2 h^{-1}, giving 100% conversion into hydrocarbons. The distribution arising from cracking giving the relative fractions of liquids/gases/and char was not discussed in the paper [52].

Hydrocracking has also been scaled up from the laboratory bench. A larger scale test of hydrocracking of fresh sunflower oil and used cooking oil was carried out by Bezergianni and colleagues [6]. The cracking process was carried out over a number of days until steady state was reached, and then an analysis was performed on the products. A presulfided commercial HDO catalyst was used, with sulfur in the form of dimethyldisulfide and nitrogen as tetra-butyl amine added to maintain the activity. The liquid hourly space velocity (LHSV) was 1.5 h^{-1} and the H_2-to-liquid ratio was 1098 Nm^3/m^3 (at 14 MPa). The difference in performance between the conversion of used and new oil was very small, with high yields of product in the diesel fuel boiling point range (70-80%). Less cracking to small molecules was observed at lower temps, i.e., 350°C, than 390°C, which is desirable for fuel manufacture.

The effect of sulfur on hydrodeoxygenation is of interest because it relates to the performance in an oil refinery with hydrodesulfurization (HDS) as well as HDO processing. Experiments over Pt/H-Y, Pt/H-ZSM-5, and sulfidedNiMo/γ-Al_2O_3 have been carried out in a batch reactor over a temperature range of 300-400°C and initial hydrogen pressures from 5 to 11 MPa. The reaction time was limited to 3 h [53]. Investigation of the performance of CoMo/Al_2O_3 at different sulfur levels (1% to < 10 mg kg^{-1}) and its effect on the HDO of sunflower oil were done under the following conditions: 300-380°C, 20-80 bar, 1-3 h^{-1}, and H_2/oil volume 200-800 Nm^3/m^3 [54]. Up to 75% of the target product C18-paraffins were made at the highest temps and lowest LHSV. At higher H_2 pressures, more hydrocracking occurred, forming light gases such as propane. Adding presulfided catalysts got better yields (5-8%) under less severe conditions. But the addition of sulfur can produce H_2S, which needs to be removed from product and recycled. Sulfur can react to form mercaptans, which are corrosive, adding cost and complexity to the process. H_2S has been found not to prevent catalyst deactivation as was anticipated. The effect of sulfur has also been investigated for the HDO of aliphatic ester model compounds [55].

Hydrogenation and deoxygenation to n-paraffins followed by isomerization is expensive and complicated, but now is performed on an industrial scale [41]. A Finnish company, Neste Oil, has built and operated three NexBTL plants around the world, in Singapore, Rotterdam, and Porvoo in 2009, to convert 190,000 t/a of C12-C16 triglycerides, fatty acid esters, and fatty acids to green diesel. The conversion involves hydrotreatment followed by isomerization to produce green diesel and is described in a Neste patent [56]. The Neste Singapore plant is rated at 800,000 t/a of palm oil, used oil and waste animal fat. A plant in Rotterdam started production in mid 2011 that uses a variety of feedstocks.

Another company, UOP/ENI S.p.A., has a process to produce green diesel that involves a number of catalytic steps to achieve deoxygenation and conversion to branched hydrocarbons. The process, based on hydrodeoxygenation, produces fuel that can be blended directly with petroleum, or added to an input stream in an oil refinery [56]. Emerald Biofuels plans to build a 85×10^6 gallon production facility based on the UOP technology (licensed by Honeywell) at a Dow Chemical site in Plaquemine, LA. Dynamic Fuels is already in production (75×10^6 gal) and Diamond Green Diesel (137×10^6 gal) also has a plant under construction at the mouth of the Mississippi, to take advantage of the proximity to shipping and petroleum refineries. The existing capacity, along with the operating Neste plants, currently produces 600×10^6 gal/a [57].

ExxonMobile is building a hydrotreating plant in Singapore to deliver bio-derived low sulfur diesel of up to 16×10^6 L/d, and has similar plants planned for Baytown and Baton Rouge, LA, in the USA [58].Other planned green biodiesel projects include sites at Norco LA (Darling International, Diamond Green Diesel, LLC, and Valero Energy Corp) to produce 137×10^6 gal/a from waste oil, and animal fats. KiOR plans construction of a plant to produce refinery intermediates in Columbus MS. Joule Unlimited will be constructing a biofuels demonstration plant (75×10^6 gal/a green diesel) in New Mexico [57].

Although hydrodeoxygenation is fairly mature, with industrial-scale production, technical problems remain that would benefit from further research. Minimization of hydrogen used in the conversion of biomass must happen to make the process economically viable [59]. A large fraction of the products from hydrodeoxygenation are linear paraffinic hydrocarbons, which tend to form waxes that can cause cold flow problems. Research into isomerization reactions and selection of catalysts to promote branched alkanes would be beneficial. In the case of soybean oil, however, thermal cracking has produced a preponderance of aromatic compounds, suggesting that mixing of fractions produced through different pathways may give rise to a fungible fuel. A third area of interest is the effect of acylglycerides and HDO products on catalysts that are used in oil refinering, especially if the bio derivatives are to be introduced into the feed stream along with petroleum. In particular, there is a concern that the acylglycerides may affect the performance of catalysts such as CoMo, used in hydrode-sulfurization [41].

The activities undertaken by industry show that the underlying drivers for biofuel production low sulfur requirements in diesel, low net CO_2 emission during production and use, and potentially disruptive oil supply disruptions are pushing major companies to make investments in this area. Although the feedstock streams for biorefineries are not specific to soybean oil, the engineering efforts contribute to the general knowledge of producing green diesel from a variety of sources. Yet, the industry is sensitive to changes in feedstock and oil prices, and smaller initiatives have lost traction during changes in the market, particularly during the last few years. Currently, green diesel remains a niche player in the larger petroleum refining industry.

5. Oil feedstocksfor materials

Soybean plants have been used to fabricate a variety of materials and products. Soybean derived materials have been used in the development of bio-based fibers and yarns [60]. In particular soybean protein fiber has been identified having potential uses [61] in the manufacturing of fabrics. Materials production may not require the degree of chemical conversion and breakdown of the triglyceride that hydrocarbon production requires, but modification is still required to give the desired properties. For instance, soybean protein separated through precipitation after the oil has been removed from the seed requires further processing to crosslink the derived fibers and reduce brittleness and degradation [62, 63]. Soybean straw, available after harvesting of the beans, can be converted to technical fibers through alkali processing. The straw-derived fibers have a higher lignin content than cotton or linen, but after processing the cellulose content is comparable to these other agricultural sources, and could represent a bioresource estimated to be 55 million tons derived from 220 million tons of straw [64]. Soybean fiber left over from oil and protein extraction can be converted to ethanol through a two-step process: (i) pretreatment with aqueous ammonia to remove lignin (by 74% after 12 hours), and (ii) simultaneous saccharification and fermentation, giving 0.25g ethanol per gram of fiber [65]. The straw can also be converted to a bio-oil through fast pyrolysis [66].

6. Oil feedstocksfor chemicals

Soybean oil can be used to manufacture a number of different compounds including surfactants, fuel additives, detergents, polymers such as polyurethanes [67], and adhesives [68]. Polymer production from biomaterials has recently been reviewed by Lligadas and colleagues [69]. Fatty acids can be converted to a polyurethane through a di-isocyanate intermediate [70].Resin alternatives have been prepared from a number of different plant-derived materials, including soybean oil. To achieve the physical properties required for a thermo-plastic polymer, that is having sufficient rigidity and tensile strength to be used in fabrication, soybean oil derived resins must be crosslinked via epoxidation (introduction of epoxy groups into double bonds on the fatty acid chains) or mixed with petroleum-based materials. Good results obtained without incorporating conventional polymers used a combination of epoxidized soybean oil and an anhydrided soyate. The linking process was catalyzed with hexamethylenediamine, and gave a fiber with a tensile strength of >10 MPa [71]. Adhesives made of renewable polymers have been made from mixing of dimeric fatty acids and diols with maleic-anhydrided soybean triglycerides. The gel is formed from cross linking of the esters and extending the fatty acid chains within the structure [72]. Hybrid coatings have been prepared from mixing blown soybean oil and sol-gel precursors (titanium and zirconium peroxides) to improve properties such as tensile strength, adhesion, flexibility, hardness and impact resistance [73].

The conversion of plant-based acylglycerides to nitrogen containing compounds has been reviewed by Biswas and colleagues [74].Properties of various products from palmitic acid have been predicted based on a combinatorial approach, and then linked to an optimization routine to select the product of choice based on predefined criteria (lubricity, critical micelle concentration, or hydrophilic-lipophilic balance) [75]. The authors, Carmada and Sunderesani, wanted to refocus the synthesis paradigm. They developed a method to choose a chemical structure that would give thermophysical properties optimized to a particular application. The structure then determined which synthetic method would be needed to produce the desired end product.

7. Methyl ester feedstocks for chemicals

Methyl esters derived from soybean oil can also be used as starting materials for the production of hydrocarbons for fuels or chemical feedstocks. Various catalytic pathways from oxygenated precursor to hydrocarbons include: pyrolysis [76], deoxygenation and hydrogenation [77, 78], and hydrotreatment [79]. The focus of many of these studies has been production of fuels that are miscible or fungible with petroleum products, e.g., the work published by the group of Daniel Resasco at the University of Oklahoma [80], for fuel production rather than chemicals. In addition, much of the published literature focuses on simpler chemical representatives of the methyl esters from soybean oil; but these results are directly applicable to the production of chemical feedstocks, such as the synthesis of ethylbenzene that can be used for a variety of products: polymers, solvents, and reagents [77]. Although differences in the product distribution would be expected from TAG and single methyl ester conversions, comparison studies carried out by Kubatova and colleagues [46-48] on individual acylglycerides, as well as soybean mixtures, showed similar conversion chemistry.

Because it appears as if the products from these deoxygenation processes will require further processing to make fuels or chemical feedstocks, the FCC of triglycerides has been studied by groups such as Melero and colleagues [81]. In particular, the effect of soy-based biomass on the FCC process is of interest. Melero investigated both the cracking of 100% soybean oil, and a mixture of 30% soybean oil with petroleum, the latter being a typical feedstock for an oil refinery. The feed injection was held at 70°C, but the FCC itself was carried out at 560°C, representing typical conditions in a refinery. The boiling point range for the soybean biodiesel was from 545.7 to 636.0°C in comparison with the diesel range from 200-330°C [82]. Because of the unsaturated chains in the soybean oil, the aromatic fraction in the product was enhanced relative to that of pure petroleum, but the PAH was reduced. Saturated fatty acid fragments gave rise to light alkanes, light oil, and diesel fractions. There could be some concern about enhanced corrosion in the cracker because of the presence of the FFA, but research suggests that the lifetime of these compounds is very short at these temperatures, and so they may not present a problem. Many of the FCC products are similar from acylglycerides and petroleum, although the composition distributions are different. In the case the biomass, of FCC products come from reactions in the cracker and in the case of petroleum, most hydrocarbons are present in the original feedstock. Issues such as gum formation or the effect of impurities, such as entrained alkali metals that could be present in biological materials, were not studied by Melero [81].

8. Conclusions

Although many chemical pathways have been demonstrated in the laboratory, the scale-up to handle large quantities of bio-derived material presents a number of challenges in comparison with petroleum refining. These range from additional transportation costs because of distributed feedstock production, to catalyst cost and regeneration. Seasonal variations in the cultivation and harvesting of soybeans and production of oil can result to chemical changes in the feedstock material and minor components. However, it appears as if the chemical modification processes are robust to minor changes in FFA distribution. Impurities and their impact on catalyst performance and lifetime may be significant and difficult to test outside of an industrial setting. Impurity and effects of minor components are highly dependent on unpredictable phenomena such as feedstock composition or process variability. Thus, these effects may not appear, much less be quantified, in a bench-scale operation using laboratory grade chemicals. Hence, operation of pilot and demonstration scale facilities will be very informative. The feasibility of the production of hydrocarbons from soybean triglycerides or methyl esters derived from these triglycerides is often dependent on the availability of low cost hydrogen. Other technical hurdles include the optimization of interfacial reactions and separations before soybean oil can make a significant contribution to the hydrocarbon economy. The question of whether feedstocks from soybean oil should be introduced into a stream in an oil refinery, or converted in a small scale refinery to fungible products depends on the final application and cost issues. However, once converted to hydrocarbons, separations to commodity chemicals or fuel should be analogous to handling conventional petroleum products.

Nomenclature and abbreviations

ASTM - American Society of Testing and Materials

BHA - butylatedhydroxyanisole

BHT - butylatedhydroxytoluene

FAME - Fatty Acid Methyl Esters

FCC - Fluidized Catalytic Cracking

FFA - Free Fatty Acid (monoglyceride)

HDO - Hydrodeoxygenation

LHSV - Liquid hourly space velocity

PAH - Polyaromatic Hydrocarbon

TAG- Triacylglyercol or triglyceride

TBHQ- Tert-butyl hydroquinone

WHSV - Weight hourly space velocity

Acknowledgements

Oak Ridge National Laboratory is managed by UT-Battelle, LLC, for the U.S. Department of Energy under contract DE-AC05-00OR22725.

Author details

Joanna McFarlane

Energy and Transportation Science Division, Oak Ridge National Laboratory, Oak Ridge, USA

References

[1] A. McIlroy and G. McRae, *Basic Research Needs for Clean and Efficient Combustion of 21st Century Transportation Fuels*. 2006.

[2] USEPA. *RFS2 – Federal Policy Drivers for Increased Biofuels Usage*. US Environmental Protection Agency, January 2009, accessed August 1, 2012;

[3] EIA. *Monthly Biodiesel Production Report*. May 2012, accessed August 1 2012;

[4] API. *US Oil Shale: Our Energy Resource, Our Energy Security, Our Choice*. 2011, accessed Aug 1, 2012; Available from: http://www.api.org/~/media/Files/Oil-and-Natural-Gas/Oil_Shale/Oil_Shale_Factsheet_1.ashx.

[5] D.J. Murphy, *Oil Crops as Potential Sources of Biofuels*, in *Technological Innovations in Major World Oil Crops*, S.K. Gupta, Editor. 2012, Springer: NY. p. 269-284.

[6] S. Bezergianni, S. Voutetakis, and A. Kalogianni, "Catalytic Hydrocracking of Fresh and Used Cooking Oil".*Industrial & Engineering Chemistry Research*, 2009. 48(18): p. 8402–8406.

[7] J. McFarlane, *Processing of Soybean Oil into Fuels*, in *Recent Trends for Enhancing the Diversity and Quality of Soybean Products*, D. Krezhova, Editor. 2011, InTech.

[8] J. McFarlane, C. Tsouris, J.F.J. Birdwell, Jr., D.L. Schuh, H.L. Jennings, A.M. Pahmer-Boitrago, and S.M. Terpstra, "Production of Biodiesel at Kinetic Limit Achieved in a Centrifugal Reactor/Separator".*Industrial and Engineering Chemistry Research* 2010. 49: p. 3160–3169.

[9] X.-W. Zhang and A. McWhirter, *Conserving Fossil Fuel Resources Using Reaction Simulations to Maximize Biodiesel Production*. 2012, Oak Ridge High School: Oak Ridge.

[10] M.S. Manic, V. Najdanovic-Visak, M.N. da Ponte, and Z.P. Visak, "Extraction of Free Fatty Acids from Soybean Oil Using Ionic Liquids or Poly(ethyleneglycol)S".*Aiche Journal*, 2011. 57(5): p. 1344-1355.

[11] C. Komintarachat and S. Chuepeng, "Solid Acid Catalyst for Biodiesel Production from Waste Used Cooking Oils".*Industrial & Engineering Chemistry Research*, 2009. 48(20): p. 9350-9353.

[12] MatLab. 2010, Mathworks: Natick, MA.

[13] A. Baig and F.T.T. Ng, "A Single-Step Solid Acid-Catalyzed Process for the Production of Biodiesel from High Free Fatty Acid Feedstocks".*Energy & Fuels*, 2010. 24: p. 4712-4720.

[14] V.K. Chakravarthy, J. McFarlane, C.S. Daw, Y. Ra, J.K. Griffin, and R. Reitz, "Physical Properties of Soy Bio-Diesel and N-Heptane: Implications for Use of Bio-Diesel in Diesel Engines".*SAE 2007 Transactions, Journal of Fuels and Lubricants*, 2008. 116: p. 885-895.

[15] Y. Ra, R.D. Reitz, J. McFarlane, and C.S. Daw, "Effects of Fuel Physical Properties on Diesel Engine Combustion Using Diesel and Bio-Diesel Fuels".*SAE International Journal of Fuels and Lubricants*, 2009. 1(1): p. 703-718.

[16] G.W. Mushrush, J.H. Wynne, H.D. Willauer, C.T. Lloyd, J.M. Hughes, and E.J. Beal, "Recycled Soybean Cooking Oils as Blending Stocks for Diesel Fuels".*Industrial & Engineering Chemistry Research*, 2004. 43(16): p. 4944-4946.

[17] G.W. Mushrush, J.H. Wynne, J.M. Hughes, E.J. Beal, and C.T. Lloyd, "Soybean-Derived Fuel Liquids from Different Sources as Blending Stocks for Middle Distillate Ground Transportation Fuels".*Industrial & Engineering Chemistry Research*, 2003. 42(11): p. 2387-2389.

[18] D.L. Purcell, B.T. McClure, J. McDonald, and H.N. Basu, "Transient Testing of Soy Methyl Ester Fuels in an Indirect Injection, Compression Ignition Engine".*Journal of the American Oil Chemists Society*, 1996. 73(3): p. 381-388.

[19] R. Ashen, K.C. Cushman, and J. McFarlane, *Chemical Kinetic Simulation of the Combustion of Bio-Based Fuels*. in press, Oak Ridge National Laboratory: Oak Ridge TN.

[20] B. Bunting, M. Bunce, T. Barone, and J. Storey, *Fungible and Compatible Biofuels: Literature Search, Summary, and Recommendations*. 2010, Oak Ridge National Laboratory: Oak Ridge, TN.

[21] J. Risher, *Toxilogical Profile for Fuel Oils*. 1995, Sciences International, Inc: Research Triangle Institute

[22] T.A. Foglia, K.C. Jones, and J.G. Phillips, "Determination of Biodiesel and Triacylglycerols in Diesel Fuel by Liquid Chromatography".*Chromatographia*, 2005. 62(3-4): p. 115-119.

[23] ASTM, *Specification for Diesel Fuel Oil, Biodiesel Blend (B6 to B20)*. 2008, American Society of Testing and Materials.

[24] D. Mackay, *Illustrated Handbook of Physical Chemical Properties and Environmental Fate for Organic Chemicals*. Vol. 5. 1997, Boca Raton, FL: Lewis Publishers.

[25] J. Sangster, "Octanol-Water Partition Coefficients". *Journal of Physical and Chemical Reference Data*, 1989. 18(3): p. 1111-1230.

[26] H.B. Krop, M.J.M. van Velzen, J.R. Parsons, and H.A.J. Govers, "N-Octanol-Water Partition Coefficients, Aqueous Solubilities and Henry's Law Constants of Fatty Acid Esters". *Chemosphere*, 1997. 34(1): p. 107-119.

[27] USEPA. *Chemicals in the Environment: OPPT Chemical Fact Sheets*. October 5, 2010; Available from: http://www.epa.gov/chemfact/.

[28] G.R. Stansell, V.M. Gray, and S.D. Sym, "Microalgal Fatty Acid Composition: Implications for Biodiesel Quality". *Journal of Applied Phycology*, 2012. 24(4): p. 791-801.

[29] C.P. Tan and Y.B.C. Man, "Differential Scanning Calorimetric Analysis for Monitoring the Oxidation of Heated Oils".*Food Chemistry*, 1999. 67(2): p. 177-184.

[30] G. Knothe, "Some Aspects of Biodiesel Oxidative Stability".*Fuel Processing Technology*, 2007. 88(7): p. 669-677.

[31] R.C. Simas, D. Barrrera-Arellano, M.N. Eberlin, R.R. Catharino, V. Souza, and R.M. Alberici, "Triacylglycerols Oxidation in Oils and Fats Monitored by Easy Ambient

Sonic-Spray Ionization Mass Spectrometry". *Journal of the American Oil Chemists Society*, 2012. 89: p. 1193-1200.

[32] T.D. Crowe and P.J. White, "Adaptation of the AOCS Official Method for Measuring Hydroperoxides from Small-Scale Oil Samples". *Journal of the American Oil Chemists Society*, 2001. 78(12): p. 1267-1269.

[33] E.N. Frankel, "Lipid Oxidation: Mechanisms, Products and Biological Significance". *Journal of the American Oil Chemists Society*, 1984. 61(12): p. 1908-1917.

[34] N. Canha, P. Felizardo, J.C. Menezes, and M.J.N. Correira, "Multivariate near Infrared Spectroscopy Models for Predicting the Oxidative Stability of Biodiesel: Effect of Antioxidants Addition".*Fuel*, 2012. 97: p. 352-357.

[35] T.S. Yang, Y.H. Chu, and T.T. Liu, "Effects of Storage Conditions on Oxidative Stability of Soybean Oil". *Journal of the Science of Food and Agriculture*, 2005. 85(9): p. 1587-1595.

[36] S. Schober and M. Mittelbach, "The Impact of Antioxidants on Biodiesel Oxidation Stability". *European Journal of Lipid Science and Technology*, 2004. 106(6): p. 382-389.

[37] G. Knothe, "A Technical Evaluation of Biodiesel from Vegetable Oils Vs. Algae. Will Algae-Derived Biodiesel Perform?". *Green Chemistry*, 2011. 13(11): p. 3048-3065.

[38] I. Kriston, A. Orban-Mester, G. Nagy, P. Staniek, E. Foldes, and B. Pukanszky, "Melt Stabilisation of Phillips Type Polyethylene, Part I: The Role of Phenolic and Phosphorous Antioxidants".*Polymer Degradation and Stability*, 2009. 94: p. 719-729.

[39] D. Lomonaco, F.J.N. Maia, C.S. Clemente, J.P.F. Mota, A.E.J. Costa, and S.E. Mazzetto, "Thermal Studies of New Biodiesel Antioxidants Synthesized from a Natural Occurring Phenolic Lipid".*Fuel*, 2012. 97: p. 552-559.

[40] A. Günther, T. König, W.D. Habicher, and K. Schwetlick, "Antioxidant Action of Organic Sulphites-I. Esters of Sulphurous Acid as Secondary Antioxidants ". *Polymer Degradation and Stability*, 1997. 55: p. 209-216.

[41] T.V. Choudhary and C.B. Phillips, "Renewable Fuels Via Catalytic Hydrodeoxygenation". *Applied Catalysis A: General*, 2011. 397(1-2): p. 1-12.

[42] T. Morgan, D. Grubb, E. Santillan-Jimenez, and M. Crocker, "Conversion of Triglycerides to Hydrocarbons over Supported Metal Catalysts". *Topics in Catalysis*, 2010. 53(11-12): p. 820-829.

[43] D. Jovanovic, R. Radovic, L. Mares, M. Stankovic, and B. Markovic, "Nickel Hydrogenation Catalyst for Tallow Hydrogenation and for the Selective Hydrogenation of Sunflower Seed Oil and Soybean Oil". *Catalysis Today*, 1998. 43(1-2): p. 21-28.

[44] C. Perego and A. Bosetti, "Biomass to Fuels: The Role of Zeolite and Mesoporous Materials". *Microporous and Mesoporous Materials*, 2011. 144(1-3): p. 28-39.

[45] V. Calemma, S. Peratello, and C. Perego, "Hydroisomerization and Hydrocracking of Long Chain N-Alkanes on Pt/Amorphous Sio2–Al2o3 Catalyst". *Applied Catalysis A: General*, 2000. 190(1-2): p. 207-218.

[46] A. Kubatova, J. St'avova, W.S. Seames, Y. Luo, S.M. Sadrameli, M.J. Linnen, G.V. Baglayeva, I.P. Smoliakova, and E.I. Kozliak, "Triacylglyceride Thermal Cracking: Pathways to Cyclic Hydrocarbons". *Energy & Fuels*, 2012. 26(1): p. 672-685.

[47] Y. Luo, I. Ahmed, A. Kubatova, J. St'avova, T. Aulich, S.M. Sadrameli, and W.S. Seames, "The Thermal Cracking of Soybean/Canola Oils and Their Methyl Esters". *Fuel Processing Technology*, 2010. 91(6): p. 613-617.

[48] A. Kubatova, Y. Luo, J. St'avova, S.M. Sadrameli, T. Aulich, E. Kozliak, and W. Seames, "New Path in the Thermal Cracking of Triacylglycerols (Canola and Soybean Oil)".*Fuel*, 2011. 90(8): p. 2598-2608.

[49] D. Kubicka and L. Kaluza, "Deoxygenation of Vegetable Oils over Sulfided Ni, Mo and Nimo Catalysts".*Applied Catalysis A: General*, 2010. 372: p. 199-208.

[50] S. Kovács, T. Kasza, A. Therneszb, I.W. Horváth, and J. Hancsók, "Fuel Production by Hydrotreating of Triglycerides on Nimo/Al2o3/F Catalyst". *Chemical Engineering Journal*, 2011. 176-177: p. 237-243.

[51] O.V. Kikhtyanin, A.E. Rubanov, A.B. Ayupov, and G.V. Echevsky, "Hydroconversion of Sunflower Oil on Pd/SAPO-31 Catalyst". *Fuel*, 2010. 89(10): p. 3085-3092.

[52] T.M. Sankaranarayanana, M. Banub, A. Pandurangana, and S. Sivasanker, "Hydroprocessing of Sunflower Oil–Gas Oil Blends over Sulfided Ni–Mo–Al–Zeolite Beta Composites". *Bioresource Technology*, 2011. 102(22): p. 10717-10723.

[53] R. Sotelo-Boyas, Y. Liu, and T. Minowa, "Renewable Diesel Production from the Hydrotreating of Rapeseed Oil with Pt/Zeolite and Nimo/Al$_2$O$_3$ Catalysts". *Industrial & Engineering Chemistry Research*, 2010. 50: p. 2791-2799.

[54] M. Krár, S. Kovács, D. Kalló, and J. Hancsók, "Fuel Purpose Hydrotreating of Sunflower Oil on Como/Al$_2$O$_3$ Catalyst". *Bioresource Technology*, 2010. 10(23): p. 9287–9293.

[55] O.İ. Şenol, T.-R. Viljava, and A.O.I. Krause, "Effect of Sulphiding Agents on the Hydrodeoxygenation of Aliphatic Esters on Sulphided Catalysts". *Applied Catalysis A: General*, 2007. 326(2): p. 236-244.

[56] J. Myllyaja, P. Aalto, and E. Harlin, *Process for the Manufacture of Diesel Range Hydrocarbons*, WPTO, Editor. 2007, Neste Oil OYJ: Finland.

[57] J. Lane. *Renewable Diesel Roundup*. 2012 May 9, accessed August 1, 2012; Available from: http://www.altenergystocks.com/archives/2012/05/renewable_diesel_roundup_1.html.

[58] eco-business.com. *Singapore: Exxonmobil to Raise 'Green' Diesel Output*. 2010 November 10, accessed August 1, 2012; Available from: Nov. 24 2010, The Business Times,

http://www.eco-business.com/news/singapore-exxonmobil-raise-green-diesel-output/.

[59] J. McFarlane, J.A. Gluckstein, M. Hu, M. Kidder, C. Narula, and M. Sturgeon, "Investigation of Catalytic Pathways and Separations for Lignin Breakdown into Monomers and Fuels". *Separation Science and Technology*, 2012. in press.

[60] B. Ozgen, "New Biodegradable Fibres, Yarn Properties and Their Applications in Textiles: A Review". *Industrial Textila*, 2012. 63(1): p. 3-7.

[61] S. Huda, N. Reddy, D. Karst, W.J. Xu, W. Yang, and Y.Q. Yang, "Nontraditional Biofibers for a New Textile Industry". *J. Biobased Materials and Bioenergy*, 2007. 1(2): p. 177-190.

[62] Y. Zhang, S. Ghasemzadeh, A.M. Kotliar, S. Kumar, S. Presnell, and L.D. Williams, "Fibers from Soybean Protein and Poly(Vinyl Alcohol)". *J. Appl. Polymer Sci.*, 1999. 71(1).

[63] Y. Li, N. Reddy, and Y. Yang, "A New Crosslinked Protein Fiber from Gliadin and the Effect of Crosslinking Parameters on Its Mechanical Properties and Wataer Stability". *Polymer International*, 2008. 57: p. 1174-1181.

[64] N. Reddy and Y. Yang, "Natural Cellulose Fibers from Soybean Straw". *Bioresource Technology*, 2009. 100(14): p. 3593-3598.

[65] B. Karki, D. Maurere, S. Box, T.H. Kim, and S. Jung, "Ethanol Production from Soybean Fiber, a Co-Product of Aqueous Oil Extraction, Using a Soaking in Aqueous Ammonia Pretreatment". *Journal of the American Oil Chemists Society*, 2012. 89(7): p. 1345-1353.

[66] A.A. Boateng, C.A. Mullen, N.M. Goldberg, K.B. Hicks, T.E. Devine, I.M. Lima, and J.E. McMurtrey, "Sustainable Production of Bioenergy and Biochar from the Straw of High-Biomass Soybean Lines Via Fast Pyrolysis". *Environmental Progress & Sustainable Energy*, 2010. 29(2): p. 175-183.

[67] D.A. Babb, *Polyurethanes from Renewable Resources*, in *Synthetic Biodegradable Polymers*, B. Rieger, et al., Editors. 2012. p. 315-360.

[68] B.K. Ahn, S. Kraft, D. Wang, and X.S. Sun, "Thermally Stable, Transparent, Pressure-Sensitive Adhesives from Epoxidized and Dihydroxyl Soybean Oil". *Biomacromolecules*, 2011. 12(5): p. 1839-1843.

[69] G. Lligadas, J.C. Ronda, M. Galia, and V. Cadiz, "Plant Oils as Platform Chemicals for Polyurethane Synthesis: Current State-of-the-Art". *Biomacromolecules*, 2010. 11(11): p. 2825-2835.

[70] L. Hojabri, X.H. Kong, and S.S. Narine, "Fatty Acid-Derived Dilsocyanate and Bio-based Polyurethane Produced from Vegetable Oil: Synthesis, Polymerization, and Characterization". *Biomacromolecules*, 2009. 10(4): p. 884-891.

[71] P. Tran, D. Graiver, and R. Narayan, "Biocomposites Synthesized from Chemically Modified Soy Oil and Biofibers". *J. Appl. Polymer Sci.*, 2006. 102(1): p. 69-75.

[72] R. Vendamme, K. Olaerts, M. Gomes, M. Degens, T. Shigematsu, and W. Eevers, "Interplay between Viscoelastic and Chemical Tunings in Fatty-Acid-Based Polyester Adhesives: Engineering Biomass toward Functionalized Step-Growth Polymers and Soft Networks". *Biomacromolecules*, 2012. 13(6): p. 1933-1944.

[73] G.H. Teng, J.R. Wegner, G.J. Hurtt, and M.D. Soucek, "Novel Inorganic/Organic Hybrid Materials Based on Blown Soybean Oil with Sol-Gel Precursors". *Progress in Organic Coatings*, 2001. 42(1-2): p. 29-37.

[74] A. Biswas, B.K. Sharma, J.L. Willett, S.Z. Erhan, and H.N. Cheng, "Soybean Oil as a Renewable Feedstock for Nitrogen-Containing Derivatives". *Energy & Environmental Science*, 2008. 1(6): p. 639-644.

[75] K.V. Camarda and P. Sunderesan, "An Optimization Approach to the Design of Value-Added Soybean Oil Products". *Industrial & Engineering Chemistry Research*, 2005. 44(4361-4367).

[76] K.D. Maher and D.C. Bressler, "Pyrolysis of Triglyceride Materials for the Production of Renewable Fuels and Chemicals". *Bioresource Technology*, 2007. 98: p. 2351-2368.

[77] T. Danuthai, S. Jongpatiwut, T. Rirksomboon, S. Osuwan, and D.E. Resasco, "Conversion of Methylesters to Hydrocarbons Over an H-ZSM5 Zeolite Catalyst H-Zsm5 Zeolite Catalyst". *Applied Catalysis A: General*, 2009. 361: p. 99-105.

[78] B. Donnis, R.G. Egeberg, P. Blom, and K.G. Knudsen, "Hydroprocessing of Bio-Oils and Oxygenates to Hydrocarbons. Understanding the Reaction Routes". *Topics in Catalysis*, 2009. 52(3): p. 229-240.

[79] G.W. Huber, S. Iborra, and A. Corma, "Synthesis of Transportation Fuels from Biomass: Chemistry, Catalysts, and Engineering". *Chemical Reviews*, 2006. 106(9): p. 4044-4098.

[80] T. Sooknoi, T. Danuthai, L.L. Lobban, R.G. Mallinson, and D.E. Resasco, "Deoxygenation of Methylesters over CsNaX". *Journal of Catalysis*, 2008. 258: p. 199-209.

[81] J.A. Melero, M.M. Clavero, G. Calleja, A. Garcia, R. Miravalles, and T. Galindo, "Production of Biofuels Via the Catalytic Cracking of Mixtures of Crude Vegetable Oils and Nonedible Animal Fats with Vacuum Gas Oil". *Energy & Fuels*, 2010. 24: p. 707-717.

[82] R.W. Hurn and H.M. Smith, "Hydrocarbons in the Diesel Boiling Range". *Industrial & Engineering Chemistry*, 1951. 43(12): p. 2788-2793.

Chapter 7

Effect of Dietary Plant Lipids on Conjugated Linoleic Acid (CLA) Concentrations in Beef and Lamb Meats

Pilar Teresa Garcia and Jorge J. Casal

Additional information is available at the end of the chapter

http://dx.doi.org/10.5772/52608

1. Introduction

Beef and lamb, are a food category with positive and negative nutritional attributes. Ruminant meats are major sources for many bioactive compounds including iron, zinc and B vitamins. However they are associated with nutrients and nutritional profiles that are considered negative including high levels of saturated fatty acids (SFA) and cholesterol. It is well know that the low PUFA/SFA and high n-6/n-3 ratio of meats contribute to the imbalance in the in the fatty acid intake of today consumers [1]. Consumers are becoming more aware of the relationships between diet and health and this has increased consumer interest in the nutritional value of foods. Nutritionist advisers recommended a higher intake of polyunsaturated fatty acids (PUFA), especially n-3 PUFA at the expense of n-6 PUFA.

The nutritional beef and lamb profile could be further improved by addition of potentially health promoting nutrients. There are many references of improved fatty acid composition in grass fed beef. Besides the beneficial effects of n-3 fatty acids on human health one fatty acid that has drawn significant attention for its potential health benefits in the last two decades is conjugated linoleic acid (CLA). Conjugated linoleic acids (CLA) are implicated as anti-carcinogenic, anti-atherosclerosis, and anti-inflammatory agents in a variety of experimental model systems. It has been shown that in ruminants grazing have potential beneficial effects on PUFA/SFA and n-6/n-3 ratios, increasing the PUFA and CLA content and decreasing the SFA concentration of beef [2].

The total CLA content of beef varies from 0.17 to 1.35% of fat [3]. This wide range is related to the type feed, breed differences, and management strategies used to raise cattle [3, 4]. Grazing beef steers on pasture or increasing the amounts of forage (grass or legumes hay) in the diet has been shown to increase the CLA content in the fat of cattle. Also, supplementing

© 2013 Garcia and Casal; licensee InTech. This is a paper distributed under the terms of the Creative Commons Attribution License (http://creativecommons.org/licenses/by/3.0), which permits unrestricted use, distribution, and reproduction in any medium, provided the original work is properly cited.

high-grain diets of beef cattle with oils (e.g., soybean oil, linseed oil, sunflower oil) may increase the CLA content of beef [3, 5].

There has been an increased interest in the substitution of animal fat sources with vegetable oils in animal nutrition. Vegetable oils have been attributed with reducing the level of saturation in monogastric animal tissues due to their unsaturated fatty acid concentration when compared with animal fat. In ruminants, dietary lipids were undergo two important transformations in the rumen. The initial transformation is the hydrolysis of the ester bond by microbial lipases. This initial step is a pre requisite for the second transformation, the biohydrogenation of unsaturated fatty acids [6, 7].

Several factors influence the CLA content of beef as breed, sex, seasonal variation, type of muscle, production practices but diet plays the most important role. Dietary CLA from beef can be increased by manipulation of animal diets. CLA concentration in beef can be influenced by dietary containing oils or oilseeds high in PUFA, usually linoleic or linolenic fatty acids.These dietary practices can increase CLA concentrations up to 3 fold [5, 8]. Moreover, trans-11 18:1 (vaccenic acid,VA) is the precursor of cis-9,trans-11 18:2 (rumenic acid, RA) is the major CLA isomer in animal and humans and, therefore, it might be considered as a fatty acid with beneficial properties.

Soybean oil is one of the few plant sources providing ample amounts of both essential fatty acids 18:2 n-6 and 18:3 n-3. The fatty acid content of soy foods is often unrecognized by health professionals, perhaps because there is so much focus on soy proteins. Soybeans are used in cattle, poultry and pigs diets and could be a more important source of 18:3 n-3 for animal nutrition and also increase 18:3 n-3 and its fatty acids metabolites in meats. Genomics, specifically marker assisted plant breeding combined with recombinant DNA technology, provided powerful means for modifying the composition of oilseeds to improve their nutritional value and provide the functional properties required for various food oils [9].

Thus, the manipulation of the fatty acid composition in ruminant meat to reduce SFA content and the n-6/n-3 ratio whilst, simultaneously increasing the PUFA and CLA contents, is the major importance in meat research. The supplementation of ruminant diets with PUFA rich lipids is the most effective approach to decrease saturated FA and promote the enrichment of CLA and n-3 PUFA.

2. CLA structure, biosynthesis and potential beneficial effects on human health

The CLA acronym refers to a group of positional and geometric isomers of linoleic acid, in which the double bands are conjugated. At least twenty four different CLA isomers have been reported as occurring naturally in food, especially from ruminant origin [10]. Isomerisation and incomplete hydrogenation of PUFA in the rumen produce several of octadecenoic, octadecadienoic and octadecatrienoic isomeric fatty acids [11] and, at least some of them, have powerful biological properties. The formation of conjugated dienes in the rumen during biohydrogenation of lipids in feed was observed previously, however, the anticarcinogenic effect of beef extracts was first observed and later identified [12, 13, 14].

The dominant CLA in ruminant meats is the cis-9, trans-11 isomer (RA) which has being identified as possessing a range of health promoting biological properties including antitumoral and anticarcinogenic activities [15]. The rumenic acid is mostly produced in tissues by delta 9 desaturation of trans-11 18:1, (VA) and by ruminal biohydrogenation of dietary PUFA. The higher deposition of CLA in the neutral lipid fraction, 88% of total CLA relatively to phospholipid fraction, has been reported [16].The majority of the main natural isomer cis-9,trans-11 CLA does not originate directly from the rumen. Instead, only small amounts of CLA escape the rumen and trans-18:1 isomers are the main biohydrogenation intermediates available. El absorbed trans-11 18:1 is desaturated in the tissues by Δ9-desaturase to form RA [17]. Stearoyl-CoA (SCD) is a rate-limiting enzyme responsible for the conversion of SFA into monounsaturated fatty acids (MUFA). This enzyme, located in the endoplasmic reticulum, inserts a double band between carbons 9 and 10 into SFA and affects the fatty acid composition of membrane phospholipids, triglycerides and cholesterol esters [18]. SCD is also a key enzyme in the endogenous production of the cis-9,trans-11 isomer of conjugated linoleic acid (CLA). Trans octadecenoates (trans 18:1) are the major intermediates formed during rumen biohydrogenation of C18 PUFA. High trans-10 18:1 have been observed in tissues of concentrated-fed ruminants, whereas vaccenic acid is consistently associated with forage feeding [11, 19]. Evidence is accumulating that different trans 18:1 isomers have differential effects on plasma LDL cholesterol. Trans-9 and trans-10 18:1 are more powerful in increasing plasma LDL cholesterol than trans-11 18:1 [20]. Comparison of antiproliferative activities of different CLA isomers present in beef on a set of human tumour cells demonstrates that all CLA isomers possess antiproliferative properties. It appears that important to determine the variations of the distribution of CLA isomers in beef since these proportions could influence the biological properties of bioformed CLA [21].

3. Factors influencing CLA concentrations on beef lipids

Amounts of CLA in beef vary mainly with feeding conditions, nature and quality of forages, proportions between forage and concentrate, oil-seed supplementations, but also with intrinsic factors such as breed and sex and age of animals [22].

3.1. Breed, sex and age (Table 1)

Breed or genotype and production system are determinant factors of the fatty acid composition of the ruminant meats. Breed affects the fat content of meat and fat content itself is a factor determining fatty acid composition. Genetic variability relates to differences between breeds or lines, variation due to the crossing of breeds and variation between animals within breeds reported that it can be difficult to assess the real contribution of genetics to variation in the CLA content.

	CLA	Reference
Breed		
LD Limousin	2.24 g/100g	[29]
LD Angus	1.96 g/100g	[29]
LD Angus	0.51 b% FAME	[25]
LD Charolais x AA	0.57 a% FAME	[25]
LD Holando x AA	0.58a% FAME	[25]
LD Nguni grass	0.34% FA	[28]
LD Bonsmara grass	0.31% FA	[28]
LD Angus grass	0.33% FA	[28]
LD Holstein grass	0.84 % FA	[24]
LD Simmental grass	0.87% FA	[24]
LD Holstein concentrate	0.75% FA	[24]
LD Simmental concentrate	0.72% FA	[24]
SM Pasture and Silage Steers Longhorn	6.75a mg/100g	[23]
SM Pasture and Silage Steers Charolais	3.29b mg/100g	[23]
SM Pasture and Silage Steers Hereford	2.93b mg/100g	[23]
SM Pasture and Silage Steers B. Gallowey	5.09a mg/100g	[23]
SM Pasture and Silage Steers Beef Shorton	4.01ab mg/100g	[23]
Sub Pasture and Silage Steers Longhorn	1210a mg/100g	[23]
Sub Pasture and Silage Steers Charolais	651b mg/100g	[23]
Sub Pasture and Silage Steers Hereford	584b mg/100g	[23]
Sub Pasture and Silage Steers B. Gallowey	796 b mg/100g	[23]
Sub Pasture and Silage Steers Beef Shorton	808b mg/100g	[23]
Mertolenga PDO beef	0.39ab g/100g FA	[27]
Mertolenga PDO veal	0.46a g/100g FA	[27]
Vitela Tradicional do Montado PGI veal	0.35b g/100g FA	[27]
LT & LL Veal Limousin	1.09% FAME	[26]
LT & LL Veal Tudanka x Charolais	1.00% FAME	[26]
Sex and age		
LL bulls 14 month	0.37 % FA	[31]
LL bulls 18 month	0.39% FA	[31]
LL heifers 14 month	0.44% FA	[31]
LL heifers 18 month	0.41 % FA	[31]
L lumborum steers	0.20 % FA	[32]
L.lumborum bulls	0.21 % FA	[32]

Table 1. Conjugated linoleic acid (CLA) concentrations on beef according to breed, sex and age .a b Indicates a significant differences (at least $p < 0.05$) between breed, sex or age reported within each respective study. Abbreviations LD: Longissimus dorsi ; SM: Semimembranosus ; LL Longissimus lumborum; LT Longissimus thoracis; Sub: Subcutaneous fat.

Significant between-breed differences in CLA content were observed in both muscle and subcutaneous adipose tissue of five breeds of cattle with the highest values in Longhorn and with the lowest in Hereford [23]. German Holstein bulls accumulated a higher amount of CLA compared with German Simmental bulls [24]. CLA percentages were affected by breed with the low values for Angus beef compared with Charolais x Angus and Holstein Argentine steers [25]. The content of trans -10 C18:1 isomer tended to be higher in Limousin compared to Tudanca meat when expressed as mg/100g of meat, and the difference was only significant when expressed in terms of relative percent. The higher level of trans-10 C18:1 was consistent with the greater consumption of concentrate by Limousin calves [26]. Within a similar production system the age/weight, gender and crossbreeding practices have minor effects on muscle FA composition but Mertolenga-PDO veal has higher total CLA contents that PDO beef and PGI veal [27]. On the contrary the cis-9, trans-11 CLA levels among steers of Nguni, Bonsmara and Angus breeds raised on natural pasture were similar [28]. Similar results were found comparing the CLA content of Limousin and Aberdeen Angus beef [29].

Sex and age differences in muscle FA contents are often be explained by the degree of fatness and associated changes in the triacylglycerol/phospholipid ratio [30]. Sex-dependent differences in the FA composition of muscle an adipose tissue from cattle slaughtered at different ages were demonstrated [31]. Concentration CLA in meat beef not affected by castration [32].

3.2. Type muscle and anatomical location (Table 2)

Little work has been conducted to assess the effects of slaughter season and muscle type on meat CLA profile. The type of muscles strongly influenced proportions of total CLA and of all CLA isomers classes in intramuscular fatty acids (Table 2). CLA is mainly associated to the triacylglycerol fraction which is linked to the fat content of tissues [21]. VA and CLA percentages were lower in lean muscle than subcutaneous fat or marbling [33]. The CLA content of steaks differs depending on the location of the fat, CLA level was almost doubled in outer subcutaneous fat compared to lean muscle [34]. There was significant differences in the concentration of CLA among depot sites through-out a bovine carcass. The brisket contained a higher concentration of cis-9, trans-11 CLA but no significant differences in the concentrations of trans-10, cis-12 CLA among the locations [35].

3.3. Season and pasture type (Table 2)

No differences between dietary grass silage and red clover silage were detected on CLA content of LD muscle of dairy cull cows [36]. Total CLA content was lower (p< 0.05) in intensively produced beef than in Carnalentejana-PDO meat, which did not show significant differences (p<0.05) when the slaughter season was compared. Furthermore *Longissimus thoracis* (LT) muscle had a higher (p<0.001) total CLA content relative to *Longissimus dorsi* (LD) muscle. In addition no significant differences (p<0.05) regarding specific CLA content were observed when slaughter season, production system and muscle type were analyzed [37]. Significant interactions between the slaughter season and muscle type were obtained for several fatty acid and CLA isomers and for total lipid and CLA. Mirandesa –PDO veal showed seasonal differences in the levels of CLA isomers but the CLA content was affected by much more influence by the muscle type [38]. The variation of CLA milk fat content during pasture season might be related to the alfa-linolenic/linoleic acid ratio in the pasture. The ratio in the average pasture sample decreased from 4.36 in May to 1.97 in August, and subsequently it increased to 3.14 in September, thus close to that at the beginning of pasture season. Thus the seasonal variation of the ratio in pasture were directly proportional to the corresponding content of CLA in ewe milk fat [39].

	CLA	Reference
Muscle and adipose tissue location		
Steak muscle	0.30b % FA	[33]
Steak marbling	0.50a % FA	[33]
Outer subcutaneous fat	0.50a % FA	[33]
Inner subcutaneous fat	0.50a% FA	[33]
Seam	0.40ab % FA	[33]
Adipose tissue brisket	0.70a g/100g FA	[35]
Adipose tissue chuck	0.62ab g/100g FA	[35]
Adipose tissue flank	0.56b g/100g FA	[35]
Adipose tissue loin	0.53b g/100g FA	[35]
Adipose tissue plate	0.57b g/100g FA	[35]
Rib	0.52b g//100g FA	[35]
Round	0.63ab g/100g FA	[35]
Sirloin	0.57b g/100g FA	[35]
LT concentrate	4.45 mg/g fat	[37]
ST concentrate	3.88 mg/g fat	[37]
Season		
LT Autumn	5.07 mg/g fat	[37]
LT Spring	4.92 mg/g fat	[37]]
ST Autumn	3.82 mg/g fat	[37]
ST Spring	5.06 mg/g fat	[37]
L L Spring	0.30 a g/100g FA	[34]
L L Autumn	0.31a g/100g FA	[34]
ST Spring	0.23 b g/100g FA	[34]
ST Autumn	0.19b g/100g FA	[34]]
Pasture type		
LD Tall fescue	0.28%	[91]
LD Alfalfa	0.37%	[91]
LD Red clover	0.30%	[91]
LD cull cows grass silage	0.22 % TFA	[36]
LD cull cows red clover silage	0.17 % TFA	[36]

Table 2. Conjugated linoleic acid (CLA) concentrations on beef according to muscle and adipose tissue location, season and grass composition a b Indicates a significant differences (at least p<0.05) between anatomical location, season or pasture type reported within each respective study. Abbreviations LD: Longissimus dorsi ; SM: Semimembranosus ; LL Longissimus lumborum; LT Longissimus thoracis

3.4. Grass vs. concentrate (Tables 3 & 5)

A direct linear relation between grass percentage in cattle diet and meat CLA content has been described by [2] although the mechanism remains controversial. They suggested that grass in the diet enhances the growth of ruminal bacterium *Butyrivibrio fibrisolvens* which convert 18:2 n-6 into cis-9, trans-11 CLA isomer through the action of a linoleic acid isomerase. Others [40] proposed that the increased content of CLA in animals fed forage- based diets is associated with an increase in trans-11 18:1, which is the substrate of stearoyl-CoA desaturase in tissues. It is generally accepted that the concentrations of CLA can be increased in beef by increasing the forage to concentrate ratio, and by feeding fresh grass instead of grass silage [4, 22] (Table 3). Beef contains both of the bioactive CLA isomers, namely, cis-9, trans-11 and trans-10, cis-12. Many reports demonstrated that cis-9, trans 11 CLA is a major fatty acid in tissue and little or no trans-10, cis 12 CLA was detected [5,41]. High trans-10 18:1 have been observed in tissues of concentrated-fed ruminants, whereas vaccenic acid is consistently associated with forage feeding [11, 19]. Significantly higher contents of trans-18:1 were found in animals fed on concentrate diets relative to the pasture diet. This is mainly due to the trans-6, trans-8, trans-9 and trans-10 isomers, since the trans-11 and trans-12 18:1 remains unaffected by the dietary treatments. The feeding systems, pasture only, pasture feeding followed by 2 or 4 months of finishing on concentrate, and concentrate only, had a major impact on the concentration of CLA isomers from bull LD muscles. Beef fat from pasture-fed animals had a higher nutritional quality relative to that from concentrate-fed bulls and the feeding regimen had a major impact on the CLA isomeric distribution of beef affecting 10 of 14 CLA isomers. The CLA isomeric profile showed a clear predominance of the cis-9, trans-11 isomer for all diets [42]. The grass silage diets increased the proportions of trans-11 18:1 and cis-9, trans-11 18:2. Feeding a high forage diet may therefore have increased the rate of appearance of trans-11 18:1 in the rumen, providing more substrate for the endogenous production and deposition of CLA in bovine tissues [43]. This hypothesis is consistent with an increase of trans-11 18:1 concentration with no effect on cis-9, trans-11 in duodenal content of Hereford steers fed increasing levels of grass hay [44]. The relative flow of PUFA through the major biohydrogenation pathways, trans-10 or trans-11, 18:1, can be judged by the 11t-/10t- 18:1 ratio with a higher ratio denoting an improvement in its healthfulness to its human consumers [45]. Backfat composition was compared in steers fed either a control (barley grain based) diet or diets containing increasing levels of corn or wheat derived dried distillers'grains with solubles (DDGS). Back fat from control and wheat derived DDGS fed steers had lower levels of trans-18:1 and a higher 11trans/10 trans 18:1 ratio compared to back fat from corn derived DDGS fed steers [45]. The explanation might be found in ruminal biohydrogenation pathway of LA and ALA. Most of the cis-9, trans-11 CLA isomer present in tissues derive from endogenous desaturation of trans-11, 18:1, which originates during biohydrogenation of 18:2 n-6 and 18:3 n-3. The CLA concentrations in three different muscles of pasture- or feedlot-finished cattle were greater from pasture-finished than from cattle feedlot-finished [46]. The absolute cis-9, trans-11 CLA was about twice as high in Asturiana de la montaña (AV) and Asturiana del Valle (AV) animals than in other AV genotypes, probably due to the much higher fat content of the AM and AV animals [47]. This effect was also found in other studied were cis-9, trans-11 content variation was influenced by the total lipid content, and hence with variation in the neutral lipid fraction [48]. A linear correlation between VA and cis-9, trans-11 CLA was observed in several studies [8] and in other studied no significant correlation was found [49]. Breed or genotype effects could act by enhancing or inhibiting the Δ9-desaturase activity. The major isomers in beef fed a high barley diet is trans-10, 18:1 rather than trans-11, 18:1. In feedlot finished beef fed a diet containing 73% barley was found 2.13 % of trans-10 18:1 and only 0.77% of trans-11 18:1 in subcutaneous fat [19, 50]. Feeding ruminants diets with high levels of barley (low fiber, high starch) reduces rumen pH, alters the bacterial flora and causes a shift in the biohydrogenation pathway towards producing trans-10 18:1 instead of trans-11 18:1 [51]. Subcutaneous fat is quite sensitive to changes in diet and rumen function. This is due to adipose tissue having a high proportion of neutral lipids which accumulate greater levels of PUFA biohydrogenation products relative to polar lipids [24]. In addition, subcutaneous fat is easily accessible, inexpensive and levels of trans-18:1 have been reported to be linearly related to those found in muscle [52]. Vaccenic acid made up the greatest concentration of total trans fats in grass-fed beef, whereas CLA accounted for approximately 15% of the total trans fats [53].

	CLA	Reference
LD Grazing	10.8a mg/g fat	[2]
LD Concentrate-fed	3.7b mg/g fat	[2]
LD Grazing	5.3a mg/g fat	[90]
LD Concentrate-fed	2.5 b mg/g fat	[90]
LD Pasture	0.72 % FAME	[25]
LD Pasture +0.7% corn	0.61 % FAME	[25]
LD Pasture+1.0 %corn	0.58% FAME	[25]
LD Feedlot	0.31 % FAME	[25]
LD Grass silage (GS)	3.62% FA	[43]
LD GS +Low concentrate	2.50% FA	[43]
LD GS+ High concentrate	2.72% FA	[43]
LT Semi-intensive 12 month	0.49a %	[92]
LT Semi-intensive 14 month	0.49a %	[92]
LT Intensive 12 month	0.25b %	[92]
LT Intensive 14 month	0.29b %	[92]
Ground control	0.50b g/100g	[53]
Ground grass	0.94a g/100g	[53]
Steaks control	0.38b g/100g	[53]
Steaks grass	0.66a g/100g/	[53]
Control	0.82 % FA	[45]
Back fat 20% DDGS corn	0.88 % FA	[45]
Back fat 20% DDGS wheat	0.88 % FA	[45]
Back fat 40%DDGS corn	0.97 % FA	[45]
Back fat 40% DDGS wheat	0.81 % FA	[45]
LT concentrate	4.45 mg/g fat	[37]
ST concentrate	3.88 mg/g fat	[37]

Table 3. Conjugated linoleic acid (CLA) concentrations on beef under dietary grass or concentrate a b Indicates a significant differences (at least p<0.05) between dietary grass or concentrate reported within each respective study. Abbreviations LD: Longissimus dorsi ; Longissimus thoracis ; ST: Semitendinosus

3.5. Oil supplementation (Tables 4 & 5)

The most common method of enhancing the CLA and VA content of ruminant meat and dairy products is to provide the animal with additional dietary unsaturated fatty acids, usually from plants oils such as soybean oil (SBO), for use as substrates for ruminal biohydrogenation [4]. Steers fed a corn-based diet supplemented with SBO may enhance TVA without impacting CLA, while reducing the MUFA content of lean beef [54]. Both oilseed and free oils affect CLA content in a similar manner. Free plant oils with high PUFA concentrations are normally not included in ruminant diets as high levels of dietary fat disturb the rumen environment and inhibit microbial activity. The main sources of supplementary fatty acids in ruminant rations are plant oils and oilseeds, fish oils, marine algae and fat supplements. Since dietary inclusion of fatty acids must be restricted to avoid impairment of rumen function, the capacity to manipulate the fatty acid composition by use of ruminally available fatty acids is limited [55]. Many researchers have found higher CLA content in muscle lipids by supplementing with different oils. However, some studies reported no significant differences in CLA content due to oil supplementations. The differences in responses to plant oils were probably due to variations in stage of growth of cattle, levels of oil supplementation, levels of oil in total ration and amount of linoleic acid in oils. Researchers have successfully increased CLA content by supplementation of different oils [4,48,56]. Others [3] supplementing with 4% SBO to diets did not affect the CLA. Similar to [41] who reported that feeding 5% SBO no affected CLA but increased trans10-cis-12 CLA. The addition of different vegetable oils to the bulls diet (soybean or linseed, either protected or not protected from rumen digestion) increased the CLA content, with an average CLA value of 0.72 %. The increase of CLA was also due to the addition of oils presenting large quantities of its precursor LA in diets with unprotected soybean and linseed oils [57]. Diets containing silage and concentrate or sugarcane and sunflower seeds fed Canchim- breed animals, produce an improvement in CLA levels (0.73g/100g vs. 0.34g/100) [58]. Rapeseed oil and whole rapeseed do not seem to have positive effects. Of the three studied none showed increased CLA concentrations in the LDi after supplementation with 6% rapeseed oil [41]. Soybean oil (SBO) has been used as a source de LA throughout the finishing period to promote greater CLA accretion in lean tissues with equivocal results [56, 41]. and where CLA accretion was increased with SBO addition, growth performance was reduced [56]. Fed steers with 5% of soybean oil in a finishing experiment for 102 days had no effects in meat cis-9, trans, 11 CLA [41]. In a study with steers, supplementation of 4% soybean oil to a finishing diet based on concentrate and forage (80:20) resulted in a depression of the CLA deposition in muscle tissues (2.5 vs. 3.1 mg/g FAME) compared to the same diet without soybean oil On the other hand, comparing 4% with 8% added soybean oil in a 60:40 concentrate : forage diet showed a numerical increase of the CLA content with the higher soybean supplementation (2.8/3.1 mg/g FAME) [59]. The inclusion of sunflower oil in the diets (80% barley, 20% barley silage) of finishing cattle at 0%, 3%, or 6% increased the CLA content of the beef by 75% when cattle were fed 6% sunflower oil [4]. Although supplementation with oil or oil seeds increased CLA content in muscle, the inclusion of linoleic acid –rich oil or oilseeds such as safflower or sunflower, in the diet of ruminants appears to be the most effective [60]. Supplementation of cattle with a blend of oils rich in n-3 PUFA and linoleic acid results in a synergistic accumulation of ruminal and tissue concentrations of TVA [61]. VA is the substrate for Δ9 –desaturase- catalyzed de novo tissue tissue synthesis of cis-9 trans-11 isomer of CLA. However, despite increases in its substrate, muscle tissue concentrations of cis-9, trans-11 CLA have not increased by using this strategy [62]. Inclusion of extruded linseed in the diet of Limousin and Charolais cattle, increase CLA [63]. The importance of the contribution of TVA to total CLA intake is further reinforced by a French study [64] in which a huge 233% increase of VA was shown, along with 117% increase of RA, which was caused by adding extruded linseeds into the animal fodder. Several authors reported that diets containing proportionally high levels of linolenic acid, such as fresh grass, grass silage, and concentrates containing linseed, resulted in increased deposition of the cis-9, trans-11 CLA isomer in muscle [65]. The biohydrogenation by rumen microorganism does not include the cis-9, trans-11 CLA isomer as an intermediate. The trans-11 18:1 is the common intermediate during the biohydrogenation of dietary linoleic acid and linolenic acid to stearic acid [6]. Since only a relatively small percentage the cis-9, trans-11 CLA isomer, formed in the rumen, is available for deposition on the muscles, the major source of this isomer in muscle results from the endogenous synthesis involving Δ 9 desaturase and vaccenic acid [17].. Hereford steers cannulated in the proximal duodenum were used to evaluated the effects of forage and sunflower oil level on ruminal biohydrogenation and conjugated linoleic acid. Flow of trans-10 18:1 decreased linearly as dietary forage level increased whereas trans -11 18:1 flow to the duodenum increased linearly with increased dietary forage. Dietary forage or sunflower oil levels did not alter the outflow of cis-9, trans-11 CLA [44]. Linseed supplementation was an efficient way to increase CLA proportion in beef (+22% to 36%) but was highly modulated by the nature of the basal diet, and by intrinsic factors as breed, age/sex, type of muscle, since these ones could modulate CLA proportions in beef from 24% to 47% [21]. Soybean oil, which is rich in linoleic acid, has been found in several studied [66,67,] to be more efficient than linseed oil, which is rich in linolenic acid, in increasing the CLA content of milk. In beef cattle the addition of 3% and 6% sunflower oil to a barley based finishing diet results in increased CLA content in LD muscle: 2.0 vs 2.6 vs. 3.5 mg/g lipid for control, 3%,, and 6% sunflower oil, respectively. A more substantial increase in the CLA concentration was found when sunflower oil was added to both the growing and finishing diet of beef cattle.[68,69]. 4.3, 6,3 and 9.1 mg CLA / g FAME in LD muscle lipids of heifers, were found, after supplementing the feed with 0, 55, and 110 g sunflower oil per kg of the diet for 142 days before slaughter [48]. Supplementation of a high forage fattening diet with either soybean oil or extruded full fat soybeans at a level of 33g added oil per kg of diet DM resulted in a 280-410 % increase in the concentration of CLA in the intramuscular and subcutaneous lipid depots of fattening Friesian bull calves The content of VA in both lipid depots were also increased about three-fold by this oil supplementation [70].

	CLA	Reference
Concentrate IMF fat	3.4 b mg/g fat	[70]
Soybean oil IMF fat	13.0 a mg/g fat	[70]
Extruded soybean IMF fat	15.4 a mg/g fat	[70]

	CLA	Reference
Concentrate Sub fat	5.2 c mg/g fat	[70]
Soybean oil Sub fat	20.3 b mg/g fat	[70]
Extruded soybean Sub fat	26.6 a mg/g fat	[70]
LD concentrate / silage	0.41d % FA	[93]
LD Grass	0.70c % FA	[93]
LD grass +sunflower oil	1.34a % FA	[93]
LD grass +linseed oil	0.93b% FA	[93]
LD Wagyu Control	0.27 b % FA	[68]
LD Wagyu 6% sunflower oil	1.29a % FA	[68]
LD Limousin Control	0.28b % FA	[68]
LD Limousin 6% sunflower oil	1.19a % FA	[68]
LM grass	0.73c % FA	[48]
LM grass+ sunflower oil	1.78a % FA	[48]
LM grass+linseed oil	1.26b % FA	[48]
LM Corn oil 0%	0.68b % FA	[94]
LM Corn oil 0.75%	0.85a % FA	[94]
LM Corn oil 1.5%	0.81ab % FA	[94]
LT Control	0.33% FA	[95]
LT Control + Vit E	0.36 % FA	[95]
LT Control	0.34 % FA	[95]
LT Control+ flaxseed	0.34 % FA	[95]
LD Control	0.35 c mg/100g FA	[57]
LD Soybean oil	0.94a mg/100g FA	[57]
LD Linseed oil	0.80a mg/100g FA	[57]
LD Protected linseed oil	0.55b mg/100g FA	[57]
LM grass NL	0.78c g/100g FA	[48]
LM grass+sunflower oil NL	1.90a g/100g FA	[48]
LM grass+linseed oil NL	1,35b g/100g FA	[48]
LM grass PL	0.32c g/100g FA	[48]
LM grass+sunflower oil PL	0.71a g/100g FA	[48]
LM grass+linseed oil PL	0.51b g/100g FA	[48]

Table 4. Conjugated linoleic acid (CLA) concentrations on beef under dietary oils supplementation. a b Indicates a significant differences (at least p<0.05) between dietary oil supplementations reported within each respective study. Abbreviations LD: Longissimus dorsi ; SM: Semimembranosus ; LT Longissimus thoracis; Sub: Subcutaneous fat; NL: neutral lipids; PL: Phospholipids.

Diet P vs. C	Trans- 11 18:1	Trans- 10 18:1	Reference
C	0.92	1.21b	[42]
P+4month C	1.10	0.81b	[42]
P+2month C	1.15	0.98b	[42]
P	1.35	0.20a	[42]
Ground control	1.14	2.69	[53]
Ground grass	4.14	0.75	[53]
Steaks control	0.51	3.60	[53]
Steaks grass	2.95	0.60	[53]
Grass silage (GS)	2.03a	Na	[43]
GS +Low C	1.37b	Na	[43]
GS+ High C	1.15b	Na	[43]
Control	0.65	2.02	[45]
20% DDGS corn	0.78	2.37	[45]
20%DDGS wheat	0.74	1.60	[45]
40% DDGS corn	0.92	3.16	[45]
40%DDGS wheat	0.69	1.33	[45]
LT et LL Tudanca x Charolais	2.68	0.36b	[26]
LT et LL Limousin	2.24	1.01 a	[26]

Table 5. Trans-11 and trans 10 C18:1 isomer proportions on beef under different conditions. a b Indicates a significant differences (at least $p<0.05$) between trans-10 C18:1 and trans-11 C18:1 reported within each respective study. "na" indicates that the value was not reported in the original study. Abbreviations LD: Longissimus dorsi ; LL Longissimus lumborum LT: Longissimus thoracis.

4. Factors influencing CLA concentrations on lamb lipids

In lamb production, more than other species, each country or region has its own specific weight/age and type of carcass criteria, depending on the culture and the customs of the people. Many factors including breed, gender, age/body weight, fatness, depot site, environmental condition, diet and rearing management influence lamb fat deposition and composition. Further studied are needed to understand how animal circadian rhythms, diurnal rumination patterns and daily changes in herbage chemical composition could affect lamb fatty composition [71].

4.1. Production system (Tables 6 & 8)

No differences were detected in the muscle CLA/ trans-11 18:1 index of herbage or concentrate –fed lambs but the supplementation of tanino produced strong effects on the accumulation of fatty acids which are involved in the biohydrogenation pathway [72]. During two years (Y1 and Y2) lambs were under four diets. Only silage both pre and post weaning (SS), only silage until weaning, silage plus concentrate thereafter (SC), silage plus concentrate both pre and post weaning (CC) and silage plus concentrate before weaning, only silage after (CS). Treatment differences for trans-11 18:1 were presented only in Y1, with muscle from the lamb fed silage before weaning having the highest levels. The same groups has the highest levels of cis-9, trans 11 CLA in Y1. Similar in Y2 the group SS has the highest CLA level, while the CC group has the lowest [73]. The feeding strategy around parturition influence the CLA and VA content of lamb meat. Pre-partum grazing, regardless of post-partum feeding, can improve the fatty acid composition, increasing the CLA content in lamb meat [74]. The meat of lambs slaughtered at Christmas has a higher CLA content than those reared in winter (slaughtered at Easter) as a result of the traditional feeding system which provided that lambs born and reared in autumn receive milk from ewes permanently pastured while those reared in winter are suckled by ewes permanently stall-fed [75]. The grazing on *T.subterraneum* as monoculture, associated with *L. multiflorum* in the proportion T/L=66/33 incremented cis-9,trans-,11 CLA of *L. dorsi* muscle of lambs [76]. The meat fatty acid profile was affected by the grazing management: compared to a morning-grazing or to a whole day-grazing management. Allowing lambs to gaze in the afternoon resulted in a meat fatty acid profile richer in CLA. In particular, in the 4hPM meat there is a greater proportion of those fatty acids arising from ruminal biohydrogenation, among them the CLA [71].

	CLA mg/g fat	Reference
LD Grass pellets	1.29a % FA	[40]
LD Concentrate diet	1.02a % FA	[40]
LD Concentrate diet *Ad libitum*	0.74b % FA	[40]
Muscle Concentrate+concentrate CC	0.46b g/100g lipids	[73]
Muscle Silage+concentrate SC	0.61a g/100g lipids	[73]
Muscle Concentrate+silage CS	0.45b g/100g lipids	[73]
Muscle Silage+silage SS	0.65a g/100g lipids	[73]
LT Pre-partum hay	1.42b % FA	[74]
LT Pre-partum grazing	1.66a % FA	[74]
LT Post-partum hay	1.35b % FA	[74]
LT Post –partum grazing	1.73a % FA	[74]
LD Grassed 9 am to 5 pm	1.85b g/100g FAME	[71]
LD Grassed 9 am to 1 pm	1.45b g/100g FAME	[71]
LD Grassed 1 pm to 5 pm	2.39a g/100g FAME	[71]
LL Sucking lamb Autumn	1.10% IM Fat	[75]
LL Sucking lamb Winter	0.56 % IM Fat	[75]

	CLA mg/g fat	Reference
LD Grazing subterraneous clover	0.46a % FA	[76]
LD Grazing Italian rye grass	0.26 b % FA	[76]
Pasture LD	0.90b % FAME	[96]
Pasture Leg muscles	1.27a % FAME	[96]
Pasture LD total lipids	0.90 % total FAME	[96]
Pasture LD Triacylglycerols	0.62 %total FAME	[96]
Pasture LD Phospholipids	0.11 % total FAME	[96]

Table 6. Conjugated linoleic acid (CLA) concentrations on lamb meat according to concentrate, pasture, muscle type and season. a b Indicates a significant differences (at least p<0.05) between values reported within each respective study. Abbreviations LD: Longissimus dorsi ; ST: Semitendinosus ; LL Longissimus lumborum; ; LT: Longissimus toracsis; SM: Semimembranosus.

4.2. Oil supplementation (Tables 7& 8).

Several strategies have been tested in recent years to improve CLA isomers in meat of inten-sively–reared lambs, keep indoors and fed high-concentrate diets rich linoleic acid and poor in linolenic. Incorporating linseed rich, in linolenic acid, the proportion of trans-11, 18:1 and cis-9, trans-11 18:2 were higher in the muscle and in the adipose tissues of linseed –fed lambs than in control lambs [77]. This increased is in contrast to results of [78] but in agree-ment with [79]. Discrepancias between these studies may due to differences in the level of intake the linoleic and linolenic acids or the different level of Δ9- desaturase inhibition as it has been shown that Δ9 desaturase is inhibited by PUFA with increasing inhibition as the degree of fatty acid unsaturation increases. Fed lambs from weaning to slaughter with diets that contained 5% supplemental from high oleic acid safflower or normal safflower in-creased the meat cis-9,trans,11 CLA compared with the control group [80] In lambs inclu-sion of 8% of soybean oil to a lucerne hay-based diet resulted in an intramuscular (M. *Longissimus thoracis*) CLA content of 23.7 compared with 5.5 mg/g FAME in the control group [81]. Feeding soybean and linseed oils to lambs pre and post weaning did not increase CLA content of muscle, whereas post weaning oil supplementation minimally increased CLA concentration in subcutaneous fat [82]. Conflicting results have been reported on alter-ing FA content of meat supplementing ruminant diets with lipid sources high in linoleic and linolenic acids. Some research suggests supplementing CLA, linoleic or linolenic acids in high concentrate fed to lambs can increase CLA content in muscle [83], whereas supplemen-tation of linoleic in finishing diets fed to cattle had no effects on CLA in adipose or muscle tissue [3,41]. Feeding lipid sources rich in linoleic and linolenic increases the cis-9,trans-11 18:2 content of ruminant meats [21,81, 83,84]. However, feeding linseed oil, rich in linolenic acid, seems to be less effective in the increases of cis-9, trans-11 18:2 in muscle than sunflow-er oil, rich in linoleic acid [45,48]. Seems to be that a blend of sunflower and linseed oils may be a good approach to obtain an enrichment in CLA in lamb meat. Maximun CLA concen-trations (42.9 mg/100 g fresh lamb tissue) was observed with 100% of sunflower, decreasing linearly at 78% by sunflower oil with linseed oil replacement [86]. A consistent significant increase in CLA content in lamb tissues was observed with dietary supplementation with

6% of safflower oil. The CLA concentration in several lamb tissues was increased by more than 200% [39]. These results indicated that supplementation of lamb feedlot diets with a source of LA was a successful method of increasing CLA content of tissues. Merino Branco ram lambs initially fed with concentrate showed a lower proportions of cis-9,trans-11 18:2 CLA (0.98% vs. 1.38% of total fatty acids) than lambs initially fed with Lucerne. Initial diet did not compromise the response to the CLA promoting diet (dehydrated lucerne plus 10% soybean oil) and the proportion of cis-9,trans-11 C18:2 CLA in intramuscular fat increased with the duration of time on the CLA-promoting diet (1.02% vs. 1.34% of total fatty acids) [87]. Supplementation of oilseed with different levels of oleic (rapeseed), linoleic (sunflower and safflower seeds), and linolenic acid (linseed) on trans-11 18:1 and CLA isomers on ewe different tissues showed that the percentage of trans-11 C18:1 averaged around 4.56 % of total fatty acids for all supplements and tissues [88]. Increasing dietary forage and soybean oil did not change the sheep mixed ruminal microbes concentration of vaccenic acid but increased rumenic acid [89].

	CLA mg/g fat	Reference
LD Sunflower oil	2.13a mg/100 g muscle	[86]
LD Sunflower oil+ 33% linseed oil	2.06 a mg/100 g muscle	[86]
LD Sunflower oil + 66% linseed oil	1.84b mg/100g muscle	[86]
LD Linseed oil	1.56 c mg/100g muscle	[86]
Leg control	1.78a mg/g fat	[97]
Leg CLA	1,50a mg/g fat	[97]
Leg Safflower oil	4.41b mg/g fat	[97]
Adipose tissue Control	2.77 b mg/g fat	[97]
Adipose tissue CLA	2.60 b mg/g fat	[97]
Adipose tissue Safflower oil	7.33 a mg/g fat	[97]
LD control	3955 b ppm in muscle	[98]
LD Control + 5% sunflower oil	8491a ppm in muscle	[98]
Fat control	4947b ppm in fat	[98]
Fat control +5% sunflower oil	11313a ppm in fat	[98]
LT Control	0.75c % of FA	[87]
LT Control+Lucerne+10% soybean oil	1.21b % of FA	[87]
LT Lucerne	1.28 b% of FA	[87]
LT Lucerne+Lucerne+10% soyben oil	1.47 a% of FA	[87]
Muscle Control no fat	0.05 b mg/g muscle	[77]
Muscle control+wheat+linseed	0.11 a mg/g muscle	[77]
Muscle control+corn+linseed	0.12 a mg/g muscle	[77]
LL Control	0.60b g/100g FAME	[88]
LL Linseed	0.72b g/100g FAME	[88]
LL Rapeseed	0.70b g/100g FAME	[88]
LL Safflower seed	0.96a g7100g FAME	[88]
LL Sunflower seed	0.98a g/100g FAME	[88]

Table 7. Conjugated linoleic acid (CLA) concentrations on lamb meat according to dietary oil. a b Indicates a significant differences (at least p<0.05) between valus reported within each respective study. Abbreviations LD: Longissimus dorsi ; LL Longissimus lumborum; LT Longissimus thoracis

	Trans- 11 18:1	Trans- 10 18:1	Reference
LD Grazing subterraneous clover	4.22	Na	[76]
LD Grazing Italian rye grass	3.65	Na	[76]
LD Grass 9 am to 5 pm	1.55a	Na	[71]
LD Grass 9 am to 1 pm	1.06b	Na	[71]
LD Grass 1 pm to 5 pm	1.60a	Na	[71]
LD grass pellets	2.25a	0.38b	[40]
LD concentrate	1.39b	1.54a	[40]
LD concentrate ad lib	0.85b	1.73a	[40]
LD concentrate	79.4	Na	[72]
LD Herbage	31.4	Na	[72]
Silage-silage	1.54a	Na	[73]
Silage-concentrate	1.45a	Na	[73]
Concentrate-concentrate	1.08b	Na	[73]
Concentrate-silage	1.14b	Na	[73]
LM Pre Control	0.25	5.13b	[82]
LM Pre Control +oil	0.29	6.02a	[82]
LM Post Control	0.25	4.30b	[82]
LM Post Control+ oil	0.29	6.81a	[82]
Sub Pre Control	0.29	9.85b	[82]
Sub Pre Control +oil	0.31	8.25b	[82]
Sub Post Control	0.28	7.09b	[82]
Sub Post Control+ oil	0.33	11.01a	[82]

Table 8. Conjugated linoleic acid (CLA) isomer proportionson lamb under different condicions. a b Indicates a significant differences (at least p<0.05) between values reported within each respective study. "na" indicates that the value was not reported in the original study. Abbreviations LD: Longissimus dorsi ; LM: Semimembranosus; Sub: Subcutaneous

5. Conclusions

Several factors influence the CLA content of ruminant meats as breed, sex, seasonal variation, type of muscle, production practices but diet plays the most important role. CLA concentration in beef and lamb can be influenced by dietary containing oils or oilseeds high in PUFA, usually linoleic or linolenic fatty acids. The supplementation of ruminant diets with PUFA rich lipids is the most effective approach to decrease saturated FA and promote the enrichment of CLA and n-3 PUFA. The differences in responses to plant oils were probably due to variations in stage of growth of animals, levels of oil supplementation, levels of oil in total ration and amount of linoleic acid in oils. Thus, the manipulation of the fatty acid composition in ruminant meat to reduce SFA content and the n-6/n-3 ratio whilst, simultaneously increasing the PUFA and CLA contents, is the major importance in meat research.

Author details

Pilar Teresa Garcia[1*] and Jorge J. Casal[2]

*Address all correspondence to: pgarcia@cnia.inta.gov.ar

1 Area Bioquimica y Nutricion. Instituto Tecnologia de Alimentos. Centro de Investigacion en Agroindustria. Instituto Nacional Tecnologia Agropecuaria. INTA Castelar, CC77(B1708WAB) Moron, Pcia Buenos Aires, Argentina

2 Universidad de Moron. Facultad de Ciencias Agroalimentarias. Cabido 137. Moron , Pcia Buenos Aires, Argentina

References

[1] Russo, G. L. (2009). Dietary n-6 and n-3 polyunsaturated fatty acids: From biochemical to chemical implications in cardiovascular prevention. Biochemical Pharmacolog , 77, 937-946.

[2] French, P., Stanton, C., Lawless, F., O`, Riordan. E.g, Monahan, F. J., Caffrey, P. J., & Moloney, A. P. (2000). Fatty acid composition, including conjugated linoleic acid, of intramuscular fat from steers offered grazed grass, grass silage or concentrate-based diets. *Journal Animal Science*, 78, 2849-2855.

[3] Dhiman, T. R., Nam, S. H., & Ure, A. L. (2005). Factors affecting conjugated linoleic acid content in milk and meat. *Crit. Rev Food Sci. Nutr.*, 45(6), 463-482.

[4] Mir, P. S., Mc Allister, T. A., Scott, S., Aalhus, J., Baron, V., Mc Cartney, D., et al. (2004). Conjugated linoleic acid-enriched beef production. *American Journal of Clinical Nutrition*, 79, 1207S-1211S.

[5] Madron, M. S., Peterson, D. G., Dwyer, D. A., Corl, B. A., Baumgard, L. H., Beermann, D. H., & Bauman,d, E. (2002). Effect of extruded full-fat soybeans on conjugated linoleic acid content of intramuscular, intermuscular, and subcutaneous fat in beef steers. *Journal Animal Science*, 80, 1135-1143.

[6] Harfoot, C. G., & Hazelwood, G. P. (1997). Lipid metabolism in the rumen. *In The Rumen Microbial Ecosystem, 2nd ed. P.N. Hobson, ed. Elservier Science Publishing Co., Inc., New York, NY.*

[7] Jenkins, T. C., Wallace, R. J., Moate, P. J., & Mosley, E. E. (2008). Board-invited review: Recent advances in biohydrogenation of unsaturated fatty acids within the rumenmicrobial ecosystem. *Journal Animal Science*, 86, 397-412.

[8] Enser, M., Scollan, N. D., Choi, N. J., Kurt, E., Hallett, K., & Wood, J. D. (1999). Effect of dietary lipid on the content of conjugated linolenic acid (CLA) in beef muscle. *Animal Science, 69*, 149-156.

[9] Jimenez, J. J., Bernal, J. L., Nozal, M. J., Toribio, L., & Bernal, J. (2009). Profile and relative concentrations of fatty acids in corn and soybean seed from transgenic and isogenic crops. *Journal of Chromatography, 1216*, 7288-7295.

[10] Sehat, N., Rickert, R., Mossoba, M. M., Kramer, J. K., Yurawez, M. P., Roach, J. A., et al. (1999). Improved separation of conjugated fatty acids methyl esters by silver ion-high performance liquid chromatography. *Lipids, 34*, 407-413.

[11] Bessa, R. J. B., Santos-Silva, J., Ribeiro, J. M. R., & Portugal, A. V. . (2000). Reticulo-rumen biohydrogenation and the enrichment of ruminant edible products with linoleic acid conjugated isomers. *Livestock Production Science, 63*, 201-211.

[12] Pariza, M. W., Ashoor, S. H., Chu, F. S., & Lund, D. B. (1979). Effects of temperature and time on mutagen formation in pan-fried hamburger. *Cancer Letters, 17*, 63-69.

[13] Pariza, M. W., Park, Y., & Cook, M. E. (2001). The biologically active isomers of conjugated linoleic acid. *Progress in Lipid Research, 40*, 283-298.

[14] Ha, Y. L., Grimm, N. K., & Pariza, M. W. (1987). Anticarcinogens from fried ground beef: heat- altered derivatives of linoleic acid. Carcinogenesis , 8, 1881-1887.

[15] Belury, M. A. (2002). Dietary conjugated linoleic acid in health: physiological effects and mechanisms of action. *Annual Review of Nutrition, 22*, 505-531.

[16] Wood, J. D., Enser, M., Fisher, A. V., Nute, G. R., Sheard, P. R., Richardson, R., et al. (2008). Fat deposition, fatty acid composition, and meat quality: A review. *Meat Science, 78*, 343-358.

[17] Griinari, J. M., & Bauman, D. E. (1999). Biosynthesis of conjugated linoleic acid and its incorporation in meat and milk of ruminants. In Advances in Conjugated Linoleic Acid Research. M.P. Yur awcez, Mossoba, M.M.,, 1

[18] Ntambi, J. M., & Miyazaki, M. (2004). Regulation of stearoyl-CoA desaturases and role in metabolism. *Progress Lipid Research, 43*, 91-104.

[19] Dugan, M. E., Kramer, R., J. K., G., Robertson, W. M., Meadus, W. J., Aldai, N., & Rolland, D. C. . (2007). Comparing subcutaneous adipose tissue in beef and muskus-with emphasis on trans 18:1 and conjugated linoleic acids. *Lipids, 42*, 509-518.

[20] Willet, W. C. (2005). The scientific basis for TFA regulation- is it sufficient? *In Proceedings of the first international symposium on "Trans Fatty Acids and Health", Rungstedgaard, Denmark*, 11-13 September, 24.

[21] De la Torre, A., Gruffat, D., Durand, D., Micol, D., Peyron, A., Scilowski, V., & Bauchard, D. (2006). Factor influencing proportion and composition of CLA in beef. *Meat Science, 73*, 258-268.

[22] Raes, K., De Smet, D., & Demeyer, D. (2004). Effect of dietary fatty acids on incorporation of long chain polyunsaturated fatty acids and conjugated linoleic acid in lamb,beef and pork meat: A review. *Animal Feed Science and Technology*, 113, 199-221.

[23] Dance, I. J. E., Matthews, K. R., & Doran, G. (2009). Effect of breed on fatty acid composition and stearoyl-CoA desaturase protein expression in the Semimembranosus muscle and subcutaneous adipose tissue of cattle. *Livestock Science*, 125, 291-297.

[24] Nuernberg, K., Dannenbrger, D., Nuernberg, G., Ender, K., Voigt, J., Scollan, N. D., Wood, J. D., Nute, G. R., & Richardson, R. I. (2005). Effect of grass-based and a concentrate feeding system on meat quality characteristics and fatty acid composition of longissimus muscle in different cattle breeds. *Livestock Production Science,,* 94, 137-147.

[25] Garcia, P. T., Pensel, N. A., Sancho, A. M., Latimori, N. J., Kloster, A. M., Amigone, M. A., & Casal, J. J. (2008). Beef lipids in relation to animal breed and nutrition in Argentina. *Meat Science*, 79, 500-508.

[26] Aldai, N., Lavin, P., Kramer, J. K. G., Jaroso, R., & Mantecon, A. R. . (2012). Beef effect on quality veal production in mountain areas: emphasis on meatfatty acid composition. *Meat Science, (in press)*.

[27] Monteiro, A. C. G., Fontes, M. A., Bessa, R. J. B., Prates, J. A. M., & Lemos, J. P. C. (2012). Intramuscular lipids of Mertolenga PDO beef, Mertolenga PDO veal and "Vitela Tradicional do Montado" PGI veal. *Food Chemistry*, 132, 1486-1494.

[28] Muchenje, v., Hugo, A., Dzama, K., Chimonyo, M., Strydom, P. E., & Raats, J. G. (2009). Cholesterol levels and fatty acid profiles of beef from three cattle breeds raised on natural pasture. *Journal of Food Composition and Analysis*, 22, 354-358.

[29] Ward, R. E., Wood, B., Otter, N., & Doran, O. (2010). Relationship between the expression of key lipogenic enzymes, fatty acid composition, and intramuscular fat content of Limousin and Aberdeen Angus cattle. *Livestock Science*, 127, 20-29.

[30] De Smet, S., Raes, K., & Demeyer, D. (2004). Meat fatty composition as affected by fatness and genetic factors: A review. *Anim. Res.*, 53, 81-98.

[31] Barton, L., Bures, D., Kott, T., & Rehak, D. (2011). Effects of sex and age on bovine muscle and adipose tissue fatty acid composition and stearoyl-CoA desaturase mRNA expression. *Meat Science*, 89, 444-450.

[32] Monteiro, A. C. G., Santos-Silva, J., Bessa, R. J. B., Navas, D. R., & Lemos, J. P. C. (2005). Fatty acid composition of intramuscular fat of bulls and steers. *Livestock Production Science*, 99, 13-19.

[33] Jiang, T., Busboom, J. R., Nelson, M. L., O`, Fallon. J., Ringkob, T. P., Joos, D., & Piper, K. (2010). Effectof sampling fat location and cooking on fatty acid composition of beef steaks. *Meat Science*, 84, 86-92.

[34] Pestana, J. M., Costa, A. S. H., Alves, S. P., Martins, S. V., & Alfaia, C. M. (2012). Seasonal changes and muscle type effect on the nutritional quality of intramuscular fat in Mirandesa-PDO veal. *Meat Science*, 90, 819-827.

[35] Turk, S. N., & Smith, S. B. (2009). Carcass fatty acid mapping. *Meat Science*, 81, 658-663.

[36] Lee, M. R. F., Evans, P. R., Nute, G. R., Richardson, R. I., & Scollan, N. D. (2009). A comparison between red clover silage and grass silage feeding on fatty acid composition, meat stability and sensory quality of the M. Longissimus muscle of dairy cull cows. *Meat Science*, 81, 738-744.

[37] Alfaia, C. M. M., Ribeiro, V. S. S., Lourenco, M. R. A., Quaresma, M. A. G., Martins, S. I. V., Portugal, A. P. V., et al. (2006). Fatty acid composition, conjugated linoleic acid isomers and cholesterol in beef from crossbred bullocks intensively produced and from Alentejana purebred bullocks reared according to Carnalentejana-PDO specifications. *Meat Science*, 72, 425-436.

[38] Alfaia, C. P. M., Castro, M. L. F., Martins, S. I. V., Portugal, A. P. V., Alves, S. P. A., Fontes, C. M. G. A., et al. (2007). Influence of slaughter season and muscle type on fatty acid composition, conjugated linoleic acid isomeric distribution and nutritional quality of intramuscular in Arouquesa-PDO veal. *Meat Science*, 76, 787-795.

[39] Mel'uchova, B., Blasko, J., Kubinec, R., & Gorova, R. (2008). Seasonal variations in fatty acid composition of pasture forage plant and CLA content in ewe milk fat. *Small Ruminant Research*, 78, 56-65.

[40] Daniel, Z. C. T. R., Wynn, R. J., Salter, A. M., & Buttery, P. J. (2004). Differing effects of forage and concentrate diets on the oleic acid and the conjugated linoleic acid content of sheep tissues. *Journal of Animal Science*, 82, 747-758.

[41] Beaulieu, A. D., Drackley, J. K., & Merchen, N. R. (2002). Concentrations of conjugated linoleic acid (cis-9, trans-11-octadecadienoic acid) are not increased in tissue lipids of cattle fed a high concentrate diet supplement with soybean oil. *Journal Animal Science*, 80, 847-861.

[42] Alfaia, C. P. M., Alves, S. P., Martins, S. I. V., Costa, A. S. H., Fontes, C. M. G. A., Lemos, J. P. C., Bessa, R. J. B., & Prates, J. A. M. (2009). Effect of the feeding system on intramuscular fatty acids and conjugated linoleic acid isomers of beef cattle, with emphasis on their nutritional value and discriminatory ability. *Food Chemistry*, 114, 939-946.

[43] Faucitano, L., Chouinard, P. Y., Fortin, J. J., Mandell, I. B., Lafreniere, C., Girard, C. L., & Berthiaume, R. (2008). Comparison of alternative beef production systems based on forrage finishing or grain-forage diets with or without growth promotans: 2. Meat quality, fatty acid composition, and overall palatability. *Journal Animal Science*, 86, 1678-1689.

[44] Sackman, J. R., Duckett, S. K., Gillis, M. H., Realini, C. E., Parks, A. H., & Eggeslton, R. B. (2003). Effect of forage and sunflower oil levels on ruminal biohydrogenation of fatty acids and conjugated linoleic acid formation in beef steers fed finishing diets. *Journal Animal Science*, 81, 3174-3181.

[45] Aldai, N., Dugan, M. E. R., Juarez, M., Martinez, A., & Koldo, O. (2010). Double-musculin carácter influences the trans-18:1 conjugated linoleic acid profile in concentrate-fed yearling bulls. *Meat Science*, 85, 59-65.

[46] Rule, R. D., Mac, Neil. M. D., & Short, R. E. (1997). Influence of sire growth potential, time on feed, and growing-finishing strategy on cholesterol and fatty acids of the ground carcass and Longissimus muscle of beef steers. *Journal of Animal Science*, 75, 1525-1533.

[47] Aldai, N., Murray, B. E., Olivan, M., Martinez, A., Troy, D. J., Osoro, K., & Najera, A. I. . (2005). The influence of breed and mh-genotype on carcass conformation, meat physic-chemical characteristics, and the fatty acid profile of muscle from yearling bulls. *Meat Science*, VER.

[48] Noci, F., French, P., Monahan, F., & Napier, J. A. . (2007). The.J.& Moloney, A.P.The fatty acid composition of muscle fat and subcutaneous adipose tissue of grazing heifers supplemented with plant oil-enriched concentrates. *Journal Animal Science*, 85, 1062-1073.

[49] Raes, K., & Smet, S. D. (2001). Effect of double-muscling in begian blue young bulls on the intramuscular fatty acid composition with emphasis on conjugated linoleic acid and polyunsaturated fatty acids. *Animal Science*, 73, 253-260.

[50] Bauman, D. E., Baumgard, L. H., Corl, B. A., & Griinari, J. M. (1999). Biosynthesis of conjugated linoleic acids in ruminants. *Proceedings American. Society Animal Science.*, VER.

[51] Dugan, M. E. R., Rollan, D. C., Aalhus, J. L., Aldai, N., & Kramer, J. K. G. (2008). Subcutaneous fat composition of youthful and mature Canadian beef. *Canadian Journal of Animal Science*, 88, 591-599.

[52] Basarab, J. A., Mir, P. S., Aalhus, J. L., Shah, M. A., Baron, B. S., Okine, E. K., & Robertson, W. M. (2007). Effect of sunflower seed supplementation on the fatty acid composition of muscle and adipose tissue of pasture-fed and feedlot finished beef. *Canadian Journal Animal Science*, 87, 71-86.

[53] Leheska, J. M., Thompson, L. D., Howe, J. C., Hentges, E., Boyce, J., & Brooks, J. C. (2008). Effects of conventional and grass-feeding systems on the nutrient composition of beef. *Journal Animal Science*, 86, 3575-3585.

[54] Ludden, P. A., Kucuk, O., Rule, D. C., & Hess, B. W. (2009). Growth and carcass fatty acid composition of beef steers fed soybean oil for increasing duration before slaughter. *Meat Science*, 82, 185-192.

[55] Scollan, N. D. (2005). Effect of a grass-based and a concentrate feeding system on meat quality characteristics and fatty acids composition of longissimus muscle in different cattle breeds. *Livestock Production Science*, 94, 137-147.

[56] Engle, T. E., Spears, J. W., Fellner, V., & Odle, J. (2000). Effects of dietary soybean oil and dietary copper on ruminal and tissue lipid metabolism in finishing steers. *Journal Animal Science*, 78, 2713-2721.

[57] Oliveira, E. A., Sampaio, A. A. M., Henrique, W., Pivaro, T. M., Rosa, B. L., Fernandes, A. R. M., & Andrade, A. T. (2012). Quality traits and lipid composition of meat from Nellore young bulls fed with different oils protected or unprotected from rumen degradation. *Meat Science*, 90, 28-35.

[58] Fernandez, A. R. M., Sampaio, A. A. M., Henrique, W., Oliveira, E. A., Oliveira, R. V., & Leonet, F. R. . (2009). Composicao em acidos e qualidade da carne de tourinhos Nelore e Canchim alimentados com dietas a base de cana-de-azucar edos niveis de concentrado. *Revista Brasileña de Zootecnia*, 38, 328-337.

[59] Griswold, K. E., Apgar, G. A., Robinson, R. A., Jacobson, B. N., Johnson, D., & Woody, H. D. (2003). Effectiveness of short-term feeding strategies for altering conjugated linoleic acid content of beef. *Journal. Animal Science*, 61, 1862-1871.

[60] Cassut, M. M., Scheeder, M. R., Ossowski, D. A., Sutter, F., Sliwinski, B., Danilo, A. A., et al. (2000). Comparative evaluation of rumen protected fat, coconut oil and various oilseeds supplemented to fattening bulls. *Archiv der Tierernahrung*, 23, 25-44.

[61] Abu, Ghagazaleh. A. A., Schingoethe, D. G., Hippen, A. R., Kalscheur, K. F., & Whitlock, L. A. . (2002). Fatty acid profile of milk and rumen digesta from cows fed fish oil, extruded soybean or their blend. *Journal Dairy Science.*, 85, 2266-2276.

[62] Kenny, D. A., Kelly, J. P., Monahan, F. J., & Moloney, A. P. (2007). Effect of dietary fish and soya oil on muscle fatty acid concentrations and oxidative lipid stability in beef cattle. *Journal Animal Science*, 85(1), 911.

[63] Barton, L., Marounek, M., Kudrna, V., Bures, D., & Zahradkova, R. (2007). Growth performance and fatty acid profiles of intramuscular and subcutaneous fat from Limousin and Charolais heifers fed extruded linseed. *Meat Science*, 76, 517-523.

[64] Weill, P., Schmitt, B. C. G., Chesneau, G., Daniel, N., Safraou, F., & Legrand, P. (2002). Effects of introducing linseed oil in livestock diet on blood fatty acid composition of consumers of animal products. *Annals of Nutrition and Metabolism*, 46, 182-191.

[65] Dannenberger, D., Nuernberg, K., Nuernberg, G., Scollan, N., Steinhart, H. y., & Ender, K. (2005). Effect of pasture vs. concentrate diet on CLA isomers distribution in different tissue lipids of beef cattle. *Lipids*, 40, 589-598.

[66] Dhiman, T. R., Satter, L. D., Pariza, M. W., Galli, M. P., Albright, K., & Tolosa, M. X. . (2000). Conjugated linoleic acid (CLA) content of milk from cows offered diets rich in linoleic and linolenic acid. *Journal Dairy Science*, 83, 1016-1027.

[67] Chouinard, P. Y., Corneau, L., Butler, W. R., Chilliard, Y., Drackley, J. K., & Bauman, D. E. . (2001). Effect of dietary lipid source on conjugated linolei acid concentrations in milk fat. Journal Dairy Science , 84, 680-690.

[68] Mir, P. S., Mir, Z., Kuber, P. S., Gasking, C. T., Martin, E. L., Dodson, M., Elias, V., Calles, J. A., Johnson, K. A., Busboom, J. R., Wood, A. J., Pittenger, G. L., & Reeves, J. J. . (2002). Growth, carcass characteristics, muscle conjugated linoleic acid (CLA) con-tent, and response to intravenous glucose challenge in high percentage Wagyu, Wa-gyu x Limousin, and Limousin steers fed sunflower oil-containing diets. *Journal Animal Science*, 80, 2996-3004.

[69] Mir, P. S., Mc Allister, T. A., Zaman, S., Jones, S. D. M., He, M. L., Aalhus, J. L., Jere-miah, L. E., Goonewardene, L. A., Weselake, R. J., & Mir, Z. (2003). Effect of dietary sunflower oil and vitamin E on beef cattle performance, carcass characteristics and meat quality. *Canadian Journal Animal Science*, 83, 53-66.

[70] Aharoni, Y., Orlow, A., Brosh, A., Granit, R., & Kanner, J. (2005). Effects of soybean supplementation of high forage fattening diet on fatty acid profiles in lipid depots of fattening bull calves, and their levels of blood vitamin E. *Animal Feed Science and Technology*, 119, 191-202.

[71] Vasta, V., Pagano, R. I., Luciano, g., Scerra, M., Caparra, P., Foti, F., et al. (2012). Ef-fect of morning vs. afternoon on intramuscular fatty acid composition in lamb. *Meat Science*, 90, 93-98.

[72] Vasta, V., Priolo, A., Scerra, M., Hallet, K. G., Wood, J. D., & Doran, O. (2009). Δ9 de-saturase protein expression and fatty acid composition of longissimus dorsi muscle in lamb fed green herbage or concentrate with or without added tannins. *Meat Sci-ence*, 82, 357-364.

[73] Bernes, G., Turner, T., & Pickova, J. (2012). Sheep fed only silage supplement with concentrates. 2. Effects on lamb performance and fatty acid profile of ewe milk and lamb meat. *Small Ruminant Research*, 102, 114-124.

[74] Joy, M., Ripoll, G., Molino, F., Dervishi, E., & Alvarez, Rodriguez. J. (2012). Influence of the type of forage supplied to ewes in pre- and post-partum periods on the meat fatty acid of suckling lambs. *Meat Science*, 90, 775-782.

[75] Mazzone, G., Giammarco, M., Vignola, G., Sardi, L., & Lambertini, L. (2010). Effects of rearing season on carcass and meat quality of suckling Apennine light lambs. *Meat Science*, 86, 474-478.

[76] Chiofalo, B., Simonella, S., Di Grigoli, A., Liotta, L., Frenda, A. S., Lo, Presti. V., Bo-nanno, A., & Chiofalo, V. (2010). Chemical and acidic compostion of Longissimus dorsi of Comisana lambs fed with Trifolium subterraneum and Lolium multiflorum. *Small Ruminant Research*, 88, 89-96.

[77] Berthelot, V., Bas, P., & Schmidely, P. (2010). Utilization of extruded linseed to modify fatty acid composition of intensively-reared lamb meat: Effect of associated cereals (wheat vs. corn) and linoleic acid content of the diet. *Meat Science*, 84, 114-124.

[78] Bas, P., Berthelot, V., Pottier, E., & Normand, J. (2007). Effect of linseed on fatty acid composition of muscles and adipose tissues of lamb with emphasis on trans fatty acids. Meat Science , 77, 678-688.

[79] Wachira, A. M., Sinclair, L. A., Wilkinson, R. G., Enser, M., Wood, J. D., & Fisher, A. V. (2002). Effects of dietary fat source and breed on the carcass composition, n-3 polyunsaturated fatty acid and conjugated linoleic acid content of sheep and adipose tissue. *Journal of Nutrition*, 88, 697-709.

[80] Bolte, M. R., Hess, B. W., Means, W. J., Moss, G. E., & Rule, D. C. (2002). Feeding lambs high-oleate or high-linoleate safflower seeds differentially influences carcass fatty acid composition. *Journal Animal Science*, 80, 609-616.

[81] Santos-Silva, J., Bessa, R. J. B., & Mendes, I. A. (2003). The effect of supplementation with expanded sunflower seed on carcass and meat quality of lambs raised on pasture. *Meat Science*, 65, 1301-1308.

[82] Radunz, A. E., Wickersham, L. A., Loerch, S. C., Fluharty, F. L., Reynolds, C. K., & Zerby, H. N. (2009). Effects of dietary polyunsaturated fatty acid composition in muscle and subcutaneous adipose tissue of lamb. *Journal Animal Science*, 87, 4082-4091.

[83] Bessa, R. J. B., Portugal, V. P., Mendes, I. A., & Santos-Silva, J. (2005). Effect of lipid supplementation on growth performance, carcass and meat quality and fatty acid composition of intramuscular lipids of lambs fed dehydrated Lucerne or concentrate. *Livestock Production Science*, 96, 185-194.

[84] Bessa, R. B., Alves, S. P., Figueredo, R., Texeira, A., Rodriguez, A., Janeiro, A., et al. (2006). Discrimination of production system and origin of animal products using chemical markers. In Animal Products from the Mediterranean area. EAAP Publication The Netherland: Wageningen Academic Publishers.(119), 231-240.

[85] Bessa, R. J. B., Alves, S. P., Jeronimo, E., Alfaia, C. M., Prates, J. A. M., & Santos-Silva, J. (2007). Effect of lipid supplements on ruminal biohydrogenation intermediates and muscle fatty acids in lambs. *European Journal of Lipid Science Technology*, 109, 868-878.

[86] Jeronimo, E., Alves, S. P., Prates, J. A. M., Santos-Silva, J., & Bessa, R. J. B. (2009). Effect of dietary replacement of sunflower oil with linseed oil on intramuscular fatty acids of lamb meat. *Meat Science*, 84, 499-505.

[87] Bessa, R. J. B., Lourenco, M., Portugal, P. V., & Santos-Silva, J. (2008). Effects of previous diet and duration of soybean oil supplementation on light lamb carcass composition, meat quality and fatty acid composition. Meat Science , 80, 1100-1105.

[88] Peng, Y. S., Brown, M. A., Wu, J. P., & Liu, Z. (2010). Different oilseed supplements alter fatty acid composition of different adipose tissues of adult ewes. *Meat Science*, 85, 542-549.

[89] Kucut, O., Hess, B. W., & Rule, D. C. (2008). Fatty acid composition of mixed ruminal microbes isolated from sheep supplemented with soybean oil. *Research in Veterinary Science*, 84, 213-224.

[90] Realini, C. E., Duckett, S. K., Brito, G. W., Rizza, M., & De Mattos, D. (2004). Effect of pasture vs. concentrate feeding with or without antioxidants on carcass characteristics, fatty acid composition, and quality of Uruguayan beef. *Meat Science*, 66, 567-577.

[91] Dierking, R. M., Kallenbach, R. L., & Grun, I. U. (2010). Effect of forage species on fatty acid content and performance of pasture-finished steers. *Meat Science*, 85, 597-605.

[92] Humada, M. L., Serrano, E., Sañudo, C., Rolland, D. C., & Dugan, M. E. R. (2012). Production system and slaughter age effects on intramuscular fatty acids from Young Tudanca bulls. *Meat Science*, 90, 678-685.

[93] Sarries, M. V., Murray, B. E., Moloney, A. P., Troy, D., & Berian, M. J. (2009). The effect of cooking on the fatty acid composition of Longissimus dorsi muscle from beef steers fed rations designed to increase the concentration of conjugated linoleic acid in tissue. *Meat Science*, 81, 307-312.

[94] Pavan, E., & Duckett, S. K. (2007). Corn oil supplementation to steers grazing endophyte-free tall fescue. II Effects on Longissimus muscle and subcutaneous adipose fatty acid composition and stearoyl-CoA desaturase activity and expression. Journal Animal Science , 85, 1731-1740.

[95] Juarez, M.., Dugan, M. E. R., Aalhus, J. L., Aldai, N., Basarab, J. A., Baron, B. S., & Mc Allister, T. A. (2011). Effects of vitamin E and flaxseed on rumen-derived fatty acid intermediates in beef intramuscular fat. *Meat Science*, 88, 434-440.

[96] Garcia, P. T., Casal, J. J., Fianuchi, S., Magaldi, J. J., Rodriguez, F. J., & Nancucheo, J. A. (2008). Conjugated linolenic acid (CLA) and polyunsaturated fatty acids in muscle lipids of lamb from the Patagonian area of Argentina. *Meat Science*, 79, 541-548.

[97] Mir, Z., Rushfeldt, M. I., Paterson, L. J., & Weselake, P. (2000). Effects of dietary supplementation with CLA or linoleic rich oil in the lamb content of lamb tissues. *Small Ruminant Research*, 36, 25-31.

[98] Kott, R. W., Hatfield, P. G., Bergman, J. W., Flyn, C. R., Van Wagoner, H., & Boles, J. A. (2003). Feedlot performance, carcass composition, and muscle and fat CLA concentrations of lambs fed diets supplemented with safflower seeds. *Meat Science*, 49, 11-17.

Chapter 8

Value - Added Products from Soybean: Removal of Anti-Nutritional Factors *via* Bioprocessing

Liyan Chen, Ronald L. Madl, Praveen V. Vadlani,
Li Li and Weiqun Wang

Additional information is available at the end of the chapter

http://dx.doi.org/10.5772/52993

1. Introduction

Soybean is the second largest acreage crop in the United States (29%), right after corn (35%) according to the American Soybean Association [1]. Soybean is widely consumed in the world, particularly in Asian countries. The various soybean products could be separated into non-fermented and fermented soybean products. The non-fermented soybean products include soymilk, tofu, yuba, soybean sprouts, okara, roasted soybeans, soynuts and soy flour, immature soybeans, cooked whole soybeans, and the fermented oriental soybean products include soy paste (Jiang and Miso), soy sauce, Tempeh, Natto, soy nuggets (Douchi), sufu. In the United States, soy oil is often used for food and biodiesel production. The soybean processing process is shown in Figure 1. After the oil extraction, the residue – flaked soy meal, is usually produced into four products (textured soy flour, soy protein concentrate and soy protein isolate, 48% soy meal, soluble soy carbohydrate). The textured soy flour could be used in bakery products, meat products, infant food etc. Soy protein concentrate and isolate could be used in baby food, bakery products, cereals, lunch meat etc. SSPS (soluble soybean polysaccharides) functions as a dispersing agent, stabilizer, emulsifier, and has good adhesion properties [2]. The 48% soy meal is used for animal feed. The portions of different animal usages are poultry (48%), swine (26%), beef (12%), dairy (9%), pets (2%), others (3%) [1]. Poultry and swine usages account for74%.

The popular usage of soybean for food and feed is due to its nutritional profile. Soybean is a good protein source and the only dietary isoflavone source together with other legumes. Anti-nutritional factors in soybean, such as phytic acid, oligosaccharides, trypsin inhibitor etc, limit its usage. Fermentation with GRAS (generally recognized as safe) microorganisms has been used to help degrade these anti-nutritional factors. The nutritional value of fer-

© 2013 Chen et al.; licensee InTech. This is a paper distributed under the terms of the Creative Commons Attribution License (http://creativecommons.org/licenses/by/3.0), which permits unrestricted use, distribution, and reproduction in any medium, provided the original work is properly cited.

mented soybean and soy meal products with additional nutritional factors is then largely enhanced.

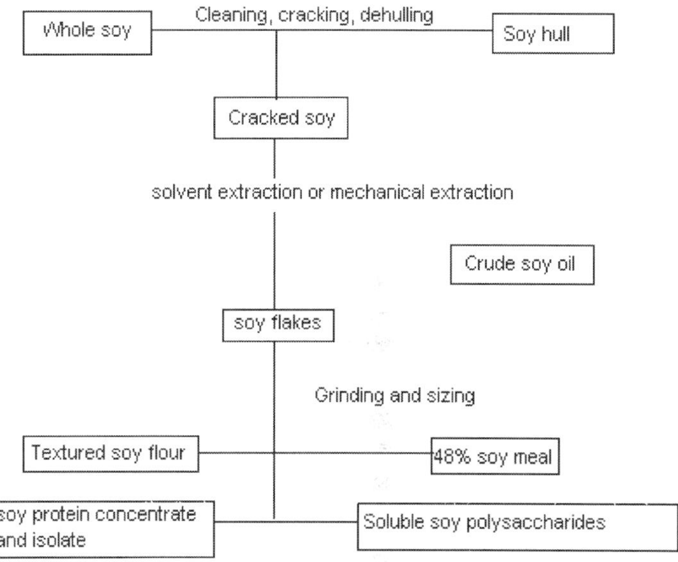

Figure 1. Soybean processing and products

2. Nutritional enhancement of soybean and soy meal via fermentation

2.1. Soy Protein

Protein content in soybean and soy meal are around 40%- 50% respectively. The high protein content makes soybean and soy meal a rich protein source for food and feed. As food source, the Protein Digestibility Corrected Amino Acid Score (PDCAAS), which is adopted by FDA and FAO/WHO, for isolated soy protein is 0.92, soy protein concentrate is 0.99, comparing with beef (0.92) and egg white (1.00). The human subject studies show that well-processed soy protein could serve as the sole source of protein intake for human beings [3]. FDA claims that diets containing 25 g of soy protein can reduce levels of low-density lipoproteins by as much as 10 percent and have considerable value to heart health. The specific reason for the heart protection function is unclear, for there are hundreds of protective compounds in soybean. As feed source, soy protein is high in lysine, but low in sulfur-containing amino acids, with methionine being the most limiting amino acid, followed by threonine [4]. The complementation of soy and corn for lysine and methionine makes them a valuable feed when combined.

2.1.1. Fermentation increases protein and amino acid content, and degrades protein into small functional peptides.

During fermentation, microorganisms digest the carbohydrates in soybean or soy meal and use for their own growth. The decreased dry matter and increased microorganisms weight ratio result in enhanced protein content[5-7]. In reference [5], fermented soy meal with *S.cerevisae* increased its protein level from 47% to 58%, while with *L. plantarum* and *B. lactis*, protein level increased to 52.08% and 52.14%. Microorganisms used for soybean fermentation have been reported to secret protease during fermentation [8-11]. In Cheonggukjang, the *Bacillus subtilis* fermented traditional soybean food in Korea, the acidic protease activity level could be as high as 590.24±2.92 µg/ml. Neutral protease activity level could achieve 528.13±3.11 µg/ml [9]. Because of protein degradation during fermentation, fermented soybean products are easier to digest.

Four parameters have been often used to evaluate the protein degradation of fermented soybean products. They are trichloroacetic acid (TCA) soluble nitrogen, degree of protein hydrolysis, SDS-PAGE profile, and amino acid content. Usually peptides having 10 or fewer amino acids would dissolve in TCA[12]. During fermentation, the degree of protein hydrolysis increases because of protease hydrolysis [13-14]. Meanwhile, TCA soluble nitrogen and peptide contents could also be enhanced [6, 9, 13-14]. SDS-PAGE analysis shows less large (>70 kDa) and medium (20-60 kDa) peptides and more small (<10 kDa) peptides in soy meal after fermentation of *Lactobacillus plantarum, Bifidobacterium lactis, Sccharomyces cerevisiae*, or *Aspergillus oryzae* [5, 7, 15]. Reference [13] showed that after 24 hr *Bacillus subtilis* fermentation, soy protein with molecular weight above 20 Kd disappeared from the electrophoretograms. The total amino acid content increased significantly (p<0.05) in fermented soy meal or soybean with *Lactobacillus plantarum, Bifidobacterium lactis, Sccharomyces cerevisiae, Bacillus subtilis, Aspergillus oryzae, Rhizopus oryzae, Actinomucor elegans, Rhizopus oligosporus* et al. [5, 7-8, 10-11, 13, 16]. *L.plantarum* fermentation of soy flour led to an increase in sulfur amino acids (Met plus Cys), Phe, Tyr, Lys, and Thr [16].

2.1.2. Functional biopeptide

Fermentation degrades large protein molecules into small peptides and amino acids. The biologically active peptides from soybean play an important role as angiotenisin converting enzyme (ACE) inhibitor [17] and as antioxidants [18]. In this section, we will discuss the ACE inhibitor. The antioxidant activity of biopeptides will be discussed in the antioxidant section.

Angiotensin I-converting enzyme (ACE, EC 3.4.15.1) is a dipeptidylcarboxypeptidase associated with the regulation of blood pressure as well as cardiovascular functions [19]. ACE-inhibitory substances are used to lower the blood pressure of hypertensive patients. Various ACE inhibitory peptides have been isolated from traditional fermented soybean foods, like natto, doujiang, soy sauce, and miso paste [20-23].

ACE inhibitory activity of peptides generated by protease is greatly dependent upon fermentation time. ACE inhibitory activity in Textured Vegetable Protein (TVP) fermented by

Bacillus subtilis for 24 and 72hr showed IC50 values of 2.20 and 3.80 mg/ml, respectively [24]. The initial fermentation of TVP resulted in production of effective peptides, but longer fermentation time produced less active peptides as ACE inhibitor. Peptide with ACE inhibitory activity consisted of low molecular weight. Molecular weight of 500-1,000 Da shows the highest ACE inhibitory activity [24]. In [25], oligopeptides generated from soy hydrolysate and fermented soy foods though endoprotease digestion, demonstrated a range of biological activities – angiotensin converting enzyme (ACE) inhibitory, anti-thrombotic, surface tension and antioxidant properties.

2.1.3. Fermentation decreases soy immunoreactivity

Soybean is defined as one of the "big 8" food allergens in the United States [16].The estimated prevalence of soybean allergies is about 0.5% of the total U.S. population [16]. Patients with soy allergy could react subjectively and objectively with 0.21 and 37.2 mg of soy protein, respectively [5]. The principle for food allergy is that epitopes in allergenic protein bind to the immunoglobulin E (IgE) molecules residing in the mast cells and basophils, causing them to release inflammatory mediators, including histamine. Alpha- (72 kDa) and beta- (53 kDa) conglycinin subunits, P34 fraction, and glycinin basic (33 kDa) and acidic (22 kDa) subunits, and trypsin inhibitor (20 kDa) are the main protein components causing plasma immunoreactivity [16]. Glycine was found to stimulate local and systemic immune responses in allergic piglets and had negative effects on piglet performance [26]. The severity of the immune reactions depends on the dose of glycinin; higher doses cause more severe symptoms. The effect of purified beta-conglycinin on the growth and immune responses of rats were investigated [27]. Results showed that purified beta-conglycinin possesses intrinsic immune-stimulating capacity and can induce an allergic reaction. Also, newly weaned pigs with limited stomach acid and enzymatic secretions in the small intestine can have difficulty digesting proteins with complex structures and large molecular weights [28].

Studies have confirmed degradation of soybean allergens during fermentation by microbial proteolytic enzymes in fermented soybean products, such as soy sauce, miso and tempeh [5, 29]. In the fermented soy products, soy protein has been hydrolyzed into smaller peptides and amino acids;therefore the structure of antigen epitopes might be altered, becoming less reactive. The IgE binding potential is reduced and therefore the immunoreactivity is lowered. In soy sauce, proteins are completely degraded into peptides and amino acids after fermentation and allergens are no longer present [29]. The reduction of immunoreactivity by nature and induced fermentation of soy meal with *Lactobacillus plantarum, Bifidobacterium lactis, Saccharomyces cereuisiae* were evaluated [5]. *S.cerevisiae, B.lactis* and *L. plantarum* reduced the immune response77.2%, 77.2%, 78.0%, when using 97.5 kUA/l human plasma and 88.7%, 86.3%, 86.9%, respectively, when the pooled human plasma was used. All three fermented soy meal products showed fewer large (>70 kDa) and medium (20-60 kDa) peptides, and more small (<10 kDa) peptides. Protein hydrolysis reduction of soy protein immunoreactivity was also confirmed through enzyme hydrolysis conducted in reference [30], which showed that after hydrolyzing with three food-grade proteases (Alcalase, Neutrase, Corolase PN-L), no residual antigenicity was observed in resulting soy whey.

Animal experiments have also confirmed the hypoallergenic properties of fermented soybean or soy meal products. With regard to the soybean allergy, fermentation of soy meal decreased the immune response to soy protein in piglets and the level of serum IgG decreased by 27.2% [31]. Antigenic soybean proteins in the diet of early weaned pigs provoke a transient hypersensitivity associated with morphological changes including villi atrophy and crypt hyperplasia in the small intestine [31]. All of these morphological changes can cause a malabsorption syndrome [26, 33], growth depression, and diarrhea [34, 35]. Differences of the villi condition in such pigs fed soy meal and fermented soy meal were investigated by using scanning electron microscopy [36]. Piglets fed soy meal had shorter, disordered, and broader villi, whereas piglets fed fermented soy meal had long, round, regular, and tapering villi that could better digest and absorb nutrients.

2.2. Isoflavone

One of the acknowledged bioactive compounds in soybean is isolavone. Isoflavones generally exist in soybeans and soy foods as aglycones (daidzein, genistein, and glycitein), beta-glycosides (daidzin, genistin, and glycitin), acetylglycosides (6″-O-acetyldaidzin, 6″-O-acetylgenistin, and 6″-O-acetylglycitin), and malonylglycosides (6″-O-malonyldaidzin, 6″-O—malonylgenistin, and 6″-O-mlonylglycitin). The structures of the 12 isomers are shown in figure 2.Isoflavones physiological effects include their estrogenic activity, antioxidant and antifungal activity, and more importantly, to act as anti-carcinogens. Isoflavones may also help to reduce blood serum cholesterol levels [4].

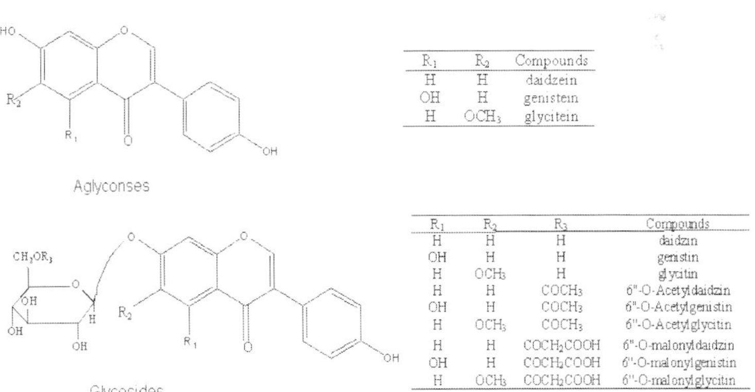

R_1	R_2	Compounds
H	H	daidzein
OH	H	genistein
H	OCH_3	glycitein

Aglyconses

R_1	R_2	R_3	Compounds
H	H	H	daidzin
OH	H	H	genistin
H	OCH_3	H	glycitin
H	H	$COCH_3$	6″-O-Acetyldaidzin
OH	H	$COCH_3$	6″-O-Acetylgenistin
H	OCH_3	$COCH_3$	6″-O-Acetylglycitin
H	H	$COCH_2COOH$	6″-O-malonyldaidzin
OH	H	$COCH_2COOH$	6″-O-malonylgenistin
H	OCH_3	$COCH_2COOH$	6″-O-malonylglycitin

Glycosides

Figure 2. Chemical structures of 12 isoflavone isomers in soybean [4]

Among soy isoflavones, the relative abundance of genistein including respective derivatives, is the highest – about 50% of isolavone content, followed by daidzein (40%) and glycitein (10%). However, glycitein has been shown to be more bio-available than other isoflavones, followed by genistein [37]. Most of the isoflavones are associated with proteins in soy, with very little present in the lipid fraction. In their natural state, the majority of isoflavones exists as inactive glycosides (genistin, daidzin, and glycitin) and the remaining as their active aglycone forms (genistein, daidzein, and glycitein, respectively). Glycoside forms are heat sensitive, being converted into malonyl-beta-glycosylated isoflavone when heated. The aglycone forms are quite stable at high temperature [38].

Isoflavones when ingested are metabolized extensively in the intestinal tract, absorbed, transported to the liver, and undergo enterohepatic recycling. Intestinal bacterial glucosidases cleave the sugar moieties and release the biologically active isoflavones as aglycones. Aglycones could be directly absorbed in the adult and these can be further biotransformed by microorganisms to the specific metabolites, equol, desmethylangolensin, and p-ethylphenol. All of these phyto-oestrogens are then eliminated, mainly by the kidney, and therefore share the physiological features and behavior of endogenous estrogens. Among the glycones, beta-glucosides are easier to be hydrolyzed than 6''-O-malonylglucosides and 6''-O-acetylglucosides.

Some microorganisms have been reported to secrete beta-glucosidase, which could convert isoflavones to aglycones. The quantities of melonyl, acetyl, and glycosidyl isoflavonoids decrease during fermentation but those of isoflavonoid aglycones, daidzein, and genistein increase by over 10 to 100 fold. In Meju (long term fermented soybeans), compared with unfermented samples, the total glycosides in 60 hr fermented samples decreased from 1827 ug/g to 487 ug/g, while total aglycones increased from 22 ug/g to 329 ug/g, with daidzein increasing from trace to 152 ug/g, genistein from 16 ug/g to 170 ug/g. However, the quantity of glycitein was not increased [39]. In another study about meju fermented by *Aspergillus oryzae*, malonyl glycosides that initially accounted for 57.2-72.2% in the different soybeans markedly decreased to 7.4-19.9%, while aglycones originally accounted for only 1.1-2.8% in the soybeans, but markedly increased to 34.1-53.2% in miso [40]. In reference [38], total aglycones increased from 12.27 ug/g in whole soy flour to 446.90 ug/g after 48 hr fermentation by *Aspergillus oryzae*(ATCC 22876). Its percentage of total isoflavones increased from 2.67% to 75.51%. Daidzein content increased from trace to 133.07 ug/g, glycitein from trace to 35.56 ug/g, genistein from 12.27 ug/g to 278.27 ug/g. In *Bacillus subtilis* fermented soy paste ChungGuklang (CGJ), about 85% of isoflavones were in the aglycone form in the CGJ, 14% in the glucoside form and acetylglucoside and malonylglucoside forms contributed less than 1% [41]. In *Bacillus pumilus* HY1 fermented Cheonggkjang, the beta glucosidase increased to 24.8 U/g until 36 h. The glycosides and malonylglycosides decreased throughout the fermentation to about 80% - 90% of their starting amount at 60 hr [42]. Part of isoflavonoid aglycones are broken down into secondary metabolites, so the total quantity of isoflavonoids decreased [38, 42].

Aglycones could alleviate the symptoms of type2 diabetes, which the beta-glycosides, acetylglycosides and malonylglycosides forms are not able to do. Type 2 diabetes mellitus emerges from uncompensated peripheral insulin resistance that is associated with unregulated nutrient homeostasis, obesity, peripheral insulin resistance and progressive beta-cell failure. The effect of isoflavones in Meju on alleviating the symptoms of the type2 diabetes was investigated and four mechanismswere found [39]. Isoflavonoid aglycones could improve insulin-stimulated glucose intake. Also, they could induce PPAR-γ activation to increase insulin-stimulated glucose uptake. The PPAR-γ is the central regulator of insulin and glucose metabolism. It could help improve insulin sensitivity in type 2 diabetic patients and in diabetic rodent models. Besides, aglycones have strong effects for insulin/IGF – 1 signaling through IRS2, which plays a crucial role in beta-cell growth and survival. In this study, aglycones in meju increased GLP-1 secretion. GLP-1 is one of the incretins secreted from enteroendocrine L-cells that augments insulin secretion after the oral intake of glucose and free fatty acids. The induction of its secretion can prevent and/or relieve diabetic symptoms.

2.3. Antioxidant activity

Oxidative stress has been found to be the primary cause of many chronic diseases as well as the aging process itself. Antioxidants could help to delay or prevent oxidative stress. Epidemiological studies show that antioxidants could lower the risk of cardiovascular disease, cancer (overall risk reduction between 30 – 50%), diabetes, neurological diseases, immune diseases, eye diseases et al. [43]. Antioxidant compounds play an important role as a health protecting factor. It is beneficial to eat antioxidant enriched food. Products prepared through the fermentation of soybean including various traditional oriental fermented products of soybean such as natto, tempeh, miso and other fermented products, have been found to exhibit a significantly higher antioxidant activity than their respective non-fermented soybean substrate.

Di(phenyl)-(2,4,6-trinitrophenyl)iminoazanium(DPPH) radical scavenging activity, Fe2+ chelating activity and reducing property have been used to quantify antioxidant activity. DPPH is a stable nitrogen-centered, lipophilic free radical that is used in evaluating the antioxidant activities in a short time. Ferrous ionsare the most effective pro-oxidants in food system. Metal chelation agents prevent metal-assisted homolysis of hydroperoxides and block the formation of chain initiators. IC50 and the relative scavenging effect are the parameters to describe the DPPH radical scavenging and Fe2+ chelating abilities. IC50 is the inhibition concentration of extracts required to decrease initial DPPH radical or Fe2+ concentration by 50%. The relative scavenging effect is calculated by dividing the extract content with the IC50 of the respective extract, and then compare with the scavenging effect of control samples. The reducing property indicates that these compounds are electron donors and can reduce the oxidized intermediates, and therefore, can act as primary or secondary antioxidants. The reduction of Fe3+/ferricyanide complex to ferrous form in presence of antioxidants has been used to test the reducing activity of samples [44].

Antioxidant activity of methanol extracts of soybean koji fermented with *Asp. oryzae*, *Asp. sojae*, *Actinomucor taiwanensis*, *Asp. awamori*, and *Rhizopus* spp. have been investigated [44]. Methanol extracts of soybean koji, which mainly contained phenolic acid, have higher relative DPPH – scavenging effect and Fe2+ chelating effect than the unfermented steamed soybeans. The koji methanol extracts had a relative DPPH - scavenging effect of 2.3-8.9 compared with that of the non-fermented soybean, which was assigned as 1.0. Among them, the *Aspergillus awamori*-soybean koji exhibited the highest DPPH-scavenging effect, at a level approximately 9.0 fold than that exhibited by the non-fermented steamed soybean [44].The Fe2+ ion chelating ability of soybean increased by 2.1 – 6.7 fold after fermentation, depending on starter organism employed [44]. In Cheonggukjang fermented by *Bacillus pumilus HY1*, the level of DPPH radical scavenging activity increased from 54.5% to 96.2% by 60 hr [42]. In *Bacillus subtilis* fermented soybean kinema, when the methanol extract concentration was 50 mg/ml, 60% DPPH radical scavenging was observed. In the same product with a concentration of 10 mg/ml, the methanolic extract of kinema exhibited 64% metal chelation which was much higher than the activity shown by cooked non-fermented soybean (22%) [45]. Similar findings of enhanced reducing power of fermented bean and bean products were reported from *Bacillus subtilis* fermented soybean kinema[45] and from *Aspergillus oryzae* fermented soybean koji [44].

The enhanced antioxidant activities in fermented soybean products may be due to the increased phenolic compounds contents. Phenolic compounds have been demonstrated to exhibit a scavenging effect for free radicals and metal-chelating ability [46]. Most phenolic acids in cereals primarily occur in the bound form, as conjugates with sugars, fatty acids, or proteins [47].Isoflavones are the predominant phenolics in soybean, and the glucoside form of isoflavones represents 99% of the total isoflavones in soybean [45].This condition lowers the antioxidant activity since the availability of free hydroxyl groups in the phenolic structure is an important characteristic for the resonance stabilization of free radicals. The enhancement of bioactive phenolic compounds by enzymatic hydrolysis from different cereals was reported by Wojdylo [47] and Yuan [48]. Different enzymes during bacteria or fungi fermentation, such as alpha-amylase, alpha and beta-glucosidase, beta-glucuronidase, cellulase et al., have been reported to be involved in the lignin remobilization and phenolic compounds contents enhancement [49]. Fermented soybean products have higher amount of phenolic compounds [42, 44]. In Cheonggukjang fermented by *Bacillus pumilus HY1*, total phenolics increased markedly from the starting amount of 253 g/kg to 9586 g/kg at the end of fermentation (60hr) [42]. In *Bacillus subtilis* fermented kinema, total phenol content of kinema was 144% higher than that of cooked non-fermented soybean. The total phenol content was positively correlated (p<0.001) with radical- scavenging, reducing power, metal-chelating activity in *Bacillus subtilis* fermented kinema [45].

The other reason is the short chain peptides generated by fungi or bacterial protease. Antioxidant activities of peptides have been reported [50-52]. In [51], five different proteases were used to hydrolyze soybean β-conglycinin and the hydrolysates from three of them had significant enhanced antioxidative activities. Peptide antioxidant activity is related, but not limited, to the amino acid composition and its sequence. In reference [51], peptides isolated from the antioxidative beta-conglycinin hydrolysate contain histidine, proline or tyrosine residue in their sequences and hydrophobic amino acids, valine or leucine at the N terminus. The constituent amino acids had no antioxidant activity when mixed at the same concentration as the peptides. Anti-oxidative activities of peptides with different amino acid composition or sequences have different antioxidant mechanisms. Reference [52] studied the anti-oxidative properties of combinatorial tri-peptides. Tri-peptides Tyr-His-Tyr, Xaa-Xaa-

Trp/Tyr, and Xaa-Xaa-Cys (SH) had a strong synergistic effect with phenolic antioxidants, a high radical scavenging activity, and a high peroxynitrite scavenging activity, respectively.

2.4. Phytic acid

Phytate is the calcium-magnesium-potassium salt of inositol hexaphosphoric acid commonly known as phytic acid [Figure 3]. Phytate and phytic acid are also referred to as phytin in some literature.

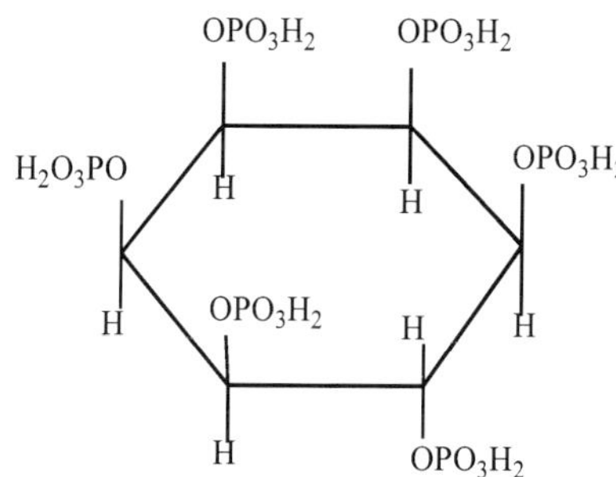

Figure 3. Basic structure of phytic acid [52]

Phytate is the main storage form of phosphorous in soybean. Its content in soybean ranged from 1.00 to 1.47% on a dry matter basis [4].Phytate is known to be located in the protein bodies, mainly within their globoid inclusions. Phosphorous in the phytate form could not be absorbed by monogastric animals, because they lack phytase, the digestive enzyme required to release phosphorus from the phytate molecule. Phytic acid could form protein-phytate or protein-phytate-protein complexes; these have more resistance to digestion by proteolytic enzymes; thus, utilization of dietary protein is reduced. Also, phytic acid has a strong binding affinity to important minerals such as calcium, magnesium, iron, and zinc. When a mineral binds to phytic acid, it becomes insoluble, precipitates, and is not absorbable in the intestines. In food industry, the presence of phytic acid in high concentration is undesirable. In feed industry, the unabsorbed phytate passes through the gastrointestinal tract of monogastric animals, elevating the amount of phosphorus in the manure. Excess phosphorus excretion can lead to environmental problems such as eutrophication. With the pressure on the swine industry to reduce the environmental impact of pork production, it is important to use feed ingredients that can minimize this influence.

The ability of the molds for oriental fermented soybean food to produce phytase has been investigated. Phytase is the enzyme hydrolyzing phytic acid to inositol and phosphoric acid and thereby removing the metal chelating property of phytic acid. There are two strains of *Rhizopus oligosporus* used for tempeh fermentation, two strains of *Aspergillus oryzae* used for soy sauce and six strains of *Aspergillus oryzae* used for miso fermentation, all the ten strains were reported to be able to secret both extracellular and intracellular phytases [54]. The phytic acid content of soybeans was reduced by about one-third as a result of *Rhizopus oligosporus* NRRL 2710 fermentation [55-56]. That was from original 1.27% to 0.61% after 48 hr fermentation [56]. Animal test showed that fermentation of soy meal increased phosphorus availability [57-58] and zinc availability [59] and reduced phosphorus excretion without affecting growth of chicks. Using fermented soy meal as substitute for regular soy meal saved 0.2% of dietary inorganic phosphorous [60].

2.5. Oligosaccharides

Galacto-oligosacchrides (GOS) generally represent approximately 4 to 6% of soybean dry matter. In soy meal produced at 10 commercial processing plants in the United States, concentrations of stachyose, raffinose, and verbascose ranged from 41.0 to 57.2, 4.3 to 9.8, and 1.6 to 2.4 mg/g DM, respectively [61]. Oligosaccharides in the carbohydrate fraction, particularly raffinose and stachyose [figure 4], could lead to flatulence and abdominal discomfort for monogastric animals [62-63] because of the deficiency of alpha-galactosidase.

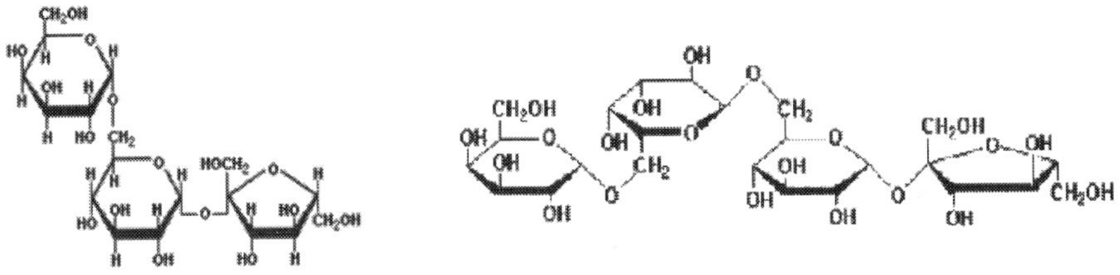

(1)

(2)

Figure 4. Structure of raffinose and stachyose (1) raffinose; (2) stachyose

Galacto-oligosacchrides are digested to some extent in the small intestine (76 to 88% for stachyose, 31 to 65% for raffinose, and 32 to 55% for verbascose), resulting in the production of carbon dioxide and hydrogen [64]. In some cases, the accumulation of flatulent rectal gas provokes gastrointestinal distress such as abdominal pain, nausea, and diarrhea. Weanling pigs fed a GOS-free diet supplemented with 2% stachyose or fed a diet containing soy meal had increased incidence of diarrhea compared with pigs fed a control diet [63]. Additionally, fermentation of GOS has been implicated to have negative effects on nutrient digestibilities and energy availability of soy meal. Roosters fed soy meal with low oligosaccharide concentrations had higher total net metabolizable energy values (2931 kcal/kg dm) than those fed conventional soy meal (2739 kcal/kg DM) [65]. The removal of polysaccharides from soy foods and feed is, therefore, a major factor in improving their nutritive value. To reduce non-digestible oligosaccharides, fermentation with fungi, yeast, and bacteria with alpha-galactosidase secreting ability has been attempted over the years.

The enzyme alpha-D-Gal (alpha-D-galactoside galactohydrolase, EC 3.2.1.22) is of interest for hydrolyzing the raffinose-type sugars found in soybeans. *Rhizopus oligosporus, Lactobacil-*

lus curvatus R08, Leuconostoc mesenteriodes, Lactobacillus fermentum, Bifidobacterium sp. et al. have been reported to be able to produce alpha-galactosidase [56, 66-70]. These microorganisms have been applied for soybean fermentation to reduce the oligosaccharides [56, 67-68, 71]. In [56],stachyose and raffinose decreased by 56.8% and 10%, respectively, in soybean during 48 hr fermentation by *Rhizopus oligosporus*. In *Leu.mesenteriodes JK55* and *L.curvatus R08* fermented soymilk, the non-digestive oligosaccharides were completely hydrolyzed after 18-24 h of fermentation [67].

Alpha-galactosidase has been isolated from plant and microbial sources, and its properties have been documented. In general, alpha-galactosidase acts upon gal-gal bonds in the tetrasaccharide stachyose, releasing galactose and raffinose, and also acts upon gal-glu bonds with the release of sucrose. Sucrose is, in turn, split by invertase, producing fructose and glucose [72]. So, α-galatosidase activity is noticeably dependent on the type of sugar. The type and concentration of the carbon source are known to be nutritional factors that regulate the synthesis of bacterial galactosidase. Reference [67] found that the existence of glucose and fructose inhibit the alpha-galatosidase expression both in *Lactobacillus curvatus R08* and *Leucomostoc mesenteriodes*.

2.6. Trypsin inhibitor

Protease inhibitors constitute around 6% of soybean protein [73]. Two protein protease inhibitors have been isolated from soybeans. The Kunitz trypsin inhibitor has a specificity directed primarily toward trypsin and a molecular weight of about 21.5 kDa. The Bowman-Birk (BB) inhibitor is capable of inhibiting both trypsin and chymotrypsin at independent reactive sites and has a molecular weight of 7.8 kDa [74]. Trypsin and chymotrypsin, the two major proteolytic enzymes produced in the pancreas, belong to the serine protease class.

Trypsin inhibitors present in soybeans are responsible for growth depression by reducing proteolysis and by an excessive fecal loss of pancreatic enzymes rich in sulfur-containing amino acid which can not be compensated by dietary soy protein [75]. Trypsin inhibitors account for 30-50% of the growth inhibition effect [76]. Rats fed a raw soybean extract from which trypsin inhibitors had been inactivated showed improved growth performance when compared with control rats fed diets containing raw soybeans from which inhibitors had not been inactivated [77].

Diets with a trypsin inhibitor concentration of 0.77 mg/g and less did not reduce the growth of pigs according to reference [7]. And, research showed that after fermentation with *Aspergillus oryzae GB-107*, the trypsin inhibitor in soy meal was reduced from 2.70 mg/g to 0.42 mg/g [7]. In the *in vivo* experiment of reference [15], the activities of total protease and trypsin at the duodenum and jejunum of piglets fed with fermented soy meal increased because of the inactivation of trypsin inhibitor. Just as was mentioned above, protease produced during fermentation could degrade protein molecules into peptides and amino acids. The trypsin inhibitor may be degraded or modified during fermentation and lose its activity binding to trypsin.

2.7. Vitamin

The increased content of some vitamins or provitamins, both water-soluble and fat-soluble vitamins, such as riboflavin, niacin, vitamin B6, β-carotene et al., which are due to fungal metabolic activities,is one of the healthy and nutritional advantages of fermented soybean products and has been extensively examined.

Vitamin or provitamin formation during tempeh fermentation by *Rhizopus oligosporus, R.arrhizus, R. oryzae* and *R.slolonifer*, respectively, were studied [11, 78]. In ref [11], all of the fourteen Rhizopus strains used in the research could form riboflavin, vitamin B6, nicotinic acid, nicotinamide, ergosterol, with isolates of *R.oligosporus* the best vitamin formers. In ref [78], six of 14 Rhizopus strains were able to form β-carotene in significant amounts. Five of these six strains belonged to the species of *R.oligosporus* and one was identified as *R.stolonifer*. A newly fourfold increase in β-carotene from 0.6 ug/g dw to 2.2 ug/g dw could be detected between 34 and 48hr in fermentations with *R.oligosporus* strain. During this period the content of total carotenoids increased from 9.1 ug/g dw to 11.2 ug/g dw in the fermentations with *R.oligosporus* strain. Ergosterol is the vitamin D2 precursor. Vitamin D can be derived from two naturally occurring compounds: ergocalciferol (D2) and cholecalciferol (D3). Both forms have equal biological activities in humans. The fourteen strains could produce ergosterol. The ergosterol concentration could be up to 1610 ug/g dw after 96 hr fermentation.

Vitamin K is an essential cofactor for the posttranslational conversion of glutamic acid residues of specific proteins in the blood and bone into γ-carboxyglutamic acid (Gla). There are two naturally occurring forms of vitamin K, vitamins K1 and K2. Vitamin K1 (phylloquinone) is formed in plants. Vitamin K2 (menaquinone, MK) is primarily synthesized by bacteria [79].Menaquinone (MK) plays an important role in blood coagulation and bone metabolism [80]. Japanese fermented soybean product, Natto, has been regarded as a high content of MK source (about 6- 9 ug/g) and is found in everyday products. Natto produced by a mutated *B. subtilis* strain showed a much higher content of MK up to 12.98 ug/g [81].Aromatic amino acids (phenylalanine, tyrosine, and trypstophan) could slow down the MK synthesis rate in cheonggukjang by using *Bacillus amyloliquefaciens KCTC11712BP*, while the supplement of 4% glycerol could signicantly increase its yield [82].

3. Conclusion

After fermentation by GRAS microorganisms, the anti-nutritional factors in soybean or soy meal are totally degraded, including oligosaccharides, trypsin inhibitor and phytic acid. Fermentation could also degrade large soy protein into peptides and amino acids, therefore, removing the allergenic effect of soy protein. Nutritional factors are formed during fermentation along with removal of undesirable factors. Functional peptides, such as peptides with ACE inhibitory activity are created by protein degradation. Isoflavones are converted to their functional forms, the aglycones. Antioxidant activity is enhanced, contributed by the increase of short chain peptides and phenolic compounds. Certain vitamins or provitamins are formed such as riboflavin, β-carotene, vitamin K2 and ergosterol. Total nutrition-al profiles of soybean and soy meal are greatly enhanced by fermentation.

Acknowledgements

The authors are grateful to the Kansas Soybean Commission and the Department of Grain, Science and Industry, Kansas State University, for funding this project. This chapteris contri-bution no 13-041-B from the Kansas Agricultural Experiment Station, Manhattan, KS 66506

Author details

Liyan Chen[1], Ronald L. Madl[1], Praveen V. Vadlani[1*], Li Li[2] and Weiqun Wang[3]

*Address all correspondence to: vadlani@ksu.edu

1 Bioprocessing and Renewable Energy Laboratory, Department of Grain Science and Indus-try, Kansas State University, Kansas, USA

2 Department of Food Science, South China University of Technology, Guangzhou, China

3 Department of Human Nutrition, Kansas State University, Manhattan, Kansas, USA

References

[1] American Soybean Association. (2012). Soy Stats. http://www.soystats.com/2012, ac-cessed 1st May 2012).

[2] Maeda, H., & Nakamura, A. (2009). Soluble soybean polysaccharide. In: Phillips GO, Williams PA., *Handbook of Hydrocolloids*, New York: CRC Press;, 693-709.

[3] Young, V. R. (1991). Soy protein in relation to human protein and amino acid nutri-tion. *Journal of American Dietetic Association*, 91(7), 828-835.

[4] Liu, K. (1997). Soybeans: Chemistry, Technology, and Utilization. New York: Chap-man & Hall Press;.

[5] Song, Y. S., Frias, J., Martinez-Villaluenga, C., et al. (2008). Immunoreactivity reduc-tion of soybean meal by fermentation, effect on amino acid composition and antige-nicity of commercial soy products. *Food Chemistry*, 108, 571-581.

[6] Chen, C. C., Shih, Y. C., Chiou, P. W. S., et al. (2010). Evaluating nutritional quality of single stage-and two stage-fermented soybean meal. *Asian-Australasian Journal of Ani-mal Science*, 23(5), 598-606.

[7] Hong, K. J., Lee, C. H., & Kim, S. W. (2004). Aspergillus oryzae GB-107 Fermentation Improves Nutritional Quality of Food Soybeans and Feed Soybean Meals. *Journal of Medical Food*, 7(4), 430-435.

[8] Kim, J., Hwang, K., & Lee, S. (2010). ACE Inhibitory and hydrolytic enzyme activities in textured vegetable protein in relation to the solid state fermentation period using Bacillus subtilis HA. *Food Science and Biotechnology*, 19(2), 487-495.

[9] Kim, M., Han, S., Ko, J., & Kim, Y. (2012). Degradation characteristics of proteins in Cheonggukjang (fermented unsalted soybean paste) prepared with various soybean cultivars. *Food Science and Biotechnology*, 21(1), 9-18.

[10] Omafuvbe, B. O., Abiose, S. H., & Shonukan, O. O. (2002). Fermentation of soybean (Glycine max) for soy-daddawa production by starter cultures of Bacillus. *Food Microbiology*, 19, 561-566.

[11] Bisping, B., Hering, L., Baumann, U., Denter, J., et al. (1993). Tempeh fermentation: some aspects of formation of γ-linolenic acid, proteases and vitamins. *Biotechnology Advances*, 11, 481-493.

[12] Low, A. G. (1980). Nutrient absorption in pigs. *Journal of the Science of Food and Agriculture*, 31, 1087-1130.

[13] Weng, T., & Chen, M. (2010). Changes of protein in Natto (a fermented soybean food) affected by fermenting time. *Food Science and Technology Research*, 16(6), 537-542.

[14] Weng, T., & Chen, M. (2011). Effect of two-step fermentation by Rhizopus oligosporusrus and Bacillus subtilis on protein of fermented soybean. *Food Science and Technology Research*, 17(5), 393-400.

[15] Feng, J., Liu, X., Xu, Z. R., et al. (2006). The effect of Aspergillus oryzae fermented soybean meal on growth performance, digestibility of dietary components and activities of intestinal enzymes in weaned piglets. *Animal Feed Science and Technology*, 134, 295-303.

[16] Frias, J., Song, Y. S., Martinez-Villaluenga, C., et al. (2008). Immunoreactivity and amino acid content of fermented soybean products. *Journal of Agricultural and Food Chemistry*, 56, 99-105.

[17] Zhang, J. H., Tatsumi, E., Ding, C. H., & Li, L. T. (2006). Angiotensin converting enzyme inhibitory peptides in douchi, a Chinese traditional fermented soybean product. *Food Chemistry*, 98, 551-557.

[18] Kim, S. K., Choi, Y. R., Park, P. J., et al. (2000). Purification and characterization of antioxidative peptides from enzymatic hydrolysate of cod teiset protein. *J. Korean Fish. Soc*, 33, 198-204.

[19] Saxena, P. R. (1992). Interaction between the renin-angiotensin-aldosterone and sympathetic nervous systems. *Journal of Cardiovascular Pharmacology*, 19, 6-8.

[20] Okamoto, A., Hanagata, H., Kawamura, Y., & Yanagida, F. (1995). Antihypertensive substances in fermented soybean, natto. *Plant Foods for Human Nutrition, 47,* 39-47.

[21] Shin, Z. I., Nam, H. S., Lee, H. J., Lee, H. J., & Moon, T. H. (1995). Fractionation of angiotensin converting enzyme (ACE) inhibitory peptides from soybean paste. *Korean Journal of Food Science and Technology, 27,* 230-234.

[22] Oka, S., & Nagata, K. (1974). Isolation and characterization of neutral peptides in soy sauce. *Agricultural and Biological Chemistry, 38,* 1185-1194.

[23] Teranaka, T., Ezawa, M., Matsuyama, J., Ebine, H., & Kiyosawa, I. (1995). Inhibitory effects of extracts from rice-koji miso, barley-koji miso, and soybean-koji miso on activity of angiotensin I converting enzyme. *Nippon Nogeik. Kaishi, 69,* 1163-1169.

[24] Kim, J. E., Hwang, K., & Lee, S. P. (2010). ACE inhibitory and hydrolytic enzyme activities in textured vegetable protein in relation to the solid state fermentation period using Bacillus subtilis HA. *Food Science and Biotechnology, 19*(2), 487-495.

[25] Gibbs, B. F., Zougman, A., Masse, R., & Mulligan, C. (2004). Production and characterization of bioactive peptides from soy hydrolysate and soy-fermented food. *Food Research International, 37*(2), 123-131.

[26] Sun, P., Li, D. F., Dong, B., et al. (2008a). Effects of soybean glycinin on performance and immune function in early weaned pigs. *Archives of Animal Nutrition, 62,* 313-321.

[27] Guo, P., Piao, X., Ou, D., et al. (2007). Characterization of the antigenic specificity of soybean protein b-conglycinin and its effects on growth and immune function in rats. *Archives of Animal Nutrition, 61*(3), 189-200.

[28] Kim, S. W. (2010). Bio-fermentation technology to improve efficiency of swine nutrition. *Asian-Australasian Journal of Animal Science, 23*(6), 825-832.

[29] Kobayashi, M. (2005). Immunological functions of soy sauce: hypoallergenicity and antiallergicantivity of soy sauce. *Journal of Bioscience and Bioengineering, 100*(2), 144-151.

[30] Penas, E., Restani, P., Ballabio, C., et al. (2006). Assessment of the residual immunoreactivity of soybean whey hydrolysates obtained by combined enzymatic proteolysis and high pressure. *European Food Research and Technology, 222,* 286-290.

[31] Liu, X., Feng, J., Xu, Z., et al. (2007). The effects of fermented soybean meal on growth performance and immune characteristics in weaned piglets. *Turkish Journal of Veterinary and Animal Sciences, 31*(5), 341-345.

[32] Dreau, D., & Lalles, J. P. (1999). Contribution to the study of gut hypersensitivity reactions to soybean proteins in preruminant calves and early-weaned piglets. *Livestock Production Science, 60,* 209-218.

[33] Gu, X., & Li, D. (2004). Effect of dietary crude protein level on villous morphology, immune status and histochemistry parameters of digestive tract in weaning piglets. *Animal Feed Science and Technology, 114,* 113-126.

[34] Dreau, D., Lalles, J. P., Philouzerome, V., et al. (1994). Local and systemic immune-responses to soybean protein ingestion in early-weaned pigs. *Journal of Aanimal Science*, 72, 2090-2098.

[35] Sun, P., Li, D. F., Li, Z. J., et al. (2008b). Effects of glycinin on IgE-mediated increase of mast cell numbers and histamine release in the small intestine. *The Journal of Nutritional Biochemistry*, 19, 627-633.

[36] Feng, J., Liu, X., Xu, Z. R., et al. (2007). Effect of fermented soybean meal on intestinal morphology and digestive enzyme activities in weaned piglets. *Digestive Diseases and Sciences*, 52, 1845-1850.

[37] Masilamani, M., Wei, J., & Sampson, H. A. (2012). Regulation of the immune response by soybean isoflavones. *Immunologic Research*, in press.

[38] Da, Silva. L. H., Celeghini, R. M. S., & Chang, Y. K. (2011). Effect of the fermentation of whole soybean flour on the conversion of isoflavones from glycosides to aglycones. *Food Chemistry*, 128, 640-644.

[39] Kwon, D. Y., Hong, S. M., Ahn, S., et al. (2011). Isoflavonoids and peptides from meju, long-term fermented soybeans, increase insulin sensitivity and exert insulinotropic effect in vitro. *Nutrition*, 27, 244-252.

[40] Yamabe, S., Kobayashi-Hattori, K., Kaneko, K., Endo, H., & Takita, T. (2007). Effect of soybean varieties on the content and composition of isoflavone in rice-koji miso. *Food Chemistry*, 100, 369-374.

[41] Chung, I., Seo, S., Ahn, J., et al. (2011). Effect of processing, fermentation, and aging treatment to content and profile of phenolic compounds in soybean seed, soy curd and soy paste. *Food Chemistry*, 127, 960-967.

[42] Cho, K. M., Hong, S. Y., Math, R. H., Lee, J. H., et al. (2009). Biotransformation of phenolics (isoflavones, flavanols and phenolic acids) during the fermentation of cheonggukjang by Bacillus pumilus HY1. *Food Chemistry*, 114, 413-419.

[43] Willcox, J. K., Ash, S. L., & Catignani, J. L. (2004). Antioxidants and prevention of chronic disease. *Critical reviews in Food Science and Nutrition*, 44, 275-295.

[44] Lin, C., Wei, Y., & Chou, C. (2006). Enhanced antioxidative activity of soybean koji prepared with various filamentous fungi. *Food Microbiology*, 23, 628-633.

[45] Moktan, B., Saha, J., & Sarkar, P. K. (2008). Antioxidant activities of soybean as affected by Bacillus-fermentation to kinema. *Food Research International*, 41, 986-593.

[46] Shahidi, F., Janitha, P. K., & Wanasundara, P. D. (1992). Phenolic antioxidants. *Critical Reviews in Food Science and Nutrition*, 32, 67-103.

[47] Wojdylo, A., & Oszmainski, J. (2007). Comparison of the content phenolic acid, alpha-tocopherol and the antioxidant activity in oat naked and weeded. *Electronic Journal of Environmental, Agricultural and Food Chemistry*, 6(4), 1980-1988.

[48] Yuan, X., Wang, J., & Yao, H. (2006). Production of feruloyl oligosaccharides from wheat bran insoluble dietary fibre by xylanases from Bacillus subtilis. *Food Chemistry*, 95, 484-492.

[49] Mc Cue, P., & Shetty, K. (2003). Role of carbohydrate-cleaving enzymes in phenolic antioxidant mobilization from whole soybean fermented with Rhizopusoligosporus. *Food Biotechnology*, 17(1), 27-37.

[50] Amadou, I., Le Shi, G., et, Y., & al, . (2011). Reducing, radical scavenging, and chelation properties of fermented soy protein meal hydrolysate by Lactobacillus plantarum LP6. *International Journal of Food Properties*, 14, 654-665.

[51] Chen, H., Muramoto, K., & Yamauchi, F. (1995). Structural analysis of antioxidative peptides from soybean β-conglycinin. *Journal ofAgricutural and Food Chemistry*, 43, 574-578.

[52] Saito, K., Jin, D., Ogawa, T., & Muramoto, K. (2003). Antioxidative properties of tripeptide libraries prepared by the combinatorial chemistry. *Journal of Agricultural and Food Chemistry*, 51, 3668-3674.

[53] Oatway, L., Vasanthan, T., & Helm, J. H. (2001). Phytic acid. *Food Reviews International*, 17(4), 419-431.

[54] Wang, H. L., Swain, E. W., & Hesseltine, C. W. (1980). Phytase of molds used in oriental food fermentation. *Journal of Food Science*, 45, 1262-1266.

[55] Sudarmadji, S., & Markakis, P. (1977). The phytate and phytase of soybean tempeh. *Journal of the Science of Food and Agriculture*, 28, 381-383.

[56] Egounlety, M., & Aworth, O. C. (2003). Effect of soaking, dehulling, cooking and fermentation with Rhizopus oligosporus on the oligosaccharides, trypsin inhibitor, phytic acid and tannins of soybean (Glycine max Merr.), cowpea (Vignaunguiculata L. Walp) and groundbean (Macrotyloma geocarpa) Harm . *Journal of food engineering*, 56, 249-254.

[57] Kim, S. S., Galaz, G. B., Pham, et., & al, . (2009). Effects of dietary supplementation of a meju, fermented soybean meal, and Aspergillus oryzae for Juvenile Parrot Fish (Oplegnathusfasciatus). *Asian-Australasian Journal of Animal Science*, 22(6), 849-856.

[58] Matsui, T., Hirabayashi, M., Iwama, Y., et al. (1996). Fermentation of soya-bean meal with Aspergillus usami improves phosphorus availability in chicks. *Animal Feed Science and Technology*, 60(1-2), 131-136.

[59] Hirabayashi, M., Matsui, T., & Yano, H. (1998a). Fermentation of soybean meal with Aspergillus usami improves zinc availability in rats. *Biological Trace Element Research*, 61(2), 227-234.

[60] Hirabayashi, M., Matsui, T., Yano, H., et al. (1998b). Fermentation of soybean meal with Aspergillusususamii reduces phosphorus excretion in chicks. *Poultry Science*, 77, 552-556.

[61] Grieshop, C. M., Kadzere, C. T., Clapper, G. M., et al. (2003). Chemical and nutritional characteristics of United States soybeans and soybean meals. *Journal of Agricultural and Food Chemistry*, 51, 7684-7691.

[62] Krause, D. O., Easter, R. A., & Mackie, R. I. (1994). Fermentation of stachyose and raffinose by hindgut bacteria of the weanling pig. *Letters in Applied Microbiology*, 18, 349-352.

[63] Zhang, L. Y., Li, D. F., Qiao, S. Y., et al. (2003). Effects of stachyose on performance, diarrhoea incidence and intestinal bacteria in weanling pigs. *Archives of Animal Nutrition*, 57, 1-10.

[64] Karr-Lilienthal, L. K., Kadzere, C. T., Grieshop, C. M., et al. (2005). Chemical and nutritional properties of soybean carbohydrates as related to nonruminants: a review. *Livestock Production Science*, 97, 1-12.

[65] Parsons, C. M., Yhang, Y., & Araba, M. (2000). Nutritional evaluation of soybean meals varying in oligosaccharides content. *Poultry Science*, 79, 1127-1131.

[66] Nout, M. J. R., & Rombouts, F. M. (1990). Recent developments in tempeh research. *Journal of Applied Bacteriology*, 69(5), 609-633.

[67] Yoon, M. Y., & Hwang, H. (2008). Reduction of soybean oligosaccharides and properties of alpha-D-galactosidase from Lactobacillus curvatus R08 and Leuconostoc mesenteriodes JK55. *Food Microbiology*, 25, 815-823.

[68] Wang, Y. C., Yu, R. C., Yang, H. Y., & Chou, C. C. (2002). Growth and survival of bifidobacteria and lactic acid bacteria during the fermentation and storage of cultured soymilk drinks. *Food Microbiology*, 19, 501-508.

[69] Garro, M. S., de Valdez, G. F., Oliver, G., & de Giori, G. S. (1996). Purification of alpha-galactosidase from Lactobacilus ferment. *Journal of Biotechnology*, 45, 103-109.

[70] Leblanc, J. G., & Garro de, Giori. G. S. (2004a). Effect of pH on Lactobacillus fermentum growth, raffinose removal, alpha-galactosidase activity and fermentation products. *Applied Microbiology and Biotechnology*, 65, 119-123.

[71] Le Blanc, J. G., Garro, Silvestroni. A., Connes, C., Piard, J. C., Sesma, F., & de Giori, G. S. (2004b). Reduction of alpha-galactooligosaccharides in soyamilk by Latobacillus-fermentum CRL 722: in vitro and in vivo evaluation of fermented soyamilk. *Journal of Applied Microbiology*, 97, 876-881.

[72] Mulimani, V. II., Thippeswamy, S., & Ramalingam, S. (1997). Enzymatic degradation of oligosaccharides in soybean flour. *Food Chemistry*, 59, 279-282.

[73] Friedman, M., & Brandon, D. L. (2001). Nutritional and health benefits of soy proteins. *Journal of Agricultural and Food Chemistry*, 49(3), 1069-1086.

[74] Hocine, L. L., & Boye, J. I. (2007). Allergenicity of soybean: new developments in identification of allergenic proteins, cross-reactivities and hypoallergenization technologies. *Critical Reviews in Food Science and Nutrition*, 47, 127-143.

[75] Rackis, J. J. (1981). Gumbmann MR. Protease inhibitors: physiological properties and nutritional significance. In: Ory RL. (ed.) Antinutrrents and natural toxicants in foods., *USA: Food & Nutrition Press*, Inc;, 203-237.

[76] Isanga, J., & Zhang, G. (2008). Soybean bioactive components and their implications to health- a review. *Food reviews International*, 24, 252-276.

[77] Kakade, M. L., Hoffa, D. E., & Liener, I. E. (1973). Contribution of trypsin inhibitors to the deleterious effects of unheated soybeans fed to rats. *Journal of Nutrition*, 103, 1172.

[78] Denter, J., Rehm, H., & Bisping, B. (1998). Changes in the contents of fat-soluble vitamins and provitamins during tempeh fermentation. *International Journal of Food Microbiology*, 45, 129-134.

[79] Sato, T., Yamada, Y., Ohtani, Y., et al., & 200, . (2001). Production of Menaquinone (Vitamine K2)- 7 by Bacillus subtilis. *Journal of Bioscience and Bioengineering*, 91(1), 16-20.

[80] Schurgers, I. J., Knapen, M. H. J., & Vermeer, C. (2007). Vitamin K2 improves bone strength in postmenopausal women. *International Congress Series*, 1297, 179-187.

[81] Tsukamoto, Y., Ichisse, H., Kakuda, H., et al. (2000). Intake of fermented soybean (natto) increases circulating vitamin K2 (menaquinone-7) and γ-carboxylatedosteocalcin concentration in normal individuals. *Journal of Bone and Mineral Metabolism*, 18, 216-222.

[82] Wu, W., & Ahn, B. (2011). Improved Menaquinone (Vitamin K2) production in Cheonggukjang by optimization of the fermentation conditions. *Food Science and Biotechnology*, 20(6), 1585-1591.

Chapter 9

Soybean and Isoflavones – From Farm to Fork

Michel Mozeika Araújo,

Gustavo Bernardes Fanaro and

Anna Lucia Casañas Haasis Villavicencio

Additional information is available at the end of the chapter

http://dx.doi.org/10.5772/52609

1. Introduction

1.1. General

Soybean (*Glycine max* L. Merrill) were first grown as a crop in China about 5000 years ago [1] and have been widely consumed as folk medicines in China, India, Japan and Korea for hundreds of years [2]. Today is a major source of plant protein (70%) and oil (30%) and become a globally important crop [3,4]. Its nutrients become basic for humans consumption, beyond its by-products, that offer great diversities of products to the food industry. Soybean oil is highly consumed world-wide and soy milk is often used as a milk substitute to people who have lactose intolerance [5]. In addition soybean has phytoestrogens which can be used in replacement to women hormone [6]. Usage oil can be reused in several forms, including as a fuel source [7,8].

There are many kinds of soybean cultivars with different biological composition and economic values. According to the consensus recommendations of the Organization for Economic Cooperation and Development, soybean nutrients (such as amino acids, fatty acids, isoflavones) and antinutrients (such as phytic acid, raffinose and stachyose) are important markers in assessment the nutritional quality of soybean varieties [9].

Soybean production has expanded to most of continents and 90% of the world's soybean production is concentrated in tropical and semi-arid tropical regions which are characterized by high temperatures and low or erratic rainfall. In tropics, most of the crops are near their maximum temperature tolerance [10].

© 2013 Araújo et al.; licensee InTech. This is a paper distributed under the terms of the Creative Commons Attribution License (http://creativecommons.org/licenses/by/3.0), which permits unrestricted use, distribution, and reproduction in any medium, provided the original work is properly cited.

USA is the largest soybean producer in the world, following by Brazil and Argentina. World soybean harvest production reached 264.25 million metric tons in 2010/2011. USA, Brazil and Argentina were responsible to about 92.75% of all production [11].

Soybeans have been appreciated by consumers as a health-promoting food [12]. Soybeans and soy products are wide consumed by Asian populations and are encouraged for western diets because of its nutritional benefits. Soy products are abundant in traditional Asian diets which daily intake is from 7–8 g/day in Hong-Kong and China, up to 20–30 g/day in Korea and Japan. Most of Europeans and North Americans, however, consume less than 1 g/day [13].

The soybean consumption will depend on grains characteristics, for example large seed sizes with high sucrose content are desirable for the production of vegetable soybean, which is harvested at immature stage, also called edamame. On the other hand, cultivars with small seed size and low calcium contents are desirable for natto, a traditional fermented soy food in Japan with a firmer texture. For soymilk and tofu production, soybeans with light hilum color, large seed size, high water absorption, high protein and sucrose contents and low oligosaccharides contents are desirable [14].

1.2. Composition

Soybean's unique chemical composition place this food products as one of the most economical and valuable agricultural commodity. Among cereal and other legume species, it has the highest protein content (around 40%) and the second highest (20%) oil content among all food legumes, after peanuts. Other valuable components found in soybeans include phospholipids, vitamins and minerals. Furthermore, soybeans contain many minor substances known to be biologically active including oligosaccharides, phytosterols, saponins and isoflavones. The actual composition of the whole soybean and its structural parts depends on many factors, including varieties, growing season, geographic location and environmental stress [12,15].

1.2.1. Macronutrients

On the average, oil and protein together constitute about 60% of dry soybeans. The remaining dry matter is composed of mainly carbohydrates (about 35%) and ash (5%) [15].

Soybeans store their lipids in an organelle known as oil bodies or lipid-containing vesicles. Their lipids are mainly in the form of triglycerides or triacylgycerols, with varying fatty acids as part of their structure. Triglycerides are neutral lipids, each consisting of three fatty acids and one glycerol that bound to three acids. Both saturated and unsatured can occur in the glycerides of soybean oil, however the fatty acids of soybean oil are primarily unsaturated [15,16]. The highest percentage of fatty acid in soybean oil is linoleic acid, following in decreasing order by oleic, palmitic, linolenic and stearic acid. Soybean oil contains some minor fatty acids, including arachidic, behenic, palmitoleic and myristic acid [16].

Protein content varies between 36% and 46% depending on the variety [17-19]. This component is present in the greatest amount in soybeans. Seed proteins are usually classified based on biological function in plants (metabolic proteins and storage proteins) and on solubility patterns. According to their functionality, metabolic proteins (such as enzymatic and structural) are concerned in normal cellular activities, including even the synthesis of storage proteins. Storage proteins, together with reserve of oils, are synthesized during soybean seed development and provide a source of nitrogen and carbon skeletons for the development seedling. In soybeans, most of proteins are storage type. A solubility pattern divides proteins into those soluble in water (albumins) and in salt solution (globulins). Globulins are further divided in legumins and vicilins. Under this classification system, most of soy protein is globulin. Certain soy proteins have their trivial names, as glycinin (legumins) and conglycinin (vicilins). Others, particularly those with enzymatic function, are based on the biological function of proteins themselves. Examples include hemagglutinin, trypsin inhibitors and lipoxygenases. A solubility classification can pose problems because an association with other proteins can change their solubility profile. Thus, a more precise means of identifying proteins has been based on approximate sedimentation coefficient using ultracentrifugation to separate seed proteins. Under appropriate buffer conditions, soy proteins exhibit four fractions after centrifugation. These fractions are designed as 2, 7, 11 and 15S. The major portion of the protein component is formed by storage proteins such as 7S globulin (β-conglycinin) and 11S globulin (glycinin), which represent about 80% of the total protein content [17]. Other proteins or peptides present in lower amounts include enzymes such as lipoxygenase, chalcone synthase and catalase.

On average, moisture-free soybeans contain about 35% of carbohydrates and defatted dehulled soy grits and flour contain about 17% soluble and 21% insoluble carbohydrates. Therefore, they are the second largest component in soybeans. However, the economical value of soy carbohydrates is considered much less important than soy protein and oil. A limited use of soybeans in human diet is due the flatucence produced by soluble carbohydrates such as raffinose and stachyose. Humans lack the enzymes to hydrolyze the galactosidic linkages of raffinose and stachyose to simpler sugars, so the compounds enter the lower intestinal tract intact, where they are metabolized by bacteria to produce flatus [16]. As a result, relatively fewer efforts have been made to study soy carbohydrates and their potential utilization. The principal use of soybean carbohydrate has been in animal feeds (primarily ruminant because they can digest the compound better than monogastric animals) where it contributes calories to the diet. Food processing can alter the carbohydrates composition making them more digestible to human organism. Although the presence of these oligosaccharides is generally considered undesirable because of their flatus activity, some studies shown some beneficial effects of dietary oligosaccharides in humans, mainly due to a increasing population of indigenous bifidobacteria in colon, such as: suppressing the activity of putrefactive bacteria by antagonist effect; preventing pathogenic and autogenous diarrhea; anti-constipation due to the production of high levels of short-chain fatty acids; producing nutrients such as vitamins. Other positive effects are: toxic metabolites and detrimental enzymes reduction; protecting liver function due to reduction of toxic metabolites; blood pressure reduction; anticancer effects [15,16].

1.2.2. Micronutrients

Soybean also contains a wide range of micronutrients and phytochemicals including minerals, vitamins, phytic acid (1.0–2.2%), sterols (0.23–0.46%) and saponins (0.17–6.16%) [20].

The primary inorganic compounds of the soybeans are minerals. Potassium is found in the highest concentration, followed by phosphorus, magnesium, sulfur, calcium, chloride and sodium, which vary in concentration to the variety, growing location and season. Minor minerals include silicon, iron, zing, manganese, copper and other [15,16].

Both water-soluble and oil-soluble vitamins are present in soybeans. Water-soluble vitamins in soybeans include thiamin, riboflavin, niacin, panthotenic acid and folic acid. They are not substantially lost during oil extraction and subsequently toasting of flakes. The oil-soluble vitamins present in soybeans are vitamins A and E. Vitamin E is especially unstable during soybean processing. During solvent extraction of soybeans, vitamin E goes with oil [15,16].

Phytate is the calcium-magnesium-potassium salt of inositol hexaphosphoric acid commonly known as phytic acid. As in the most seeds, phytate is the principal source of phosphorus in soybeans. His content depends on not only variety, but also growing conditions and assay methodology [15,16].

One important group of minor compounds present in soybean that has received considerable attention is a class of phytoestrogen called the isoflavones. Phytoestrogens are non-steroidal compounds that bind to and activate estrogen receptors (ERs) α and β, due to the fact that they mimic the conformational structure of estradiol. Phytoestrogens are naturally occurring plant compounds found in numerous fruits and vegetables, and are categorized into three classes: the isoflavones, lignans and coumestans [21,22]. Isoflavone compounds have been considered as nonnutrients, because they neither yield any energy nor function as vitamins. However, they play significant roles in the prevention of several diseases, so they may considered health-promoting substances.

Isoflavones belong to a group of compounds that share a basic structure consisting of two benzyl rings joined by a three-carbon bridge, which may or may not be closed in a pyran ring. This group of compounds is known as flavonoids, which include by far the largest and most diverse range of plant phenolics [15,16].

Isoflavones are present in just a few botanical families, because of the limited distribution of the enzyme chalcone isomerase, which converts 2(R)-naringinien, a flavone precursor, into 2-hydroxydaidzein. The soybean is unique in that it contains the highest amount of isoflavones, being up to 3 mg.g-1 dry weight [15,23].

The isoflavones in soybeans and soy products are of three types, with each type being present in four chemical forms. Therefore there are twelve isomers of isoflavones.

Isoflavones in soybean are mainly found as aglycones (daidzein, genistein, glycitein) (Figure 1), β-glucosides (daidzin, genistin, glycitin), malonyl-β-glucosides (6″-O-malonyldaidzin, 6″-O-malonylgenistin, 6″-O-malonylglycitin) and acetyl-β-glucosides (6″-O-acetyldaidzin, 6″-O-acetylgenistin, 6″-O-acetylglycitin) (Figure 2) [24].

Compounds	R₁	R₂
Daidzein	H	H
Glycitein	H	OCH₃
Genistein	OH	H

Figure 1. Chemical structure of soy aglycones.

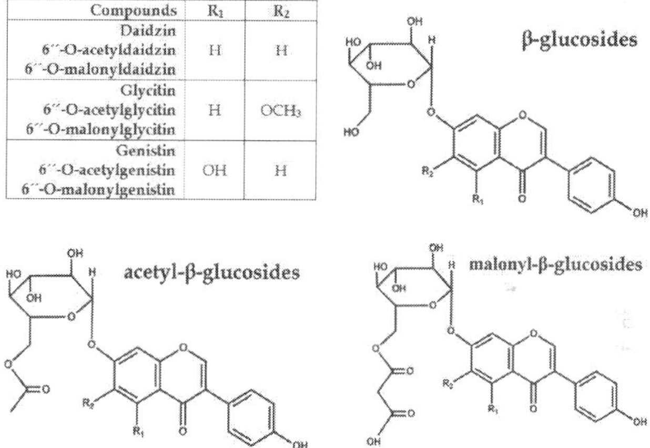

Compounds	R₁	R₂
Daidzin		
6″-O-acetyldaidzin	H	H
6″-O-malonyldaidzin		
Glycitin		
6″-O-acetylglycitin	H	OCH₃
6″-O-malonylglycitin		
Genistin		
6″-O-acetylgenistin	OH	H
6″-O-malonylgenistin		

Figure 2. Chemical structure of β-glucosides soybeans isoflavones.

Aglycones are flavonoid molecules without any attached sugars or other modifiers. Among the different forms of isoflavones, aglycones are especially important because they are readily bioavailable to humans [24]. Flavonoid β-glucosides also may carry additional small molecular modifiers, such as malonyl and acetyl groups. Sugar-linked flavonoids are called glycosides. The term glucoside only applies to flavonoids linked to glucose [25]. The malonyl-β-glucosides are the predominant form in conventional raw soybean [24].

Isoflavones concentration and composition vary greatly with structural parts within soybean seed. The concentration of total isoflavones in soybean hypocotyl is 5.5-6 times higher than that in cotyledons. Glycitein and its three derivatives occur extensively in the hypocotyl. Isoflavones are almost absent in seed coats [15]. The isoflavone content varies among soybean varieties, but most varieties contain approximately 100–400 mg per 100 g dry basis [26].

1.3. Isoflavones importance in human health

Isoflavones have received much attention because of their weak estrogenic property and other beneficial functions [24]. A large number of researchers have reported the positive aspects of isoflavones on human health, such as the ability to reduce the risk of cardiovascular, atherosclerotic and haemolytic diseases; alleviation of osteoporosis, menopausal and blood-cholesterol related symptoms; inhibition of the growth of hormone-related human breast cancer and prostate cancer cell lines in culture; and increased antioxidant effect in human subjects [27-30]. Isoflavones are also known for having anticancer activity and an effect on cell cycle and growth control [31-33].

Isoflavones in glycosides forms are poorly absorbed in the small intestine, due to their higher molecular weight and hydrophilicity. However bacteria in the intestine wall can biologically activate by action of β-glucosidase to their corresponding bioactive aglycone forms. Once hydrolyzed, aglycone forms are absorbed in the upper small intestine by passive diffusion. Nevertheless, pharmacokinetic studies confirm that healthy adults absorb isoflavones rapidly and efficiently. The average time to ingested aglycones reach peak plasma concentrations is about 4–7 h, which is delayed to 8–11 h for the corresponding β-glycosides. Despite the fast absorption, isoflavones or their metabolites are also rapidly excreted [13].

1.4. Food processing effect in isoflavones

Several investigations have been performed during the last years on soybeans consumption and their benefits, however the effects of seed processing and soybean processing into foods on the distribution of soy isoflavones are sparse. Processing significantly affects the retention and distribution of isoflavone isomers in soyfoods. The conversion and loss of isoflavones during processing significantly affect the nutraceutical values of soybean. Post-harvest changes in isoflavones in soybeans are influenced by processing methods however genotype has an effect on isoflavones profiles during seed development [34] and the environment has a greater e•ect than the genotype [26,35-37]. Distribution of isoflavones in soybean can be altered during various processing steps including fermentation, cooking, frying, roasting, drying and storage [24]. These effects in isoflavones can be even accelerated by heat, acid, alkaline and enzymes [38-40] (Figure 3). The effects of several food processing techniques on soybeans isoflavones will be reviewed and discussed in the following section.

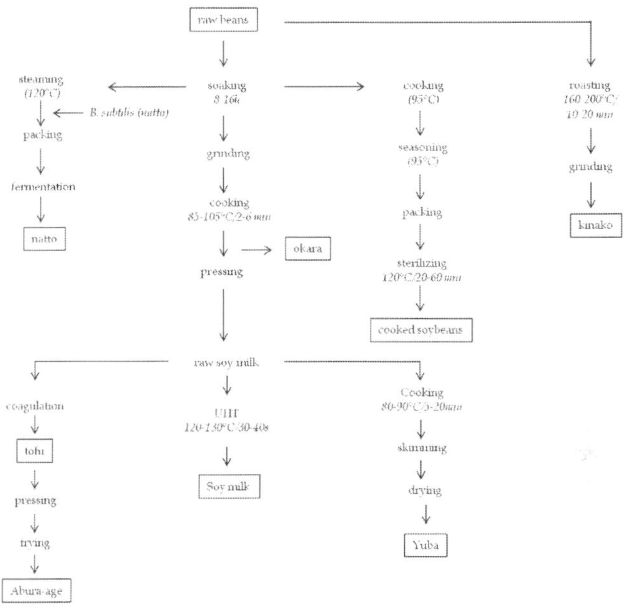

Figure 3. Flow diagram illustrating the processing of soybeans to commercial soybeans products. Adapted from Toda *et al.* [41].

1.4.1. Harvesting

Harvesting soybean seeds after their development and maturation is a critical step in profitable soybean production. Although most soybeans are harvested at the dry mature stage, a very small portion is harvested at the immature stage in certain regions. Immature soybeans refer to soybeans harvested at 80% maturity. The immature seed is used as a vegetable or as ingredient recipes. Soybeans are considered dry mature when seed moisture reduces to less than 14% in the field [15,42]. At this stage, seeds are ready for harvesting. The exact harvesting date depends on the variety, growing regions, plating date and local weather conditions [15].

1.4.2. Drying

Drying is an important process to extend the shelf life or to prepare food, including soybean, for subsequent production [43]. After harvest, if moisture content is more than 14%, soybeans need to be dried immediately in order to the following reasons: meet the quality standard of soybean trading; retain maximum quality of the grain; reach a level of moisture that does not allow the growth of bacteria and fungi and; prevent germination of seeds. Drying could be done naturally or artificially. Sun drying, or natural drying, soybeans are spread on the threshing for 2-3 days with frequently turning between top to bottom layers.

Once dried, seeds are transferred to storage facilities. Sun drying is not suitable for large quantities of soybeans or under humid and cloudy weather conditions. Artificially drying is carried out with various mechanical driers, including low temperature driers, on-the-floor driers, in-bin driers, medium temperature driers, tray driers, multiduct ventilated driers, countercurrent open-flame grain driers and solar driers. Regardless of which driers are used, caution is required so as to avoid too rapid drying; rapid drying hardens outlayers of seeds and seals moisture within the inner layer. Although the temperature of soybeans must be raised sufficiently to achieve the desired moisture content during drying process, excess heating (not exceed 76%) should be avoided to protect beans from discoloration and beans proteins from denaturation [15].

It is well known that drying significantly affects the quality and nutrients of dried food products. Drying temperatures affect the activity and stability due to chemical and enzymatic degradation which can for example alter significantly the isomeric distribution of the 3 aglycones [44]. All forms of isoflavones are generally lost due to thermal degradation and oxidation reactions during processing [45]. Conventional thermal treatment decreases malonyl derivatives into β-glucosides via intra-conversion while aglycones have higher heat resistance. Dry heat treatment such as frying, toasting, or baking process increases the formation of acetyl derivatives of isoflavones through decarboxylation from malonyl derivatives [24,41].

A comparison between freeze-drying and drum-drying of germ flour demonstrated that the former contained higher isoflavone aglycones than the latter. However, isoflavone glucoside contents in freeze-dried germ flour were lower than those of drum-dried germ flour. The content of isoflavone glucosides was significantly lower in processed (drum- and freeze-dried) germ flours compared with that of unprocessed germ flour because of conversion of isoflavone glucosides to isoflavone aglycones. Total isoflavone contents of drum-dried and freeze-dried germ flours were comparable but more than that of unprocessed germ flour [46].

Different drying methods (hot-air fluidized bed drying, HAFBD; superheated-steam fluidized bed drying, SSFBD; gas-fired infrared combined with hot air vibrating drying, GFIR-HAVD) were compared at various drying temperatures (50, 70, 130 and 150 °C). Higher drying temperatures led to higher drying rates and higher levels of β-glucosides and antioxidant activity, but to lower levels of malonyl-β-glucosides, acetyl-β-glucosides and total isoflavones. Comparing different drying methods to each other, at the same drying temperature GFIR-HAVD resulted in the highest drying rates and the highest levels of β-glucosides, aglycones and total isoflavones, antioxidant activity as well as α-glucosidase inhibitory activity of dried soybean. A drying temperature of 130 °C gave the highest levels of aglycones and α-glucosidase inhibitory activity in all cases [47].

Dehydration also helps to achieve longer shelf life and easier transportation and storage, enabling wider distribution of a product. Fermentation of soymilk with microorganisms with β-glucosidase activity promotes the biotransformation of isoflavone glucosides into bioactive aglycones, brings down the contents of aldehydes and alcohols responsible for the beany flavor in soy milk [48] and reduces the content of indigestible oligosaccharides. A spray-drying technology has been also applied to fermented soymilk. Daidzein and genistein retention were of about 87.6% and 85.3%, respectively. Retention was better at the lower inlet air temperatures. Coarser droplets formed at higher feed rates helped in higher retention of isoflavones because it reduced the incidence of direct exposure of isoflavones to higher temperature [49].

1.4.3. Storage

Soybeans are stored at farms, elevators and processing plants in various types of storage structures (steel tanks or concrete silos) before being channeled to next destination, and finally to processing. Loss in quality of soybeans during storage results from the biological activities of seeds themselves, microbiological activities and attacks of insects, mites and rodents. Quality loss is characterized by reduced seed viability and germination rate, coloration, reduced water absorption, compositional changes and ultimately reduced quality of protein and oil. Heat damage is a major cause of quality loss. Characterized by darkening of seed color coat, it results mainly from the improper control of temperature and moisture during storage and transportation. The excess presence of foreign matter can also cause heat buildup. Thus, cleaning soybeans before drying minimizes heat damage. Although minor losses are inevitable, major losses can be prevented by carefully control of storage temperature and humidity. Any biological activity requires a certain level of moisture present. Higher moisture content (or high moisture humidity) not only promotes bacteria and mold infection but also speeds up biological activity of seed themselves. Excessive moisture may also leed to seed germination. Generally, moisture content of 13.5% or below is considered to ensure storage stability of soybeans over reasonably long periods. However, this is true only when temperatures are kept below certain levels [15].

Storage conditions have also an important effect in the composition of soybeans, including isoflavone, anthocyanin, protein, oil and fatty acid.

Variation in isoflavone contents of different soybean cultivars were evaluated under different location and storage duration. Total isoflavone contents of soybeans stored for 1 year were only slightly higher than those of soybeans stored for 2 or 3 years. However, the concentrations of individual isoflavones, especially 6''-O-malonyldaidzin and 6''-O-malonylgenistin, decreased markedly in soybeans stored for 2 or 3 years, probably due to high temperatures during storage and oxidative reactions which transformed malonylated type to glycoside and aglycone groups, increasing their amounts with longer storage. They also pointed that the effect of crop year seemed to have a much greater influence on the variation in isoflavone content than did the location because of weather differences from one year to another. Differences among cultivars have been already expected [35]. Similar results were found comparing cropping year and storage for 3 years of soybeans seeds. Malonyldaidzin and malonylgenistin concentrations also decreased and the concentration of glucosides increased slightly over the 3 years [50]. Storage effect in soybeans isoflavones have been exhaustedly studied, and a markedly decreased in isoflavones content is proportional to storage periods, whereas protein, oil, and fatty acid of black soybeans showed a slight decrease over storage at room temperature [51].

A combined effect of storage temperature and water activity was evaluated on the content and profile of isoflavones of soy protein isolates and defatted soy flours. Storage for up to 1 year of soy products, at temperatures from -18 to 42 °C, had no effect on the total content of isoflavones, but the profile changed drastically at 42 °C, with a significant decrease of the percentage of malonyl glucosides with a proportional increase of β-glucosides. A similar effect was observed for soy protein isolates stored at aw = 0.87 for 1 month. For defatted soy flours, however, there was observed a great increase in aglycons (from 10 to 79%), probably due to the action of endogenous β-glucosidases [52].

1.4.4. Fermentation

Processing and fermentation of soybean has been reported to influence the forms isoflavones take. Studies have shown that the fermentation process of soybean promotes changes in the phytochemical compounds, causing changes in the isoflavone forms, hydrolysing the proteins and reducing the antinutritional factors, by reducing trypsin inhibitor content [53-55]. Fermentation process of soy leads to manufacturing different soy fermented foods, such as tempeh, soy extract, miso and natto.

Differences on isoflavones content between non-fermented and fermented soybean products have been extensively describe in literature in the last years. Isoflavone glucosides were the major components in soybean and non-fermented products, while isoflavone aglycones were abundant in sufu and partially in miso of soybean fermented products [56].

Tempeh is a traditional fermented soybean food product from Indonesia. It is normally consumed fried, boiled, steamed or roasted. A way to processing soybeans into tempeh occurs by fungus mediated fermentation. Several authors have already studied the effect of fermentation on isoflavones content. It was reported an increase of aglycones amount with fermentation time of tempeh, approximately two-fold higher after 24 h fermentation [57]. Later on, similar results were found by Haron *et al.* [58] who reported higher values of aglycone forms in raw tempeh. In addition, these researchers showed that fried tempeh had its total isoflavone content reduced in almost 50% during frying processing (reduction from 205 to 113 mg in 100 g of fried tempeh). A combined process of fermentation and refrigeration was evaluated by Ferreira *et al.* [25]. They quantified isoflavones content of two different soybean cultivars (low isoflavone content versus high isoflavone content cultivar) during processing of tempeh combined to refrigeration at 4 °C for 6, 12, 18, and 24 hours. After 24 hours fermentation, isoflavone glucosides were 50% reduced, and the aglycone forms in the tempeh from both cultivars was increased. The malonyl forms reduced 83% after cooking. Refrigeration process up to 24 hours did not affect the isoflavone profile of tempeh from either cultivar. The fermentation process improves the nutritional value of tempeh by increasing the availability of isoflavone aglycones. Fermented soy foods, which are usually prepared by mixing soy with other components such as barley, rice and wheat, contained isoflavones at lower concentrations. In addition these fermented soy foods contain predominantly isoflavone aglycones [15]. Fermentation with microorganisms or natural products containing high β-glucosidase activity converts β-glucosides into corresponding aglycones by breaking the carbohydrate bond [59,60]. These aglycones exist in smaller amounts in other nonfermented soy products such as tofu and soymilk [37,61]. Other soybean products or by-products showed a similar behavior. Soy pulp is generated as a by-product during tofu or soymilk production and is sometimes used as animal food but is mainly burnt as waste. Fermentation increased isoflavone aglycone contents in black soybean pulp. Genistein concentrations in black soybean pulp were 6.8 and 7.2 fold higher than controls respectively after 12 h and 24 h of fermentation with *L. acidophilus*. Fermentation with *B. subtilis* showed a similar genistein concentration increase [62]. The effect of fermentation of whole soybean flour was investigated and also showed a conversion of isoflavone glycosides to the aglycone form [63].

Sufu, a fermented tofu product, showed ambiguous changes in isoflavone contents during manufacturing. Sufu manufacturing procedure promoted a significant loss of isoflavone content mainly attributed to the preparation of tofu and salting of pehtze. The isoflavone composition was altered during sufu processing. The initial fermentation corresponded to the fastest period of isoflavone conversion. Aglycones levels increased while the corresponding levels of glucosides decreased. The changes in the isoflavone composition were significantly related to the activity of β-glucosidase during sufu fermentation, which was inhibited by the NaCl content [64]. These influence of processing and NaCl supplementation on isoflavone contents was also investigated during douchi manufacturing. Douchi is a popular Chinese fermented soybean food. These results indicated that 61% of the isoflavones in raw soybeans were lost when NaCl content was 10%. Indeed, changes in isoflavone isomer distribution were found to be related to β-glucosidase activity during fermentation, which was affected by NaCl supplementation [65].

Other fermented soybean products are miso and soy sauce. Their production involves the application of pressurized steaming following to fermentation by bacteria for a lengthy period. During the fermentation of miso and soy sauce, β-glucosides are also reported to be hydrolyzed to aglycons by β-glucosidase of bacteria [41].

Fermentation shows the same pattern on isoflavones transformation, no matter the fermented soybean product. The fungi-fermented black soybeans (koji) also contained a higher content of aglycone, the bioactive isoflavone, than did the unfermented black soybeans [66]. However, it was later realized that the contents of various isoflavone isomers in black soybean koji may reduce during after 120 days of storage. Although the retention of isoflavone varied with storage temperature, packaging condition enabled black soybean koji to retain the highest residual of isoflavone [67].

1.4.5. Non-fermentation processing

Tofu is a popular nonfermented soy food. Processing of tofu involves soaking and heating procedures as well as the addition of protein coagulants such as calcium sulfate to soymilk to coagulate to make tofu. This soybean product has been also target of several studies during its manufacturing. Results of the stability of isoflavone during processing of tofu showed that the concentrations of the three aglycones increased with increasing soaking temperature and time, while a reversed trend was found for the other nine isoflavones. Tofu produced with 0.3% calcium sulfate was found to contain the highest total isoflavones yield (2272.3 µg/g) whereas a higher level (0.7%) of calcium sulfate resulted in a lower yield

(1956.6 µg/g) of total isoflavones in tofu. In the same study, authors showed that during processing of soymilk, an increase of concentration for β-glucosides and acetyl genistin, whereas malonyl glucosides exhibited a decreased tendency and the aglycones did not show significant change [68]. Previous reports have already demonstrated that during soaking of soybean malonyl glucosides can be converted to acetyl glucosides, which can be further converted to glycosides or aglycones depending on soaking temperature and time [34,69].

Regarding soymilk isoflavones, a research was done on the transformation of isoflavones and the β-glucosidase activity in soymilk during fermentation. Regardless of employing a lactic acid bacteria or a bifidobacteria as starter organism, fermentation causes a major reduction in the contents of glucoside, malonyl glucoside and acetyl glucoside isoflavones along with a significant increase of aglycone isoflavones content. Indeed, the increase of aglycones and decrease of glucoside isoflavones during fermentation coincided with the increase of β-glucosidase activity observed in fermented soymilk [70].

1.4.6. Heating processing

Distribution of isoflavones in soybeans and soybeans products are significantly affected by the method, temperatures and duration of heating.

Toasted soy flour and isolated soy protein had moderate amounts of each of the isoflavone conjugates. Pananum *et al.* [44] evaluated the effect of longer toasting of defatted soy flakes at 150 °C. They stated that toasting led to higher aglycone concentration, which increased the total phenolic recovery. Apparently, malonyl glucoside conjugates are thermally unstable and are converted to their corresponding isoflavone glycosides at high temperatures [15]. The chemical modification of isoflavones in soy foods during cooking and processing studies showed interesting results on isoflavones stability. Baking or frying of textured vegetable protein at 190 °C and baking of soy flour in cookies did not alter total isoflavone content. However, there was a steady increase in β-glucoside conjugates at the expense of 6''-O-malonylglucosides conjugates [71]. The de-esterifying reaction was presumably a result of transesterification of the ester linkage between the malonate or acetate carboxyl group and the 6''-O-hydroxyl group of the glucose moiety, yielding methyl malonate or methyl acetate and the isoflavone glucoside [15].

Roasting has been used to deactivate anti-nutritional components in soybeans and to give characteristic flavour and brown color to final products [43]. Kinako is a soybean product produced by roasting raw soybeans around 200 °C for 10-20 min, following by grinding. Roasting also promotes changes in isoflavones contents. Soybeans roasted upon at 200 °C without prior soaking in water showed a change in isoflavones profile. It was found that at the first 10 min of roasting caused an increase in 6''-O-acetyl-β-glucosides while 6''-O-malonyl-β-glucosides decreased drastically. Continued roasting showed a slightly decreased proportion of the 6''-O-acetyl-β-glucosides. These authors proposed that most 6''-O-malonyl-β-glucosides were decarboxylated and changed to 6''-O-acetyl-β-glucosides when roasted at a high temperature. Besides, β-glucosides and aglycons also increased gradually over time [41]. A similar trend was found roasting soybeans at 200 °C for 7, 14 and 21 min. Malonyl derivatives decreased drastically and acetyl-β-glucosides and β-glucosides increased significantly [24].

Similar to the roasting process, explosive puffing caused a significant decrease in malonyl derivatives and significant increase in acetyl derivatives and β-glucosides through 686 kPa explosive puffing treatment. Otherwise, aglycones did not increase during the explosive puffing process. This fact was suggested due that temperature of explosive puffing may not be high enough to cleave glycosidic linkage between β-glucopyranose and aglycones [24].

Some soy products are prepared by frying process. Abura-age is one of them being produced by frying tofu in oil. As pointed by Toda *et al.* [41], In comparison with tofu, heating in oil at a high temperature to produce abura-age resulted in a smaller proportion of 6″-O-malonyl-β-glucosides and a higher proportion of 6″-O-acetyl-β-glucosides.

Simonne *et al.* [34] evaluated the retention and changes of soy isoflavones in immature soybeans seeds during processing and found total isoflavone retention percentages means of 46% after boiling, 53% after freezing and 40% after freeze-drying. They assumed that probably the loss of isoflavones could be due to their leaching into the cooking water, however these authors did not analyze isoflavones in the cooking water. In the same study, they noted that boiling process also caused a substantial increase in daidzin, genistin, and genistein. In a previous work on soy milk production and cooking of dry soy products, it was proposed that hydrolysis of the malonyl and acetyl glucosides during boiling probably contributes to the conversion of isoflavone forms [71]. Changes in isoflavone compositions of different soybean foods during cooking process were performed later and supported this proposition [41]. Concerning to a possible leaching effect of isoflavones during blanching or boiling, Wang *et al.* [72] reported about 26% retention of isoflavones during the production of soy protein isolate. This soybean product is made by extracting soy flour under slightly alkaline pH, followed by precipitation, washing, and drying. Soy protein isolate showed a different isoflavone profile in comparison to soy flour. The former contained much more aglycones (genistein and daidzein), while the latter had almost none [12,72]. The high content of aglycones in soy protein isolate was probably due to the hydrolysis of glycosides. The percentages of total isoflavones lost during extraction, precipitation, and washing were 19, 14, and 22%, respectively. Washing was the step where most isoflavones were lost [72]. This statement was supported by another study on thermal processing of tofu. It was demonstrated a significant total isoflavone content decrease, most likely due to leaching of isoflavones into the water [69].

An approach on the stability of isoflavones in soy milk stored at temperatures ranging from 15 to 90 °C showed that genistin in soy milk is labile to degradation during storage. Although the loss rate was low at ambient temperatures, authors highlighted the potential loss of genistin when estimating shelf life of soy milk products [73].

Influence of thermal processing such as boiling, regular steaming and pressure steaming were also investigated in yellow and black soybeans. Again, all thermal processing caused significant increases in aglycones and β-glucosides of isoflavones, but caused significant decreases in malonyl glucosides of isoflavones for both kinds of soybeans. The malonyl glucosides decreased dramatically with an increase in β-glucosides and aglycones after thermal processing [74].

1.4.7. Irradiation

Food irradiation is a process in which food is exposed to ionizing radiations such as high energy electrons and X-rays produced by machine sources or gamma rays emitted from the radioisotopes ^{60}Co and ^{137}Cs. Depending on the absorbed radiation dose, various effects can be achieved resulting in reduced storage losses, extended shelf life and/or improved micro-biological and parasitological safety of foods [75,76]. Food irradiation is one of the most effective means to disinfect dry food ingredients. Disisfestation is aimed at preventing losses caused by insects in stored grains, pulses, flour, cereals, coffee beans, dried fruits, dried nuts, and other food products. The dosage required for insect control is fairly low, of the order of 1 kGy or less [77,78].

Soybeans have been processed by ionizing radiation in order to improve their properties. An important improvement in microbiological properties, such as insect disinfestation and microbial contamination, can be achieved by radiation treatment. Physical properties are also enhanced, such as reduction of soaking and cooking time. Indeed, higher radiation doses may break glycosidic linkages in soybean oligosaccharides to produce more sucrose and decrease the content of flatulence causing oligosaccharides [79].

Gamma irradiation at 2.5-10.0 kGy caused the reduction of soaking time in soybeans by 2-5 hours and the reduction of cooking time by 30-60% compared to non-irradiated control samples. The irradiation efficacy on physical quality improvement was also recognized in stored soybeans for one year at room temperature [80,81].

Influence of radiation processing on isoflavones content has been also studied in the last years. Gamma-radiated (0.5-5.0 kGy) soybeans showed a radiation-induced enhancement of antioxidant contents. Interestingly, a decrease in content of glycosidic conjugates and an increase in aglycons were noted with increasing radiation doses. These results suggested a radiation-induced breakdown of glycosides resulting in release of free isoflavones. Whereas the content of genistein increased with radiation dose, that of daidzein showed an initial increase at a dose of 0.5 kGy and then decreased at higher doses. Degradation of daidzein beyond 0.5 kGy could thus be assumed. Glycitein appears to be the least stable among the three aglycons as its content decreased at all of the doses studied [82]. Gamma irradiation also induced an enhancement in isoflavones content of varying seed coat colored soybean up to a radiation dose of 0.5 kGy. However, the genotypes showed decrease in total isoflavone content at a higher radiation dose of 2.0 and 5.0 kGy [83].

Such as gamma-irradiation, an enhanced effect on soy germ isoflavones was found after electron beam processing. Interestingly, this study showed that applied radiation doses ranging from 1.0 to 20.0 kGy showed an increase in the amount of both glucosides and aglycones simultaneously [84].

Acknowledgements

We thank CAPES, IPEN and CNPq.

Author details

Michel Mozeika Araújo[1,2], Gustavo Bernardes Fanaro[1] and
Anna Lucia Casañas Haasis Villavicencio[1*]

*Address all correspondence to: mmozeika@yahoo.com

1 Nuclear and Energy Research Institute (IPEN), Brazil

2 Strasbourg University (UdS), France

References

[1] Liu, X., Jin, J., Wang, G., & Herbert, S. J. (2008). Soybean yield physiology and development of high-yielding practices in Northeast China. *Field Crops Research*, 105, 157-171.

[2] Jeng, T. L., Shih, Y. J., Wua, M. T., & Sung, J. M. (2010). Comparisons of flavonoids and anti-oxidative activities in seed coat, embryonic axis and cotyledon of black soybeans. *Food Chemistry*, 123, 1112-1116.

[3] Chan, C., Qi, C., Li, M. W., Wong, F. L., & Lam, H. M. (2012). Recent Developments of Genomic Research in Soybean. *Journal of Genetics and Genomics*, 39, 317-324.

[4] Tacarindua, C. R. P., Shiraiwa, T., Homma, K., Kumagai, E., & Sameshima, R. (2012). The response of soybean seed growth characteristics to increased temperature under near-field conditions in a temperature gradient chamber. *Field Crops Research*, 131, 26-31.

[5] Rosenthal, A., Deliza, R., Cabral, L. M. C., Cabral, L. C., Farias, C. A. A., & Domingues, A. M. (2003). Effect of enzymatic treatment and filtration on sensory characteristics and physical stability of soymilk. *Food Control*, 14, 187-192.

[6] Baber, R. (2010). Phytoestrogens and post reproductive health. *Maturitas*, 66, 344-349.

[7] Diya'uddeen, B. H., Abdul, Aziz. A. R., Daud, W. M. A. W., & Chakrabarti, M. H. (2012). Performance evaluation of biodiesel from used domestic waste oils: A review. *Process Safety and Environmental Protection*, 90, 164-179.

[8] Viola, E., Blasi, A., Valerio, V., Guidi, I., Zimbardi, F., Braccio, G., & Giordano, G. (2012). Biodiesel from fried vegetable oils via transesterification by heterogeneous catalysis. *Catalysis Today*, 179, 185-190.

[9] Jiao, Z., Si, X. X., Zhang, Z. M., Li, G. K., & Cai, Z. W. (2012). Compositional study of different soybean (Glycine max L.) varieties by 1H NMR spectroscopy, chromatographic and spectrometric techniques. *Food Chemistry*, 135, 285-291.

[10] Thuzar, M., Puteh, A. B., Abdullah, N. A. P., Lassim, M. B. M., & Jusoff, K. (2010). The Effects of Temperature Stress on the Quality and Yield of Soya Bean [(Glycine max L.) Merrill.]. *Journal of Agricultural Science*, 2, 172-179.

[11] United States Department of Agriculture (USDA). (2012). World Agricultural Supply and Demand Estimates:. *WASDE- 504*.

[12] Shao, S., Duncan, A. M., Yanga, R., Marconec, M. F., Rajcand, I., & Tsaoa, R. (2009). Tracking isoflavones: From soybean to soy flour, soy protein isolates to functional soy bread. *Journal of Functional Foods I*, 119-127.

[13] Cederroth, C. R., Zimmermann, C., & Nef, S. (2012). Soy, phytoestrogens and their impact on reproductive health. *Molecular and Cellular Endocrinology*, 355, 192-200.

[14] Saldivar, X., Wanga, Y. J., Chen, P., & Hou, A. (2011). Changes in chemical composition during soybean seed development. *Food Chemistry*, 124, 1369-1375.

[15] Liu, K. (1997). Soybeans: chemistry, technology, and utilization. *New York: Chapman & Hall*.

[16] Erickson, D. R. (1995). Practical Handbook of soybean processing and utilization. *USA: AOCS*.

[17] Garcia, M. C., Torre, M., Marina, M. L., & Labord, F. (1997). Composition and characterization of soyabean and related products. *Critical Reviews in Food Science and Nutrition*, 361-391.

[18] Grieshop, C. M., & Fahey Jr, G. C. (2001). Comparison of quality characteristics of soybeans from Brazil, China and the United States. *Journal of Agricultural and Food Chemistry*, 49(5), 2669-2673.

[19] Grieshop, C. M., Kadzere, C. T., Clapper, G. M., Flickinger, E. A., Bauer, L. L., Frazier, R. L., & Fahey Jr, G. C. (2003). Chemical and nutritional characteristics of United States soybeans and soybean meals. *Journal of Agricultural and Food Chemistry*, 51(26), 7684-7691.

[20] Kang, J., Badger, T. M., Ronis, M. J., & Wu, X. (2010). Non-isoflavone phytochemicals in soy and their health effects. *Journal of Agricultural and Food Chemistry*, 58(14), 8119-8133.

[21] Kuiper, G. G. J. M., Carlsson, B., Grandien, K., Enmark, E., Häggbla, J., Nilsson, S., & Gustafsson, J. A. (1997). Comparison of the Ligand Binding Specificity and Transcript Tissue Distribution of Estrogen Receptors α and β. *Endocrinology*, 138(3), 863-870.

[22] Kuiper, G. G., Lemmen, J. G., Carlsson, B., Corton, J. C., Safe, S. H., van der Saag, P. T., van der Burg, B., & Gustafsson, J. A. (1998). Interaction of estrogenic chemicals and phytoestrogens with estrogen receptor beta. *Endocrinology*, 139(10), 4252-4263.

[23] Rostagno, M. A., Palma, M., & Barroso, C. G. (2004). Pressurized liquid extraction of isoflavones from soybeans. *Analytica Chimica Acta*, 522, 169-177.

[24] Lee, S. W., & Lee, J. H. (2009). Effects of oven-drying, roasting, and explosive puffing process on isoflavone distribution in soybeans. *Food Chemistry*, 112, 316-320.

[25] Ferreira, M. P., Oliveira, M. C. N., Mandarino, J. M. G., Silva, J. B., Ida, E. I., & Car-rão-Panizzi, M. C. (2011). Changes in the isoflavone profile and in the chemical composition of tempeh during processing and refrigeration. *Pesquisa Agropecuária Brasileira*, 46(11), 1555-1561.

[26] Hoeck, J. A., Fever, W. R., Murphy, P. A., & Grace, A. W. (2000). Influence of genotype and environment on isoflavone concentrations of soybean. *Journal of Agricultural and Food Chemistry*, 40, 48-51.

[27] Kwak, C. S., Lee, M. S., & Park, S. C. (2007). Higher antioxidant properties of chungkookjang, a fermented soybean paste, may be due to increased aglycone and malonylglycoside isoflavone during fermentation. *Nutrition Research*, 27, 719-727.

[28] Lee-J, S., Kim-J, J., Moon-I, H., Ahn-K, J., Chun-C, S., Jung-S, W., et al. (2008). Analysis of isoflavones and phenolic compounds in Korean soybean [Glycine max (L.) Merrill] seeds of different seed weights. *Journal of Agricultural and Food Chemistry*, 56, 2751-2758.

[29] Scambia, G., Mango, D., Signorile, P. G., Anselmi, Angeli. R. A., Palena, C., Gallo, D., et al. (2000). Clinical effects of a standardized soy extract in postmenopausal women: A pilot study. *Menopause*, 7, 105-111.

[30] Zhang, X., Shu, X. O., Gao, Y. T., Yang, G., Li, Q., Li, H., et al. (2003). Soy food consumption is associated with lower risk of coronary heart disease in Chinese women. *Journal of Nutrition*, 133, 2874-2878.

[31] Gourineni, V. P., Verghese, M., & Boateng, J. (2010). Anticancer effects of prebiotics Synergy1® and soybean extracts: Possible synergistic mechanisms in caco-2 cells. *International Journal of Cancer Research*, 6, 220-233.

[32] Katdare, M., Osborne, M., & Telang, N. T. (2002). Soy isoflavone genistein modulates cell cycle progression and induces apoptosis in HER-2/neu oncogene expressing human breast epithelial cells. *International Journal of Oncology*, 21, 809-815.

[33] Kim, M. H., Gutierrez, A. M., & Goldfarb, R. H. (2002). Different mechanisms of soy isoflavones in cell cycle regulation and inhibition of invasion. *Anticancer Research*, 22(6C), 3811-3817.

[34] Simonne, A. H., Smith, M., Weaver, D. B., Vail, T., Barnes, S., & Wei, C. I. (2000). Retention and changes of soy isoflavones and carotenoids in immature soybean seeds

(Edamame) during processing. *Journal of Agricultural and Food Chemistry, 48,* 6061-6069.

[35] Lee, S. J., Chung, I. M., Ahn, J. K., Kim, J. T., Kim, S. H., & Hahn, S. J. (2003). Variation in isoflavones of soybean cultivars with location and storage duration. *Journal of Agricultural and Food Chemistry, 51,* 3382-3389.

[36] Lee, S. J., Yen, W., Ahn, J. K., & Chung, I. M. (2003). Effects of year, site, genotype, and their interactions on various soybean isoflavones. *Field Crop Research, 81,* 181-192.

[37] Wang, H. J., & Murphy, P. A. (1994). Isoflavone composition of American and Japanese soybeans in Iowa- Effects of variety, crop year, and location. *Journal of Agricultural and Food Chemistry, 42(8),* 1674-1677.

[38] Beddows, C. G., & Wong, J. (1987). Optimization of yield and properties of silken tofu from soybeans. I. *The water:bean ratio. International Journal of Food Science and Technology, 22,* 15-21.

[39] Riedl, K. M., Zhang, Y. C., Schwartz, S. J., & Vodovotz, Y. (2005). Optimizing dough proofing conditions to enhance isoflavone aglycones in soy bread. *Journal of Agricultural and Food Chemistry, 53,* 8253-8258.

[40] Vaidya, N. A., Mathias, K., Ismail, B., Hayes, K. D., & Corvalan, C. M. (2007). Kinetic modeling of malonylgenistin and malonyldaidzin conversions under alkaline conditions and elevated temperatures. *Journal of Agricultural and Food Chemistry, 55,* 3408-3413.

[41] Toda, T., Sakamoto, A., Takayanagi, T., & Yokotsuka, K. (2000). Changes in isoflavone compositions of soybean during soaking process. *Food Science and Technology Research, 6,* 314-319.

[42] Rubel, A., Rinne, R. W., & Canvin, D. T. (1972). Protein, oil, and fatty acid in developing soybean seeds. *Crop Science, 12,* 739-741.

[43] Im, M. H., Choi, J. D., & Choi, K. S. (1995). The oxidation stability and flavor acceptability of oil from roasted soybean. *Journal of Agricultural and Food Chemistry and Biotechnology, 38(5),* 425-430.

[44] Pananun, T., Montalbo-Lomboy, M., Noomhorm, A., Grewell, D., & Lamsal, B. (2012). High-power ultrasonication-assisted extraction of soybean isoflavones and effect of toasting. *Food Science and Technology, 47,* 1999-2007.

[45] Chien, J. T., Hsieh, H. C., Kao, T. H., & Chen-H, B. (2005). Kinetic model for studying the conversion and degradation of isoflavones during heating. *Food Chemistry, 91,* 425-434.

[46] Tipkanon, S., Chompreeda, P., Haruthaithanasan, V., Prinyawiwatkul, W., No, H. K., & Xu, Z. (2011). Isoflavone content in soy germ flours prepared from two drying methods. *International Journal of Food Science and Technology, 46,* 2240-2247.

[47] Niamnuy, C., Nachaisin, M., Laohavanich, J., & Devahastin, S. (2011). Evaluation of bioactive compounds and bioactivities of soybean dried by different methods and conditions. *Food Chemistry*, 129, 899-906.

[48] Blagden, T. D., & Gilliland, S. E. (2005). Associated with the "Beany" flavor in soymilk by Lactobacilli and Streptococci. *Journal of Food Science*, 70(3), M186-M189.

[49] Telang, A. M., & Thorat, B. N. (2010). Optimization of process parameters for spray drying of fermented soy milk. *Drying Technology*, 28, 1445-1456.

[50] Kim, S. H., Jung, W. S., Ahn, J. K., & Chung, I. M. (2005). Analysis of isoflavone concentration and composition in soybean [Glycine max (L.)] seeds between the cropping year and storage for 3 years. *European Food Research and Technology*, 220, 207-214.

[51] Lee, J. H., & Cho, K. M. (2012). Changes occurring in compositional components of black soybeans maintained at room temperature for different storage periods. *Food Chemistry*, 131, 161-169.

[52] Pinto, M. S., Lajolo, F. M., & Genovese, M. I. (2005). Effect of storage temperature and water activity on the content and profile of isoflavones, antioxidant activity, and in vitro protein digestibility of soy protein isolates and defatted soy flours. *Journal of Agricultural and Food Chemistry*, 53, 6340-6346.

[53] Kim, H., Peterson, T. G., & Barnes, S. (1998). Mechanism of action of the soy isoflavone genistein: emerging role for its effects via transforming growth factor signaling pathways. *American Journal of Clinical Nutrition*, 68, 1418S-1425S.

[54] Molteni, A., Brizio-Molteni, L., & Persky, V. (1995). In-vitro hormonal effects of soybean isoflavones. *Journal of Nutrition*, 125, 751S-756S.

[55] Zhu, D., Hettiarachchy, N. S., Horax, R., & Chen, P. (2005). Isoflavone contents in germinated soybean seeds. *Plant Foods for Human Nutrition*, 60, 147-151.

[56] Chen, T. R., & Wei, Q. K. (2008). Analysis of bioactive aglycone isoflavones in soybean and soybean products. *Nutrition and Food Science*, 38, 540-547.

[57] Nakajima, N., Nozaki, N., Ishihara, K., Ishikawa, A., & Tsuji, H. (2005). Analysis of isoflavone content in tempeh, a fermented soybean, and preparation of a new isoflavone-enriched tempeh. *Journal of Bioscience and Bioengineering*, 100, 685-687.

[58] Haron, H., Ismail, A., Azlan, A., Shahar, S., & Peng, L. S. (2009). Daidzein and genestein contents in tempeh and selected soy products. *Food Chemistry*, 115, 1350-1356.

[59] Murphy, P. A., Song, T., Buseman, G., Barua, K., Beecher, G. R., Trainer, D., & Holden, J. (1999). Isoflavones in Retail and Institutional Soy Foods. *Journal of Agricultural and Food Chemistry*, 47(7), 2697-2704.

[60] Yang, S. O., Chang, P. S., & Lee, J. H. (2006). Isoflavone distribution and beta-glucosidase activity in Cheonggukjang, a traditional Korean whole soybean-fermented food. *Food Science and Biotechnology*, 15(1), 96-101.

[61] Astuti, M., & Dalais, F. S. (2000). Tempeh, a nutritious and health food from Indonesia. *Asia Pacific Journal of Clinical Nutrition, 9*, 322-325.

[62] Hong, G. E., Mandal, P. K., Lim, K. W., & Lee-H, C. (2012). Fermentation increases isoflavone aglycone contents in black soybean pulp. *Asian Journal of Animal and Veterinary Advances, 6*, 502-511.

[63] Silva, L. H., Celeghini, R. M. S., & Chang, Y. K. (2011). Effect of the fermentation of whole soybean flour on the conversion of isoflavones from glycosides to aglycones. *Food Chemistry, 128*, 640-644.

[64] Li-Jun, Y., Li-Te, L., Zai-Gui, L., Tatsumi, E., & Saito, M. (2004). Changes in isoflavone contents and composition of sufu (fermented tofu) during manufacturing. *Food Chemistry, 87*, 587-592.

[65] Wang, L., Yin, L., Li, D., Zou, L., Saito, M., Tatsumi, E., & Li, L. (2007). Influences of processing and NaCl supplementation on isoflavone contents and composition during douchi manufacturing. *Food Chemistry, 101*, 1247-1253.

[66] Lee, I. H., & Chou, C. C. (2006). Distribution profiles of isoflavone isomers in black bean kojis prepared with various filamentous fungi. *Journal of Agricultural and Food Chemistry, 54*, 1309-1314.

[67] Huang-Y, R., & Chou-C, C. (2009). Stability of isoflavone isomers in steamed black soybeans and black soybean koji stored under different conditions. *Journal of Agricultural and Food Chemistry, 57*, 1927-1932.

[68] Kao, T. H., Lu, Y. F., Hsieh, H. C., & Chen, B. H. (2004). Stability of isoflavone glucosides during processing of soymilk and tofu. *Food Research International, 37*, 891-900.

[69] Grün, I. U., Adhikari, K., Li, C., Li, Y., Lin, B., Zhang, J., & Fernando, L. N. (2001). Changes in the profile of genistein, daidzein, and their conjugates during thermal processing of tofu. *Journal of Agricultural and Food Chemistry, 49*, 2839-2843.

[70] Chien-L, H., Huang-Y, H., & Choua-C, C. (2006). Transformation of isoflavone phytoestrogens during the fermentation of soymilk with lactic acid bacteria and bifidobacteria. *Food Microbiology, 23*, 772-778.

[71] Coward, L., Smith, M., Kirk, M., & Barnes, S. (1998). Chemical modification of isoflavones in foods during cooking and processing. *American Journal of Clinical Nutrition, 68*, 1486S-1491S.

[72] Wang, C., Ma, Q., Pagadala, S., Sherrard, M. S., & Krishnan, P. G. (1998). Changes of isoflavones during processing of soy protein isolates. *Journal of the American Oil Chemists' Society, 75*, 337-341.

[73] Eisen, B., Ungar, Y., & Shimoni, E. (2003). Stability of Isoflavones in Soy Milk Stored at Elevated and Ambient Temperatures. *Journal of Agricultural and Food Chemistry, 51*, 2212-2215.

[74] Xu, B., & Chang, S. K. C. (2008). Total phenolics, phenolic acids, isoflavones, and anthocyanins and antioxidant properties of yellow and black soybeans as affected by thermal processing. *Journal of Agricultural and Food Chemistry*, 56, 7165-7175.

[75] Farkas, J. (2006). Irradiation for better foods. *Trends in Food Science and Technology*, 17, 148-152.

[76] Farkas, J., & Mohácsi-Farkas, C. (2011). History and future of food irradiation. *Trends in Food Science and Technology*, 22, 121-126.

[77] World Health Organization (WHO). (1994). Safety and Nutritional Adequacy of Irradiated Food. *Geneva, Austria*, 23-39.

[78] Hayashi, T., Takahashi, Y., & Todoriki, S. (1998). Sterilization of foods with low-energy electrons ("Soft-Electrons". *Radiation Physics and Chemistry*, 52, 73-76.

[79] Dixit, A. K., Kumar, V., Rani, A., Manjaya, J. G., & Bhatnagar, D. (2011). Effect of gamma irradiation on lipoxygenases, trypsin inhibitor, raffinose family oligosaccharides and nutritional factors of different seed coat colored soybean (Glycine max L.). *Radiation Physics and Chemistry*, 80, 597-603.

[80] Byun-W, M., Kwon-H, J., & Mori, T. (1993). Improvement of physical properties of soybeans by gamma irradiation. *Radiation Physics and Chemistry*, 42, 313-317.

[81] Pednekar, M., Das, A. K., Rajalakshmi, V., & Sharma, A. (2010). Radiation processing and functional properties of soybean (Glycine max). *Radiation Physics and Chemistry*, 490-494.

[82] Variyar, P. S., Limaye, A., & Sharma, A. (2004). Radiation-induced enhancement of antioxidant contents of soybean (Glycine max Merrill). *Journal of Agricultural and Food Chemistry*, 52, 3385-3388.

[83] Dixit, A. K., Bhatnagar, D., Kumar, V., Rani, A., Manjaya, J. G., & Bhatnaga, D. (2010). Gamma irradiation induced enhancement in isoflavones, total phenol, anthocyanin and antioxidant properties of varying seed coat colored soybean. *Journal of Agricultural and Food Chemistry*, 58, 4298-4302.

[84] Park, J. H., Choi, T. B., Kim, S. W., Hur, M. G., Yang, S. D., & Yu, K. H. (2009). A study on effective extraction of isoflavones from soy germ using the electron beam. *Radiation Physics and Chemistry*, 78, 623-625.

Chapter 10

Evaluation of Soybean Straw as Litter Material in Poultry Production and Substrate in Composting of Broiler Carcasses

Valeria Maria Nascimento Abreu,

Paulo Giovanni de Abreu,

Doralice Pedroso de Paiva, Arlei Coldebella,

Fátima Regina Ferreira Jaenisch,

Taiana Cestonaro and Virginia Santiago Silva

Additional information is available at the end of the chapter

http://dx.doi.org/10.5772/54068

1. Introduction

The litter has never been a subject of extensive studies or considered a priority in large poultry company. However, due to the increasing lack of good litter material, more attention has been given to proper litter management, litter reuse, and to the search for new litter materials. In this context, the use of crop residues as poultry litter material seems to be promising. There is a current trend in poultry production to use alternative litter materials, that is, other than wood shavings. Despite being demonstrated by several authors that the use of these alternative materials do not interfere with flock live performance, most agree that they are more difficult to manage and may result in a higher incidence of carcass lesions. The effects of several many materials used as poultry litter substrate on poultry performance have been evaluated. It was shown that litter made of rice husks not only does not impair performance, but also reduced foot-pad and breast lesions [1, 2 – 9]. A study on the use of soybean crop residues as poultry litter did not show any influence of this material on broiler performance or on its agronomic value [10]. However, these bedding materials have not been evalatuated as to the evolution of darkling beetle and enteric parasite populations. *Alphitobius diaperinus* (darkling beetle) adults and larvae are considered a problem in intensive broiler and turkey production.

© 2013 Abreu et al.; licensee InTech. This is a paper distributed under the terms of the Creative Commons Attribution License (http://creativecommons.org/licenses/by/3.0), which permits unrestricted use, distribution, and reproduction in any medium, provided the original work is properly cited.

These beetles replicate in the litter, becoming potential vectors of pathogens and parasites both on site and to the neighboring farms. Those insects have been associated to many pathogenic agents, and there are reports they carry *Escherichia coli, Salmonella spp*, avian leucosis virus, as well as internal parasites, such as coccidian, avian tapeworms, and helminths [11,12], enterobacteria [13] and *Clostridium perfringens* [14]. Attempts to control darkling beetles have been made by changing the litter pH using hydrated lime [15]) or applying insecticides in the entire poultry house during downtime [16] or as a management complement during rearing, with application in specific spots, as suggested by [17]. The continuous contact of birds with the excreta in the litter poses a risk of infection with parasites, and coccidiosis caused by *Eimeria spp* is one of the most significant diseases in this production system. The evolution of oocyst population may determine the need to change the litter, and an evaluation may aid decision-making regarding this need. Litter reutilization for more than one flock is practiced in several countries. However, aspects relative to the potential health risk posed by these alternative litter materials have been discussed, and their use may limit the international chicken meat trade due to the requirements of showing equivalence of production processes practiced among exporting and importing countries [18]. The study was divided into three parts, where the use of soybean straw as litter in the poultry production was compared with rice husk under two ventilation conditions, as follows.

2. The study

The experiment was carried out at the experimental field of Suruvi, belonging to Embrapa Swine &Poultry, Concórdia, Santa Catarina, Brazil. Four 12m×10m broiler houses were internally divided in four pens each (total of 16 pens), at a density of 200 birds/pen (28kg meat/m^2), totaling 3,200 birds/flock. Four consecutive flocks were followed up. Each flock was reared to 42 days of age, and an interval between flocks (downtime) of 15 days was applied. Two ventilation systems (stationary or oscillating), reaching a distance of 10m, and two litter materials (soybean straw or rice husks) were tested. Rice husks and ventilation system using stationary fans are considered as standards as they are commonly used in broiler production. Fans were activated by a thermostat when the environmental temperature reached 25°C, and were equipped with a potentiometer and speed regulator matching the broiler house size. Treatments were distributed as follows (Figure 1): house 1 – stationary ventilation system, pens 2 and 3 with soybean straw; pens 1 and 4 with rice husks; house 2 – oscillating ventilation system, pens 2 and 3 with soybean straw; pens 1 and 4 with rice husks; house 3 - oscillating ventilation system, pens 1 and 4 with soybean straw; pens 2 and 3 with rice husks; house 4 – stationary ventilation system, pens 1 and 4 with soybean straw; pens 2 and 3 with rice husks. Birds and feeds were weekly weighed and the following parameters were evaluated: body weight, weight gain, feed intake and feed conversion ratio when birds were 21, 35 and 42 days of age. Performance data were analyzed according to the theory of mixed models for repeated measures, considering the effects of flock, ventilation system, litter material, bird age, and the interaction of these parameters up to third order, and 16 types of variance and covariance matrices, using PROC MIXED procedure of SAS statistical package [19], according to [20].

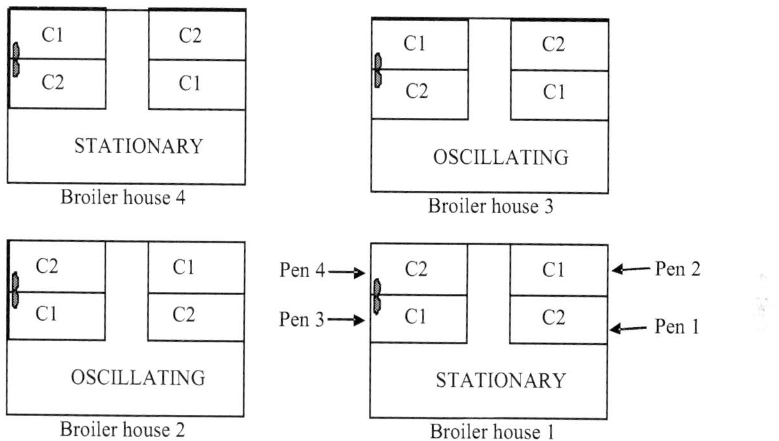

Figure 1. Fan (stationary and oscillating) and litter material (C1 – soybean straw and C2 – rice husks) distribution in the experimental facilities

The variance and covariance structure used for analysis was chosen based on the lowest value of the Akaike Information Criterion (AIC). The estimation method was that of restricted maximum likelihood. Mortality was daily recorded and assigned to ascitis, sudden death, or other causes. Total mortality was also evaluated. Breast and foot-pad lesions were evaluated by gross examination the last time birds were weighed, and scored as present or absent. Because mortality and presence of foot-pad lesion data have binomial distribution, these data were analyzed by logistic regression, using the LOGISTIC procedure of SAS statistical package [19], and considering the effects of flock, litter material, ventilation system, and their interactions. The overdispersion of the presence of foot-pad lesion was adjusted by the dispersion parameter estimated by Pearson's χ^2 statistics divided by degrees of freedom. Litter samples were collected at each flock, when chicks were housed (day 0) and when they were removed (day 42), and submitted to enterobacteria quantitative exam. Litter samples weighing 10g were diluted at 1:10 in PBS (phosphate buffer saline solution) and serially diluted to 10^{-7}. Aliquots of 100 µL of 10^{-3} to 10^{-7} dilutions were seeded in Mac Conkey agar and incubated at 37ºC for 48h, and submitted to colony-forming unit (CFU) counting in plates containing 30 to 300 colony-forming units. CFU data analysis considered the effects of flock, litter material, ventilation system, and their interactions and used the theory of mixed models for repeated measures and nine variance and covariance matrix structures applied to the PROC MIXED procedure of SAS statistical package [19]. The following psychrometric parameters were recorded at the center of each pen and in the external environment: dry- and wet-bulb temperature, black globe temperature, and air velocity. Temperatures were collected using copper-constantan thermocouples connected to a potentiometer and a 20-channel selecting key. The wet bulb thermometer was characterized by securing a cotton wick attached to the thermocouple terminal to the mercury bulb and immersing it in a flask with distilled water. Black globe temperature was collected by placing thermocouples inside a hollow 15cm-diameter PVC sphere painted with black mat spray paint. air velocity was recorded using a digital anemometer (Pacer® Model DA40V), with a resolution of 0.01m/s. Data were collected every three hours, from 08:00 to 18:00h, when broilers were 4, 5, and 6 weeks. Based on the data collected at each time, wet bulb globe temperature (WBGT) and radiant heat load and a radiant heat load (RHL). Litter temperature was also measured using an infrared thermometer Raytec® in five different spots in each pen (two near the lateral pen wall, two near the central aisle, and at the geometrical center of the pen) every three hours, from 08:00 to 18:00 when broilers were 6 weeks of age. Based on the average data in each spot, isothermal maps of litter temperature using the kriging method of the SURFER software program were built. Internal thermal environmental parameters were evaluated as to the effects of flock, ventilation, litter, week, hour, and the interactions among the last four factors using mixed models for repeated measures and 19 variance and covariance matrix structures, applying PROC MIXED procedure of SAS statistical package [19], according to [20]. The structure used for analysis was chosen based on the lowest value of the Akaike Information Criterion (AIC). The estimation method was that of restricted maximum likelihood. The effect of hour was detailed using orthogonal polynomial analysis up to the polynomial of the third degree. For the external environment, parameter means were calculated as a function of hour and week in order to compare the internal with the external thermal environment. The soybean straw was chopped into approximately 3cm long particles. Litter was initially 10cm high and was reused for four consecutive flocks. Litter quality was evaluated in terms of physical-chemical composition, moisture, and compactness. The development of the population of darkling beetles and the evolution of the number of parasite eggs/oocysts was also observed. In order to evaluate darkling beetles population, three traps per pen made with 20 × 5cm PVC tubes filled with 20 × 50cm rolled-up corrugated cardboard, totaling 48 traps [21]. Traps were placed under the litter in three sites in each pen: at the center, between the external wall and the first line of feeders, and the other two between the feeders, equally distant from the drinkers. Traps remained in the pens for seven days and were removed after catching of each flock. Traps were then placed in 1L thick plastic bags, closed with thin plastic-coated wire, and submitted to the laboratory. *Alphitobius diaperinus* adults and larvae were identified and counted, and the total number of individuals was recorded. Litter samples were collected to count number of endoparasite eggs/oocysts per gram of litter (epg), and also submitted to physical-chemical analysis. Two samples were collected per flock: on the day chicks were housed and after catching. In each pen, 15 litter samples, weighing 50g each, were collected on the surface and under the surface of the litter centrally to the feeders, drinkers, and external and internal limits of the pens. In the laboratory, after homogenization, 50g aliquots were used for endoparasite counting. The remaining sample was submitted to the physical-chemical analyses lab to determined dry matter, ashes, and phosphorus contents. Phosphorus was colorimetrically determined by the molybdovanadate method [22]. Pre-dry matter or moisture was determined at 65ºC. Copper, zinc, calcium, manganese, and iron contents were determined by flame atomic absorption spectrometry after nitro-perchloric digestion [23]. Nitrogen was determined by the Kjeldahl method. Litter pH was measured according to the method described by [24]. Organic carbon was titered after chemical oxidation with sulfochromic solution [25]. Potassium was determined by flame photometry [24]. Litter temperature was recorded using an infrared thermometer (Raytec®) in five different spots in each pen (two near the lateral pen wall, two near the central aisle, and at the geometrical center of the pen). Darkling beetle counts were transformed into log (y+1) and submitted to analysis of variance using a model that considered the effects of litter material, ventilation, flock, and the interaction among these factors. The GLM procedures of SAS statistical package were used [19]. Parasite egg and oocyst counts were used to characterize the presence of absence of parasite in the pens. Parasite presence was analyzed by logistic regression considering the effects of litter material, ventilation and flock, using LOGISTIC procedures of SAS statistical package [19]. Litter quality data were analyzed by mixed models for repeated measures, considering the effects of litter material, ventilation, and the interaction between litter material and ventilation (plot), sample sampling and respective interactions (subplot, and period (subplot), and three matrix structures of variance and covariance, using the PROC MIXED procedure of SAS statistical package [19], according to [20]. The structure used for analysis was chosen based on the lowest value of the Akaike Information Criterion (AIC). The estimation method was that of restricted maximum likelihood.

2.1. Evaluation poultry production using different ventilation systems and litter material: I – general performance

According to statistical analysis results, flock, litter, and age significantly (p<0.05) influenced all evaluated parameters, whereas ventilation system had no effect on any variable. Two interactions significantly affected almost all parameters: litter x age and flock x ventilation x age. The details of the litter x age interactions indicated better results for rice husks litter as compared to soybean straw litter in all studied parameters at all ages (Table 1). These results are similar to those described by Mizubuti et al. (1994), who evaluated rice husks, guinea grass, and napier grass as litter material. On the other hand, opposite results were obtained by [27, 5, 27, 28, 9, 8, 10], who tested different broiler litter materials and did not find any differences in body weight or feed intake. Moreover, [29] studied five broiler litter materials and observed significant reduction of body weight, feed intake, and antibody titers in broilers reared on rice husks litter. [30] compared the use of rice husks, coconut hulls, and wood shavings as broiler litter material and low- and high-density diets, and concluded that weight gain and feed intake during the total experimental period were higher in broilers reared on coconut hulls litter. The evaluation of the details of the triple interaction among flock, ventilation and age indicated that there was no consistent effect of ventilation system on performance parameters. The analysis of the presence of foot-pad lesion showed a significant influence (p<0.0001) of the interaction between flock and litter and of the main effects litter and flock. Independently from the interaction, soybean straw litter caused higher incidence of foot-pad lesions as compared to rice husks, in all flocks (Figure 2).

Litter material	Bird age (days)		
	21	35	42
	Feed intake (g)		
Rice husks	1196 ± 3.90a	3373 ± 8.88a	4764 ± 13.15a
Soybean straw	1179 ± 3.90b	3310 ± 8.88b	4692 ± 13.15b
	Feed conversion ratio		
Rice husks	1.278 ± 0.004a	1.529 ± 0.003a	1.666 ± 0.004a
Soybean straw	1.306 ± 0.004b	1.543 ± 0.003b	1.683 ± 0.004b
	Weight gain (g)		
Rice husks	894 ± 3.75a	2164 ± 6.05a	2819 ± 9.44a
Soybean straw	861 ± 3.75b	2104 ± 6.05b	2749 ± 9.44b
	Body weight (g)		
Rice husks	937 ± 3.76a	2207 ± 6.05a	2862 ± 9.45a
Soybean straw	903 ± 3.76b	2146 ± 6.05b	2791 ± 9.45b

Means followed by different letters in the same columns are different ($p<0.05$) by the F test.

Table 1. Feed intake, feed conversion ratio, weight gain and body weight of broiler reared on two different litter materials and two different ventilation systems

These results are opposite to those found in other studies evaluating litter materials that concluded that litter material has not effect on carcass lesions [31]. The obtained results indicate that litter material causes foot-pad lesions, resulting in carcass condemnation in the processing plant and consequent economic losses. [1] also found higher incidence of foot-pad lesions when napier grass and coast-cross grass hays were used as litter material. However, [7] used sunflower crop residue and Brachiaria hay as litter, but did not find any effects on breast, hock, and foot-pad lesions. In the present study, it was also observed that the percentage of foot-pad lesions markedly increased from the first to the second flock, decreased in the third flock, and that this reduction was maintained in the fourth flock. We believe that, as the number of flocks increased, caked litter was removed and the remaining litter was turned, making litter softer, and thereby, reducing leg lesions. Mortality was influenced by flock ($p<0.0001$) and by the interaction between flock and litter ($p<0.05$). However, mortality was only significantly affected by litter material in the third flock in favor of soybean straw. Evaluating the use of wood shavings, rice husks, sugarcane residue and carnauba palm residue as broiler litter material, [32] did not find any differences in mortality, although in absolute numbers, values were higher in broilers reared on wood shavings and sugarcane residue. [5] and [29] reported similar results, that is, no mortality differences in studies on alternative broiler litter materials. According to [33], bacteria present in the litter may have different effects. Many Gram-positive bacteria, such as lactobacilli and bifidobacteria, are present in broiler excreta and in the litter, but are not necessarily related to problems. On the other hand, the frequent presence of pathogens in the litter, particularly of enterobacteria and zoonotic bacteria in general, is a reason of concern due to possible diseases transmitted both to the broiler flock itself and to consumers.

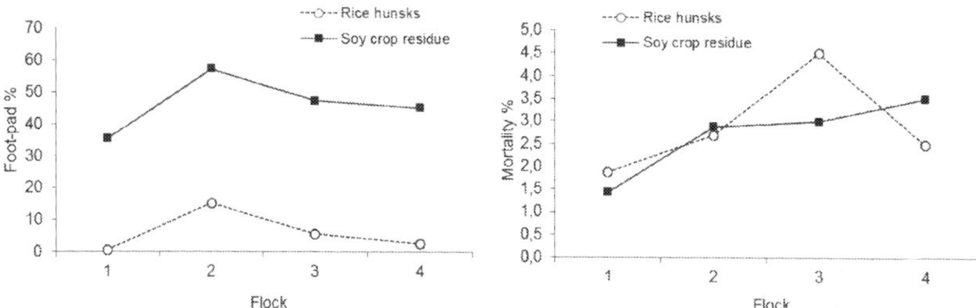

Figure 2. Presence of foot-pad lesions and mortality rate of broilers reared on two different litter materials for four consecutive flocks

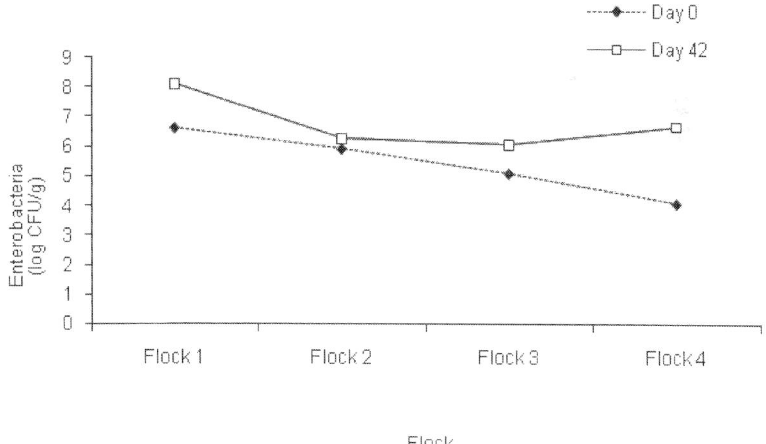

Figure 3. Enterobacteria (log CFU/g) in rice husks litter reused for four broiler flocks (flock housing and removal – Days 0 and 42)

Moreover, according to [33], the composition of the bacterial population in the litter is usually very similar to the composition of the physiological microbiota of the ileum of broilers, and consists approximately of 70% lactobacilli, 11% *Clostridium* spp., 6.5% *Streptococcus* spp., and 6.5% *Enterococcus* spp. The litter presents, in average, 10 times less bacteria than the digesta, but this is still a high concentration of microorganisms. The digest contains between 10^8 and 10^{10} of Gram-positive bacteria and 10^6 to 10^7 Gram-negative bacteria per gram. As the concentration of bacteria in the litter can increase 10 times per reared flock, it may achieve the same levels as the digesta. From the practical perspective, it may be assumed that bacterial concentration in the litter of broilers is similar to that in feces. In the present study, total enterobacteria counts (expressed as colony-forming units, CFU) per gram of litter for each treatment were carried out when each of the four flocks was removed (day 42) and when the following flocks were housed (day 0), aiming at evaluating the effect of downtime on litter bacterial load. Significant effects ($p < 0.05$) of flock, evaluation day (0 or 42), and their interaction was observed, whereas litter material (rice husks or soybean straw) and ventilation system had no influence on litter enterobacterial load (Figures 3 and 4). The evaluation carried out on day 42 always presented better results as compared to that performed on day 0, as expected. However, the difference increased as the number of flocks increased (Figure 5), as shown by the lowest UFC count on day 42 (immediately after flock removal) in the third flock, whereas on day 0 (day of housing, after downtime), bacterial load the lowest only in the fourth flock. This suggests that the reduction of the load of enterobacteria after downtime (day 0), may continue to occur in subsequente flocks, as the reduction was linear and did not achieve a stabilization point.

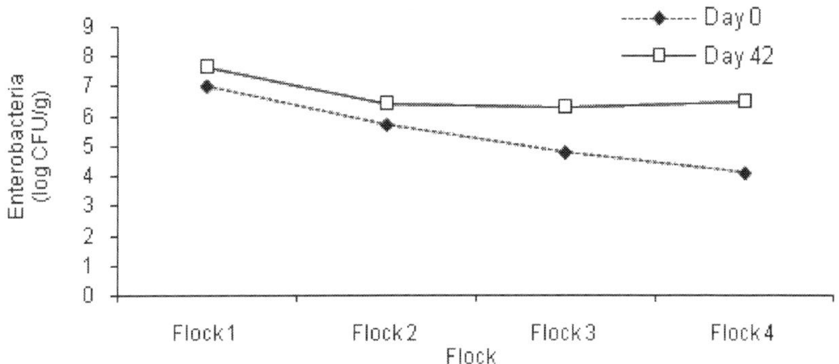

Figure 4. Enterobacteria (log CFU/g) in soybean straw litter reused for four broiler flocks (flock housing and removal – Days 0 and 42)

It must be mentioned that, during downtime, flame-gun was used twice on the litter: immediately after bird removal and before the following flock was housed. However, during the 15-d downtime, litter was not submitted to any management practice for the reduction or control of undesirable bacteria, including potentially pathogenic bacteria. Despite the reduction in the bacterial load along the four flocks, enterobacteria counts were high in all flocks, and therefore some method of litter treatment during downtime is recommended to reduce pathogen load. The initial enterobacteria counts in the litters used in the experiment before the first flocks were housed were considered high. When evaluating litter treatment methods for bacterial load reduction, [18] observed that average loads of enterobacteria and total mesophilles were high in new litters. According to those authors, these results call the attention to the quality of new litter, as high bacterial loads in that material are associated to their origin, possibly due to production, preservation, storage, and transport conditions to the broiler house. In addition, these high bacterial loads are a hazard to birds that will be housed in that environment, particularly considering that these will be day-old chicks. [34] determined bacterial levels in pine shavings and sand used as broiler litter, and found a marked increase in bacterial counts after birds were housed. Pine shavings reached a level of 10_8 CFU/g of aerobic bacteria in the second week, and this level remained stable up to six weeks, after which ir increased approximately 1 log the next week, and remained at this level until birds were removed at the seventh week. Enteric bacteria in pine shavings reach a plateau at 10_7 to 10_8 CFU/g in the second week, and these values showed little variation until the seventh week. In the present study, enterobacteria counts were higher, in general, than those obtained in untreated wood shavings litter by [18], which were about 10^5 CFU/g. Therefore, the need to treat the litter during downtime is stressed, independently of the material used.

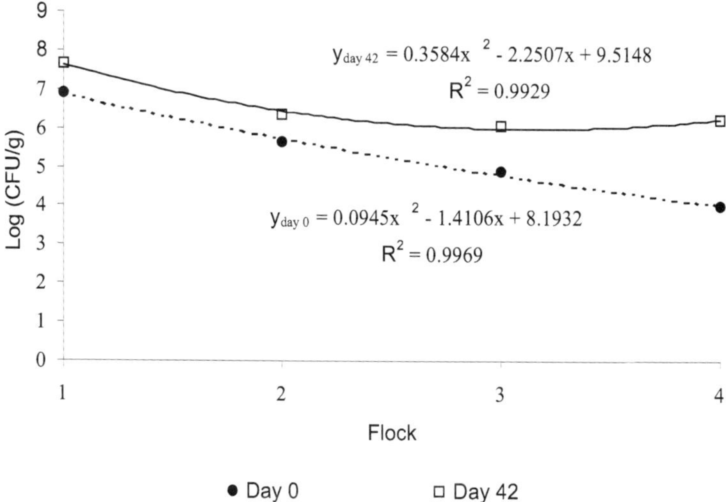

Figure 5. Enterobacteria load on days 0 and 42 of the evaluated flocks, considering all treatments

2.2. Poultry production using different ventilation systems and litter materials: II – thermal comfort

The main effects of flock, week and hour were significant (p<0.05) for all evaluated parameters. Litter material significantly influenced (p<0.05) only air relative humidity, whereas ventilation system did not affect any of the parameters. The hour × week interaction significantly affected all parameters, whereas the effect of the ventilation × hour interaction was significant only for air temperature and WBGT. Heat load inside the broiler house did not show much variation, and the results were more favorable to the birds as compared to the external environment. When rice husks litter was used, higher air relative humidity values were detected (Figure 6). According to [35] optimal air temperature values are between 23ºC and 26ºC; 20ºC and 23ºC; and 20ºC for 4-, 5-, and 5-week-old broilers, respectively. On the other hand, the recommended air relative humidity is 60 to 70%, regardless bird age. In the present study, air relative humidity remained higher than the recommendations for broiler production in all studied weeks, times, and litter materials (Figure 6). Controlling air and litter humidity is important to reduce pathogens, ammonia, and parasite, such as coccidia, in the poultry house environment [36]. [37] did not find significant differences in breast blisters as a function of bird density (10 birds/m² or 14 birds/m²), litter material (wood shavings or saw dust) or by their interaction, as this part of the broiler body is not in permanent contact with the litter, and therefore is no influenced by litter humidity. Litter temperature presented the same behavior as air temperature, as expected, with a linear correlation of 0.95. At all evaluated times, litter temperature was, in average, 2°C to 3ºC higher than air temperature. It must be noted that litter temperature was not affected by the used material, that is, independently of using soybean straw or rice husks, litter temperature remained similar (Figure 6). According to [38], temperature measured 5cm below litter surface was 23.5°C and 31.3°C at housing densities of 19 and 40 kg/m², respectively, whereas air temperature 1m above litter surface was 22°C, indicating that litter temperatures were 1.5°C and 9.3°C higher than the environmental temperature. The higher litter temperature at higher bird densities may be explained by different effects. As bird density increases, nitrogen and humidity levels increase in the litter, allowing higher microbial activity and the heat transference from the litter surface to the environmental air is prevented when the litter surface area is covered by birds, particularly as birds reach market weight.

2.3. Poultry production using different ventilation systems and litter materials: III – effect of litter reuse on the populations of *Alphitobius diaperinus* and intestinal parasites

Considering the composition of rice husks and soybean straws before its utilization and the values established in [39] for simple organic fertilizers (Table 2), both materials are very different, except for organic carbon content, and do not comply with the recommendations of IN-23 for organic fertilizers. Rice husks humidity and pH values are different from those described by [40], who evaluated the reutilization of rice husks as litter material by 18 broiler flocks and reported 9.4% humidity, pH 7.0, 0.47% nitrogen, 0.03% phosphorus, and 0.27% potassium in the beginning of the experiment, that is, composition before utilization. According to [10], considering that plant nutrient requirements vary as a function of cultivar, soil, expected yield, etc., and exceeding supplied quantities remain in the soil and are susceptible to leaching and percolation, it is essential to balance soil and litter compositions. Based on these considerations, the knowledge on the quality (physical-chemical composition, compactness, and reutilization) of litter materials used as alternative to wood shavings is essential as the disposal of such material is part of good production practices. Litter quality was evaluated according to chemical element levels and physical parameters. Litter chemical composition parameters were significantly affected ($p<0.05$) by litter material, except for humidity and zinc. Sampling affected only ashes and potassium content, whereas period significantly influenced ($p<0.05$) all parameters. The interaction litter material x period x sampling affected organic carbon, copper, iron, potassium, pH and zinc. In order to study litter chemical composition, samplings were performed in three different periods, as follows: period 1 – 15-day interval (downtime) between removal of flock 1 and housing of flock 2; period 2 - 15-day interval (downtime) between removal of flock 2 and housing of flock 3; and period 3 - 15-day interval (downtime) between removal of flock 3 and housing of flock 4.

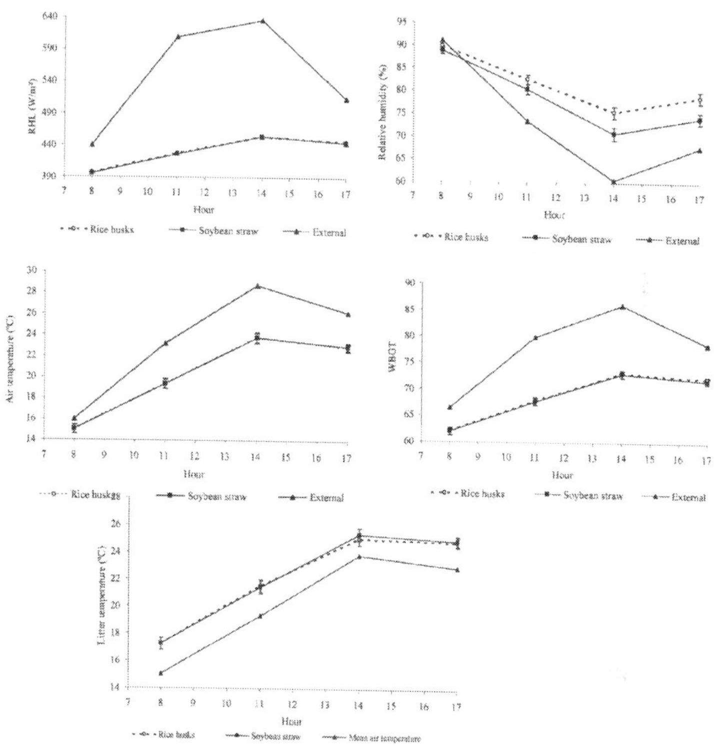

Figure 6. Air temperature, wet bulb globe temperature (WBGT), radiant heat load (RHL), air relative humidity and litter temperature inside broiler houses with two different litter materials

	Rice husks	Soybean straw	IN-23
Ashes (%)	15.18	4.33	
Organic carbon (%)	35.13	33.33	≥ 20
Copper (%)	traces	0.00065	
Iron (mg/kg)	433.25	607.63	
Potassium (%)	0.075	0.108	
Manganese (mg/kg)	169.85	32.75	
Humidity (%)	10.35	14.33	≤ 30
Nitrogen (%)	0.344	0.662	≥ 1
Phosphorus (%)	0.056	0.092	
pH	6.56	7.60	a.d.*
Zinc (mg/kg)	11.43	14.70	

* c.d. – as declared.

Table 2. Chemical composition of rice husks and soybean straw before their utilization as litter and values established in Normative Instruction n. 23

The chemical composition of rice husks and soybean straw after utilization (Table 3) indicates a differentiation pattern of chemical element values. In general, values were high after the removal of flock, and decreased after 15 days of downtime in all studied periods. An exception was pH, which was higher at the end of downtime. Humidity values at flock removal were high, exceeding the recommendations both for poultry rearing and for organic fertilizer; however, at the end of downtime, values returned to acceptable levels. [40] observed that the moisture of rice husks litter varied between 23.4 and 29.1%, averaging 26.4%. Those authors found that litter pH increased from 7.05 to 8.59 after the first flock, and significantly increased between the first and the second flock, after which only minor changes were observed, with an average litter pH of 8.80 after 18 flocks. Similar pH values were obtained by [41] for rice husks litter reused by four flocks (pH 8.75); by [42], with pH values of 8.4-8.5; and by [10], with an average pH of 8.79. For soybean straw, [10] obtained an average pH of 8.97. Ammonia and nitrates are the most common chemical forms of nitrogen found in poultry waste. Nitrates can significantly contaminate underground waters when excessive levels of broiler litter as used as crop fertilizer [43]. Moreover, according to that author, phosphorus is found in large amounts in poultry excreta, and its excessive application for crop fertilization may exceed soil and plant capacity to use that nutrient, resulting in leaching and subsequent contamination of underground waters. [10] studied six different litter materials reused for six consecutive flocks and found nitrogen values of 2.46 and 2.63 and 0.84 and 1.00 of phosphorus in rice husks and soybean straw, respectively. [40] showed that nitrogen, phosphorus, and potassium contents significantly increased in the first seven to eight flocks. Nitrogen, phosphorus, and potassium values obtained in the litter of the fourth flock of the study of [40] were 3.56, 1.59, and 3.12%, respectively, and are used for the comparison with those obtained in the present study.

Period	Rice husks		Soybean straw	
	Removal	Housing	Removal	Housing
Ashes (%DM)				
1	20.86 ± 0.78	-	15.71 ± 0.35	-
2	20.84 ± 0.17 aA	21.13 ± 0.23 aA	17.29 ± 0.32 bA	17.04 ± 0.35 bA
3	21.40 ± 0.20 aA	21.77 ± 0.12 aA	18.95 ± 0.29 bA	18.39 ± 0.30 bA
Organic carbon (%DM)				
1	32.93 ± 0.32 bB	36.43 ± 0.24 bA	36.08 ± 0.43 aB	38.69 ± 0.33 aA
2	32.72 ± 0.45 bA	34.95 ± 1.53 aA	35.92 ± 0.28 aA	32.02 ± 0.64 bB
3	32.14 ± 1.41	-	28.45 ± 0.66	-
Copper (%DM)				
1	45.19 ± 0.84 bA	39.33 ± 1.33 bB	63.31 ± 1.26 aA	49.44 ± 1.64 aB
2	56.05 ± 1.09 bA	52.99 ± 1.65 bB	67.55 ± 0.91 aA	62.72 ± 1.20 aB
3	63.27 ± 1.34 bA	59.13 ± 0.98 bB	74.75 ± 1.22 aA	73.03 ± 1.43 aA
Iron (mg/kg – DM)				
1	911 ± 26 bA	608 ± 29 bB	1492 ± 57 aA	1469±117 aA
2	1014 ± 28 bA	738 ± 16 bB	1452 ± 44 aA	1361± 59 aA
3	1035 ± 21 bA	507 ± 18 bB	1364 ± 49 aA	736± 44 aB
Potassium (%DM)				
1	2.04 ± 0.05 bA	1.97 ± 0.05 bA	2.78 ± 0.05 aA	2.73 ± 0.05 aA
2	2.35 ± 0.04 bA	2.00 ± 0.06 bB	2.81 ± 0.04 aA	2.78 ± 0.04 aA
3	2.16 ± 0.05 bB	2.57 ± 0.08 aA	2.69 ± 0.07 aA	2.62 ± 0.03 aA
Manganese (mg/kg – DM)				
1	351 ± 8 aA	356 ± 7 aA	311 ± 6 bA	252 ± 12 bB
2	427 ± 9 aB	448 ± 9 aA	386 ± 6 bA	375 ± 12 bA
3	420 ± 8 aA	421 ± 7 aA	404 ± 7 aA	383 ± 8 bB
Nitrogen (%DM)				
1	2.31 ±0.09 bA	1.77 ± 0.05 bB	2.92 ± 0.02 aA	2.30 ± 0.06 aB
2	2.67 ± 0.04 bA	2.32 ± 0.02 bB	3.26 ± 0.04 aA	2.83 ± 0.03 aB
3	2.90 ± 0.07 bA	2.44 ± 0.03 bB	3.32 ± 0.04 aA	2.88 ± 0.05 aB
pH				
1	8.59 ± 0.08 aB	9.20 ± 0.01 bA	8.39 ± 0.09 aB	9.35± 0.02 aA
2	8.39 ± 0.04 bA	8.98 ± 0.02 bA	8.68 ± 0.05 aB	9.11± 0.03 aA
3	8.54 ± 0.06 aB	8.86 ± 0.01 bA	8.71± 0.04 aB	8.97± 0.02 aA
Phosphorus (%DM)				
1	1.38 ± 0.07 aA	0.88 ± 0.05 aB	1.54 ± 0.07 aA	1.06 ± 0.07 aB
2	1.65 ± 0.07 bA	1.43 ± 0.07 bB	1.91 ± 0.10 aA	1.66 ± 0.04 aB
3	1.43 ± 0.05 aA	1.35 ± 0.06 bA	1.46 ± 0.09 aA	1.54 ± 0.08 aA
Humidity (%)				
1	42.71 ± 3.22 aA	25.56 ± 1.31 a B	44.09 ±1.01 aA	27.38 ± 0.94 aB
2	31.12 ± 1.01 aA	16.51 ± 0.50 b B	33.85 ± 0.98 aA	19.94 ± 0.49 aB
3	33.44 ± 0.79 aA	16.55 ± 0.21 b B	35.35 ± 0.93 aA	18.62 ± 0.22 aB
Zinc (mg/kg DM)				
1	227 ± 19 bA	145 ± 15 aB	286 ± 12 aA	80.57 ±9.87 bB
2	266 ± 5 aA	254 ± 21 aA	285 ± 6 aA	254 ± 14 aA
3	266 ± 5 bA	141 ± 2 bB	299 ± 7 aA	158 ± 5 aB

Means followed by different small letters in the same row are different, within sampling, (p<0.05) by the F test.

Table 3. Chemical composition of broiler litter made of rice husks or soybean straw

There is little information on copper (Cu), iron (Fe), zinc (Zn), and manganese (Mn) levels in different poultry litter materials. According to [43], poultry feeds are rich in iron, and high levels of this mineral are commonly found in broiler litter. Excessive levels of those mineral in the soil affect plant root growth. That author presented a table with findings of several authors relative to the levels of those trace minerals in poultry litter, and those values (in mg/kg, dry weight) are used to compare with the results of the present study. That table presents average concentrations and ranges of 77 and 58-100 for copper, 1,625 and 1,026-2,288 for iron, 348 and 125-667 for manganese, and finally 315 and 106-669 for zinc. It must also be mentioned that the chemical characteristics of the litter materials after three flocks comply, in terms of their nutritional aspects, with the legislation relative to simple organic fertilizer. However, it is recommended that the litter removed from the poultry house is distributed in rows for an additional composting period in order to eliminate or reduce health risks. Despite the effect of litter material on the evaluated parameters, with higher averages promoted by soybean straw, litter made of these crop residues were similar and presented similar characteristics to those described in literature for wood shavings [44], showing that after three flocks, both materials presented excellent fertilizing characteristics. Relative to the physical aspects, when the first flock was removed, soybean straw litter was more compacted and caked as compared to rice husks litter, and this condition remained for the following three flocks, requiring labor interference to break the caked parts, including during the rearing period. At flock removal, soybean straw required more labor to allow its reuse due to the formation of a caked layer on the top of the litter. When the fourth and last flock was removed, the litter made with soybean straw had reached its maximum limit of reutilization, with decomposition of the lower layer, presenting fiber breakdown and humic matter formation. Rice husks litter, on the other hand, presented reuse conditions after the removal of the fourth flock. [45] recommended monitoring insect populations in poultry housed as a routine procedure of the management program, independently from the strategy used for insect control. However, according to [46], it is difficult to carry out darkling beetle population studies because their population is usually very high in poultry houses, and they have cryptic behavior. In the present study, one of the objectives was to know which litter material was more favorable to the dissemination of darkling beetles, as well as the evolution of this population as the litter was reused by several consecutive flocks. Darkling beetle count was significantly influenced ($p < 0.05$) by litter material, flock, and the interactions between litter material and ventilation and between litter material and flock. Rice husks presented lower darkling beetle count ($p < 0.05$) as compared to soybean straw, with oscillating ventilation in all flocks, and after the second flock, with stationary ventilation (Figure 7). Darkling beetle count increased from the first to the third flock, and decreased in the fourth flock, independently of litter material or ventilation. [47], studying darkling beetle distribution and population dynamics, observed that in broiler houses with cement floor covered with wood shavings litter reused for four flocks and equipped with automatic feeders the average numbers of insects trapped in the first flock were 385.4 larvae and 24.5 adults, and these figures increased to 615.3 larvae and 208.7 adults in the next flock. Those authors found that the population tended to become stable in the third flock (651.3 larvae and 248 adults), and found an apparent reduction to 422 larvae and 160.2 adults per trap in the fourth flock. They also found that average litter temperature in flock 1 was 30.7ºC, which favors the multiplication of saprophyte microorganisms. Air and litter temperatures directly influence the population of darkling beetles [48]: temperatures of 22 and 31ºC determine incubation periods of 8.9 and 3 days, respectively. Below 17ºC, eggs do not hatch. The larval stage can take 70.1 days when environmental temperature is 22ºC or 33.2 dias at 28ºC, whereas the pupa stage takes 4 days at 31ºC and 9.7 days at 22ºC. Therefore, according those authors, the complete life cycle of the darkling beetle at a constant temperature of 28ºC is 42.5 days, and considering that a new flock is housed every 50 days, a new generation of darkling beetles may occur at each flock housed in the farm.

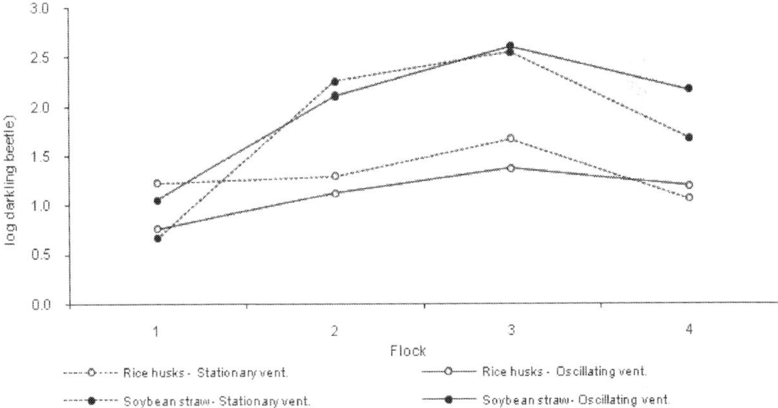

Figure 7. Average darkling beetle counts, transformed into log(y+1), in the litter of broilers reared on two different litter materials and under two ventilation systems.

In the present study, average air and litter temperatures were mild (Table 5), but nevertheless allowed the multiplication of darkling beetles. When the presence of parasites was investigated, only *Eimeria sp.* occysts were identified, and significant effects ($p<0.05$) of litter material and of the interaction between litter material and ventilation on the presence of oocystes were determined. Rice husks submitted to oscillating ventilation presented the higher percentage of contaminated pens and 18.78 more chances of being contaminated as compared to soybean straw submitted to the same ventilation (Table 4). When ventilation was stationary, no differences between litter materials were observed. When comparing ventilation types, it was found that rice husks litter contamination was significantly higher when ventilation was oscillating (odds ratio = 7.22), whereas there was no significant influence of ($p>0.05$) ventilation type on the contamination of soybean straw by *Eimeiria spp*. Other factors may have influenced litter contamination, such as the higher nutrient levels – particularly nitrogen levels – in soybean straw, explaining the lower oocyst count due to the negative effects of the release of ammonia levels that are lethal to oocysts [49]. As optimal *Eimeria* spp oocyst sporulation, which makes it infective, occurs at temperatures of 28 to

30°C [49], data were tested to verify if oocyst counts could have changed with litter temperature variations (Table 5), but the results showed that litter temperature did not have any significant influence on oocyst counts.

Litter material	Ventilation		Odds ratio	$P''/>\chi^2$
	Oscillating	Stationary		
Rice husks	81.3%	37.5%	7.22	0.0163
Soybean straw	18.8%	31.3%	0.50	0.4182
Odds ratio	18.78	1.32		
$P''/>\chi^2$	0.0012	0.7100		

Table 4. Percentage of pens contaminated with *Eimeiria spp* and odds ratio (probability) of being contaminated

Hour	Week			
	4	5	6	Média
	Air temperature (°C)			
8:00	16.15 ± 0.45	13.03 ± 0.47	15.93 ± 0.59	15.04 ± 0.30
11:00	19.86 ± 0.48	17.08 ± 0.51	21.10 ± 0.63	19.34 ± 0.33
14:00	24.30 ± 0.49	21.14 ± 0.52	25.93 ± 0.65	23.79 ± 0.33
17:00	23.23 ± 0.39	20.64 ± 0.41	24.97 ± 0.52	22.95 ± 0.27
	Litter temperature (°C)			
8:00	17.12 ± 0.36	14.86 ± 0.50	19.77 ± 0.51	17.25 ± 0.27
11:00	20.87 ± 0.38	19.39 ± 0.53	24.12 ± 0.54	21.46 ± 0.28
14:00	24.35 ± 0.37	23.48 ± 0.51	27.71 ± 0.52	25.18 ± 0.27
17:00	24.20 ± 0.27	22.84 ± 0.38	27.53 ± 0.38	24.86 ± 0.20

Means followed by different letters in the same column are different ($p \leq 0.05$) by the F test.

Table 5. Means and standard error of the parameters air temperature (°C) and litter temperature (°C) as a function of measurement week and time

According to [43], poultry production generates nutrient-rich residues that can be utilized to generate energy or to fertilize crops; however, their application to the soil must follow nutrient management plans in order to prevent environmental impacts. Considering the composition of rice husks and soybean straws before its utilization and the values established in Normative Instruction n. 23 for simple organic fertilizers, both materials are very different, except for organic carbon content, and do not comply with the recommendations of IN-23 for organic fertilizers. According to [10], considering that plant nutrient requirements vary as a function of cultivar, soil, expected yield, etc., and exceeding supplied quantities remain in the soil and are susceptible to leaching and percolation, it is essential to balance soil and litter compositions. Based on these considerations, the knowledge on the quality (physical-chemical composition, compactness, and reutilization) of litter materials used as alternative to wood shavings is essential as the disposal of such material is part of good production practices. It must also be mentioned that the chemical characteristics of the litter materials after three flocks comply, in terms of their nutritional aspects, with the legislation relative to simple organic fertilizer. However, it is recommended that the litter removed from the poultry house is distributed in rows for an additional composting period in order to eliminate or reduce health risks. Despite the effect of litter material on the evaluated parameters, with higher averages promoted by soybean straw, litter made of these crop residues were similar and presented similar characteristics to those described in literature for wood shavings [44], showing that after three flocks, both materials presented excellent fertilizing characteristics. Relative to the physical aspects, when the first flock was removed, soybean straw litter was more compacted and caked as compared to rice husks litter, and this condition remained for the following three flocks, requiring labor interference to break the caked parts, including during the rearing period. At flock removal, soybean straw required more labor to allow its reuse due to the formation of a caked layer on the top of the litter. When the fourth and last flock was removed, the litter made with soybean straw had reached its maximum limit of reutilization, with decomposition of the lower layer, presenting fiber breakdown and humic matter formation. Rice husks litter, on the other hand, presented reuse conditions after the removal of the fourth flock.

At the same time also the same comparison was made using the two materials as substrate for the composting of broliler carcasses.

2.4. Rice husks and soy straw as substrate for composting of broiler carcasses.

Considering the possibility of using rice husk and soybean straw as litter, the present study aimed to evaluate these products as substrates for broiler carcasses composting. The appropriate destination of waste from poultry production is a challenge for producers. Carcasses of broilers dead during the rearing period need to be managed so as not to cause problems like unpleasant odors and attraction of flies. One alternative for destination of carcasses considered economically and environmentally acceptable is the composting [50], a natural process of organic matter decay performed by bacteria and fungi that turn the carcasses into a useful product, the compost.

The experiment was conducted in the Experimental Field of Suruvi, of Embrapa Swine and Poultry, in Concórdia (Santa Catarina State) using a composter with six cameras, with internal measures: 0.80 m wide, 1.20 m depth and 1.50 m height of the wall. Cameras were constructed with concrete floor, wooden walls, and asbestos tiles. Two types of substrate were tested for composting: soybean straw (T1) and rice husk (T2). The experiment started with new substrates, which were reused in the composting, accompanying four flocks of broilers. Three repetitions of each treatment for each composting period were randomly selected. By the end of each flocks, each camera received 10 newly slaughtered broilers adding up a total of 60 animals per flocks. The total of 10 broilers was weighed, calculating the amount of water to be added, equal to 30% of broilers weight. The composting pile was arranged on a 30 cm-layer of new substrate, placing at the beginning 5 carcasses in a same layer, and the other 5 in a second layer, covered by a new 20 cm-layer of the same substrate. After a composting period of 15 days, pile was tumbled carcasses and substrates were weighed separately, piles were rearranged and water was added in amount corresponding to 30% of carcasses weight. All the process was performed using equipment for individual protection (rubber gloves, dust mask, boots, hat and overalls). An electronic scale with capacity of 100 kg (Toledo 2124-C5) has weighed substrates and carcasses. Carcasses were placed into thick plastic bags (20 kg) and substrates into raffia bags (60 kg). For the substrate removal, it was used a rounded tip cupped shovel and a watering can to add water. The removal of carcasses waste was made with a garden spade and a polyethylene broom. Thermocouples (S. E. Test tools A20) were installed to monitor the temperature of the composting pile, with readers inserted into each camera, in the central portion of the pile, with reading at three points (top, middle and bottom) and record of data at 7, 15, 22 and 30 days after mounting the pile. On the 30th day after the start of composting, the second weighing of carcasses and substrate was performed separately, being mounted a new pile with the same substrate and the remaining waste was divided into two layers, allowing the composting for more 15 days. The procedure was repeated for four periods, reusing the same substrate, forming from the second period, three layers of carcasses, being the bottom made up by the remaining waste of the previous batch, and the other two with five newly slaughtered broilers each. By the end of 30 days of composting, samples were collected from each camera, at nine points per layers (subsamples) and taken a pool of these points at 5 layers per camera (totaling 30 samples per batch) for analyses. It was analyzed the content of dry matter, ash, phosphorus, Cu; Zn; Mn; Fe, Nitrogen, pH; Calcium and Magnesium. For the analysis of the organic carbon, with the homogenization of part of five samples of layers of each camera, it was formed a new sample for each camera, totaling six samples per period, being the analysis performed by the titration method after chemical oxidation with sulfochromic solution [25]. The variable "Carcass decomposition" was obtained by the ratio between the difference of the initial and final weight, divided by the initial weight multiplied by 100. The variable "Substrate decomposition" was obtained by the ratio between the difference of the initial and final weight of the substrate. These data were examined by an analysis of variance for the model considering the effect of the composting cycle, substrate and the interaction, using the procedure GLM of SAS [19]. Data of composting temperature were analyzed using the mixed model theory for repeated measures, considering the effects of the composting cycle, substrate, composting week and interaction of these latter two variables and 16 types of structures of matrices of variance and covariance, using the PROC MIXED of SAS [19]. The structure employed in the analysis was chosen based on the Akaike's Information Criterion (AIC). The estimation method used was the restricted maximum likelihood.

The analysis of variance for the percentage of decomposition of carcasses and substrates, a significant effect ($p < 0.05$) was detected for the interaction between period and substrate for the variable "carcass decomposition", whereas for the "substrate decomposition", a significant effect was observed only for the period. The Table 6 presents the mean values and standard errors of the % of decomposition of carcasses and substrates according to the type of substrate and period. A significant effect (p<0.05) was observed for the carcass decomposition on the 4th period in relation to the substrate type, being the soybean straw the substrate with the highest value. The variable "substrate decomposition" had values above 100 because it was added to the substrate part of the decomposed carcasses besides the addition of water and the natural process of formation of humic matter, characterized by being a complex of several elements [51] which works in the supply of nutrients to the plants, on the structure and compatibility of the soil and water retention capacity [52]. Nevertheless, the change in the shape and color of substrate particles indicate the decomposition of the substrate. The Figure 8 shows the mean profile of composting temperature according to substrates. No significant effect was observed for the interaction between week and treatment, but a significant effect (p<0.05) of the week and treatment separately. The rice husk maintained higher temperatures than the soybean straw, and in both substrates there was an increase of temperature with the tumbling of the pile from the 2nd to the 3rd week.

Substrate	Period			
	1	2	3	4
Carcass Decomposition (%)				
Rice husk	64.29 ± 1.90 a	65.25 ± 1.90a	60.18 ± 1.90 a	55.76 ± 1.90 b
Soybean straw	63.74 ± 1.90a	59.79 ± 1.90a	64.47 ± 1.90 a	62.70 ± 1.90a
Substrate Decomposition (%)				
Rice husk		106.79 ± 3.73	96.17 ± 3.73	102.79 ± 3.73
Soybean straw		104.97 ± 3.73	91.54 ± 3.73	100.90 ± 3.73

Means followed by different letters in the columns are significantly different by the F-test (p≤0.05)

Table 6. Mean values and standard errors of the % of decomposition of carcasses and substrates according to the type of substrate and period.

Nevertheless, the absolute maximum temperature registered inside the piles was higher in the camera with soybean straw, reaching 73.3°C, while the absolute maximum of the camera with rice husk reached 65.9°C. Both were close to the temperatures reported by [53] for composted mass on the fifth day after mounting the piles (57 - 71°C). [54] in the composting experiment with agroindustrial waste have obtained temperatures within the range of 40 - 60°C, with some peaks, with the highest temperature at 71°C. Biologically, the operating limits for the temperature can be classified as: > 55ºC to maximize the sanitization; 45 - 55ºC to maximize the biodegradation rate; and between 35 - 45ºC to maximize the microbial diversity [55]. In this study, temperatures remained between 35 – 45ºC, but as abovementioned, there were temperature peaks above 55ºC promoting the sanitization of the composting mass. In the first week it was also found values indicative of the maximization of the degradation rate. [56] in a composting study using different stations obtained temperatures higher than 55ºC for over 3 consecutive days, achieving the maximum reduction of pathogenic microorganisms and indicating the biosafety of the composting. [57] recommended that the conditions of time-temperature for the compost to meet the biosafety standards should be any of the following procedures: 53ºC for 5 days, 55ºC for 2.6 days or 70ºC for 30 min. [58] with poultry carcasses composting in the climate of the United Kingdom, during the autumn and winter, for 8 weeks, have obtained positive results of carcass decomposition, as well as appropriate temperatures (60ºC - 70ºC) for the control of pathogens. [59] in studies on the biosafety of composting quoted that little attention has been given to strategies to evaluate the microbiological safety of composting systems, once there are different zones of the compost (e.g. the external edge of the pile) that usually have less organic matter and lower temperature. Thus the challenges for this type of benefit are greater than conventional. The analysis of variance for physical and chemical variables of the compost, are presented a significant effect ($p<0.05$) of the interaction between period and substrate for all variables, except for pH and K. The major effect of substrate was not significant only for the dry matter and the main effect of the period was significant for all variables. The significance effect of the interaction demonstrates that the effect of the substrate type depends on the period. The levels of physical and chemical variables measured in the substrate at the time zero, before its use as a composting substrate. In this way, it can be calculated the C/N ratio of the two substrates used, of 50.37 for the soybean straw and of 101.86 for the rice husk. Meanwhile, studies were performed using different sources of waste and residue from livestock and crop production, presenting a large variation in the initial C/N ratio, from 5/1 to 513/1 [60]. The mean values and standard errors of physical and chemical variables of composts according to periods and substrate type can be found in the Table 7. There was a significant difference between the substrate types for all variables in all periods, except for the dry matter, in the periods 2 and 4. Also there was an increase in the concentrations of the different parameters, expected given the addition of carcasses at each cycle. This increase was higher and statistically significant for the soybean straw, in the variables organic C, Cu, Fe, K, N, P, Zn and pH, besides the dry matter, although this difference had not been significant. The rice husk only presented levels of ash and Mn significantly higher than the soybean straw. The products obtained with carcass composting using two substrates tested can be classified as class D "compost organic fertilizer", according to the Normative Instruction 23 [39] meeting the requirements set concerning the minimum levels of N, organic C and moisture. But in relation to the C/N ration considered an indicative of the process maturity level [61], only the compost with soybean straw presented desired levels on the 3rd (17.75) and 4th (13.29) periods. The compost with rice husk would need to be subjected to a secondary composting to reduce this ratio and meet the requirements of the IN (maximum C/N of 18) or be used for composting new carcasses until reaching the suitable C/N ratio.

Substrate	Period			
	1	2	3	4
Ash (%)				
Rice husk	14.45 ± 0.30a	15.55 ± 0.30a	15.05 ± 0.30 a	15.46 ± 0.30a
Soybean straw	6.19 ± 0.30 b	6.61 ± 0.30b	8.52 ± 0.30b	9.67 ± 0.30b
Organic C (g/kg)				
Rice husk	306 ± 3.19b	326 ± 3.19b	301 ± 3.19 b	296 ± 3.19b
Soybean straw	345 ± 3.19a	384 ± 3.19a	355 ± 3.19a	336 ± 3.19 a
Cu (mg/kg)				
Rice husk	1.78 ± 0.33b	2.95 ± 0.33b	2.29 ± 0.33b	3.62 ± 0.33 b
Soybean straw	7.04 ± 0.33a	6.82 ± 0.33a	6.88 ± 0.33a	9.92 ± 0.33a
Fe (mg/kg)				
Rice husk	224 ± 71.37 b	256±71.37b	263±71.37b	463±71.37b
Soybean straw	1384 ± 71.37a	852±71.37a	980±71.37a	1544±71.37 a
K (mg/kg)				
Rice husk	1828 ± 373.0b	2706±373.0b	3398±373.0 b	3688±373.0 b
Soybean straw	11983 ± 373.0a	13574±373.0a	13717±373.0 a	15226±373.0 a
Dry matter (%)				
Rice husk	81.74 ± 0.59 a	86.83 ± 0.59a	80.65 ± 0.59a	82.82 ± 0.59 a
Soybean straw	78.96 ± 0.59 b	86.70 ± 0.59a	83.02 ± 0.59 b	83.71 ± 0.59a
Mn (mg/kg)				
Rice husk	211 ± 7.48a	282 ± 7.48a	196 ± 7.48a	236 ± 7.48 a
Soybean straw	56.45 ± 7.48 b	48.21 ± 7.48b	61.25 ± 7.48 b	64.04 ± 7.48b
N (mg/kg)				
Rice husk	5058 ± 711.7b	7416±711.7b	11962±711.7b	14020±711.7b
Soybean straw	10233 ± 711.7a	13491±711.7a	19993±711.7a	25281±711.7a
P (mg/kg)				
Rice husk	714 ± 166.4b	1017±166.4b	1870±166.4a	1387±166.4b
Soybean straw	1295 ± 166.4a	1757±166.4 a	1247±166.4b	2096±166.4a
pH				
Rice husk	8.88 ± 0.06b	8.99 ± 0.06b	8.64 ± 0.06b	8.33 ± 0.06b
Soybean straw	9.20 ± 0.06a	9.30 ± 0.06a	8.95 ± 0.06a	8.64 ± 0.06a
Zn (mg/kg)				
Rice husk	13.87 ± 1.16 b	14.61 ± 1.16b	20.03 ± 1.16b	24.63 ± 1.16b
Soybean straw	19.52 ± 1.16a	20.21 ± 1.16a	29.36 ± 1.16a	41.89 ± 1.16a

Means followed by different letters in the columns are significantly different by the F-test ($p \leq 0.05$)

Table 7. Mean values and standard errors of physical and chemical variables of the composts according to the periods and type of substrate.

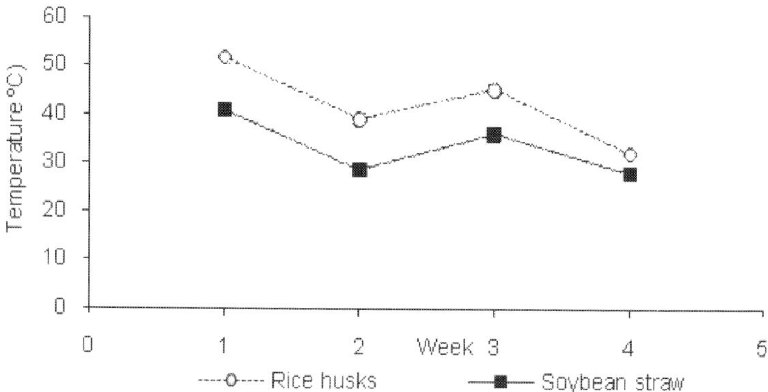

Figure 8. Mean profile of composting temperature according to substrates

Regarding the pH, both substrates presented varied values during the experimental period and reached the levels required by the IN-23 (pH 6.0) by the end of the 4th period. [62] found in the final compost in similar experiments, values of water pH in the range between 8.20 and 9.34. The levels of nitrogen increased at the end of composting periods. [54] achieved a compost with values of N, P, K between 17,710 -26,700 mg/Kg, 4,810-6,600 mg/Kg and 5,000-13,000 mg/Kg, respectively.

3. Conclusion

As compared to soybean straw, the use of rice husks as broiler litter material promotes better live performance of broilers up to 42 days of age. The use of soybean straw litter increases the incidence of footpad lesions relative to rice husks litter. Enterobacteria counts in broiler litter are reduced after downtime (day 0) when the litter is consecutively used by four flocks. Air relative humidity was higher when rice husks were used as litter material. Broiler litter used for three flocks, in average, complies with the minimal legal requirements to be traded as simple organic fertilizer, independently from the material. Soybean straw can be used as litter for rearing up to four flocks of broilers. At this same number of flocks, rice husks remain usable, whereas soybean straw is deteriorated and under humification. The number of darkling beetles was higher in the soybean straw litter and the rice husks litter presented 18.78 more chances of being contaminated with oocysts when ventilation was oscillating as compared to soybean straw litter. The soybean straw can represent an alternative of substrate for poultry carcasses composting, reaching values of C/N required by legislation with three reuses. Likewise, the rice husk can be used in composting broilers carcasses and can be reused for a greater number of times. The soybean straw presented a higher percentage of carcasses decomposition by the end of the 4th composting period (p<0.05).

Acknowledgements

The authors thank Fundação de Apoio à Pesquisa de Santa Catarina - FAPESC, for funding this study, Unifrango Agroindustrial de Alimentos Ltda., in the person of the farmer Mr. Arsênio for supplying the litter material, and Roster Ind. e Com. Ltda for lending the fans.

Author details

Valeria Maria Nascimento Abreu*, Paulo Giovanni de Abreu, Doralice Pedroso de Paiva, Arlei Coldebella, Fátima Regina Ferreira Jaenisch, Taiana Cestonaro and Virginia Santiago Silva

*Address all correspondence to: Valeria.abreu@embrapa.br

Brazilian Agricultural research Corporation, Embrapa Swine & Poultry, Concórdia, SC, Researh DT CNPq, Brazil

References

[1] Angelo, J.C.; Gonzales, E.; Kondo, N. et al. Material de cama: qualidade, quantidade e efeito sobre o desempenho de frangos de corte. Revista Brasileira de Zootecnia, v. 26, n.1, p.121-130, 1997.

[2] Benabdeljelil, K.; Ayachi, A. Evaluation of alternative litter materials for poultry. Journal Applied Poultry Science, v.5, p.203-209, 1996.

[3] Mizubuti, I.Y.; Fonseca, N.A.N.; Pinheiro, J.W. Desempenho de duas linhagens comerciais de frangos de corte, criadas sob diferentes densidades populacionais and diferentes tipos de camas. Revista da Sociedade Brasileira de Zootecnia, v.23, n.3, p. 476-484, 1994.

[4] Mouchrek, E.; Linhares, F.; Moulin, C.H.S. et al. Identificação de materiais de cama de frango de corte criados em diferentes densidades na época fria. In: Reunião Anual Da Sociedade Brasileira De Zootecnia, 29, 1992, Lavras. Anais... Lavras: SBZ, 1992. p. 344.

[5] Santos, E.C.; Teixeira, C.J.T.B.; Muniz, J.A. et al. Avaliação de alguns materiais usados como cama sobre o desempenho de frangos de corte. Ciência e Agrotecnologia, v.14, n.4, p.1024-1030, 2000

[6] Willis, W.L.; Murray, C.; Talbott, C. Evaluation of leaves as a litter material. Poultry Science, v.76, p.1138-1140, 1997.

[7]   Oliveira, M.C.; Carvalho, I.D. Rendimento e lesões em carcaça de frangos de corte criados em diferentes camas e densidades populacionais. Ciência e Agrotecnologia, v.26, n.5, p.1076-1081, 2002.

[8]   Araújo, J.C.; Oliveira, V.; Braga, G.C. Desempenho de frangos de corte criados em diferentes tipos de cama e taxa de lotação. Ciência Animal Brasileira, v.8, n.1, p.59-64, 2007.

[9]   Atapattu, N.S.B.M; Wickramasinghe, K.P. The use of refused tea as litter material for broiler chickens. Poultry Science, v.86, n.5, p.968-972, 2007.

[10]   Avila, V.S.; Oliveira, U.; Figueiredo, E.A.P. et al. Avaliação de materiais alternativos em substituição à maravalha como cama de aviário. Revista Brasileira de Zootecnia / Brazilian Journal of Animal Science, v.37, p.273-277, 2008.

[11]   Arends, J.J. Control, management of the litter beetle. Poultry Digest, v.46, n.542, p. 172-176, 1987.

[12]   Arends, J.J. External parasites and poultry pests. *In*: CALNEK, B.W. Diseases of Poultry. 9. ed. Ames: Iowa State Univ. Press. 1991. p.703-730 (710-712)

[13]   Chernaki-Leffer, A.M.; Biesdorf, S.M.; Almeida, L.M. et al. Isolamento de enterobactérias em *Alphitobius diaperinus* and na cama de aviários no oeste do Estado do Paraná, Brasil. Revista Brasileira de Ciência Avícola/Brazilian Journal of Poultry Science, v.4, n.3, p.243-247, 2002.

[14]   Vittori, J.; Schocken-Iturrino; R.P.; Trovó, K.P. et al. *Alphitobius diaperinus* como veiculador de *Clostridium perfringens* em granjas avícolas do interior paulista-Brasil. Ciência Rural, v.37, n.3, p.894-896, 2007.

[15]   Watson, D.W.; Denning, S.S.; Zurek, L. et al. Effects of lime hidrated on the growth and development of darkling beetle, *Alphitobius diaperinus*. International Journal of Poultry Science, v.2, n.2, p.91-96, 2003.

[16]   Salin, C.; Delettre, Y.R.; Vernon, P. Controling the mealworm *Alphitobius diaperinus* (Coleoptera: Tenebrionidae) in broiler and turkey houses: field trials with a combined inseticide treatment: Insect Growth Regulator an Pyrethroid. Journal of Economic Entomology, v.96, n.1, p.126-130, 2003.

[17]   Surgeoner, G.A.; Romel, K. Control of the lesser mealworm in poultry houses (Coleoptera: Tenebrionidae). University of Ghelph. Disponível em: <http://www.uoguelph.ca/pdc/ Factsheets/FactsheetList.html>. Acesso em: 10/11/05.

[18]   Silva, V.S.; Voss, D.; Coldebella, A. et al. Efeito de tratamentos sobre a carga bacteriana de cama de aviário reutilizada em frangos de corte. Concórdia: Embrapa Suínos e Aves, 2007. 4p. (Embrapa Suínos e Aves. Comunicado Técnico, 467).

[19]   Sas Institute Inc. System for Microsoft Windows: release 9.1. Cary, 2002-2003. 1 CD-ROM.

[20] Xavier, L.H. Modelos univariado e multivariado para análise de medidas repetidas e verificação da acurácia do modelo univariado por meio de simulação. 2000. Tese (Mestrado) - Escola Superior de Agricultura "Luiz de Queiroz", Piracicaba.

[21] Safrit, R.D.; Axtell, R.C. Evaluations of sampling methods for darkling beetles (*Alphitobius diaperinus*) in the litter of turkey and broiler houses. Poultry Science, v. 63, p. 2368-2375, 1984.

[22] Windham, W.R. (Ed.) Animal feed. In: CUNNIFF, P. (Ed.) Official methods of analysis of AOAC international. 16. ed. Arlington: AOAC International, 1995. v. 1, cap. 4, p.1, 4, 27 (Method 965.17).

[23] Sindicato Nacional Dos Fabricantes De Ração. Compêndio brasileiro de alimentação animal. São Paulo: SINDIRAÇÕES, 1998.

[24] BRASIL. Ministério da Agricultura, Pecuária and Abastecimento (MAPA) Manual de métodos analíticos oficiais para fertilizantes minerais, orgânicos, organominerais and corretivos. Disponível em: <http://www.agricultura.gov.br>. Acesso em: 23 out.2007

[25] Tedesco, M.J.; Gianelo, C.; Bissani, C.A. et al. Análise de solo, plantas and outros materiais. 2. ed. Porto Alegre: UFRGS/Dep de Solos, 1995.

[26] Mendes, A.A.; Patricio, I.S.; Garcia, E.A. Utilização de fenos e gramíneas como material de cama para frangos de corte. In: Congresso Brasileiro De Avicultura, 10, 1987, Natal. Anais... Campinas: UBA, 1987. p.135.

[27] Chamblee, T.N.; Yeatman, J.B. Evaluation of Rice hull ash as broiler litter. The Journal of Applied Poultry Research, v.12, n.4, p.424-427, 2003.

[28] Grimes, J.L.; Carter, T.A.; Godwin, J.L. Use of a litter material made from cotton waste, gypsum and old newsprint for rearing broiler chickens. Poultry Science, v.85, v.4, p.563-568, 2006.

[29] Toghyani, M.; Gheisari, A.; Modaresi, M. et al. Effect of different litter material on performance and behavior of broiler chickens. Applied Animal Behaviour Science, v. 122, p.48-52, 2010.

[30] Huang, Y.; Yoo, J.S.; Kim, H.J. et al. Effect of bedding tipes and different nutrient densities on growth performance, visceral organ weight, and blood characteristics in broiler chickens. The Journal of Applied Poultry Research, v.18, n.1, p.1-7, 2009.

[31] Paganini, F.J. Reutilização de cama na produção de frangos de corte: porquê, quando e como fazer. In: Conferência Apinco 2002 De Ciência E Tecnologia Avícolas, 2002, Campinas, SP. Anais... Campinas, SP: APINCO, 2002. p.194 – 206.

[32] Azevedo, A.R.; Costa, A.M.; Alves, A.A. et al. Desempenho produtivo de frango de corte de linhagem Hubbard, criados sobre diferentes tipos de cama. Revista Científica de Produção Animal, v.2, n.1, p.52-57, 2000.

[33] Fiorentin, L. Reutilização da cama de frangos e as implicações de ordem bacteriológica na saúde humana e animal. Concórdia: Embrapa suínos e Aves, 2005. 23p. (Embrapa Suínos e Aves. Documentos, 94).

[34] Macklin, K.S.; Hess, J.B.; Bilgili, S.F. et al. Bacterial levels of pine shavings na sand used as poultry litter. The Journal of Applied Poultry Research, v.14, n.2, p.238-245, 2005.

[35] Abreu P.G.; Abreu, V.M.N. Conforto térmico para aves. Concórdia: Embrapa Suínos e Aves, 2004, 5p. (Embrapa Suínos e Aves, Comunicado técnico, 365)

[36] Soliman, E.S.; Taha, E.G.; Sobieh, M.A.A. et al. The influence of ambient environmental conditions on the survival of salmonella enteric serovar typhimurium in poultry litter. International Journal of Poultry Science, v.8, n.9, p.848-852, 2009.

[37] Oliveira, M.C.; Goulart, R.B.; Silva, J.C.N. Efeito de duas densidades e dois tipos de cama sobre a umidade da cama e a incidência de lesões na carcaça de frango de corte. Ciência Animal Brasileira, v.3, n.2, p.7-12, 2002

[38] BESSEI, W. Welfare of broilers: a review. World's Poultry Science Journal, Vol. 62, September 2006, n3, pp 455-466

[39] BRASIL. Ministério da agricultura, Pecuária e Abastecimento. Secretária de Defesa Agropecuária. Instrução Normativa nº 23 de 31 agosto, 2005. Diário Oficial da República Federativa do Brasil, Brasília, 8 set. 2005. Seção 1. p. 12-15.

[40] Coufal,C.D.; Chavez, C.; Niemeyer, P.R. et al. Effect of top-dressing recycled broiler litter on litter production, litter characteristic, and nitrogen mass balance. Poultry Science, v. 85, p.392-397, 2006.

[41] Moore Jr, P.A.; Daniel, T.C; Eduards, D.R. et al. Evaluation of chemical amendments to reduce ammonia volatilization from poultry litter. Poultry Science, v.75, p.315-320, 1996.

[42] Singh, A.; Bicudo, J.R.; Tinoco, A.L. et al. Characterization of nutrientes in built-up broiler litter using trench and random walk sampling methods. The Journal of Applied Poultry Research, v.13, p.426-432, 2004.

[43] Oviedo-Rondón, E.O. Tecnologias para mitigar o impacto ambiental da produção de frangos de corte. Revista Brasileira de Zootecnia, v.37, suplemento especial, p. 239-252, 2008.

[44] Turazi, C.M.V.; Junqueira, A.M.R.; Oliveira, A.S. et al. Horticultura Brasileira, v.24, n. 1, p.65-70, 2006.

[45] Pinto, D.M.; Ribeiro, P.B.; Bernardi, E. Flutuação populacional de *Alphitobius diaperinus* (Panzer, 1879) (Coleoptera: Tenebrionidae), capturados por armadilha do tipo sanduíche, em granja avícola, no Município de Pelotas, RS. Arquivos do Instituto Biológico, v.72, n.2, p.199-203, 2005.

[46] Godinho, R.P.; Alves, L.F.A. Método de avaliação de população de cascudinho *(Alphitobius diaperinus)* panzer em aviários de frango de corte. Arquivos do Instituto Biológico, v.76, n.1, p.107-110, 2009

[47] Uemura, D.H.; Alves, L.F.A.; Opazo, M.U. et al. Distribuição and dinâmica populacional do cascudinho *Alphitobius diaperinus* (Coleoptera: Tenebrionidae) em aviários de frango de corte. Arquivos do Instituto Biológico, v.75, n.4, p.429-435, 2008.

[48] Chernaki, A.M; Almeida, L.M. Exigências térmicas, period de desenvolvimento and sobrevivência de imaturos de *Alphitobius diaperinus* (Panzer) (Coleoptera: Tenebrionidae). Neotropical Entomology, v.30, n.3, p.365-8, 2001

[49] Shirley, M.W. Epizootiologia. In: Simpósio Internacional Sobre Coccidiose, 1994, Santos. Anais... Santos: [s.n.], 1994, p.11-22.

[50] Macsafley, L. M.; DUPOLDT, C.; GETER, F. Agricultural waste management system component design. In: Krider, J. N.; Rickman, J. D. (Ed.). Agricultural waste management field handbook. Washington, D.C.: Department of Agriculture, Soil Conservation Service, 1992. p. 1-85.

[51] Diniz Filho, E. T.; Mesquita, L. X.; Oliveira, A. M.; Nunes, C. G. F.; Lira, J. F. B. A prática da compostagem no manejo sustentável de solos. Revista Verde de Agroecologia e Desenvolvimento Sustentável, v. 2, n. 2, p. 27-36, 2007.

[52] Budziak, C. R.; Maia, C. M. B. F.; Mangrich, A. S. Transformações químicas da matéria orgânica durante a compostagem de resíduos da indústria madeireira. Química Nova, v. 27, n. 3, p. 399-403, 2004.

[53] Rynk, R. ed. On farm composting handbook. Ithaca: Northeast Regional Agricultural Engineering Service, 1992. 186p. (Cooperative Extension. NRAES, 54).

[54] Fiori, M. G. S.; Schoenhals, M.; Follador, F.A. C. Análise da evolução tempo-eficiência de duas composições de resíduos agroindustriais no processo de compostagem aeróbia. Engenharia Ambiental, v. 5, n. 3, p. 178-191, 2008.

[55] Hassen, A.; Belguith, K.; Jedidi, N.; Cherif, A. Microbial characterization during composting of municipal solid waste. Bioresource Technology, v. 80, n. 3, p. 217-25, 2001.

[56] Sivakumar, K.; Kumar, V. R. S.; Jagatheesan, P. N. R.; Viswanathan, K.; Chandrasekaran, D. Seasonal variations in composting process of dead poultry birds. Bioresource Technology, v. 99, n. 2, p. 3708-3713, 2008.

[57] Haug, R. T. The practical handbook of compost engineering. Boca Raton: Lewis Publishers Press Inc., 1993.

[58] Lawson, M. J.; Keeling, A. A. Production and physical characteristics of composted poultry carcases. British Poultry Science, v. 40, n. 5, p. 706-708, 1999.

[59] Wilkinson, K. G. The biosecurity of on-farm mortality composting. Journal of Applied Microbiology, v. 102, n. 3, p. 609-618, 2007.

[60] Valente, B. S.; Xavier, E. G.; Morselli, T. B. G.A.; Jahnke, D. S.; Brum Jr., B. S.; Cabrera, B. R.; Moraes, P. O.; Lopes, D. C. N. Fatores que afetam o desenvolvimento da compostagem de resíduos orgânicos. Archivos de Zootecnia, v. 58, n. 1, p. 59-85, 2009.

[61] Reis, M. F. P.; Escosteguy, P. V.; Selbach, P. Teoria e prática da compostagem de resíduos sólidos urbanos. Passo Fundo: UPF, 2004.

[62] Kumar, V. R. S.; Sivakumar, K.; Purushothaman, M. R.; Natarajan, A.; Amanullah, M. M. Chemical changes during composting of dead birds with caged layer manure. Journal of Applied Sciences Research, v. 3, n. 10, p. 1100-1104, 2007.

Chapter 11

Acrylated-Epoxidized Soybean Oil-Based Polymers and Their Use in the Generation of Electrically Conductive Polymer Composites

Susana Hernández López and
Enrique Vigueras Santiago

Additional information is available at the end of the chapter

http://dx.doi.org/10.5772/52992

1. Introduction

1.1. Electrical properties on polymer compounds with carbon black making emphasis in the polymer matrix for controlling them

Polymer composites with electrical properties modified by the amount of conductive filler particles emerge at a half of the XX Century. This type of materials are conformed by two components, the first is usually majority and continuous component known as matrix and the second is the minority and discontinuous frequently named conductive filler. Both, matrix and filler could be constituted by one or many different materials. For example, the matrix could be only one polymer or it may well be constituted by a polymer blend, the same way the conductive component. The polymer composites are characterized for being multiphase materials in which is possible to distinguish the two phases: polymer and conductive filler. The own properties of these composites are determined by the synergic coupling between properties of the matrix and the conductive particles, this means the electrical conductivity on these composites and in turn on their practical applications are determined by the chosen polymer matrix-conductive filler couple.

Modification of the electrical properties in polymer composites based on carbon black (CB) particles could reach until 10^{11} orders of magnitude. This characteristic has allowed designing materials with applications for electrostatic protection and electromagnetic shields. It is well known that electrical properties are modified by external factors making them useful for applications as vapor, gases, toxic or inert substances sensors as well as detectors of temper-

© 2013 Hernández López and Vigueras Santiago; licensee InTech. This is a paper distributed under the terms of the Creative Commons Attribution License (http://creativecommons.org/licenses/by/3.0), which permits unrestricted use, distribution, and reproduction in any medium, provided the original work is properly cited.

ature and external mechanical strengths. For this reason from the 70´s the massive develop of polymer compounds plus conductive particles becomes important [1] and from the beginning it has been used all type of oleo-polymers and conductive particles being of major interest the carbon particles and nanoparticles. These carbonaceous particles allow obtaining polymer composites highly conductive, at great scale, at low cost, with the inherent polymer properties as lightness, processability into different shapes, sometimes reinforced in mechanical and thermal properties. The CB particles have very good conductivity thanks to a series of structural characteristics as the graphitic composition, the structure of the fundamental aggregate that could be low, middle or high, being a high structure [1] the required in order to render lower critical volume or weight percent (wt%) fractions. The particle surface is another important characteristic to take in account for having a good dispersion into the polymer. Some CB surfaces are oxidized and contain many functional groups as hydroxyls, carbonyls, esters, etc [1] which facilitate the interaction with the polymer matrix having a very good dispersion in it. Those and other characteristics as their easy synthesis at great scale make them the most studied conductive particles for preparing conductive polymer composites.

If the difference in the electrical conductivity between conductive particles and the matrix is at least six orders of magnitude then the electrical conductivity of polymer compounds in terms of the conductive particles load could be explained very well by the percolation concept [2]. According to this concept the electrical conductivity is achieved when the conductive particles reach a critical fraction which is the smallest amount of particles required to build a conductive network. This is established by the relationship 1.

$$\sigma(X) \approx (X - X_c)^{\beta} \tag{1}$$

X represents the volumetric fraction of the conductive filler, X_c the amount of conductive particles in the percolation threshold and is a critical exponent. The critical parameters Xc and β determine the transition from the dielectric to the conductive state in those polymer composites. This make an important difference in the electrical properties respect to the conventional materials in which is possible to determine the type of material according to the order of magnitude in its electrical conductivity. In composite materials the electrical conductivity is achieved when a volume or weight filler fraction is reached and it is characterized because the particles are in electric contact building conductive paths. In this conductive network the electrical interconnection is favored when the transit of electrons is allowed from one particle to another; transitions which are governed by the mechanical quantum laws [3]. The percolation model given by equation 1 fits very well to the experimental results allowing to explain the mechanism for which these composites could be considered as intelligent materials, useful for the detection of temperature gradients, pressure, displacement, solvents, gases, etc. For all those reasons the percolation model has been accepted as the well way to explain the electric conduction in terms of conductive particles fraction. However it has the important disadvantage to no predict the percolation threshold for a specific polymer-conductive particles couple.

Such disadvantage emerges from the fact that it only considers the particles interconnection probability without taking in account the polymer matrix nature and the possible interactions between particles and polymer.

The critical fraction is also associated with the potential applications of the polymer composite. In polymer composites with high critical fraction the mechanical properties could be influenced due the secondary forces between polymer chains are modified by the presence o a high fraction of particles and as a consequence the mechanical properties of the composite could detriment. On another hand in solvent sensing applications it has been demonstrate that as lower conductive particles fraction the sensibility to the detection is considerably increased due to the best interaction area between the polymer and the solvent, vapor or gas [4]. This is the reason for which one of the goals in the preparation of conductive composites is to diminish the critical fraction of conductive particles in conductive polymer composites. However there is not a generalized rule which allow predicting it considering both the polymer and the conductive particles properties. It is known that properties of the matrix polymer as density, viscosity at molten state, dielectric constant and those of the conductive particles as oxidative surface, size, shape and structure, have an important influence in the electrical composite properties in junction with the preparation method [1, 5-17]. The conditions in preparation method are especially important to achieve a homogeneous disaggregation and dispersion of the conductive particles and in turn to reach the electrical properties at a low fraction of conductive particles. A preferential and controlled distribution of those particles contributes to build conductive networks at a low critical fraction and it also guarantees a better reproducibility. Some strategies to obtain a preferential distribution have used polymer matrix derived from two or more immiscible polymers (blends) [10] in which the particles are in the interphase; the *in situ* synthesis using the emulsion polymerization [11] in which particles prefer to surround the polymer matrix nanoparticles. The polymer matrix viscosity [8] is very important to take in account for the conductive particles dispersion. In a high viscose polymer are required high shear forces in order to produce a disaggregation of the particles but also it could produce a breaking of the CB particles structure. This last effect is not convenient due are necessary more particles to interconnect them and form the conductivity network, increasing the critical fraction. Usually the strength distribution is not uniform, not a good dispersion is achieved and the conductivity is not reached a low fraction of conductive particles. In summary the polymers with low viscosity tend to reach the percolation threshold at a lower critical fraction in comparison with polymers highly viscose. The weight molecular mass [12] also influences the process of conductive particles distribution and is directly associated with the viscosity. In according with the latest discussion, as higher the molecular weight the polymer has a higher viscosity and viceverse, in such a way that conductive particles have less problems to disperse in a low molecular weight matrix. The polymer morphology [13] determines in some way the preparation process and the critical fraction of the conductive particles in the percolation threshold. It has been well established that particles tend to disperse uniformly in the amorphous region. As a consequence in semicrystalline matrixes is possible to have a preferential distribution of the conductive particles in the amorphous and interphases zones diminishing the critical fraction. In other cases the polarity of the polymer matrix has taken in account for dispersing CB particles and there are many opinions respect to its influence on the critical fraction [9, 14,15]. It has been mentioned that a major polarity the critical fraction increases [9]. However other authors claim [14, 15] that a major polarity the particles dispersion is better due to the interactions between the groups on the surface of the CB particles and those on or pending on the polymer chain favoring their dispersion. Recently in a systematic study [16] has been demonstrated that critical fraction of CB particles included in some polystyrene-based polymers with different dielectric constant, tend to diminish as the polarity (dielectric constant) of the polymer matrix increases.

In other recently studies, has been evaluated that polymer derivate from vegetable oils as the soybean and linseed oil and some derivates have a very good compatibility not only with some natural fillers as fibers, henequen, cellulose, bamboo, etc, but also with CB particles and carbon nanotubes producing composites with a huge gamma of mechanical and thermal properties, similar to those offered for oleo-polymers, with a great potential to use them as engineering applications. Much less research has been focus to the electrical properties however those studies shown that critical fraction in composites based on acrylated-epoxidized soybean oil and epoxidized linseed oil with CB are much lower (lower than 4 wt% CB) than those exhibited by common oleo-polymers (usually higher than 8 wt % CB) [17]. This fact has been explained in terms of the functional groups of the polymer matrixes and on the ability to crosslink drawing paths in which the CB particles are favorable dispersed. In order to understand this phenomena, the next section will be devoted to describe the structure, characteristics and uses of vegetable oils, specifically focus to one derivate of soybean oil, the acrylated-epoxidized soybean oil (AESO) a derivate well studied as monomer and comonomer in the preparation of conductive composites with different CB particles.

1.2. Acrylated-epoxidized soybean oil: Some relevant advances and achievements

Polymeric materials prepared from renewable natural resources have become increasingly important because of their low cost, ready availability, and potential biodegradability. Vegetable oils represent one of the cheapest and most abundant biological feedstocks available in all around the world in large quantities, and their use as starting materials offers numerous advantages, such as low toxicity, biocompatibility, inherent biodegradability as well as certain excellent frictional properties e.g. good lubricity, low volatility, high viscosity index, solvency for lubricant additives, and easy miscibility with other fluids, and more recently in useful polymers and polymer composites [18-21].

In addition to their application in medicine, cosmetics, surfactants, nutrition, food industry, lubricants and as an alternative fuel for diesel, natural oils also have been used quite extensively to produce coatings, inks [22] plasticizers, lubricants, and agrochemical [23]. Structurally vegetable oils are predominantly constituted of triglyceride molecules. Triglycerides in turn are constituted of three fatty acids joined at a glycerol central structure. Most common oils contain fatty acids that vary from 14 to 22 carbons in length, with 0 to 3 double bonds per fatty acid. Because of the many different fatty acids present, these oils are composed of many different types of triglycerides with numerous levels of unsaturation. A high degree of multiple unsaturations (-C=C-) in the fatty acid (FA) chain of many vegetable oils causes poor thermal and oxidative stability and confines their use as lubricants to a modest range of temperature [24]. Although they possess double bonds, which could be used as reactive sites, they are of low reactivity and cross-linked polymers could be produced under a specific and limited reaction conditions [25-29]. Usually as triglycerides are made up of aliphatic chains, the produced materials lack of the necessary rigidity and strength required for some applications. To reach a higher level of molecular weight and crosslink density is necessary to copolymerize [30-34] or to incorporate chemical functionalities [35] easily to polymerize and that are known to improve the mechanical properties of the polymer networks. The double bonds are usually used to functionalize the triglyceride with polymerizable chemical groups, for example to convert the unsaturation to epoxy group is the most common reaction. There are a lot of published studies devoted to establish [36-40], control [41, 42] and scale [43] many different reaction conditions for partial [44-46] or complete epoxidization of the trygliceride's double bonds. By themselves the epoxidized oils are polymerizable under temperature [47-49] ultraviolet radiation [50-53], or by opening epoxy ring reaction. However is well know that the diamines [54, 55], anhydrides [56, 57], dicarboxylic acids [56], dioles [58, 59] are also used for curing epoxidized oils rendering cured resins with a gamma of properties controlled by the stoichiometry, the structure of the crosslinker and the cure degree.

Epoxidation in general is a commercially important reaction due to the high reactivity of that functional group that makes it to be readily transformed into other important functional and polymerizable groups. The most studied and important polymerizable groups are hydroxyl by complete or partial ring opening reaction of epoxy group [60] making the triglyceride capable of reaction via addition polymerization for producing polyurethanes [61-63] or hydroxylated polyesters for example with maleic anhydride which could further being cured by free radical reaction. Another important functional group is the acrylate inserted by reaction of epoxidized vegetable oils with acrylic acid [18, 64]. This last modification provides monomers which can then be blended with reactive diluents and cured by free radical polymerization. There are natural oils comprise fatty acids with these types of functionalities as the vernonia oil which contain epoxy groups and castor oil that comprise hydroxyl groups.

Soybean oil is an exemplar model of how the natural oils can be used for producing polymer and polymer composites useful in some applications. About 80% of the soybean oil produced each year is used for human food. Another 6% is exploiting for animal feed, while the remainder (14%) has nonfood uses (soap, fatty acids, lubricants, coatings, etc.). Structurally (Figure 1), soybean oil usually contains in average 11% palmitic, 4% stearic, 23% oleic, 54% linoleic and 8% linolenic acids. Palmitic acid contains 16 length carbons and no unsaturations, the remanent acids are all of them are 18 length carbons and are unsaturated acids: oleic (1 double bond), linoleic (2 unsaturations) and linolenic (3 double bonds). Soybean oil contain in average 4.5 double bonds which are thermally polymerizable under the special conditions above mentioned.

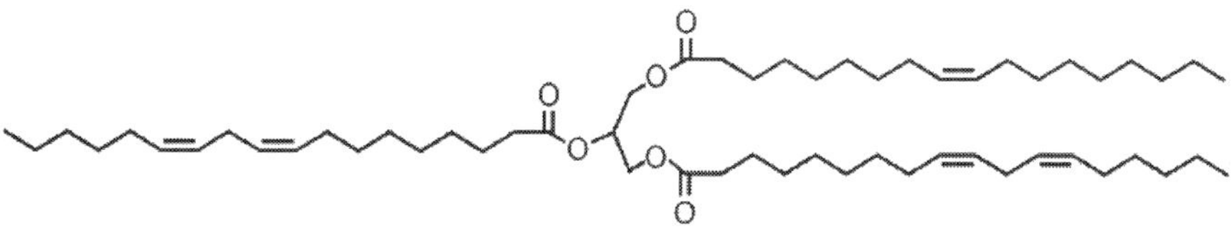

Figure 1. Representative structure of SBO showing three 18 carbon-length chains with 1 and two double bonds

For example, cationic polymerization of the soybean oil with divinylbenzene comonomer initiated by boron trifluoride diethyl etherate produces crosslinked polymers with properties from the rubbers (soft) to the thermosets (rigid) [34, 65]. These properties are dependent on the stoiquiometry of both comonomers. Some others copolymers prepared also by cationic polymerization [66] showed shape-memory effect which refers to the ability to remember a specific shape after deformation. Polymerized soybean oils have already been employed in printing inks and paints and much effort has been devoted to converting soybean oils into solid polymeric materials which now usually possess viable mechanical properties and thus may be useful as structural materials in a variety of specific applications [67]. In this context, one of the most important derivate is the epoxidized soybean oil (ESO). The epoxidized soybean oil is widely used as plasticizer instead of phtalates in the plastic industry to increase flexibility, stability and processability in PVC products. For this application the higher the epoxy degree more efficient is the stabilizer ability [41] to heat and UV-radiation. It was also studied as potential source of high-temperature lubricant [24]. It shows a low thermal and oxidative deteriorate compared with raw soybean oil and other oils due to the absence of allyl hydrogens which are the responsible of those unwanted process. Crosslinked polyesters were prepared by curing ESO with different dicarboxylic acid anhydrides [56]. The mechanical and thermal properties were evaluated and they were dependent on the type of anhydride, the type and wt% of catalyst and on the ratio ESO/anhydride. A broad range of glass transition temperatures, T_g's, from -5 to 75°C were shown and flexural modulus from 520 to 980 MPa. Another example of ESO cured with different acid anhydrides [57] in presence of tertiary amines as initiators. The dynamical mechanical properties were studied in terms of the type of anhydride, initiator and epoxidation level. Composites based on ESO resin and chicken feathers fibers (CFFs) were studied as potential applications in electronic devices such as printed circuits and boards (PCBs) [68]. Equipment based on PCBs requires high electrical resistance, relatively low electric constant and loss. Properties were compared with those of the prepared composites using the standard epoxy resin for PCBs and CFFs as reinforcement. The resistivity of CFFs composites was two to four orders of magnitude higher than E-glass fiber composites indicating that CFFs have better insulating effect. The CFF composite can be potentially used as PCBs industry. In other study ceramer coatings based on ESO were prepared using three sol-gel titanium (IV) and zirconium precursors and several coating properties were evaluated, exhibiting excellent flexibility and hardness. Ceramer coatings (inorganic/organic hybrid materials) have potential applications in protecting optical and electronic devices. In this study the properties of ELO as low volatile organic content, low price and viscosity were taken in advantage in order to prepare the mentioned ceramers [69].

Another important derivate of soybean oil (Figure 2), in which this chapter will be devoted, is the acrylated-epoxidized soybean oil (AESO), produced in two steps from soybean oil. The first one is the epoxidation of soybean oil by any described method. After, this epoxy intermediate is reacted with acrylic acid [21, 64] in presence of an acid as catalyst, being the level of acrylatation very important for the mechanical properties as studied in [70]. T_g increases linearly with the number of acrylates per triglyceride from -50 to 92°C for 0.6 to 5.8 acrylates per triglyceride and it is possible to obtain soft and rigid polymers as the level of acrylatation increases.

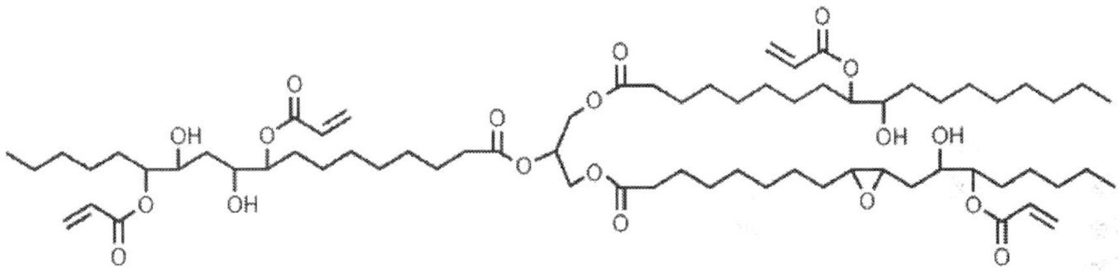

Figure 2. Representative structure for AESO

This molecule is interesting due the possibility of use the double bonds from acrylate functional group in order to polymerize/copolymerize easily via free radicals reaction under several initiator systems as thermal initiator decomposition, photoinitiators and UV or visible radiation, and high energy radiations as gamma rays. However, pure AESO polymer (poly-AESO) does not display important mechanical properties by itself. It looks like an amorphous crosslinked rubber without possibility of processing in useful shapes. However these properties were taken in advantage for modify the mechanical and tribological properties of goat leather [71]. Elastic modulus, tensile deformation, friction and wear were evaluated for goat leather before and after to graft AESO monomer by each and both faces of the leather. The grafting was made by free radicals using a photonitiator and ultraviolet light as initiator system [72]. Those properties were dependent on the irradiation dosages (360, 720 and 2700 J/m²) being the changes more evident at higher dosages and on the grafted face (inner or outer). In general friction and tensile deformation decreased (17% and 39%) respect to ungrafted leather whereas elastic modulus and wear decreasing as the dosage increased (19% and 39%, respectively).

However for engineering applications polyAESO do not have enough mechanical properties. An attempt to improve the mechanical properties of polyASEA was made using gamma radiation as initiator system [73]. In this work AESO was successfully polymerized and the polyASEA was obtained in one transparent and homogeneous piece depending on the used mold for polymerization. That piece could be cut in different shapes. Its properties as friction and scratch were evaluated respect to the dosage irradiation which was directly related to the crosslink degree. Some probes were made for preparing composites with carbon black under the same conditions, not rendering the same successfully results. In this case those composites were obtained as an unshaped mass without the possibility to processing them into specific shapes.

It is because AESO use to be copolymerized with other vinyl comonomers. The most used is the styrene comonomer which imparts stiffness and as the same time is useful as diluent in order to reduce the viscosity which is of great help when some filler is used for preparing polymer composites. In this context, a lot of interesting works have been published. These resins have proven to be comparable to commercial, oleo-based thermosetting resins. Also, AESO has been combined with several natural fibers [74] and other fillers as glass or carbon fibers and clays in order to produce new economical composites and very useful in many fields like agricultural, automotive, infrastructures, housing, and construction. Some examples are cited. Network blends of polyurethane network films were prepared using polyester urethane acrylate (PUA) having terminal double-bond functional groups and acrylated epoxidized soybean oil by a simultaneous thermal polymerization process. The weight ratios of PUA/AESO affected the thermal and mechanical properties; with an increase in AESO content the glass-transition temperature of the networks decreased from 40 to -4.8 °C, tensile strength increased from 1.2 to 9.8 MPa, and elongation at break decreased from 470 to 70% [75]. Copolymers of AESO and styrene mixed with butyrated kraft lignin as compatibilizing agent for natural fiber reinforced thermoset composites were prepared and evaluated [76]. An improvement of adhesion of the resin to the fibers was achieved and the flexural strength increased 40% for a 5 wt% butyrate lignin. The effect of different lignin concentrations on the mechanical properties of composite made of flax and wheat straw fibers, was investigated. The ultimate goal was a lignin-based, soluble additive that gives rise to a very strong resin-natural fiber interface. For construction structural panels and unit beams were manufactured from AESO and natural fibers (flax, cellulose, pulp, recycled paper and chicken feathers) as natural fiber reinforcements (20-55 wt%), E-glass fiber and closed cell structural foam achieve good mechanical strength. The goal of this work [77] was to develop monolithic structural panels that would be suitable for use as the (load-bearing) roof, floors or walls of a home. It was compared the mechanical properties. In summary, the results show that the recycled paper beam with chicken feathers and the recycled paper beam with corrugated cardboard have flexural rigidities for a commercial wood product. The beam with some E-glass is a stiff as stiff as other three of the wood commercial references. All composites beams are stronger than the woods references. In conclusion the composite beams made for recycled paper have strengths and stiffness that make them suitable for use in structural applications where wood members would normally used. Other important works are they where AESO was copolymerized with styrene in presence of cellulose fibers or corrugated cardboard boxes [78] in order to manufacture composites structures useful for residential roof construction, rendering successfully results according to the evaluated mechanical properties.

AESO/Styrene in a 65/35 wt ratio and 3 wt% of SWNT composites were prepared [79] by sonication procedure in order to study the mechanical properties. It was found that the addition of NTC results in an increasing of the flexural modulus about in a 44% and the yield strength in 9% as well as the T_g in 8%. In other work AESO/styrene (65/35 wt ratio) also has been reinforced with impure multi walled carbon nanotubes (containing 60-70% of carbon soot) with an aspect ratio of 33.3 [80]. Impure MWNT were added in 1, 3 and 5 wt% to the monomers mix and the dispersion conditions were evaluated until obtain a stable dispersion. Mixes were polymerized via free radicals using tert-butyl peroxy benzoate as initiator and after their mechanical properties were analyzed. Modulus was increased respect to the same resin composition without MWNT in a 30% for samples having 1wt% MWNT. In both cases electrical properties were not evaluated.

However, electrical properties of polymer composites based on AESO and AESO copolymers with carbon black were studied [17]. As mentioned before, crosslinked polyAESO is not adequate to processes the respective composites into convenient shapes for evaluate electrical and other properties. Nevertheless, something that was very interesting is that the percolation concentration for this type of composite was very low (4.1 wt% CB) in comparison with those obtained with the traditional oleo-polymers based composites with carbon black as polystyrene (15 - 9.4) [16, 81], low density polyethylene (17%), polybutyl methacrylate (14%) [17], etc. This behavior is one evidence that polymer matrix in conjunction with the carbon black characteristics, is another parameter that has a very important influence in reach the percolation threshold at low critical fraction of conductive particles. A possible explanation is that AESO helps to low the concentration percolation due to CB particles disperse very well thanks to the presence of the carbonyl groups on the acrylic functional groups that participate in the crosslinking of the AESO monomer, drawing for the carbon black particles a tridimensional conductive network in which they are well distributed and dispersed. To overcome the inconvenient of processing the polyAESO composites one option was to copolymerize the AESO with other acrylic monomer as butylmethacrylate (BMA). Polymer from BMA is not a rigid polymer as polystyrene or poly(metylmethacrylate) but it contains a carbonyl group as ASEA and a short alkyl chain enough to be miscible with ASEA. After studying the electrical behavior and evaluating the critical fraction of the polyAESO and polyBMA both with CB particles, rendered values of 4 and 14%CB respectively. It was established to copolymerize both monomers, ASEA and BMA, in order to improve the processing properties and to evaluate the changes in the electrical properties respect the composites based on the respective homopolymers. Considering the great difference in the critical fraction values of polyASEA and polyBMA composites it was expected that this value was dependent on the comonomers proportion. In order to establish one only proportion it was necessary to do a scanning on the composition vs resistivity property maintaining constant the CB load which was chosen as 10 wt% CB. Once the percolation curve (Figure 3 in reference 17) was built it could be appreciate that as increased the ASEA load in the range of 0 to 10%, the resistivity decreased sharply (seven orders of magnitude) until it reached a minimum and constant value. From these results it was chosen a copolymer composition of 30:70 wt% of ASEA/MAB, taking in account that a higher quantity of ASEA produces processing problems. The results will be discussed later in section 2.1.2.

Another interesting option was to copolymerize the ASEA with another commercial modified natural-derivate comonomer named (Acrylamidomethyl) cellulose acetate butyrate (ACAB). This semicrystalline polymer is a cellulosic product soluble in organic solvents and also it is crosslinkable by free radical reaction. It is well know that cellulose is the most abundant naturally occurring organic substance which is very insoluble and its products are extensively used in non-food and industrial applications mainly as textile, paper, thickeners, flocculants, etc. Many derivates have been synthesized in order to synthetic polymer replacements, being hydroxyl groups in cellulose the most available source of chemical modification that has been exploited. For example, cellulose esters and ethers, which usually are water-insoluble, have been used as plastics for molding, in modern coatings, extrusion, laminates, controlled release of actives, biodegradable plastics, composites, optical films, and membranes and related separation media [82, 83]. ACAB was synthesized in order to have not only esters and ether functionalities, but also polymerizable vinyl functionalities by react the hydroxyl cellulosic groups with N-metilolacrilamide [84].

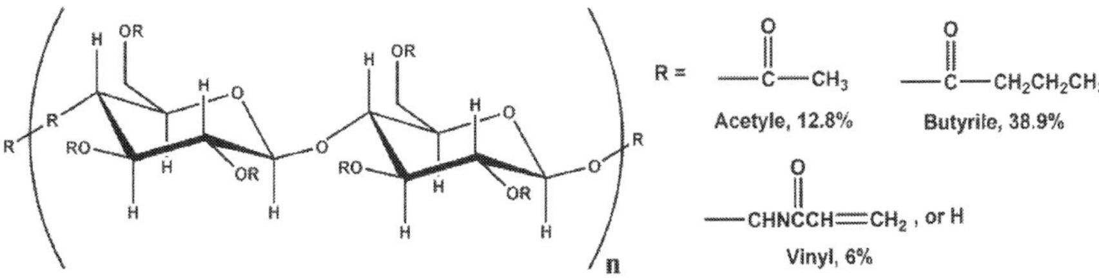

Figure 3. Structural representation and composition of commercial ACAB.

Polymerization and copolymerization could carry out by thermal, ultraviolet, gamma or electron beam radiation. It contains in average the next composition (Figure 3): 38.9% of butyril groups (~1.8 mol butyrate per mol cellulose), 12.8% of acetyl groups (~1.0 mol acetate per mol cellulose), and 0.6% of vinyl ones (~0.1 mol vinyl per mol cellulose). It is a white powder useful for produce crosslinked and insoluble resistant copolymers, hard, durables as adhesives and composites in the textile industry, to provide articles which should retain a permanent shape. Coatings are effective on wood, paper, metal, plastics and they are useful to produce containers, furniture, floors, appliances, trucks, pipe, boats, paper products, among others. Recently ACAB was used to produce columns with higher separation efficiencies of aminoacids and peptides by capillary electrochromatography [85].

Considering the good processing properties of ACAB, the miscibility with AESO, the great number of polar groups and the presence of a vinyl crosslinkable functional group, were the reasons to establish this study in which the main goal is to obtain processable conductive composites and with low critical fraction. In this study three types of CB particles were used. The properties and methodology are described in section 2.2.

The possibility to generate polymer composites from natural sources monomers with modified electrical properties is of high interest due the mentioned reasons and for the possibility of reduce the critical fraction. As mentioned before the preparation methodology is very important in order to obtain a very good CB dispersion and reproducibility of the electrical properties. The next section will be devoted to describe the preparation of the AESO-co-BMA and AESO-co-ACAB copolymers and the respective composites with CB´s as well as the characterization of products by FT-IR spectroscopy, Differential Scanning Calorimetry (DSC) and Thermogravimetric Analysis (TGA).

2. Detailed description for obtained copolymers and their structural and thermal characterization

2.1. System AESO-co-BMA

2.1.1. Reactants and equipment

The used reactants in this section are described: AESO containing in average 3.4 acrylates per molecule, and contains 4,000 ppm of monomethyl ether hydroquinone (MEHQ) as inhibitor; density 1.04 g/mL at 25°C. BMA (99%) comonomer contains 10 ppm of MEHQ; bp. 162-165°C, density 0.894 g/mL at 25°C. Potassium Bromide, KBr was spectroscopic degree; Tetrahydrofurane (THF) (99.7%), bp. 70°C; and Acetone (99%) both as solvents. All of them were purchased from Sigma-Aldrich, Co. Dibenzoyl peroxide (BPO) (> 97%), mp 102-105°C, was purchased from Merck Co; and Carbon black Vulcan X-C72 was kindly donated by Cabot Co. Before use, ASEA and BMA were surpassed by an inhibitor-remover packing (from Sigma-Aldrich, Co) in order to remove the MEHQ. Other reactants were used as received.

For Infrared characterization a FT-IR Nicolet Avatar 360 was used. All samples were mixed with previously dried KBr and pressured at 8 Tons in order to obtain transparent plates. 32 scans were recorder from 4500 to 400 cm^{-1} at a resolution of 8 cm^{-1}.

For calorimetric characterization, a SDT Q600 modulus (TGA coupled to a DSC) from TA Instruments was used. Experiments were recorder at a heating rate of 10°C/min, under nitrogen atmosphere (100 mL/min), from 23 to 500°C.

2.1.2. Synthesis and characterization of pure AESO-co-BMA copolymer, 30/70 wt%.

The pure copolymer is obtained following the next methodology, using as example 1g of total prepared compound: 0.3 g of ASEA is weighed into a 150 mL round bottom flask and 10 mL of THF are added. The mix is sonicated in an ultrasonic processor (Ultrasonik™ 28X 50/60Hz) at 5-8°C for 10 min until ASEA is completely solved. Then 0.7g (0.78mL) of BMA is added in junction with 0.05 mL of 0.5M PBO in THF and again the mix is sonicated for 2 min only, at same temperature. After, a condenser with inlet and outlet nitrogen ultrapure gas (from Infra, Co) is adapted to the flask. It was left the low nitrogen flux for 5 min and finally the temperature is increased from room temperature (RT) to 70°C in order to copolymerize. Copolymerization takes around 6 hr and it can be detected because the copolymer starts to precipitate due to the crosslinking reaction. Once the product is precipitated as several pieces, the temperature is lowered to RT, the nitrogen is turn off and the transparent lightly yellow product is filtered and washed twice with acetone to remove the possible residual monomers and initiator. The copolymer is well dried under vacuum for 24 hr. This copolymer is characterized by FTIR spectroscopy, detecting specific signals for both comonomers. Finally, the greatest piece was cut into a cylinder shapes (1.0 cm diameter x 0.2 cm thickness) and the faces were covered with silver paint in order to measure the resistivity.

For a structural characterization of the copolymer by FT-IR spectroscopy, it is necessary to know the main absorption bands of the monomers and homopolymers, in this case for MAB, polyMAB, AESO and polyAESO. For BMA, the next bands are identified: vinyl bond (=C-H) from acrylic group is found at 3100 cm⁻¹, methyl and methylene groups (-CH3-, -CH2-) between 2900 cm⁻¹ and 2800 cm⁻¹, carbonyl (C=O) from ester group at 1743 cm⁻¹, polymerizable double bond (-C=C-) from acrylic group is sited at 1640 cm⁻¹ and another vibration related with end double bonds (H2C=) also from acrylic group is sited at 950 cm⁻¹. When BMA is polymerized (Figure 3), the main changes on the respective spectrum are: the diminishing of the bands corresponding to the double bonds at 3100 cm⁻¹, 1650 cm⁻¹, and 950 cm⁻¹. These absorption bands diminish in intensity as the polymerization take place and finally they disappear when polymerization has been completed.

For AESO monomer (Figure 4) the most important identified bands are: Hydroxyl (-OH) groups at 3500 cm⁻¹, vinyl group (=C-H) from acrylic one at 3100 cm⁻¹, methyl and methylene groups (-CH3-, -CH2-) between 2900 cm⁻¹, 2800 cm⁻¹, carbonyl vibration (C=O) from esters groups is sited at 1730 cm⁻¹, double bonds (C=C) from the acrylic pendant groups are found at 1650 cm⁻¹. Finally, at 750 cm-1 is sited the typical band due to four or more continuous methylenes (-CH2CH2CH2CH2-...) from the fat acids chains. The main changes observed when ASEA is polymerized to polyASEA (Figure 4) are those related with the acrylic double bonds at 3100 cm⁻¹ and 1650 cm⁻¹. As the same as BMA polymerization, those bands decrease in intensity as polymerization takes place; at the end of polymerization they are no detected. All the other signals practically do not change.

For copolymer ASEA-co-MAB (Figure 4) the main information obtained from the FT-IR spectra are that signals of both monomers are seen overlaid and the main ones are those corresponding to the acrylic double bonds. It is logic that they are not detected by copolymerization; however there are still small signals corresponding to acrylic double bonds of the BMA at 1632 cm⁻¹, even after washing the product with THF. BMA monomer was not detected in the filtered THF after remove the solvent. That indicates that it does not correspond to an incomplete reaction of BMA (residual monomer). The small double bonds signals seen on the spectrum are attributed to a higher load than ASEA, generating segments of BMA units with disproportion ends chains.

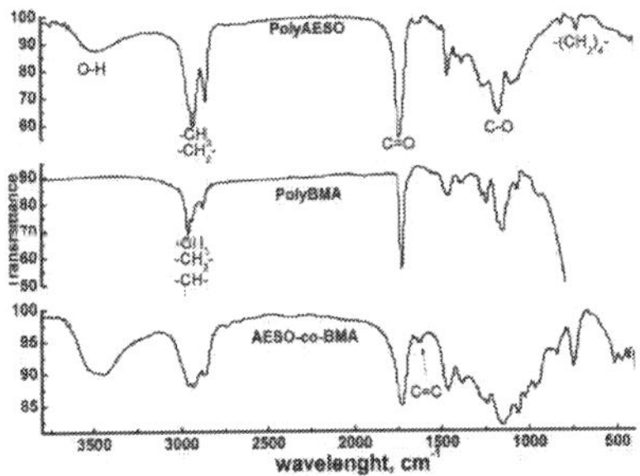

Figure 4. Comparison of the IR absorption bands for polyAESO, polyBMA and AESO-co-BMA

The FT-IR study is not determinant to corroborate if there was a copolymerization, however the thermal properties indicates that there is. DCS/TGA analysis also were made in order to found a possible transition and to determine the decomposition temperature considered it at which de polymer lost the 10 wt% (T_{10}) of its original weight percent. By DSC, ASEA monomer shows an exothermal crosslink curve with a maximal temperature at 349°C and immediately the decomposition is observed. TGA shows that ASEA's T_{10} is detected at 360°C in one step. Once the ASEA is polymerized, the DSC curve does not show any transition: no residual exothermal curve due to crosslinking, neither T_g indicating a complete polymerization under the described conditions. There is difference in the T_{10} in 7°C lower for polyASEA in comparison with ASEA monomer. This decreasing is explained in terms on that crosslink decreases the free movement of the fat acids arms on the polyASEA in such a way that heating tends to break that inter-chain tension at lower temperatures. The T_{10} of polyBMA was detected at 230°C. The T_{10} of the copolymer is 302°C which is lightly next to the polyASEA value (Figure 5) due to the crosslinking reaction between both monomers, even BMA is in major proportion. Decomposition of the copolymer occurs in only one step, indicating the product is only one compound (copolymer) without byproducts such as homopolymers or oligomers.

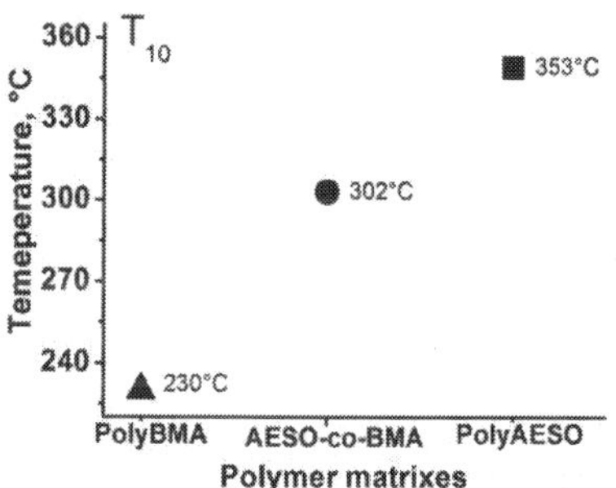

Figure 5. Decomposition temperature (T_{10}) for AESO-co-BMA copolymer and its respective homopolymers: polyAESO and polyBMA

2.1.3. Preparation and characterization of AESO-co-BMA copolymer composites with carbon black

For preparing composites based on ASEA-co-BMA and CB particles, it is followed the next methodology which is similar as the described for pure copolymer synthesis. Composites preparation starts with the dissolution of 0.3 g of ASEA monomer in 20 mL of THF and then the addition of 0.78 mL of MAB monomer. At this homogeneous mix is added the corresponding amount of CB depending on the composition to prepare. In this case were prepared the next CB loads: 0.5, 1.0, 1.2, 1.3, 1.4, 1.5, 2.0, 2.5, 3.0, 3.5, 4.0, 6.0, 8.0, 9.0, 10.0, 15.0 and 20.0 wt% CB. The mix lefts to disperse by sonication for 2 hr at 8°C. Finally the same milliliters of initiator solution is added, nitrogen flux is adapted and the composite is left to polymerize at 70°C for 4 hr and then at 95°C for 2 hr. The composites precipitate as a black unshaped mass; they are washed three times with acetone and filtered in order to remove the probable residual monomers and/or initiator. Finally they are dried under vacuum for 48 hr.

Respect to the composite characterization by FT-IF spectroscopy, the spectra are very similar to the pure copolymer with only one difference, the signals corresponding to the graphitic structure of CB, composed by double bonds C=C sited at 1633 cm^{-1} increases subtle as the CB load do. Referent to the thermal properties, T_{10} of the composites based on ASEA-co-BMA depends on the CB load. For example, for 0.5 wt% CB a T_{10} of 342°C is detected, whereas for 6 wt% CB the T_{10} is 335°C. Is a fact that CB acts as thermal reinforcing filler in this matrix however, that reinforcement decreases as the CB load increases. This is one reason for the interest in produce composites with low critical fraction, usually as the CB load increases some other properties as mechanical or processing ability tend to decrease as well the potential applications as sensing.

It was not a successfully improvement of the processability by copolymerize AESO with BMA, copolymer and the respective composites were insoluble, unmelted, and the thermomolding process was not possible, nevertheless it was obtained a very low critical fraction (1.2 wt% CB) [17].

2.2. System AESO-co-ACAB

In this section the electrical properties of polymer composites based on (Acrylamidomethyl) cellulose acetate butyrate, ACAB, and its copolymer with acrylate-epoxidized soybean oil (AESO) (50:50 wt%) are discussed. Three different commercial carbon black (CB) particles were used in this study: Raven 5000, Vulcan XC-72 and N660 which have different properties as size, structure (ramification percent), oxidize surface and conductivity (see Table 1).

2.2.1. Reactants and equipment

Commercial (Acrylamidomethyl) cellulose acetate butyrate, ACAB has a Mn of 10,000, mp 155-165 °, Tg 118°C, density 1.31 g/mL a 25°C, it is soluble on a variety of acrylic monomers. Benzophenone photoinitiador, BPN (> 99%), mp 47-51°C, both were purchased from Sigma-Aldrich, Co. Reaven 5000 and N660 CB′s particles were kindly donated by Professor Wiltold Brostow from UNT, USA. All of them were used as received. For electric contacts silver paint SPI 18DB70X was purchased from Electron Microscopy Sciences. Some properties of the carbon blacks particles are shown in Table 1.

Carbon particles have many different properties according to the raw reactants and on the synthesis method. They are into the 50 chemical products more produced in all around the world and actually, CB particles are the result of an incomplete combustion from hydrocarbons. 90% of CB is useful as reinforcement particles in elastomers for fabrication of wells, 9% as pigment and the rest 1% in other specific applications as paints and electric devices [86]. Additional to their stain power, their electric or charge action provide UV radiation protection to the polymers at low cost. The CB particles are constituted from almost spherical particles with a graphitic structure (electrically conductive) and colloidal dimensions. Their structure consists in many primary nanoparticles (from 10 to 100 nm) bonded into a bunch named "aggregate" with dimensions from 50 to 500 nm. Those aggregates use to form macro-agglomerates with more than 1 μm of size [1]. CB aggregates have different shapes, they could be 1) spherical, 2) elispsoids, 3) lineal or 4) branched. Chemically are constituted from 83% to 99% of elemental carbon and according to the fabrication method it could be founded many oxygenated functional groups as phenols, quinolics, carboxylics, etc. on the surface. This is related to the ash content which is usually less than 5%. For electrical properties, are required that CB particles have branched structures (high structure), an oxidized surface which allows to disperse them into several polymer matrix, a high diameter and good intrinsic electrical conductivity. Branched structures as in CB Vulcan X-C72 allow a large contact number or electronic sits for build the conductive paths with a lower load of conductive particles. Opposite to this, CB N660 has the lowest structure and superficial area but a particle diameter of 50 nm and a structure manly lineal. CB Raven 5000 particles show a low structure and superficial area but have a highly oxidized surface in comparison with the others. Their resistivities [see section 3.1] are not so different making interesting to study how the other mentioned properties have influence in reaching the percolation threshold at low critical fractions.

CB Particles	Average diameter size, nm	Volatile % (oxidized surface)	Branched structure, %	Resistivity (cm)
Raven 5000 Ultra II [87]	8	10.5	45.5	1.1
Vulcan XC 72 [86, 88]	32	< 2	77.3	0.08
N660 [89]	50	2.5	31.7	0.19

Table 1. Characteristics of the three used carbon black particles

The FTIR spectra for the three different types of CB, shows the same superficial functional groups, being the only difference the intensity of the signals (Figure 6). Those characteristic absorption bands are: hydroxyls groups (O-H) from 3416 cm^{-1} a 3440 cm^{-1}, methylene (C-H) from the amorphous CB composition at 2923 cm^{-1} y 2850 cm^{-1}; carbonyl bond (C=O) due to different groups as esters, carboxylic acids and quinones are seen from 1709 cm^{-1} to 1740 cm^{-1}; typical vibrations of the double bonds (C=C) that constitute the graphitic composition of the different CB′s are sited from 1578 cm^{-1} to 1632 cm^{-1}; this due there are some double conjugated bonds with carbonyl group from quinones. A combination of vibrations due to the C-O and O-H from esters and phenols, respectively are founded at 1380 cm^{-1} and 1216 cm^{-1}; finally are seen a combination of vibrations for C-O bond from hydroxyl groups in phenols at 1064 cm^{-1} y 1066 cm^{-1}. At 680 cm^{-1} is sited an out of the plane deformation vibration for O-H bond which is only seen in CB N660.

2.2.2. Synthesis and characterization of crosslinked ACAB ($_{cross}$ACAB)

This reaction was neccesary to know and stablish the copolymerization reaction conditions and the processing properties. For pure crosslinked ACAB, 1 g of ACAB is weighed into a 150 mL round bottom flask, 10 mL of THF are added and the mix is sonicated for 15 min until the complete dissolution of ACAB. At this point 0.05 mL of a 2.0M THF solution of PBO is added and shaking another 2 min. The condenser and the nitrogen atmosphere are adapted to the flask and the temperature starts to increase from RT to 70°C as a first step. The crosslinked ACAB ($_{cross}$ACAB) starts to precipitate. After 5 hr of reaction, the temperature increases from 70 to 88-90°C for 4hr in order to finish the crosslinking reaction. Heating and nitrogen gas are turn off and at RT and the transparent one piece polymer is washed twice with acetone en order to remove residual ACAB and/or initiator. Finally it was well dried under vacuum for 24 hr. The polymer was cut in small pieces and they were processed by thermo mechanic technique. This allowed preparing cylinders of 1.0 cm diameter x 0.2-0.3 mm thickness in order to measure the resistance.

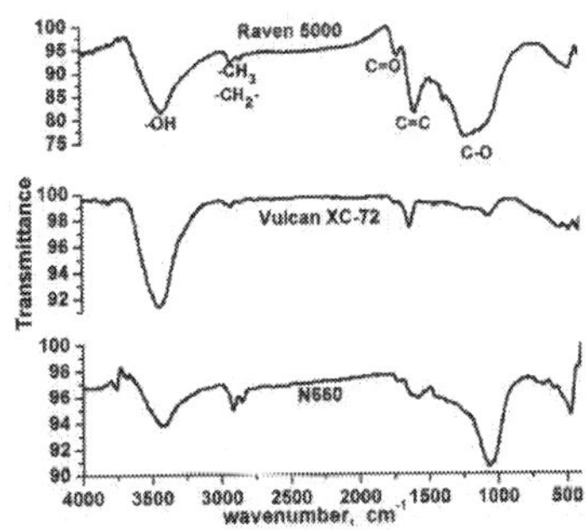

Figure 6. FI-IR spectra if the three types of used CB particles: Raven 5000, Vulcan XC-72, N660

The characterization of ACAB by FTIR spectroscopy allows distinguishing the next important signals: Hydroxyls and amide (–O-H, –N-H) vibrations appear at 3700-3250 cm⁻¹; methyl and methylene vibrations (C-H) from of the polysaccharide ring are sited at 2970 and 2879 cm⁻¹; carbonyl vibration (C=O) from ester groups (acetate and butyril) are found at 1750 cm⁻¹ and at 1679 cm⁻¹ it is possible to distinguish the carbonyl vibration from amidomethyl group. Double bond band (C=C) is sited at 1630 cm⁻¹, and the corresponding terminal double bond (C=CH₂) is at 919 cm⁻¹. Signals monitored in order to found the time in which ACAB completely crosslinks are those corresponding to the vinyl group, 1630 and 919 cm⁻¹. The $_{cross}$ACAB is characterized by FTIR spectroscopy detecting the disappearance of those mentioned signals (Figure 7).

Concerning with the thermal properties, it was interesting to note that the decomposition temperature, in this case taken as the onset temperature (T_{ons}) due some little lost before decomposition temperature due to humidity or some remnant solvents in case of $_{cross}$ACAB. Those temperatures were very similar with a difference of 2°C higher for $_{cross}$ACAB respect to the ACAB. This slightly difference could be due to the small amount of crosslink vinyl groups (only 0.6%) which not do much difference in that property. The DSC thermogram for both compounds does not show differences, it is possible to distinguish an exothermal curve with a maximal temperature at 145°C that encloses both the T_g and the T_m reported transitions. This is one reason for which polyACAB shows processability under temperature and pressure and is possible to soften or melt in order to mold it to any required shape, which was something expected.

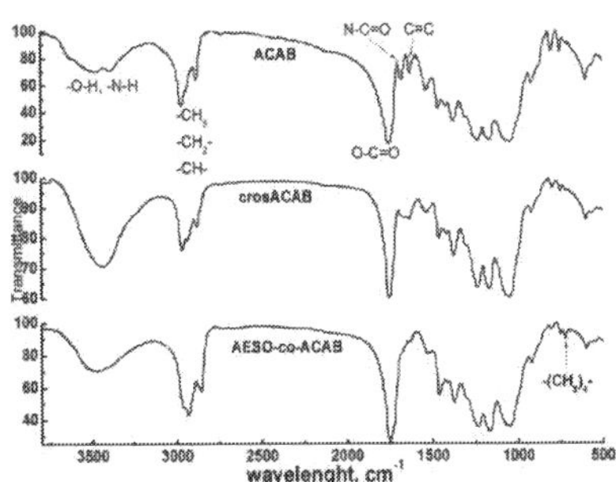

Figure 7. FT-IR spectra for ACAB, cross ACAB and AESO-co-ACAB polymer matrixes

2.2.3. Synthesis and characterization of the pure copolymer, AESO-co-ACAB (50/50 wt%)

For preparing 1g of AESO-co-ACAB (50:50 wt%), 0.5g of ACAB are weighed into a round bottom flask, 5 mL of THF are added and it was solved into an ultrasonic processor for 15 min. Then 0.5 g of ASEA and 5mL more of THF are added to the last mix and again homogenized into the ultrasonic processor for 10 min. The same quantity of PBO initiator as the described polymerizations is added and all mix is copolymerized following the same procedure described for crossACAB. The reaction for this system needs 7 hr for polymerizes at 70°C, and then the temperature is increased to 88-90°C for 4 hr. The product as a one translucent and lightly yellow homogeneous piece which is filtered and washed three times with acetone and finally dried under vacuum for 24 hr. This product, softer than crossACAB also shows processability by thermo molding being one of the main goal of this proposed system.

The FT-IR spectrum shows absorption bands of the two comonomers and the most characteristics are: 1535 cm⁻¹ (C-H and N-H), 755 cm⁻¹ (N-H) from the amide group. 1601 cm⁻¹ (C=O) it is a vibration from all esters in both monomers. The most representative band for ASEA is at 723 cm⁻¹ which is typical for more than four continuous methylene groups. There is not band sited at 1634 cm⁻¹ corresponding to double bonds indicating a complete copolymerization between both monomers.

The TGA analysis of the copolymer indicates two lost, one of 2% before the 150°C due to remnant solvent and humidity and a second at 351°C which corresponds to its decomposition. Figure 8 shows the decomposition temperature of the three pure polymer matrix. As we can see, the decomposition temperature for the copolymer is in the middle of the homopolymers. Taking in account that the decomposition occurs in only one step, it becomes and evidence that copolymerization between ACAB and ASEA took place.

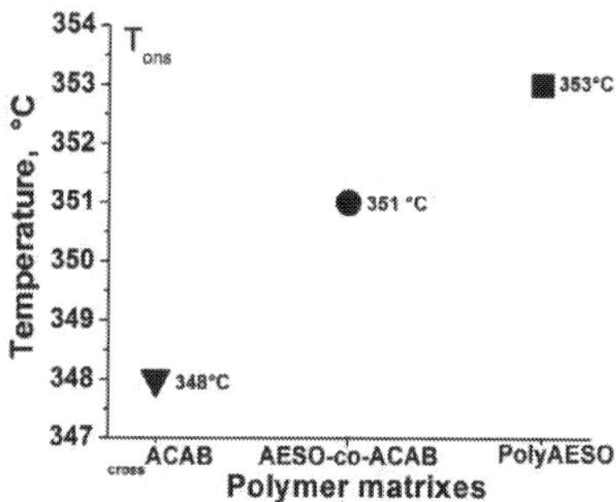

Figure 8. Decomposition temperature of the three polymer matrix: $_{cross}$ACAB, AESO and AESO-co-ACAB (50:50 wt%) copolymer

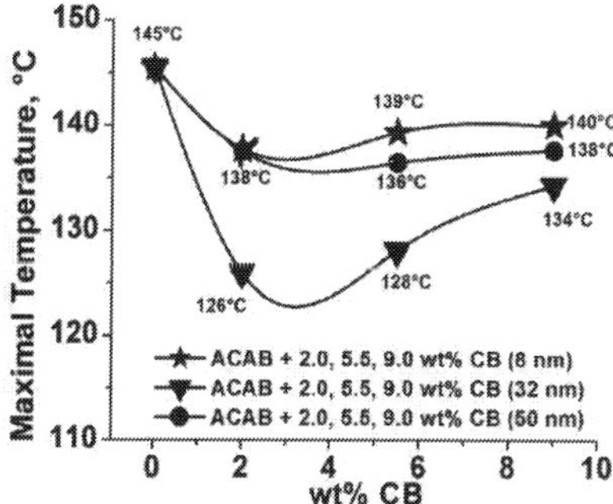

Figure 9. Changes on maximal T_{gm} according to the type and amount of CB particles

The DSC curve showed a diminishing curve due to the melting point of ACAB matrix (145°C), but it does not show any residual exothermal curve due to the ASEA cure. These events suggest that the copolymer could be processable by thermomolding thanks to the thermoplastics properties imparted by the ACAB polymer.

2.2.4. Preparation and characterization of AESO + CB particles

The CB compositions for composites based on ACAB with CB, independently of the type of CB, are: 1.0, 2.0, 2.5, 3.0, 4.0, 4.5, 5.5, 7.5 and 9.0 wt% CB. An example is described for preparing 1g of composite with 1 wt% CB. 0.99 g (90 wt%) of ACAB is solved in 30 mL of THF by sonication for 10 min into a 150 mL round bottom flask and 0.01g (1.0 wt%) of CB is added. The mix is sonicated for 2 hr until a homogeneous solution is observed. At this moment the solvent is distilled with a rotovapor and finally the solid mix is well dried under vacuum for 24 hr until a black powder is obtained. For the other compositions the methodology was the same, the only difference is the dispersion time: 2 hr for 1.0 to 4.0 wt% CB and 4hr to 4.5 to 9.0 wt% CB.

When CB is dispersed in ACAB or into de copolymer, the detection of CB particles by FTIR is the same in any case. Here is exhibited the analysis only with the copolymer and CB Vulcan XC72: the absorption band a 1632 cm^{-1} which correspond to graphitic CB zone (C=C) or 1582 cm^{-1} for 8nm-CB, is subtle increased as the CB load increases. There are other bands which overlayer the composite spectra. At 3416 cm^{-1} - 3440 cm^{-1} we have a vibration for (O-H) hydroxyl functional group, at 2923 cm^{-1} and 2850 cm^{-1} are the vibrations corresponding to the amorphous zone of the CB's (C-H). From 1709 cm^{-1} to 1740 cm^{-1} appear the vibrations of C=O bond corresponding to esters groups; at 1578 cm^{-1} a 1632 cm^{-1} are the vibrations of the graphitic composition of the CB's particles which correspond to C=C conjugated with carbonyl group from quinones. At 680 cm^{-1} we have a bending vibration of the O-H bond out of plane which is only observed in CB N660.

In Figure 9 was plotted a temperature (T_{gm}) which encloses the Tg and maximal T_m for ACAB matrix in composites with 2, 5.5, 9 wt% CB, identifying the next information: All ACAB-based composites show a diminishing on the T_{gm} respect to pure ACAB. An explanation is that CB particles act as impure or lubricant, abating both temperatures, mainly the T_m as observed for other authors [90]. The higher effect is shown when CB N660 is used.

On the other hand, respect to the decomposition temperatures, shown in Figure 10, it can be seen that the addition of any CB used particles tends to increases the decomposition temperature respect to the ACAB matrix (346°C). In Figure 10 only T_{ons} for three compositions are plotted: 2, 5.5 and 9 wt% CB. The smallest CB particles (8nm) give the composites with the higher decomposition temperatures around 351°C, followed by those based on 50nm-CB (N660) which have T_{ons} around 348°C and finally those particles that almost do not have a great effect on the decomposition temperature were the 32nm-CB particles, maintaining it around 346.5°C.

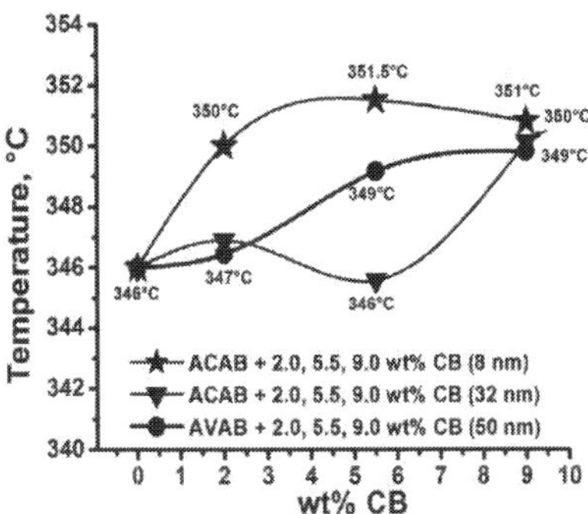

Figure 10. Decomposition temperatures for ASEA+CB composites as a function of the type and amount of the CB particles

The incorporation of the CB particles increases the decomposition temperature of the composite, it could be due to that intermolecular interaction seems to be favourable between the functional groups of the polymer matrix and those of the CB particles [1]. It is important to take in account that 8nm-CB particles have a higher oxidized superficial area implying a great number of hydroxyl and carboxyl groups that could render hydrogen bridges with those on the ACAB polymer. It is because composites prepared with 8nm-CB have a higher resistance to decompound thermally. However 50nm-CB particles have scarce superficial functional groups and less superficial area and in turn negligible interaction with the polymer matrix and a less decomposition temperatura for the respective composites. An intermediate case is for ASEA+ 32nmCB composites.

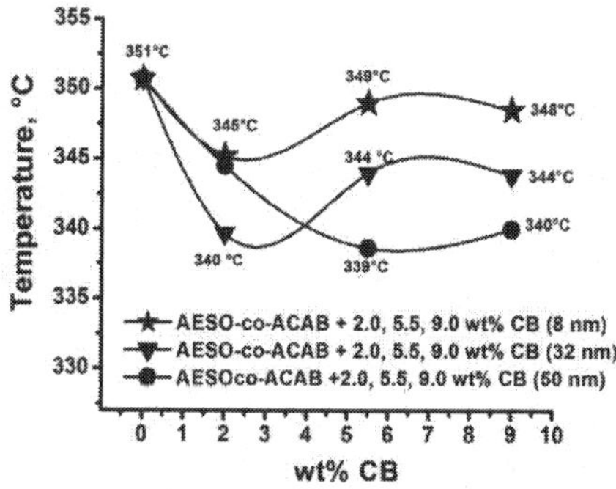

Figure 11. Decomposition temperature for the copolymer composites base don AESO-co-ACAB as a function on the type and amount of CB particles

2.2.5. Preparation and characterization of AESO-co-ACAB + CB's composites

The preparation of these composites also is also via free radicals. For 32 and 50 nm CB's, thermal initiator decomposition was used, but for CB Raven 5000 (8nm) was necessary to use the photochemical method via benzophenone/UV radiation as initiator system. The reason is that was not possible to obtain composites via thermal decomposition. A possibility is that some superficial groups on CB Raven 5000 act as inhibitors or quenching free radicals consuming them. In case of UV radiation, it is possible to reactivate the apparently inhibited free radicals. This could be supported by the fact that those copolymer composites were cured under air atmosphere, not under nitrogen flux as in thermal copolymerization.

The CB compositions prepared for these systems were the same as those described for ACAB: 1.0, 2.0, 2.5, 3.0, 4.0, 4.5, 5.5, 7.5 and 9.0 wt% CB. In the copolymer composites,the CB load was calculated considering 1 g of comonomers in which 0.5 g of ASEA and 0.5 g of ACAB are used for synthesize the copolymer matrix. The first stage of the preparation is the same independently of the type of CB used: First 0.5 g of ACAB are weighted and solved in 30 mL of THF by ultrasonic shake for 10 min. 0.5g of ASEA are added and sonicated again for 10 min in order to get a homogeneous clear solution. Then the respective amount of CB is added (depending on the composition) and the mix is sonicated for 2 or 4 hr (as same as ACAB composites). Finally, depending on the initiator system, when 8nm-CB is added, 0.01 mL of 0.2M BPN in acetone is dropped and sonicated 2 min more. The mix is poured into a glass mold covering the bottom. The solvent is slowly evaporated inside an extraction bell in such a way that composite mix forms a homogeneous layer. The glass mold is put into a CL-1000 ultraviolet-crosslinker UVP (with a maxima wavelength of 254 nm) and copolymerized at 720 J/cm^2 for 4 hr. The composites are obtained as a solid layer with a shining black aspect, which is insoluble in water and organic solvents. The composite layer is collected and reserved.

When CB of 32 and 50nm are used, 0.05 mL of 2.0M PBO in THF is added, and following the same methodology as in pure copolymer, the mix is cured at 70°C for 5h and then a 90°C for 2 hr under nitrogen flux. The cured composite precipitate as black solid pieces which are filtered and washed twice with acetone and dried under vacuum for 24 hr.

By FT-IR spectroscopy, as the same to the aforementioned for the ACAB composites, the only effect is the increasing of the band at 1632 nm which corresponds to the double bonds of the graphitic composition in CB particles. The same effect is observed in all copolymers and that signal seems to increase as the wt% CB does.

Concerning with the thermal properties of copolymer composites only was detectable the decomposition temperature, it was very difficult to detect some T_m surely due to the addition of CB particles which influence the possibility to crystallize and T_g was not evident. But TGA analysis shows interesting results, in the case of the copolymer matrix the effect on the T_{ons} is opposite to that analyzed for the ASEA polymer (Figure 10). In copolymer composites at 2, 5.5 and 9wt% CB's, the T_{ons} is diminished with the CB particles addition respect to the copolymer matrix. 8nm-CB particles seem not to have an important effect on the decomposition temperature, except for 2 wt% CB in which the temperature is reduced from 351°C (copolymer) to 345°C. At this same CB composition, 32 nm particles show the stronger effect on T_{ons} diminishing it at 340°C, but with the increasing of the CB load, the reached value for T_{ons} was in average 344°C. The 50nm-CB particles had its more critical effect at 5.5 wt% CB decreasing the T_{ons} from 351 to 338°C.

3. Electrical properties evaluation

3.1. Electrical properties of carbon black particles

In order to compare the electrical properties of the different prepared and characterized polymer composites, the electrical properties of the three types of carbon black were determined under the same conditions [91, 92]. Each type of CB was hydrostatically compressed into a system which consists of a cylinder-piston couple, as shown in Figure 12 and the electrical resistivity was calculated using the relationship 2.

$$\rho = R\frac{A}{l} \tag{2}$$

Where the electrical resistance R was measured with a LRC720 Stanford Research System. The length of the compacted particles l, and the transversal section area A were measured with a micrometer and a Vernier respectively. The same amount (100 mg) and pressure (20 Kg/cm^2) were applied to the three CB particles. Cupper electrodes of 6 mm diameter, 1cm length for the inferior and 2.5 cm length for the superior one were used. The electrical resistivities are shown on Table 1. As it can be see, the Vulcan XC-72 have the lowest resistivity and Raven 5000 the highest indicating that the most conductive CB is the Vulcan XC-72.

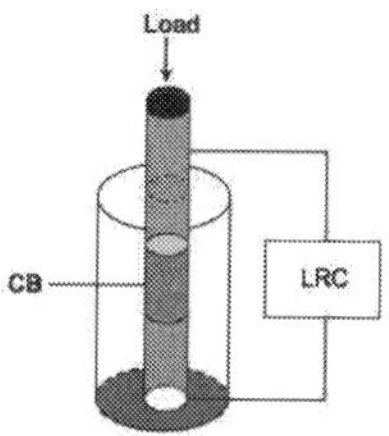

Figure 12. Design for measure the resistance of the CB particles

3.2. Percolation curves and percolation threshold calculus for composites

For composites based on AESO-co-BMA, the methodology for electrical measurements is well detailed in reference [17] and the most interesting results were the low critical fraction of both composites based on polyAESO and on the AESO-co-BMA matrixes, 4.0 and 1.2 wt% CB, respectively. These results will be discussed on section 3.3.

For the other systems (ACAB, AESO-co-ACAB) the electrical treatment is described: Polymers composites were processed by thermo-molding technique at 1.5 MPa in order to get samples of 1.2 cm diameter and 2 mm thickness [81, 93]. The processing conditions of temperature and time are summarized in Table 2 and those conditions depend on the polymer matrix. The bulk resistivity was determined using the two-points technique and silver paint (SPI de Electron Microscopy) as electric contacts were put on the parallel faces of the composite samples.

The voltage-current relationship is measured with a Keithley 6517A electrometer [17, 81, 93]. For each composition the resistivity was calculated via the relation 2 building the respective percolation curves in a width range of compositions. As an example, the percolation curves for AESO-co-ACAB are exhibited in Figure 13. From these curves and using Origin 6.0 software the percolation threshold and the critical fraction are numerically calculated fixing the experimental data to the equation 1. As we can see, the critical fraction for polymer composites based on ACAB and AESO-co-ACAB is modified due to the CB particles as well as the polymer matrix.

Polymer composite	Temperature/ °C	Compression time /min
ACAB	122	30
ASEA/ACAB	118	40

Table 2. Processing conditions of ACAB and ACAB-co-AESO composites

For the ACAB composites the critical fraction varies from 2.8 to 7.5 wt% CB whereas the AESO-co-ACAB copolymer composites the respective values are from 2.4 to 3.2 wt% CB (Figure 14). This demonstrates that polymer matrix influences the electrical properties of the final composite. However in terms of CB properties, the decreasing of the critical fraction with a increasing of the CB structure converges very well with the percolation theory. As we discussed before, when are used aggregates with small particles and low structure it tend to need more particles in order to interconnect them and to build the electrically conductive paths, in comparison with high structure and bigger particles (as Vulcan XC-72 respect to Raven 5000). Also, other CB properties as intrinsic resistivity are evident. For both systems the CB particles with the highest resistivity (Raven 5000, 8 nm) are those with a higher critical fraction.

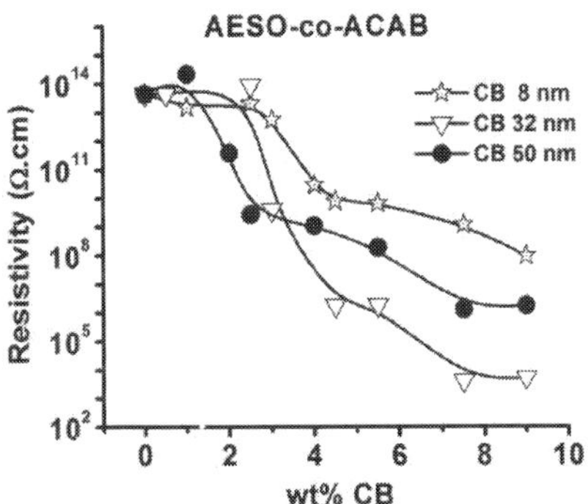

Figure 13. Percolation threshold of the AESO-co-ACAB composites for different carbon black particles

3.3. Analysis of electrical properties as a function of carbon black particles and polymer matrix nature

From the results we could discuss that the chemical nature of the polymer matrix as well the CB properties contribute to the disaggregation, dispersion and preferential distribution of the conductive particles into the polymer matrix. However it is possible to control the preferential build of conductive paths and diminishes as a consequence the critical fraction? As was mentioned before, there are some successfully attempts for controlling it but only oleo-polymer has been used for that and critical fraction lower that 3% are not obtained yet. However, AESO has become a question in to decrease the critical fraction under copolymerization. For the system AESO-co-BMA, a very low critical fraction of 1.2 wt% CB was calculated. Something

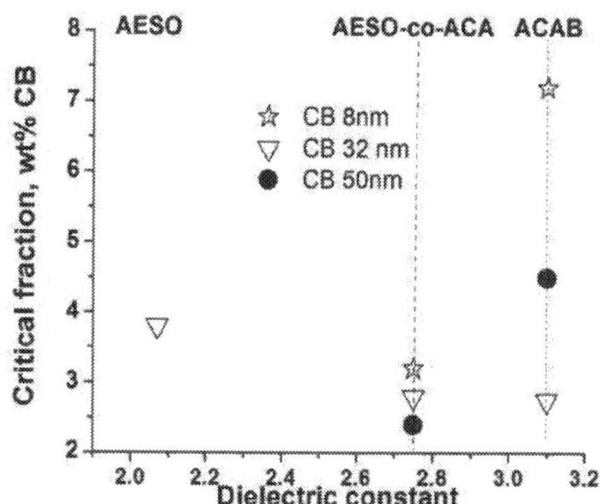

Figure 14. Critical fractions respect to the type of CB and to the polymer matrix

similar is observed for the AESO-co-ACAB composites. If we analyze the critical fractions on Figure 15, we could appreciate that copolymers composites have the lowest critical fraction independently on the type of CB particle. For oleo polymers, in a systematic work [16] was demonstrate that the dielectric constant of the polymer matrix is directly associated with the critical fraction. Due to the unprocessability of the AESO-co-BMA copolymer it was not possible to measure the dielectric constant but it was made for ACAB, AESO and the AESO-co-ACAB matrixes (Figure 15). For dielectric constant polymers were processed into disks of 2cm diameter x 0.9-1.1 mm thickness by compression molding. An Agilent 4991 A RF impedance/Material Analyzer at 450 MHz and room temperature [16].

We can realize that AESO has the lowest dielectric constant (2.1) whereas ACAB has the highest one (3.1). Even both molecules have polar groups; the AESO has large alkyl chains also, whereas ACAB has oxygen as constituent of different functional groups being carbonyl and hydroxyl the main. Unsurprising constant dielectric value for the copolymer was obtained in 2.7. The tendency with CB Vulcan XC-72 is the same as the reported results on [16]: at a highest dielectric constant, a diminishing in the critical fraction is observed. The amazing results are in the low critical fractions for the copolymer composites (from 2.4 to 3.2) with the three types of CB particles in comparison with the ACAB composite which possess the highest dielectric constant. These results claim that the dielectric constant is important to take in account however the role of the AESO is completely understood but definitively it tends to low the critical percolation. From the structural point of view, polyAESO is an amorphous and 3-D crosslinked polymer while AESO is a semicrystalline and lineal one. However when AESO crosslinks builds a bonds network with polar groups thanks to the reaction of the acrylic groups. This network draw preferential paths (in blue, Figure 16) for a good dispersion and distribution of the CB particles without have to fill the entire matrix. In conjunction with the polar functional groups of the other comonomer, very low critical fractions are reached under copolymerization with BMA and ACAB. The differences on the critical fraction will surely dependent on the CB properties as the size, structure, surface and the intrinsic conductivity.

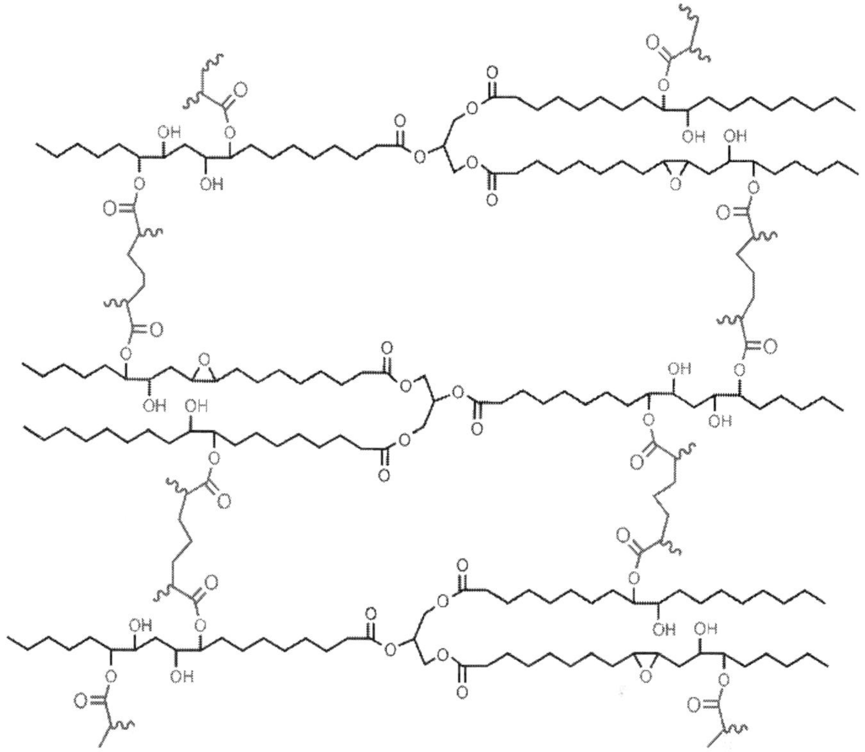

Figure 15. Representation of the 3D- polar paths (in blue) in which the CB particles surely disperse and preferentially distribute in all matrixes containing AESO comonomer.

4. Conclusions and remarks

It has been proved that polarity of the polymer matrix is very important in order to have a better distribution and dispersion of the conductive particles. However, AESO is a very interesting monomer that renders very low critical fractions under copolymerization with other polar comonomers as BMA and ACAB. The explanation we have is that AESO provides preferential distribution to the CB conductive particles due the 3D-crosslinking paths which are bonded to polar functional groups. This is the reason for thinking that conductive polymer composites based on AESO have a promising future in some applications as solvent, gases, vapors or pressure sensors. There are some previous results (unpublished) about the capacity for sensing solvents as well as pressure changes indicating they are very reliable and reproducible materials. The sensibility to pressure changes could be dependent on the crosslinking degree, the proportion and on the comonomer chemical structure. Finally is considered that polymers derived from renewable sources could be a better alternative for reducing the oleopolymers use in some specific applications as discussed in this chapter.

Acknowledgements

Authors thank to the Universidad Autónoma del Estado de México under the projects 3135/2012FS and 3198/2012U for the financial support.

Author details

Susana Hernández López* and Enrique Vigueras Santiago*

*Address all correspondence to: eviguerass@uaemex.mx

Research and Development of Advanced Materials Laboratory, Chemistry Faculty of the State of Mexico Autonomous University, Toluca, Mexico

References

[1] Donnet, J. B., Bansal, R. C., & Wang, MJ. (1993). Carbon Black. *Science and Technology.*, USA: Marcel Dekker,.

[2] Kirkpatrick, S. (1973). Percolation and Conduction. *Reviews of Modern Physics*, 45(4), 574-578.

[3] Balberg, I. (1987). Tunneling and Nonuniversal Conductivity in Composite Materials. *Physical Review Letters*, 59(12), 1305-1308.

[4] Sisk, B. C., & Lewis, N. S. (2006). Vapor Sensing Using Polymer/Carbon Black Composites in the Percolative Conduction Regime. *Langmuir*, 22(18), 7928-7935.

[5] Yu, J., Zhang, L. Q., Rogunova, M., Summers, J., Hiltner, A., & Baer, E. (2005). Conductivity of Polyolefins Filled with High-Structure Carbon Black. *Journal of Applied Polymer Science*, 98(4), 1799-1805.

[6] Gaurav, R., Kasaliwal, Goldel. A., Potschke, P., & Heinrinch, G. (2011). Influences of Polymer Matrix Melt Viscosity and Molecular Weight on MWCNT agglomerate dispersion. *Polymer*, 52(4), 1027-1036.

[7] Cheng, G. S., Hu, J. W., Zhang, M. Q., Li, M. W., & Xiao, Rong. M. Z. (2004). Electrical Percolation of Carbon Black Filled Poly (ethylene oxide) Composites in Relation to the Matrix Morphology. *Chinese Chemical Letters*, 15(12), 1501-1504.

[8] Sumita, M., Abe, H., Kayati, M., & Miyasaka, K. (2006). Effect of Melt Viscosity and Surface Tension of Polymers on the Percolation Threshold of Conductive-Particle-Filled Polymeric Composites. *Journal of Macromolecules Science*, 25(1-2), 171 EOF-184 EOF.

[9] Tchoudakov, R., Breuer, O., & Narkis, Siegmann. A. (1996). Conductive Polymer Blends with Low Carbon Black Loading: Polypropylene/Polyamide. *Polymer Engineering and Science*, 36(10), 1336-1346.

[10] Sumita, M., Sakata, K., Asai, S., Miyasaka, K., & Nakagawa, H. (1991). Dispersion of Fillers and the Electrical Conductivity of Polymer Blends Filled with Carbon Black. *Polymer Bulletin*, 25(2), 265-271.

[11] Moriarty, G. P., Whittemore, J. H., Sun, K. A., Rawlins, J. W., & Grunlan, J. C. (2011). Influence of Polymer Particle Size on the Percolation Threshold of Electrically Conductive Latex-Based Composites. *Journal of Polymer Science*, Part B: Polymer Physics, 49(21), 1547-1554.

[12] Kasaliwal, G. R., Goldel, A., Potschke, P., & Heinrinch, G. (2011). Influences of Polymer Matrix Melt Viscosity and Molecular Weight on MWCNT Agglomerate Dispersion. *Polymer*, 52(4), 1027-1036.

[13] Cheng GS, Hu JW,. Zhang MQ, Li MW, Xiao DS, Rong MZ. (2004). Electrical Percolation of Carbon Black Filled Poly (Ethylene Oxide) Composites in Relation to the Matrix Morphology. *Chinese Chemical Letters*, 15(12), 1501-1504.

[14] Carmona, F. (1989). Conducting Filled Polymers. *Phisica A.*, 157(1), 461-469.

[15] Li, Y., Wang, S., Zhang, Y., & Zhang, Y. (2005). Electrical Properties and Morphology of Polypropylene/Epoxy/Glass Fiber Composites Filled with Carbon Black. *Journal of Applied Polymer Science*, 98(3), 1142-1149.

[16] Castro-Martínez, M., Hernández, López. S., & Vigueras, Santiago. E. (2012). Relationship Between Polymer-Chemical Structure and Electrical Properties in Conductive Carbon Black Composites. Submitted to e-Polymers.

[17] Hernández-López, S., Vigueras-Santiago, E., Mercado-Posadas, J., & Sánchez-Mendieta, V. (2006). Electrical Properties of Acrylated-Epoxidized Soybean Oil Polymers-Based Composites. *Advances in Technology of Materials and Materials Processing Journal*, 8(2), 214-219.

[18] Guner, F. S., Yagci, Y., & Erciyes, A. T. (2006). Polymer from Triglyceride Oils. *Progress in Polymer Science*, 31(7), 633-670.

[19] Sharma, V., & Kundu, P. P. (2006). Addition Polymers from Natural Oils- A Review. *Progress in Polymer Science*, 31(11), 983-1008.

[20] Meier, M. A. R., Metzger, J. O., & Shubert, U. S. (2007). Plant Oil Renewable Resources as Green Alternatives in Polymer Science. *Chemical Society Reviews*, 36(11), 1788-1802.

[21] Khot, S. N., Lascala, J. J., Can, E., Morye, S., Williams, G. I., Palmese, G. R., Kusefoglu, S. H., & Wool, R. P. (2001). Development and Application of Triglyceride-based Polymers and Composites. *Journal of Applied Polymer Science*, 82(3), 703-723.

[22] Blayo, A., Gandini, A., & Le Nest-F, J. (2001). Chemical and Rheological Characterizations of Some Vegetable Oils Derivates Commonly Used in Printing Inks. *Industrial Crops and Products*, 14(2), 155-167.

[23] Hill, K. (2000). Fats and Oils as Oleochemical Raw Materials. *Pure and Applied Chemistry*, 72(7), 1255-1264.

[24] Adhvaryu, A., & Erhan, S. Z. (2002). Epoxidized Soybean Oil as a Potential Source of High-Temperature Lubricants. *Industrial Crops and Products*, 15(3), 247-254.

[25] Tuman, S. J., Chamberlain, D., Scholsky, K. M., & Soucek, . (1996). Differential Scanning Calorimetry Study of Linseed Oil Cured with Metal Catalysts. *Progress in Organic Coatings*, 28(4), 251-258.

[26] Refvik, M. D., & Larock, R. C. (1999). The Chemistry of Metathesized Soybean Oil. *Journal of the American Oil Chemists' Society*, 76(1), 99-102.

[27] Stemmelen, M., Pessel, F., Lapinte, V., Caillol, S., Habas-P, J., & Robin-J, J. (2011). A Fully Biobased Epoxy Resin from Vegetable Oils: From the Synthesis of the Precursors by Thiol-enc Reaction to the Study of the Final Material. *Journal of Polymer Science*, Part A: Polymer Chemistry, 49(11), 2434-2444.

[28] Erhan, S. Z., Bagby, M. O., & Nelsen, T. C. (1997). Drying Properties of Metathesized Soybean Oil. *Journal of the American Oil Chemists' Society*, 74(6), 703-706.

[29] Montero de, Espinosa. L., & Meier, M. A. R. (2011). Plants Oils: The Perfect Renewable Resource for Polymer Science. *European Polymer Journal*, 47(5), 837-852.

[30] Teng, G., Wegner, J. R., Hurtt, G. J., & Soucek, MD. (2001). Novel Inorganic/Organic Hybrid Materials Based on Blown Soybean Oil with Sol-Gel Precursors. *Progress in Organic Coatings*, 42(1), 29-37.

[31] Andjelkovic, D. D., Vakverde, M., Henna, P., Li, F., & Larock, R. C. (2005). Novel Thermosets Prepared by Cationic Copolymerization of Various Vegetable Oils-Synthesis and Their Structure-Property Relationships. *Polymer*, 46(23), 9674-9685.

[32] Cakmakli, B., Hazer, B., Tekin, I. O., Kizgut, S., Koksal, M., & Menceloglu, Y. (2004). Synthesis and Characterization of Polymeric Linseed Oil Grafted Methyl Methacrylate or Styrene. *Macromolecular Bioscience*, 4(7), 649-655.

[33] Rybak, A., & Meier, M. A. R. (2008). Cross-Metathesis of Oleyl Alcohol with Methyl Acrylate: Optimization of Reaction Conditions and Comparison of Their Environmental Impact. *Green Chemistry*, 10(10), 1099-1104.

[34] Li, F., & Larock, R. C. (2000). Novel Polymeric Materials from Biological Oils. *Journal of Polymers and the Environment*, 59 EOF-67 EOF.

[35] Biermann, U., Friedt, W., Lang, S., Luhs, W., Machmuller, G., Metzger, J. O., Rusch, gen., Klaas, M., Schafer, H. J., & Schneider, M. P. (2000). *Angewandte Chemie International Edition*, 39(13), 2206-2224.

[36] Rusch, gen., Klaas, M., & Warwel, S. (1997). Lipase-Catalyzed Preparation of Peroxy Acids and Their Use for Epoxidation. *Journal of Molecular Catalysis: A Chemical*, 117(1-3), 311-319.

[37] La Scala, J., & Wool, R. P. (2002). Effect of FA Composition Kinetics of TAG. *Journal of the American Oil Chemists' Society*, 9(4), 373-378.

[38] Okieimen FE, Bakare OI, Okieimen CO. (2002). Studies on the Epoxidation of Rubber Seed Oil. *Industrial Crops and Products*, 15(2), 139-144.

[39] Guidotti, M., Ravasio, N., Psaro, R., Gianotti, E., Coluccia, S., & Marchese, L. (2006). Epoxidation of Unsaturated FAMEs Obtained from Vegetable Source Over Ti(IV)-grafted Silica Catalysts: A Comparison between Ordered and Non-Ordered Mesoposous Materials. *Journal of Molecular Catalysis A: Chemical*, 250(1-2), 218-225.

[40] Campanella, A., Baltanas-Sánchez, Capel., Campos-Martín, M. C., Fierro, J. M., & , J. L. G. (2004). Soybean Oil Epoxidation with Hydroperoxide Using an Amorphous Ti/Si Catalysts. Green Chemistry , 6(7), 330-334.

[41] Parreira, T. F., Ferreira, M. M. C., Sales, H. J. S., & de Almeida, W. B. (2002). Quantitative Determination of Epoxidized Soybean Oil using Near-Infrared Spectroscopy and Multivariate Calibration. *Applied Spectroscopy*, 56(12), 1607-1614.

[42] Vlcek, T., & Petrovic, Z. S. (2006). Optimization of the Chemoenzymatic Epoxidation of Soybean Oil. *Journal of the American Oil Chemists' Society*, 83(3), 247-252.

[43] Hilker, I., Bothe, D., Pruss, J., & Warnecke-J, H. (2001). Chemo-Enzimatic Epoxidation of Unsatured Plant Oils. *Chemical Engineering Science*, 56(2), 427-432.

[44] Rusch, gen., Klaas, M., & Warwel, S. (1999). Complete and Partial Epoxidation of Plant Oils by Lipase-Catalyzed Perhydrolysis. *Industrial Crops and Products*, 9(2), 125-132.

[45] López-Téllez, G., Vigueras-Santiago, E., & Hernández-López, S. (2009). Characterization of Linseed Oil Epoxidized at Different Percentages. *Superficies y Vacío*, 22(1), 5-10.

[46] Muturi, P., Wang, D., & Dirlikov, S. (1994). Epoxidized Vegetable Oils as Reactive Diluents I. Comparison of Vernonia, Epoxidized Soybean and Epoxidized Linseed Oils. *Progress in Organic Coatings*, 25(1), 85-94.

[47] Tan, S. G., & Chow, W. S. (2010). Biobased Epoxidized Vegetable Oils and Its Greener Epoxy Blends: A Review. *Polymer-Plastics Technology and Engineering*, 49(15), 1581-1590.

[48] Jin F-L, Park S-J. (2007). Thermal and Rheological Properties of Vegetable Oil-Based Epoxy Resins Cured with Thermally Latent Initiator. *Journal of Industrial Engineering of Chemistry*, 13(5), 808-814.

[49] Uyama, H., Kuwabara, M., Tsujimoto, T., Nakano, M., Usuki, A., & Kobayashi, S. (2003). Green Nanocomposites from Renewable Resources: Plant Oil-Clay Hybrid Materials. *Chemistry of Materials*, 15(13), 2492-2494.

[50] Samuelsson, J., Sundell-E, P., & Johansson, M. (2004). Synthesis and Polymerization of a Radiation Curable Hyperbranched Resin Based on Epoxy Functional Fatty Acids. *Progress in Organic Coatings*, 50(3), 193-198.

[51] Chakrapani, S., & Crivello, J. V. (1998). Synthesis and Photoinitiated Cationic Polymerization of Epoxidized Castor Oil and Its Derivates. *Pure and Applied Chemistry A*, 35(1), 1-20.

[52] Crivello, J. V. (2008). Effect of Temperature on the Cationic Photopolymerization of Epoxides. *Journal of Macromolecular Science, Part A: Pure and Applied Chemistry*, 45(8), 591-598.

[53] Cheong, M. Y., Ooi, T. L., Amhad, S., Wan, M. Z. W. Y., & Kuang, D. (2009). Synthesis and Characterization of Palm-Based Resin for UV Coating. *Journal of Applied Polymer Science*, 111(5), 2353-2361.

[54] López-Téllez, G., Vigueras-Santiago, E., & Hernández, López. S. (2008). Synthesis and Thermal Cross-Linking Study of Partially-Aminated Epoxidized Linseed Oil. *Designed Monomers and Polymers*, 11(5), 435-445.

[55] Manthey, N. W., Cardona, F., Aravinthan, T., & Cooney, T. (2011). Cure Kinetics of an EpoxidIzed Hemp Oil based Bioresin System. *Journal of Applied Polymer Science*, 122(1), 444-451.

[56] Rosch, J., & Mulhaupt, R. (1993). Polymers from Renewable Resources: Polyester Resins and Blends upon Anhydride-Cures Epoxidized Soybean Oil. *Polymer Bulletin*, 31(6), 679-665.

[57] Gerbase, A. E., Petzhold, C. L., & Costa, A. P. O. (2002). Dynamical Mechanical and Thermal Behavior of Epoxy Resins Based on Soybean Oil. *Journal of the American Oil Chemists´ Society*, 79(8), 797-802.

[58] Yue, S., Hu, J. F., Huang, H., Fu, H. Q., Zeng, H. W., & Chen, H. Q. (2007). Synthesis, Properties and Application of Novel Epoxidized Soybean Oil Toughened Phenolic Resins. *Chinese Journal of Chemical Engineering*, 15(3), 418-423.

[59] Munoz, J. C., Ku, H., Cardona, F., & Rogers, D. (2008). Effects of Catalysts and Post-Curing Conditions in the Polymer Network of Epoxy and Phenolic Resins: Preliminary results. *Journal of Materials Processing Technology*, 486 EOF-492 EOF.

[60] Kiatsimkul, P-p., Suppes, G. J., Hsieh, F-h., Lozada, Z., & Tu-C, Y. (2008). Preparation of High Hydroxyl Equivalent Weight Polyols from Vegetable Oils. *Industrial Crops and Products*, 27(3), 257-264.

[61] Hofer, R., Daute, P., Grutzmacher, R. K., & Ga-H, A. A. W. (1997). Oleochemical Polyols- A New Raw Material Source for Polyurethane Coatings and Floorings. *Journal of Coatings Technology*, 69(869), 65-72.

[62] Zlatanić, A., Lava, C., Zhang, W., & Petrović, Z. S. (2004). Effect of Structure on Properties of Polyols and Polyurethanes Based on Different Vegetable Oils. *Journal of Polymer Science Part B: Polymer Physics*, 42(5), 809-819.

[63] Lligadas, G., Ronda, J. C., Galià, M., & Cádiz, V. (2010). Plant Oils as Platform Chemicals for Polyurethane Synthesis: Current State-of-the-Art. *Biomacromolecules*, 11(11), 2825-2835.

[64] La Scala, J., & Wool, R. P. (2002). The Effect of Fatty Acid Composition on the Acrylation Kinetics of Epoxidized Triacylglycerols. *Journal of the American Oil Chemistis´ Society*, 79(1), 59-6.

[65] Li, F., Hanson, M. V., & Larock, R. C. (2001). Soybean Oil-Styrene-Divinylbenzene Thermosetting Polymers: Synthesis, Structure, Properties and Their Relationships. *Polymer*, 42(4), 1567-1579.

[66] Li, F., & Larock, R. C. (2002). New Soybean Oil-Styrene-Divinylbenzene Thermosetting Copolymers V: Shape-Memory Effect. *Journal of Applied Polymer Science*, 84(8), 1533-1543.

[67] Li, F., & Larock, R. C. (2002). Novel Polymeric Materials from Biological Oils. *Journal of Polymers and the Environment*, 59 EOF-67 EOF.

[68] Zhang, M., Wool, R. P., & Xiao, J. Q. (2001). Electrical Properties of Chicken Feather Fiber Reinforced Epoxy Composites. *Composites Part A: Applied Science and Manufacturing*, 42(3), 229-233.

[69] Teng, G., & Soucek, . (2000). Epoxidized Soybean Oil-Based Ceramer Coatings. *Journal of the American Oil Chemists´ Society*, 77(4), 381-387.

[70] La Scala, J., & Wool, R. P. (2005). Property Analysis of Triglyceride-Based Thermosets. *Polymer*, 46(1), 61-69.

[71] Vigueras-Santiago, E., Hernández-López, S., Linares-Hernández, K., & Linares-Hernández, I. (2009). Mechanical Properties of Goat Leather Photo Grafted with Acrylate-Epoxidized Linseed Oil. *Advances in Technology of Materials and Materials Processing Journal*, 11(2), 43-48.

[72] Ureña-Núñez, F., Vigueras-Santiago, E., Hernández-López, S., Linares-Hernández, K., & Linares-Hernández, I. (2008). Structural, Thermal and Morphological Characterization of UV-Graft Polymerization of Acrylated-Epoxidized Soybean Oil onto Goat Leather. *Chemistry and Chemical Technology*, 2(3), 191-197.

[73] Hernández-López, S., Martín-López del, Campo., Sánchez-Mendieta, E., Ureña-Núñez, V., Vigueras, F., & Santiago, E. (2006). Gamma Irradiation Effect on Acrylated-Epoxi-

dized Soybean Oil: Polymerization and Characterization. *Advances in Technology of Materials and Materials Processing Journal*, 8(2), 220-225.

[74] Williams, G. I., & Wool, R. P. (2000). Composites from Natural Fibers and Soy Oil Resins. *Applied Composites Materials*, 7(5-6), 421 EOF-432 EOF.

[75] Oprea, S. (2010). Properties of Polymer Networks Prepared by Blending Polyester Urethane Acrylate with Acrilated Epoxidized Soybean oil. *Journal of Materials Science*, 45(5), 1315-1320.

[76] Thielemans, W., & Wool, R. P. (2004). Butyrated Kraft Lignin as Compatibilizing Agent for Natural Fiber Reinforced Thermoset Composites. *Composites Part A Applied Sciencie and Manufacturing*, 35(3), 327-338.

[77] Dweib, Hu. B., O', Donnell. A., Shenton, H. H., & Wool, R. P. (2004). All Natural Composite Sandwich Beams for Structural Applications. *Composite Structures*, 63(2), 147-157.

[78] Dweib, Hu. B., Shentin, I. I. I. H. W., & Wool, R. P. (2006). Bio-Based Composite Roof Structures: Manufacturing and Processing Issues. *Composite Structures*, 74(4), 379-388.

[79] Panhuis, M., in, H., Thielemans, W., Minett, A. I., Leahy, R., Le Foulgoc, B., Blau, W. J., & Wool, R. P. (2003). A Composite from Soy Oil and Carbon Nanotubes. *International Journal of Nanoscience*, 2(3), 185-194.

[80] Thielemans, W., Mc Aninch, I. M., Barron, V., Blau, W. J., & Wool, R. P. (2005). Impure Carbon Nanotubes as Reinforcements for Acrylated Epoxidized Soy Oil Composites. *Journal of Applied Polymer Science*, 98(3), 1325-1338.

[81] San-Farfán, Juan., Hernández-López, R., Martínez-Barrera, S., Camacho-López, G., Vigueras-Santiago, M. A., & , E. (2005). Electrical Characterization of Polystyrene-Carbon Black Composites. *Physica Status Solidi*, 2-3762.

[82] Verbanac, F. (1985). Cellulosic Organic Solvent Soluble Products. Patent USA,(4)

[83] Edgar, K. J., Buchanan, CM, Debenham, JS, Rundquist, PA, Seiler, BD, Shelton, MC, & Tindall, D. (2001). Advances in cellulose ester performance and application. *Progress in Polymer Science*, 26(9), 1605-1688.

[84] Verbanac, F. (1984). Cellulosic Organic Solvent Soluble Products. Patent number 4,490,516, Dec 25,.

[85] Shediac, R., Ngola, S. M., Throckmorton, D. J., Anex, Shepodd. T. J., & Singh, A. K. (2001). Reversed-Phase Electrochromatography of Amino Acids and Peptides Using Porous Polymer Monoliths. *Journal of Chromatography*, 925(1-2), 251 EOF-63 EOF.

[86] Carbon Black User's Guide, (2006). International Carbon Black Association,. ., 28.

[87] Columbian Chemicals Company. (2010). http://www.columbianchemicals.com/Portals/Markets%20and%20Applications/Industrial%20Carbon%20Black/Inks/CCC_lores_Raven_0505.pdf, (accessed 1 August).

[88] Cabot Corporation:. (2010). http://www.cabot-corp.com/wcm/download/en-us/sb/VULCAN_XC72-English1.pdf, (accessed 1 August).

[89] Nhumo[MR] . (2010). http://www.nhumo.com.mx/UploadedFiles/Adjunto-Standard_Carbon_Blacks-200671910496.pdf, (accessed 1 August).

[90] Brostow, W., Keselman, M., Mironi-Harpz, I., Narkis, M., & Peirce, R. (2005). Effects of Carbon Black on Tribology of Blends of Poly(Vinylidene Fluoride) with Irradiated an Non-Irradiated Ultrahigh Molecular Weight Polyethylene. *Polymer*, 46(14), 5058-5064.

[91] Marinho, B., Ghislandi, M., Tkalya, E., Koning, CE, & de With, G. (2012). Electrical Conductivity of Compacts of Graphene, Multiwalled Carbon Nanotubes, Carbon Black, and Graphite Powder. *Powder Technology*, 221(1), 351-358.

[92] Celzard, A., Marêché, J. F., Payot, F., & Furdin, G. (2002). Electrical Conductivity of Carbonaceous Powders. *Carbon*, 40(15), 2801-2815.

[93] Vigueras, Santiago. E., Hernández, López. S., & Mercado, Posadas. J. (2009). Modificación de propiedades eléctricas en compuestos poliméricos con negro de carbono. In: Camacho-López M.A, et al (ed), *Tópicos en Materiales,*, México: Universidad Autónoma del Estado de México;.

Chapter 12

Soybean and Prostate Cancer

Xiaomeng Li, Ying Xu and Ichiro Tsuji

Additional information is available at the end of the chapter

http://dx.doi.org/10.5772/ 52605

Part I: Consumption of Soybean and the incidence of Prostate Cancer in East and West

Prostate cancer is the most frequently diagnosed malignancy in men all over the world. The investigation from International Agency for Research on Cancer in World Healthy Organization of United Nations reported the data of 899, 000 new cases of prostate cancer in 2008, accounted for 13.6% of the total cancer cases. The investigation discovered the large difference of the incidence and the mortality of prostate cancer between in West and in East (Figure 1). In Western countries, the incidences and mortalities are significantly high. For example, the cases from Australia, Northern America or Europe, occupied nearly three-quarters of all the globe prostate cancer patients. In contrast, the incidences and mortalities in Asia are quite low, less than one tenth of that in Europe, the US or Australia of the West, especially in the South-Central Asia, showing the lowest incidence and mortality [1].

The significant difference of the incidences of prostate cancer addressed several important questions.

1. Why do Asian men have a much lower incidence of prostate cancer compared to men from the Western countries (the US and Europe)? Latent or clinically insignificant cancer of the prostate is found at autopsy at approximately the same rate in men from Asian countries as those from the USA (approximately 30% of men aged over 50 years), but there are large differences in the clinical incidence and mortality. Is there a strong possibility that diet and nutrition play a prominent role in accelerating or inhibiting the process by which clinically significant prostate cancer develops?

2. Is the fact that the hormone-dependent cancers of the prostate and breast show the same incidence and lifetime risk (the correlation r is 0.81 in 21 countries) related to diet?

© 2013 Li et al.; licensee InTech. This is a paper distributed under the terms of the Creative Commons Attribution License (http://creativecommons.org/licenses/by/3.0), which permits unrestricted use, distribution, and reproduction in any medium, provided the original work is properly cited.

3. East Asian countries, including Chinese and Japanese men, have the lowest incidence of prostate cancer in the world. But why, when Japanese men from those countries migrate to the North America, does their risk of developing prostate cancer increase 10-fold compared to their counterparts in Japan or China?

4. Although it is relatively rare in East, an increase in the incidence of prostate cancer has been reported in China, where the life style especially the diet structure is changing followed with the developed economy in recent years. What factors can account for the conspicuous increase?

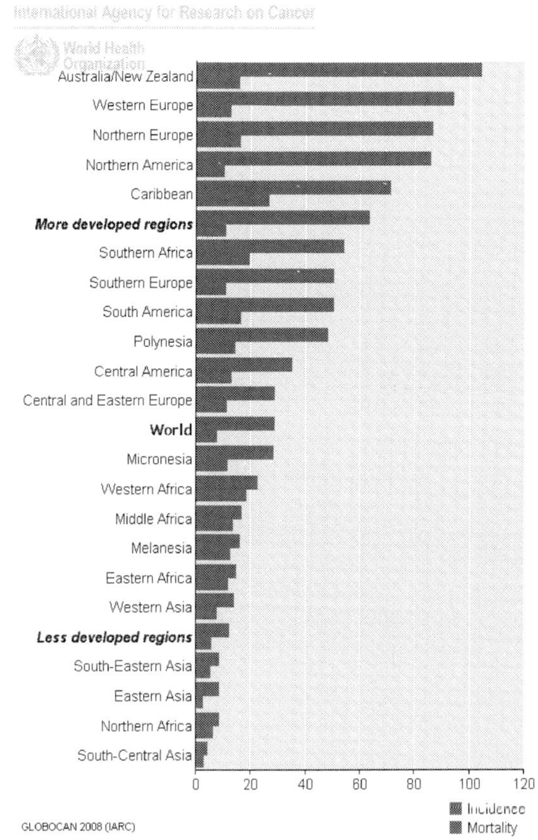

Figure 1. The incidence and mortality of Prostate cancer in the world, 2008(from International Agency for Research on Cancer)

Those questions highlight the critical roles of environmental dietary factors in the different risks of prostate cancer between East and West. Many epidemiological studies suggest that the different dietary, most probably one kind of traditional food in the East countries, soybean, may become a dominant reason for the protective effects against prostate cancer [2].

Soybean, recognized as a complete protein food, is a species of legume and an annual plant in Asia. As being a traditional food in Asia, the history of soybean in Asia is very long, more than 5000 years and even before written records. Soybean can be made to a large body of kinds of foods by fermented or non-fermented process. Typical soy foods include soy milk, bean curd and tofu skin which are non-fermented soy foods as well as miso sauce, natto and soy sauce which belong to fermented soy foods. Not only fermented soybean foods, but also non-fermented soybean products are very essential dietary ingredient for Asian people, especially Chinese, Japanese and Koreans. Some Asians often have and their fermented or non-fermented products almost everyday. Japanese nearly had the soy food daily by ingestion of miso sauce and Chinese approximately had them more than 100 grams per day.

The history of soybean used in western countries is quite short, compared to that in Asia. Soybean was introduced into America, Australia and New Zealand about in 17th century and into Canada in 1831 as a sauce named "A new dozen India Soy". Most westerners hardly consume soy foods in the diet, although soybean and its products are used largely in some other ways, for example soybean oil could be made into bio-diesel in the United States. HoweverSoy food consumption in the western countries are quite low. Soy foods have been consumed for centuries in Asian countries Japanese nearly had the soy food daily by ingestion of miso sauce and Chinese approximately had them by Tofu, soy milk and tofu skin.. The mean daily intakes of soy protein are approximately 30 grams in Japan, 20 grams in Korea, 7 grams g in Hong Kong, and 8 grams but more than 100 grams per day in some area in China [3]. While the average daily intake in the United States is less than 1 gram [4].

Country	Prostate cancer mortality	Total energy	Fish energy	Soy energy	Animal energy
Australia	29.59	3055	22	0	1019
Austria	31.23	3575	15	0	1233
Canada	29.20	3340	29	0	1297
France	31.43	3529	34	0	1343
Italy	22.57	3688	25	0	913
USA	32.19	3641	23	1	1316
Hong Kong	5.44	2771	89	36	834
Japan	5.87	2852	195	93	590
Korea	0.90	3056	67	94	269
Singapore	7.47	3165	63	29	689
Thailand	0.53	2330	37	18	152

Table 1. Prostate cancer and amount of soy food consumption in East and West, 1998 (from JNCI 1998; 90: 1637-1647)

For clear comparison of soybean in the Asian and the western diet, a table published in JNCI (Table 1), clearly clarified Prostate cancer and amount of soy food consumption in East and West. In Asian regions, the soybean energy is 36 in Hong Kong, 29 in Singapore and 18 in Thailand; and for much more dramatically difference, the soybean energy is 93 in Japanese, and 94 in Korean diet [5]. In contrast, in western countries of Australia, Austria, Canada, France, or Italia, the energy from soybean in daily dietary was zero, even in USA, the soy energy is only 1. This indicates that soybean is the dominantly different diet factor between the Eastern countries and the Western countries.

Interestingly the dramatically increased prostate cancer mortality in western countries, combined together with almost no soybean energy consumption, compare with very low mortality in Asia (only 1/60 to 1/6 of that in West), combined with high soybean energy consumption. This provided the important evidence for the consumption of soybean related to the lower risk of prostate cancer mortality. Even another point of the higher animal energy consumption in western diet, it is not that dominant as the difference of soy food.

Soy food	Study site	Finding	Study type	OR/RR	P trend	Reference
Soy food	Japan	protection	cohort	0.52(0.29–0.90)	0.010	Kurahashi 2007
Miso soup				0.65(0.39–1.11)	0.220	
Soymilk	USA	protection	cohort	0.30(0.1 – 0.9)	0.020	Jacobsen 1998
Tofu	Hawaii	protection	cohort	0.35(0.08–1.43)	0.054	Severson 1989
Soy food	UK	protection	case-control	0.52(0.30–0.91)	0.340	Heald 2007
Tofu	Japan	protection	case-control	0.47(0.20–1.08)	0.160	
All soy products				0.53(0.24–1.14)	0.110	Sonoda 2004
Soyfoods	China	protection	case-control	0.51(0.28–0.95)	0.061	
Tofu				0.58(0.35–0.96)	0.032	Lee 2003
Soyfoods	Hawaii,San Francisco,	protection	case-control	0.62(0.44–0.89)	0.060	
All legumes	Los Angeles, British Columbia and Ontario			0.62(0.49–0.80)	0.0002	Kolonel 2000
Beans/lentils/ nuts	Canada	protection	case-control	0.69(0.53–0.91)	0.030	Jain 1999
Tofu, Soybean	Canada	protection	case-control	0.80(0.60–1.10)	0.200	Villeneuve 1999
Soybean foods	China	protection	case-control	0.29(0.11-0.79)	0.02	Li 2008
Baked beans	UK	protection	case-control	0.57(0.34–0.95)	N/A	Key 1997
Garden peas	UK			0.35(0.13–0.91)	N/A	

Table 2. N/A:no adequate dataEpidemiological studies on food intake of soy products and prostate caner risk(most data summrized from Mol Nutr Food Res. 2009; 53: 217-226)

Most epidemiological studies have suggested that the consumption of soy food is associated with a reduction in prostate cancer risk in humans. Eight case-control studies and three cohort studies have reported the protective effect of soy food, with odds ratios or relative risks ranging from 0.3 to 0.80, including in China and Japan, where people consume more soybean food, tofu, soymilk and natto. Here the epidemiological data are summarized in Table 2 [6].

Some cohort studies provided the convincing data on this issue. A population based prospective study recruited 43 509 Japanese men aged 45–74 years and followed them up for 10 years (1995 through 2004). For men aged 60 years, in whom soy food were associated with a dose-dependent decrease in the risk of localized cancer, with RRs for men in the highest quartile of soy food consumption compared with the lowest obtained a protective OR of 0.52 (95% confidence interval (CI) 0.29–0.90, P trend = 0.01) [7]. A cohort study in the USA with 225 incident cases of prostate cancer in 12395 California seventh-day adventist men showed frequent consumption (more than once a day) of soymilk was associated with 70% reduction of the risk of prostate cancer (RR = 0.3, 95%CI 0.1–0.9, P trend = 0.02) [8]. In an early cohort study among 7999 men of Japanese ancestry who were first examined between 1965 and 1968 and then followed through to 1986, 174 incident cases of prostate cancer were recorded. Increased consumption of tofu did not show statistical significant association with the risk of prostate cancer (RR = 0.35, 95% CI 0.08–1.43) [9].

Much more case-control studies provided more evidence for the reduced risk of prostate cancer associated with consumption of soy foods. First we concentrate some case-control studies conducted in men living in China. Our group carried out a population-based case-control study in China to investigate the possible correlation factors for prostate cancer, 28 cases from 3940 men over 50 years old with prostate-specific antigen screening in Changchun city in China, matched them with controls of low prostate-specific antigen value (< 4.1 ng/mL) by 1: 10 according to age and place of employment. In all ten food items, the consumption of soybeans was demonstrated the only factor to decrease the risk of prostate cancer. Men who consumed the soybean product of Tofu and soymilk more than once per day had a multivariate OR of 0.29 (95% CI, 0.11–0.79) compared with men who consumed soybean products less than once per week. The P for trend was 0.02. There was no significant difference for any other dairy food (2). Another case-control study in China of 133 cases and 265 age- and residential community-matched controls from 12 cities were recruited. Results showed that the age- and total calorie-adjusted OR of prostate cancer risk was 0.58 (95% CI 0.35-0.96, P= 0.032) in the highest tertile of Tofu intake comparing to the lowest tertile. There were also statistically significant associations of intake of soy foods (OR 0.51; 95% CI 0.28–0.95, P = 0.061) [10].

The case-control studies also provide the similar evidence for that the soy intake protect against prostate cancer. A case-control study of diet and prostate cancer in Japan demonstrated the possible protective effect of traditional Japanese soybean die plays a preventive role against prostate cancer in four geographical areas (Ibaraki, Fukuoka, Nara, and Hokkaido) of Japan. All 140 cases and 140 age -matched hospital controls were analyzed to confirm the consumption of fish and natto showed significantly decreasing linear trends for risk, with RR of 0.53 (95% CI, 0.24-1.14) (P < 0.05) for all soybean products, 0.47 (95% CI, 0.20-1.08) for tofu, and 0.25 (95% CI, 0.05-1.24) for natto [11].

Not only has the case-control research conducted in Asian men supported the hypothesis that the rich soybean products may protect against prostate cancer but many case-control studies conducted in western countries also provid the epidemiological evidence for the protective effects of soy food against prostate cancer. A multiethnic case-control study carried out in Hawaii, San Francisco, and Los Angeles in the USA, and British Columbia and Ontario in Canada, 1619 cases were diagnosed during 1987–1991 and were compared to 1618 controls of African-American, white, Japanese, and Chinese men. Controls were frequency- matched to cases on ethnicity, age, and region of residence of the case, in a ratio of approximately 1:1. Intake of soy foods was inversely related to prostate cancer with OR of 0.62 (95% CI, 0.44–0.89). Results were similar when restricted to prostate-specific antigen normal controls [12]. A further four case-control studies from USA, Canada and UK showed similar results with a protective effect of consuming legumes (beans, lentils, garden peas, etc.) against prostate cancer. A case-control study in Canada for a total study population consisted of 617 incident cases of prostate cancer and 636 population controls from Ontario, Quebec, and British Columbia. To obtain a decreasing, statistically significant association was found with increasing intakes of beans/lentils/nuts (OR = 0.69, 95% CI, 0.53-0.91) [13]. Another population-based case-control study conducted in eight Canadian provinces. Risk estimates were generated by applying multivariate logistic regression methods to 1623 histologically confirmed prostate cancer cases and 1623 male controls aged 50-74 to obtained a OR 0.80 (95% CI, 0.60-1.10) of Tofu and Soybeans [14]. Not only soybean but also other legumes had the inverse relationship against prostate cancer. Oxford group in UK found the baked beans had an OR of 0.57 (95% CI, 0.34-0.95); and garden peas had an OR of 0.35 (95% CI, 0.13-0.91) [15]. And a population-based case-control study of diet, inherited susceptibility and prostate cancer was undertaken in Scotland investigated a total of 433 cases of Scottish men and 483 controls aged 50-74 years, indicates the theoretical scope for reducing the risk from prostate cancer, with the consumption of soy foods (adjusted OR 0.52, 95 % CI 0.30-0.91) [16].

All these data from both East and West, summarized in Table 2, provide the convincing epidemiological evidence that the higher intake of soy products is associated with a reduced risk of prostate cancer in human. Some experimental data in animal model also demonstrated the possible protective effects for Soybean food against prostate cancer. For example, Zhou group demonstrated that dietary soy products inhibit the experimental prostate tumor growth through a combination of direct effects on tumor cells and indirect effects on tumor neovasculature in mice, confirming that soy foods could act as a preventive factor against prostate cancer in animal models [17].

Prostate cancer has marked geographic variations between countries. The data from both western and eastern countries support the critical roles of soybean food in the protection of prostate cancer.

Part II: Correlation between prostate cancer and soybean isoflavones from diet, serum epidemiological and laboratorial data

In recent decades, much evidence from epidemiological studies support the notion that frequent consumption of soybean foods, the most different diet between East and West, is beneficial for the protection against prostate cancer.

Soybeans are high in protein, for about 35% to 40% of the dry weight. They also contain 18% polyunsaturated fats, 30% carbohydrates, and some vitamins, minerals. They are the only legume that provides ample amounts of the essential omega-3 fatty acid alpha-linolenic acid.2 Soybeans are a rich source of isoflavones (or phytoestrogens), a subclass of flavonoids that bind to estrogen receptors (though not as strongly as estrogen). Isoflavones are also discovered in peanut, alfalfa; however, soybean contains the largest amounts of isoflavones in the nature. Soybeans and soy foods such as tofu, soymilk, and miso are the only significant dietary sources of these phytoestrogens. In soybeans, isoflavones bound to a sugar molecule as glycoside, and when soybean is fermented or digested, isoflavones could be released from the bounded sugar [18].

Table 3 summarized the relationship between isoflavones levels and soy protein content in a variety of soy-containing foods. The isoflavones in soymilk are lowest, 2.5mg per 100g soymilk, may because in the fluid soymilk, the nutrition of soybeans is dissolved in water and lower concentration of nutrition in soymilk, whereas isoflavones in soy flour are highest, 131.2 to 198.9mg per 100g. In tofu, a traditional soy food in Asia, the isoflavones are 27.9mg/100g. In the traditional and very important Japanese soy food, the isoflavones in miso are 42.6mg/100g. The isoflavones in natto are 20mg/100g. The information from Table 3 clearly provides the amounted isoflavones in some other soybean foods as well. In the most frequently diet, each gram of soy protein is associated with approximately 3.6 mg of isoflavones in Tofu, 3.7 mg of isoflavones in whole bean based soymilk [19].

As the special nutrition in soybean compared to other plants, the isoflavones have many health benefits, including protection against chronic diseases, menopausal symptoms, osteoporosis and breast cancer or prostate cancer [20]. The most effective soy isoflavones are well known as genistein and daidzein. Here, the relationship between each soybean isoflavones and prostate cancer would be further illustrated.

The epidemiological evidence for the association of soy isoflavones and the prostate cancer risk is still limited. Some epidemiological data, summarized in Table 4, evaluated the effects of soybean isoflavones on prostate cancer. To date, one study suggests a causal relationship between isoflavones and prostate cancer, but two cohorts and two case-control studies suggest that soy isoflavones, genistein and daidzein, have prophylactic effects on prostate cancer. In addition,, an accumulating body of evidence from laboratory studies in recent decades has suggested that diets rich in high concentration of isoflavones are associated with anti-tumor effects in prostate cancer.

	estimated protein (g/100 g)	total isoflavones (mg/100 g)	isoflavone (mg/g protein)
soy products			
soybean chips	54.2	35	0.6
soy links frozen raw	3.9	15	3.9
natto boiled fermented	46.4	20	0.4
tofu, silken, firm	7.8	27.9	3.6
tempeh	14.9	43.5	2.9
miso	12.5	42.6	3.4
soy cheese, Cheddar	7.2	28	3.9
soy cheese, mozarella	7.7	32	4.2
soy cheese, Parmesan	6.4	36	5.6
soy milks			
soy milk, isolate based, low isoflavone	3.3	2.5	0.7
soy milk, isolate based, high isoflavone	3.3	4.4	1.3
soy milk, whole bean based, low isoflavone	3.1	2.4	0.8
soy milk, whole bean based, high isoflavone	3.1	11.6	3.7
soy milk, entire bean	3.1	3.6	1.2
soy milk	3.5	9.7	2.8
soy protein materials			
soy protein isolate, aqueous extract, type 1	85	114	1.3
soy protein isolate, aqueous extract, type 2	85	78.7	0.9
soy protein isolate, aqueous extract, type 3	85	103.4	1.2
soy protein isolate, alcohol extract	85	81.9	1.0
soy protein concentrate, alcohol extract	65	12.5	0.2
soy flour, defatted	50	131.2	2.6
soy flour, full fat, raw	35	177.9	5.1
soy flour, full fat, roasted	39	198.9	5.1

Table 3. Relationship between isoflavone levels and soy protein content in a variety of soy-containing foodsfrom J. Agric. Food Chem. 2003; 51: 4146-4155

2.1. Genistein

Genistein, firstly isolated from Genista tinctoria in 1899, was abundant in soybean. Even though a serum case-control study suggested the promotion roles of genistein for the prostate cancer risk; most studies demonstrated its protective effects against prostate cancer. Two cohort studies and two case control studies demonstrated the significant protective roles of genistein against prostate cancer, with multivariate RRs (ORs) of 0.71, 0.52, 0.58 or 0.53, summarized in Table 4.

Phynoestrogen	Study site	Findings	Study type	OR/RR	P_{trend}	Reference
genistein						
	Europe	protection	cohort	0.71(0.53-0.96)	0.030	
	Japan	protection	cohort	0.52(0.30-0.90)	0.030	
	protection	case-control		0.58(0.34-0.97)	0.040	Nagata 2007
	protection	case-control		0.53(0.29-0.97)	0.058	Lee 2003
	Japan	promotion	serum case-control			Akaza 2002
daidzein						
	Japan	protection	cohort	0.50(0.28-0.88)	0.040	Kurahashi 2007
	Japan	protection	case-control	0.55(0.32-0.93)	0.020	Nagata 2007
	China	protection	case-control	0.56(0.31-1.04)	0.116	Lee 2003
	Japan	protection	serum case-control			Akaza 2002

Table 4. Effects of photoestrogens on cacinogenesis of prostate cancer in epidemiological or animals experimental data.

The population based prospective study recruited 43 509 Japanese men aged 45–74 years and followed them up for 10 years (1995 through 2004). All 147 food items in 220 cases with organ localized cancers was investigated the relationship of isoflavones intake and the risk of prostate cancer. The increased consumption of genistein was found associated with the decreased risk of localized prostate cancer. These results were strengthened when analysis was restricted to men aged more than 60 years, in whom isoflavones and soy food were associated with a dose-dependent decrease in the risk of localized cancer, with RRs for men in the highest quartile of genistein consumption compared with the lowest of 0.52 (95% CI 0.30–0.90, P trend = 0.03) [7].

Another Japanese group examined associations between nutritional and the prevalence of prostate cancer in a case-control study of Japanese men. Two hundred patients and 200 age-matched controls were selected from 3 geographic areas of Japan. Isoflavones and their agly-cones (genistein and daidzein) were significantly associated with decreased risk, with the OR for genistein 0.58 (95% CI 0.34–0.97, P trend = 0.04), indicating that isoflavones might be an effective dietary protective factor against prostate cancer in Japanese men [21].

A case-control study in China showed an overall reduced risk of prostate cancer associated with consumption of soy foods and genistein. In this study, 133 cases and 265 age- and residential community-matched controls from 12 cities were recruited. Results showed that the age- and total calorie-adjusted OR of prostate cancer risk was 0.58 (95% CI 0.35–0.96, P trend = 0.032) comparing the highest tertile of tofu intake to the lowest tertile. There were also statistically significant associations of intake genistein (OR, 0.53; 95% CI, 0.29–0.97, P trend = 0.058) [10].

Notably, a recent prospective investigation of plasma phytoestrogens and prostate cancer in the European also found that higher plasma concentrations of genistein were associated with lower risk of prostate cancer. They examined plasma concentrations of phyto-oestrogens in relation to risk for subsequent prostate cancer in a case-control study nested in the European Prospective Investigation into Cancer and Nutrition. Concentrations of isoflavones genistein, daidzein and equol, and that of lignans enterolactone and enterodiol, were measured in plasma samples for 950 prostate cancer cases and 1042 matched control participants. Relative risks (RRs) for prostate cancer in relation to plasma concentrations of these phyto-oestrogens were estimated by conditional logistic regression. Higher plasma concentrations of genistein were associated with lower risk of prostate cancer, the RR among men in the highest vs. the lowest fifth was 0.71 (95%CI 0.53-0.96, P trend=0.03). After adjustment for potential confounders, this RR was 0.74 (95% CI 0.54-1.00, P trend=0.05). No statistically significant associations were observed for circulating concentrations of daidzein, equol, enterolactone or enterodiol in relation to overall risk for prostate cancer [22].

Laboratory studies revealed that Genistein has multiple functions in the antitumor effects against prostate cancer, with the concentration as the prominent associated factor. In 2002, an Italy group declared that at low concentration, about 1~10μM, genistein would stimulate androgen-dependent prostate cancer cell line LNCaP growth, but when genistein was at high concentration, more than 100μM, it would result in cell apoptosis in prostate caner [23]. The conclusions above were supported by the research of another Gao's group, published in the journal of the Prostate in 2004 [24].

In vitro studies in several prostate cancer cell lines have revealed that genistein directly inhibit the growth of prostate cancer cells or through inducing apoptosis, affecting the expression of a large number of genes that are related to the control of cell survival and physiologic behaviors [25,26]. As tyrosine kinase inhibitor and topoisomerase inhibitor, genistein induces cancer cell apoptosis by upregulating the expression of cyclin-dependent kinase inhibitor p21 or inhibiting the activity of nuclear factor kappa B (NF-kB) signaling pathways [27, 28]. Genistein is a potent inhibitor of protein–tyrosine kinase, which may attenuate the growth of cancer cells and decrease the level of oxidative DNA damage [29-31]. Genistein also reported to enhance the ability of endoglin, a component of the transforming growth factor beta receptor complex, in suppression of the motility of prostate cancer cell [32]. Moreover, genistein is also a potent inhibitor of angiogenesis and metastasis. It can effectively inhibit cell invasion by inhibiting transforming growth factor β-mediated phosphorylation of the p38 mitogen-activated protein kinase-activated protein kinase 2 and the 27 kDa heat shock protein [33].

The anti-tumor effects of genistein are demonstrated in the animal models. Animal experiments have also shown that dietary concentrations of genistein can inhibit metastasis of prostate cancer [34]. Lifetime consumption of isolate/isoflavones has prevented spontaneous development of metastasizing adenocarcinoma in Lobund–Wistar rat [35, 36]. Dietary genistein can also suppress the development of advanced prostate cancer in castrated transgenic adenocarcinoma of mouse prostate (TRAMP) mice [37]. Some other result also indicated biphasic role for genistein in the regulation of prostate cancer growth and metastasis. A low concentration of 500 nmol/L of genistein to 12-week-old TRAMP-FVB mice as evidenced by increased proliferation, invasion. But a pharmacologic dose (50 nmol/L) decreased proliferation, invasion, and MMP-9 activity (>2.0-fold) concomitant with osteopotin reduction [38]. With 250 mg genistein/kg diet in treatments (TRAMP) mice model, the most significant effect was seen in the TRAMP mice exposed to genistein throughout life (1-28 weeks) with a 50% decrease in poorly-differentiated cancerous lesions. In a separate experiment in castrated TRAMP mice, dietary genistein suppressed the development of advanced prostate cancer by 35% compared with controls. The data obtained in intact and castrated transgenic mice suggest that genistein may be a promising chemopreventive agent against androgen-dependent and independent prostate cancers [39]. This group further identified the associated signaling, and revealed that Genistein in the diet significantly inhibited the cell proliferation by down-regulating tyrosine kinase regulated proteins, EGFR, IGF-1R, and down-regulating the downstream mitogen-activated protein kinases, ERK-1 and ERK-2 in prostates in TRAMP mice [37].

Genistein is demonstrated to be synergy with other phyto-chemicals together to inhibit the growth of prostate cancer. For example, genistein and curcumin is reported could inhibit the growth of prostate cancer cells in a synergistic way. Genistein, together with DIM, was proved to repress the proliferation of androgen-dependent prostate cancer cell line LNCaP and androgen-independent prostate cancer cell line PC-3 [40]. Genistein, together with biochanin A, could also inhibit the growth of human prostate cancer cells [41]. Genistein is also reported acts as a radiosensitizer for prostate cancer both in vitro and in vivo inhibit metastasis of prostate cancer [34].

Based on the studies above, a conclusion is deduced that genistein showed protective effects against prostate cancer, in vivo and in vitro, at its high concentration.

2.2. Daidzein

Daidzein is another important soy isoflavone. Most of epidemiological studies prove that daidzein contributes to the reduction of prostate cancer risk and the prevention of prostate cancer. A cohort study and two case control studies demonstrated its significant preventive roles of daidzein for the prostate cancer prevention, with multivariate RRs (ORs) of 0.50, 0.55 or 0.56, summarized in Table 4.

Several studies indicated that daidzein might be an effective dietary protective factor against prostate cancer in Japanese men. One cohort study of the population based prospective study recruited 43 509 Japanese men aged 45–74 years mentioned above and followed them up for 10 years. From 147 food items in 220 cases with organ localized cancers was investigated and isoflavones intake was founded to be associated with the risk of prostate cancer. In men aged more than 60 years, RRs for men in the highest quartile of daidzein consumption compared with the lowest of 0.50 (95%CI 0.28-0.88, P trend = 0.04) [7]. Another Japanese group examined associations between nutritional and the prevalence of prostate cancer in a case-control study in Japanese men. Two hundred patients and 200 age-matched controls were selected from 3 geographic areas of Japan, demonstrated that daidzein were significantly associated with decreased risk, with the OR 0.55 (95% CI 0.32–0.93, P trend = 0.02) for daidzein [21].

A case-control study in China showed an overall reduced risk of prostate cancer associated with consumption of soy daidzein. In this study, 133 cases and 265 age- and residential community-matched controls from 12 cities were recruited. Results showed that the age- and total calorie-adjusted OR of prostate cancer risk was 0.58 (95% CI 0.35–0.96, P trend = 0.032) comparing the highest tertile of tofu intake to the lowest tertile. There were even not statistically significant associations, but the protective trend for the intake of daidzein (OR, 0.56; 95% CI, 0.31–1.04, P trend = 0.116) [10].

Daidzein displays the modest protective effect against prostate cancer from the experimental data. Daidzein could act as a radiosensitizer against prostate cancer and an inhibitor of cell growth. Daidzein inhibited cell growth and synergized with radiation, affecting APE1/Ref-1, NF-kappaB and HIF-1alpha, but at lower levels than genistein or soy [42].

In addition, the protective effect on prostate cancer was strengthened after daidzein was combined with genistein, compared to individual one. The study from Oregon State University demonstrated daidzein and genistein could also induce cell apoptosis in benign prostate hyperplasia (BPH) cells at concentration of 25 μM. Soy extractions were indicated more effective as chemopreventive agents than genistein or daidzein. A combination of active soy-derived compounds is demonstrated more efficacious and safer as chemopreventive agents than individual compound. Soy extracts also increased Bax expression in PC3 cells [43]. Isoflavones exert their chemopreventive properties by affecting apoptosis signaling TRAIL (tumor necrosis factor-related apoptosis-inducing ligand) pathways in prostate cancer LNCaP cells. The chemopreventive effects of soy foods on prostate cancer are associated with isoflavone-induced support of TRAIL-mediated apoptotic death [44].

Soy isoflavones, including daidzein and genistein, exert anticarcinogenic effects against prostate cancer, proposed that soy extracts, containing a mixture of soy isoflavones and other bioactive components, would be a more potent chemo-preventive agent than individual soy isoflavones.

Part III: The mechanisms of prevention effects of soybean phyto-estrogens against prostate cancer.

Isoflavones are the best-known phyto-estrogens, are a diverse group of naturally soy compounds, play important roles in prostate cancer inhibition. The consumptions of soy isoflavone are investigated to be an effective protection factor against certain diseases. And the products rich in isoflavones might protect against enlargement of the male prostate gland, slow prostate cancer growth and lead to prostate cancer cell death.

The chemical structure-based ligand and receptor binding is the most important primary step in generating downstream signal transduction. To reveal their mechanisms of prevention effects against prostate cancer, their specific structures for the receptor binding are illustrated. Isoflavones are classified to phyto-estrogens, because their structures are similar to estradiol (17β-oestradiol), also could mimic estrogenic effect to bind to estrogen receptors. However, their structures are similar not only to estradiol, but also to dihydrotestosterone (DHT), the most important steroid hormone in male. DHT and 17β-oestradiol, endogenous steroid hormones, have a four-ringed carbon backbone, where as non-steroid hormone thyroxin has a quite different structure from them (Figure 2). The isoflavones of genistein and daidzein are two phyto-oestrogens found at very high levels in soy formula. Flavone, another group of phyto-estrogens, is found abundant in many plants. The structures of genistein, daidzein and flavone showed similar properties: a hydrophobic core and one or two terminal polar groups, similar to 17β-oestradiol or DHT (Figure 2), have the ability to cause estrogenic or/and anti-testosterone effects through the modulation of androgen receptor (AR) transactivation.

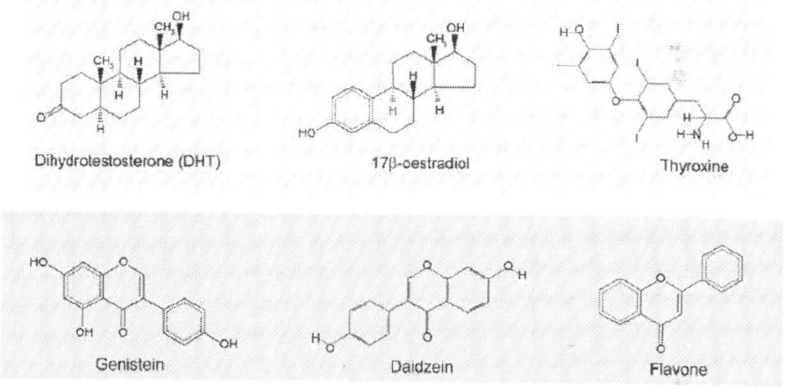

Figure 2. Molecular formulas of ligand compounds, two endogenous steroid hormones of dihydrotestosterone (DHT) and 17β-oestradiol, one endogenous non-steroidal hormone, thyroxin, and three phyto-estrogens of genistein, daidzein and flavone. from Asian J Androl. 2010; 12: 535-547

The normal development and maintenance of the prostate is dependent on androgen acting through the AR, which is driven by DHT plays a critical role in prostate cancer development and progression. AR expression is maintained throughout prostate cancer progression, and the majority of androgen-independent or hormone refractory prostate cancers with the overexpression or the mutation of AR. This progress may be affected by the soybean phyto-oestrogens, as mimic oestrogens, compete the same binding sites of AR that binds to DHT, further affect androgen-controlled AR mediated prostate cancer growth and development [45].

To reveal the mechanisms of these isoflavones for prostate cancer, our group adopted a computerized approach to examine the interaction of the human AR and isoflavone (genistein or daidzein), whereas the interaction of AR and flavone was set up as a positive control. Auto Dock method was adopted to summarize the roles of genistein, daidzein in AR activity regulation, further to evaluate the importance of isoflavones for the tumor repression against prostate cancer. Auto Dock applies a half-flexible docking method, which permits small molecular conformation changes. Based on a complex 'lock-and-key model', it is an excellent method to reveal ligand–receptor binding. The result of computer stimulation from Auto Dock contains two parts, one is the binding site of a ligand docked in the receptor and the other is the binding affinity when a ligand is docked in the receptor.

The 3D spatial structure of AR-LBD was obtained from RCSB Protein Data Bank and its PDB ID is 2ama (676-919AA). The positive-control docking result showed that 17β-oestradiol fit the ligand-binding site of AR, at the same position in AR as its natural ligand, DHT. The negative control, thyroxine, showed a quite different binding position with external docking site (Figure 3A). Comparing the three endogenous ligands, thyroxine was expected to have the weakest binding to AR and the highest affinity energy, which we measured at -5.4 kcal mol^{-1}. Very strong binding to AR with lower affinity energies was expected in the two steroid hormones, with affinity energies of -11.2 kcal mol^{-1} for DHT and -10.7 kcal mol^{-1} for 17β-oestradiol (Figure 3B) [46].

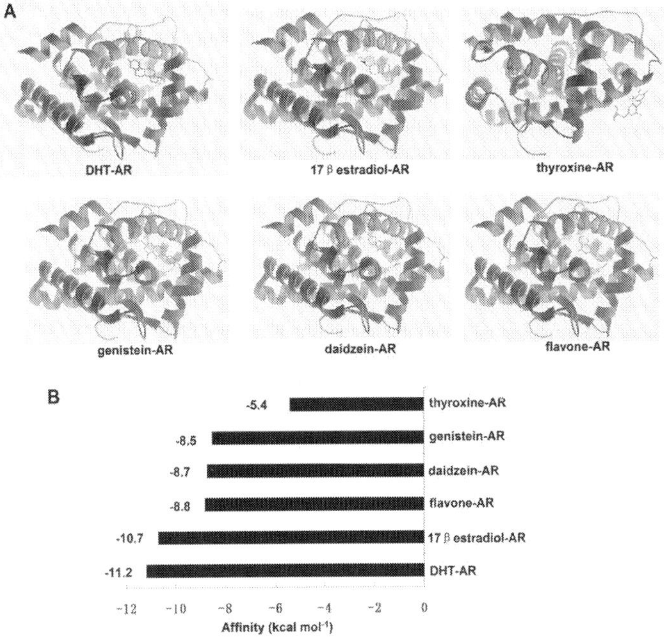

Figure 3. Auto docking analysis of endogenous hormones or phyto-estrogens binding to androgen receptor.(A) Positions of the steroid hormones of DHT or 17β-oestradiol, endogenous non-steroid hormone, thyroxin, phyto-estrogens of genistein, daidzein or flavone binding to androgen receptor.(B): The affinity energies of each ligand binding to androgen receptor. from Asian J Androl. 2010; 12: 535-547

Genistein and daidzein are abundant in soy formula and as healthy ingredient in soybean to protect against prostate cancer. To understand the role of the isoflavone compounds in prostate cancer, a computerized auto-dock system was adopted to examine the interactions between the human AR and phyto-oestrogens (genistein, daidzein, and flavone). As shown in Figure 3 genistein, daidzein and flavone fit in the middle region of the AR-LBD (Figure 3A), the same as DHT and 17β-estradiol. The affinities of them were expected to lie between the affinities of thyroxin and 17β-estradiol. Genistein and daidzein, soy isoflavones, showed affinity energies of -8.5 and -8.7 kcal mol^{-1}, respectively, which were very similar to the affinity energy of flavone of -8.8 kcal mol^{-1}. From that result, we concluded that the three (iso) flavones exhibit similar binding affinities to AR (Figure 3B). Considering their sharing of a binding site with estradiol, their affinities for AR and quantities potentially consumed in the diet, these phyto-estrogens could have significant effects on AR and AR-related cancers [46].

The phyto-estrogens (genistein, daidzein and flavone) can bind to AR in an Auto Dock model and can be regarded as androgenic effectors, suggesting important roles for them in AR-mediated cancers. Interestingly, all these phyto-estrogens are reported to be associated with prostate cancer, so we consider them AR-related phyto-estrogens. We summarized some recent data about the effects of them on AR-mediated transcriptional activity and on prostate tumorigenesis in Table 5. Genistein, daidzein and flavone, implicated as androgenic effectors in our research, indeed regulate AR-mediated PSA transcriptional activity. They have been demonstrated previously to either enhance AR-mediated transcriptional activity or inhibit DHT induced AR-mediated PSA activity. Moreover, isoflavones or soy beverage has already been shown in Phase II trials to decrease PSA levels in prostate cancer patients. It is noteworthy that two recent phase II trials showed that isoflavones or soy beverage can decrease PSA levels in prostate cancer patients, suggesting that androgen receptor target genes can be regulated by isoflavones or flavones [46].

Phynoestrogen	Findings	Study type	Reference
genistein	inhibition of R1881-induced AR mediated pPSA-luc activity	reporter assay	Davis 2002
	decrease AR binding to ARE	EMSA	
	inhibition of R1881-induced AR mediated pPSA-luc activity	reporter assay	Gao 2004
	enhancement of AR mediated pPSA/ARE/Probasin/ MMTV-luc		
daidzein	enhancement of AR (with ARA) mediated MMTV-luc	reporter assay	Chen 2007
flavone	inhibition of DHT-induced AR mediated pPSA-luc activity	reporter assay	Rosenberg 2002
soy food	decline of serum PSA	PhaseII trail	Kwan 2010 Pendleton 2008

Table 5. Effects of photoestrogens on AR mediated transcriptional activityfrom Asian J Androl. 2010; 12: 535-547

The mechanism of these phyto-estrogens against prostate cancer has been studied, some laboratory studies demonstrated that phyto-estrogens might not only mimic estrogenic activities but also interfere with other steroid hormones, for example DHT, the natural androgen in men. In human body, the activity of DHT binding with androgen receptor can be weaken in the presence of a large amount of these phyto-estrogens, genistein, daidzein and flavone.

The findings in Auto Dock interestingly demonstrate that phyto-estrogens are displayed the great binding abilities to AR, demonstrating their disrupt effects against DHT/AR binding as anti-androgenic effectors, supporting the epidemiological studies that soybean were the potential inhibitor for prostate cancer cell growth related to AR. Figure 4 illustrated clearly the mechanism of phyto-estrogens against prostate cancer. The different consumption of soy foods and the different concentrations of isoflavones possibly disrupt the endogenous DHT binding to AR in the prostate and thus inhibited DHT induced AR translocation and AR-mediated PSA transactivation, thus reduce prostate cancer risk. As the antagonists of androgen, phyto-estrogens are ale to inhibit the cell growth induced by androgen in prostate gland.

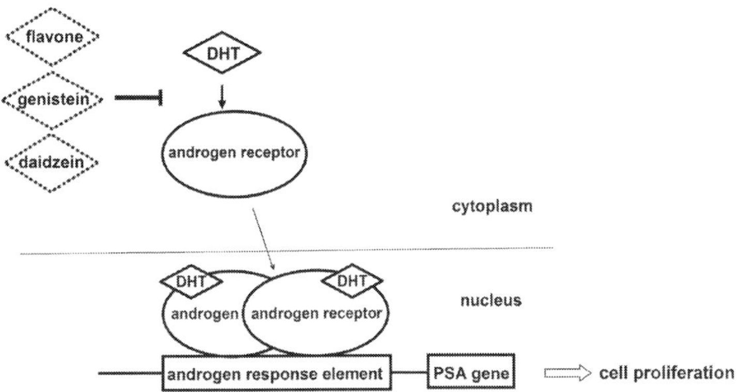

Figure 4. The mechanism of protective effect of genistein, daidzein or flavone against prostate cancer.

The reasons that Asian populations have lower rates of hormone-dependent cancers (breast, prostate) and lower incidences of menopausal symptoms and osteoporosis than Westerners are still need to be further revealed. However, the association of the large quantities of soy products consumption in Asian populations with the reduction in the risk of prostate cancer, provides a unique insight for the beneficial effects of soy foods. Soy food display completely different consumption between eastern and western populations. The geometric mean levels of plasma total isoflavonoids were demonstrated to be 7-10 times higher in Japanese men than in Finnish men. Asian immigrants living in Western nations also have increased risk of these maladies as they 'Westernize' their diets to include more protein and fat and reduce their soy intake. This provides a good explanation for much epidemiological data, indicating the significant protective effect of (iso) flavones for prostate cancer.

The much higher serum phyto-estrogen levels could hypothetically inhibit the growth of prostate cancer in Chinese and Japanese men, which may give a good explanation for the quite low incidence and mortality from prostate cancer in Japan or China. The novel important insights for soy food against prostate cancer need to be further illustrated. Soy-based food products are expected to introduce more to the western markets and to have more consumption in western daily diet.

Acknowledgements

This study was supported by Ministry of Science and Technology (No. 2010DFA31430), NCET-10-0316; the Fundamental Research Funds for the Central University (10JCXK004), and Changchun Science & Technology Department (No.2011114).

Author details

Xiaomeng Li[1*], Ying Xu[1] and Ichiro Tsuji[2]

*Address all correspondence to: lixm441@nenu.edu.cn

1 The Key Laboratory of Molecular Epigenetics of MOE, Institute of Genetics and Cytology, School of Life Sciences, Northeast Normal University, China

2 Department of Public Health, Tohoku University, Japan

References

[1] International Agency for Research on Cancer (IARC). (2008). WHO Mortality Database. *Lyon, France: IARC.*

[2] Li, X., Li, J., Tsuji, I., Nakaya, N., Nishino, Y., & Zhao, X. (2008). Mass screening-based case-control study of diet and prostate cancer in Changchun, China. *Asian J Androl*, 10(4), 551-60.

[3] Xiao, C. W. (2008). Health effects of soy protein and isoflavones in humans. *J Nutr*, 138(6), 1244S-9S.

[4] Messina, M., Nagata, C., & Wu, A. H. (2006). Estimated Asian adult soy protein and isoflavone intakes. *Nutr Cancer*, 55(1), 1-12.

[5] Hebert, J. R., Hurley, T. G., Olendzki, B. C., Teas, J., Hampl, Y., & Ma, J. S. (1998). Nutritional and socioeconomic factors in relation to prostate cancer mortality: a cross-national study. *J Natl Cancer Inst*, 90(21), 1637-47.

[6] Jian, L. (2009). Soy, isoflavones, and prostate cancer. *Mol Nutr Food Res*, 53(2), 217-26.

[7] Kurahashi, N., Iwasaki, M., Sasazuki, S., Otani, T., Inoue, M., & Tsugane, S. (2007). Japan Public Health Center-Based Prospective Study Group. Soy product and isoflavone consumption in relation to prostate cancer in Japanese men, Cancer Epidemiol. *Biomarkers Prev*, 16, 538-45.

[8] Jacobsen, B. K., Knutsen, S. F., & Fraser, G. E. (1998). Does high soy milk intake reduce prostate cancer incidence? The Adventist Health Study (USA). *Cancer Causes Control*, 9, 553-7.

[9] Severson, R. K., Nomura, A. M., Grove, J. S., & Stemmermann, G. N. (1989). A prospective study of demographics, diet, and prostate cancer among men of Japanese ancestry in Hawaii. *Cancer Res*, 49, 1857-60.

[10] Lee, M. M., Gomez, S. L., Chang, J. S., Wey, M., Wang, R. T., & Hsing, A. W. (2003). Soy and isoflavone consumption in relation to prostate cancer risk in China, Cancer Epidemiol. *Biomarkers Prev*, 12, 665-8.

[11] Sonoda, T., Nagata, Y., Mori, M., Miyanaga, N., Takashima, N., Okumura, K., Goto, K., Naito, S., Fujimoto, K., Hirao, Y., Takahashi, A., Tsukamoto, T., Fujioka, T., & Akaza, H. (2004). A case-control study of diet and prostate cancer in Japan: Possible protective effect of traditional Japanese diet. *Cancer Sci*, 95, 238-42.

[12] Kolonel, L. N., Hankin, J. H., Whittemore, A. S., Wu, A. H., Gallagher, R. P., Wilkens, L. R., John, E. M., Howe, G. R., Dreon, D. M., West, D. W., & Paffenbarger, R. S. Jr. (2000). Vegetables, fruits, legumes and prostate cancer: A multiethnic case-control study,. *Cancer Epidemiol. Biomarkers Prev*, 9, 795-804.

[13] Jain, M. G., Hislop, G. T., Howe, G. R., & Ghadirian, P. (1999). Plant foods, antioxidants, and prostate cancer risk: Findings from case-control studies in Canada. *Nutr Cancer*, 34, 173-84.

[14] Villeneuve, P. J., Johnson, K. C., Kreiger, N., & Mao, Y. (1999). Risk factors for prostate cancer: Results from the Canadian National Enhanced Cancer Surveillance System The Canadian Cancer Registries Epidemiology Research Group. *Cancer Causes Control*, 10, 355-67.

[15] Key, T. J., Silcocks, P. B., Davey, G. K., Appleby, P. N., & Bishop, D. T. (1997). A case-control study of diet and prostate cancer,. *Br J Cancer*, 76, 678-87.

[16] Heald, C. L., Ritchie, M. R., Bolton-Smith, C., Morton, M. S., & Alexander, F. E. (2007). Phyto-oestrogens and risk of prostate cancer in Scottish men,. *Br J Nutr*, 98, 388-96.

[17] Zhou, J. R., Li, L., & Pan, W. (2007). Dietary soy and tea combinations for prevention of breast and prostate cancers by targeting metabolic syndrome elements in mice. *Am J Clin Nutr*, 86(3), 882-8.

[18] Wu, Z., Rodgers, R. P., & Marshall, A. G. (2004). Characterization of vegetable oils: Detailed compositional fingerprints derived from electrospray ionization fourier

transform ion cyclotron resonance mass spectrometry. *J Agric Food Chem*, 52(17), 5322-8.

[19] Setchell, K. D., & Cole, S. J. (2003). Variations in isoflavone levels in soy foods and soy protein isolates and issues related to isoflavone databases and food labeling. *J Agric Food Chem*, 51(14), 4146-55.

[20] Food and Drug Administration. (1999). Food labeling: Health claims; soy protein and coronary heart disease. *Fed Regist.*, 64(206), 57700-33.

[21] Nagata, Y., Sonoda, T., Mori, M., Miyanaga, N., Okumura, K., Goto, K., Naito, S., Fujimoto, K., Hirao, Y., Takahashi, A., Tsukamoto, T., & Akaza, H. (2007). Dietary isoflavones may protect against prostate cancer in Japanese men. *J Nutr*, 137(8), 1974-9.

[22] Travis, R. C., Spencer, E. A., Allen, N. E., Appleby, P. N., Roddam, A. W., Overvad, K., Johnsen, N. F., Olsen, A., Kaaks, R., Linseisen, J., Boeing, H., Nöthlings, U., Bueno-de-Mesquita, H. B., Ros, M. M., Sacerdote, C., Palli, D., Tumino, R., Berrino, F., Trichopoulou, A., Dilis, V., Trichopoulos, D., Chirlaque, M. D., Ardanaz, E., Larranaga, N., Gonzalez, C., Suárez, L. R., Sánchez, M. J., Bingham, S., Khaw, K. T., Hallmans, G., Stattin, P., Rinaldi, S., Slimani, N., Jenab, M., Riboli, E., & Key, T. J. (2009). Plasma phyto-oestrogens and prostate cancer in the European Prospective Investigation into Cancer and Nutrition. *Br J Cancer*, 100, 1817-23.

[23] Maggiolini, M., Vivacqua, A., Carpino, A., Bonofiglio, D., Fasanella, G., Salerno, M., Picard, D., & Andó, S. (2002). The mutant androgen receptor T877A mediates the proliferative but not the cytotoxic dose-dependent effects of genistein and quercetin on human LNCaP prostate cancer cells. *Mol Pharmacol*, 62(5), 1027-35.

[24] Gao, S., Liu, G. Z., & Wang, Z. (2004). Modulation of androgen receptor-dependent transcription by resveratrol and genistein in prostate cancer cells. Prostate. May 1; , 59(2), 214-25.

[25] Li, Y., & Sarkar, F. H. (2002). Gene expression profiles of genisteintreated PC3 prostate cancer cells. *J. Nutr*, 132, 3623-3631.

[26] Li, Y., & Sarkar, F. H. (2002). Down-regulation of invasion and angiogenesis-related genes identified by cDNA microarray analysis of PC3 prostate cancer cells treated with genistein,. *Cancer Lett.*, 186, 157-64.

[27] Davis, J. N., Kucuk, O., & Sarkar, F. H. (1999). Genistein inhibits Nfkappa B activation in prostate cancer cells. *Nutr. Cancer.*, 35, 167-74.

[28] Raffoul, J. J., Wang, Y., Kucuk, O., Forman, J. D., Sarkar, F. H., & Hillman, G. G. (2006). Genistein inhibits radiation-induced activation of NF-kappaB in prostate cancer cells promoting apoptosis and G2/M cell cycle arrest,. *BMC Cancer*, 6, 107.

[29] Akiyama, T., Ishida, J., Nakagawa, S., Ogawara, H., Watanabe, S., Itoh, N., Shibuya, M., & Fukami, Y. (1987). Genistein, a specific inhibitor of tyrosine-specific protein kinases. *J Biol Chem.*, 262, 5592-5.

[30] Barnes, S., Peterson, T. G., & Coward, L. (1995). Rationale for the use of genistein-containing soy matrices in chemoprevention trials for breast and prostate cancer,. J. Cell Biochem. Suppl. , 22, 181-7.

[31] Djuric, Z., Chen, G., Doerge, D. R., Heilbrun, L. K., & Kucuk, O. (2001). Effect of soy isoflavone supplementation on markers of oxidative stress in men and women,. *Cancer Lett.*, 172, 1-6.

[32] Craft, C. S., Xu, L., Romero, D., Vary, C. P., & Bergan, R. C. (2008). Genistein induces phenotypic reversion of endoglin deficiency in human prostate cancer cells. *Mol. Pharmacol.*, 73, 235-42.

[33] Xu, L., & Bergan, R. C. (2006). Genistein inhibits matrix metalloproteinase type 2 activation and prostate cancer cell invasion by blocking the transforming growth factor beta-mediated activation of mitogen-activated protein kinase-activated protein kinase 2-27-kDa heat shock protein pathway,. Mol. Pharmacol. , 70, 869-77.

[34] Lakshman, M., Xu, L., Ananthanarayanan, V., Cooper, J., Takimoto, C. H., Helenowski, I., Pelling, J. C., & Bergan, R. C. (2008). Dietary genistein inhibits metastasis of human prostate cancer in mice. *Cancer Res.*, 68, 2024-32.

[35] Pollard, M., & Wolter, W. (2000). Prevention of spontaneous prostaterelated cancer in Lobund-Wistar rats by a soy protein isolate/isoflavone diet. *Prostate.*, 45, 101-5.

[36] Wang, J., Eltoum, I. E., & Lamartiniere, C. A. (2002). Dietary genistein suppresses chemically induced prostate cancer in Lobund-Wistar rats,. *Cancer Lett.*, 186, 11-8.

[37] Wang, J., Eltoum, I. E., & Lamartiniere, C. A. (2007). Genistein chemoprevention of prostate cancer in TRAMP mice. *J Carcinog.*, 6(3).

[38] El Touny, L. H., & Banerjee, P. P. (2009). Identification of a biphasic role for genistein in the regulation of prostate cancer growth and metastasis. Cancer Res. , 69, 3695-703.

[39] Wang, J., Eltoum, I. E., & Lamartiniere, C. A. (2004). Genistein alters growth factor signaling in transgenic prostate model (TRAMP). *Mol Cell Endocrinol.*, 219(1-2), 171-80.

[40] Smith, S., Sepkovic, D., Bradlow, H. L., & Auborn, K. J. (2008). Diindolylmethane and genistein decrease the adverse effects of estrogen in LNCaP and PC-3 prostate cancer cells. *J Nutr*, 138(12), 2379-85.

[41] Seo, Y. J., Kim, B. S., Chun, S. Y., Park, Y. K., Kang, K. S., & Kwon, T. G. (2011). Apoptotic effects of genistein, biochanin-A and apigenin on LNCaP and PC-3 cells by 21 through transcriptional inhibition of polo-like kinase-1. *J Korean Med Sci.*, 26(11), 1489-94.

[42] Singh-Gupta, V., Zhang, H., Yunker, C. K., Ahmad, Z., Zwier, D., Sarkar, F. H., & Hillman, G. G. (2010). Daidzein effect on hormone refractory prostate cancer in vitro and in vivo compared to genistein and soy extract: potentiation of radiotherapy. *Pharm Res.*, 27(6), 1115-27.

[43] Hsu, A., Bray, T. M., Helferich, W. G., Doerge, D. R., & Ho, E. (2010). Differential effects of whole soy extract and soy isoflavones on apoptosis in prostate cancer cells. *Exp Biol Med (Maywood)*, 235(1), 90-7.

[44] Szliszka, E., & Krol, W. (2011). Soy isoflavones augment the effect of TRAIL-mediated apoptotic death in prostate cancer cells. *Oncol Rep.*, 26(3), 533-41.

[45] Rahman, M., Miyamoto, H., & Chang, C. (2004). Androgen receptor coregulators in prostate cancer: mechanisms and clinical implications. *Clin Cancer Res.*, 10, 2208-19.

[46] Wang, H., Li, J., Gao, Y., Xu, Y., Pan, Y., Tsuji, I., Sun, Z., & Li, X. (2010). Xeno-oestrogens and Phyto-oestrogens are Alternative Ligands for the Androgen Receptor. *Asian J Androl.*, 12(4), 535-47.

Chapter 13

Soybean Meal and The Potential for Upgrading Its Feeding Value by Enzyme Supplementation

D. Pettersson and K. Pontoppidan

Additional information is available at the end of the chapter

http://dx.doi.org/10.5772/52607

1. Introduction

Soy is a crop of tremendous importance for the food industry but also in the animal feed industry. Soybeans, as well as other oilseed crops such as rape seed (canola), sunflower and palm kernel, are grown primarily for the production of vegetable oil for human consumption but the by-products after oil extraction are of similar importance as feed ingredients. Meals from these crops are obtained after the extraction of the vegetable oil from the seed. In addition, considerable amounts of cottonseed meal (a by-product of the cotton fibre production) are also available for animal feed. Soybean meal (SBM), the by-product after oil extraction of soybeans has become increasingly important as a feed component and is used in variable amounts in the feeding of all species in animal production, even to some extent in the feeding of farmed fish. On a global scale SBM is dominating the market for protein meals primarily due to the high content of good quality protein, making SBM an excellent ingredient in feed formulations. SBM is particularly important for poultry production, constituting approximately 30-40 % of a standard soy/maize diet, since broilers and layers require a high proportion of protein in their feed. It is generally estimated that approximately 46 % of all SBM produced for animal feed is used in poultry diets (broilers, layers and turkeys), while another 25 % is used for feeding pigs. In the US approximately 50 % of the SBM is used for poultry, 25 % for swine and 12 % for beef cattle. Although US is the largest producer of soybeans, Argentina is by far the largest exporter of SBM followed by Brazil and US. With very limited own production of soy the EU is one of the leading markets for import of SBM. [1]

Over the last 5 years the price for SBM has been increasing, and this trend is expected to continue in the future. Hence, with protein already being the second most expensive ingredient in e.g. poultry diets, there is a need to either replace SBM with other and cheaper pro-

© 2013 Pettersson and Pontoppidan; licensee InTech. This is a paper distributed under the terms of the Creative Commons Attribution License (http://creativecommons.org/licenses/by/3.0), which permits unrestricted use, distribution, and reproduction in any medium, provided the original work is properly cited.

tein ingredients in the diets or to increase the utilization of SBM nutrients e.g. by the use of enzymes.

2. Composition and nutritive value of soybean meal

SBM consists primarily of protein and carbohydrates in the form of indigestible neutral and acidic non-starch polysaccharides (NSP) as well as low molecular weight sugars (Table 1).

Due to the low content of easily available carbohydrates, and a high content of NSP belonging to the indigestible dietary fibre fraction [2] the apparent metabolisable energy value (AME) of SBM is low for broiler chickens and estimated to only about 9.5 MJ/kg fresh weight, while in growing pigs that, as opposed to poultry, have a high capacity for hind gut fermentation of the indigestible carbohydrate fraction, the AME value may be around 14.5 MJ/kg fresh weight [3].

	Range (g/kg dry matter)	
Component	**Low**	**High**
Crude protein	490	540
Starch	0	27
Crude fat	17	21
Low molecular weight sugars		
Mono saccharides	5	8
Sucrose	55	81
Raffinose series	53	67
Neutral non starch polysaccharides		
Rhamnose	3.7	5
Fucose	2.9	3.1
Arabinose	22	25
Xylose	15	18
Mannose	9	13
Galactose	37	40
Glucose	50	59
Acidic non starch polysaccharides		
Uronic acids	39	41

Table 1. Typical soybean meal gross chemical composition. (g/kg dry matter), compiled from literature, for crude protein, starch, crude fat and low molecular weight sugars, while the non-starch polysaccharide composition of a selection (n=6) of soybean meals were analysed at Novozymes A/S according to Theander *et al.* [4]. Data compiled from [2,3] as well as from internal analysis of neutral and acidic non starch polysaccharide constituents.

The composition of SBM may vary depending on the country of origin of the soybean, the cultivar, the processing and the inevitable year to year variation in growing conditions. Still most feeding tables contains very little, if any, information about the variability of e.g. amino acids (AA) and digestibility that may be expected although the data presented typically is based on several hundred samples collected over several years. Neither is the indigestible dietary fibre fraction well described in feed tables and at best values for neutral detergent and acid detergent fibre are provided.

2.1. Protein

Plants cultivated for their protein content are typically classified as angiosperms and belong to a number of different botanical families. Beans, peas, lupins and soybeans are all members of the leguminosae family while rape seed belong to the cruciferae family, sunflower and safflower are members of the compositae family and cotton belongs to the malvaceae [5]. On average the content of CP in the common raw oilseeds such as soybean, sunflower and rape seed ranges from 20-40 %. Due to the various processing steps and the subsequent concentration of the protein-containing fraction by solvent extraction, the average CP content of oilseed meals varies from 32 % in sunflower meal to over 50 % in some SBM [6]. SBM is used in feed rations for monogastric animals mainly due to the high protein content and also because of the superior AA profile compared to other plant protein products used as diet ingredients [3]. Poultry and swine diets are generally formulated based on AME and the level of CP. SBM has a high content of lysine, which makes it a good ingredient in poultry and swine diets as both of these species has a high requirement for this essential amino acid.

The CP fraction of SBM is made up of around 80 % storage proteins in the form of glycenin and β-conglycenin, approximately 5 % is represented as various anti nutritional factors (ANFs) and the remaining 15 % consists of other proteins. Most tables on nutrient composition of feedstuffs such as e.g. the NRC [7] operates with two types of SBM based on the crude protein CP content. One is the regular SBM with approximately 44 % CP, where a fraction of the hulls has been added back into the meal, and the other is dehulled SBM with approximately 48 % CP. The feed compound industry principally assumes that the digestible AA content of SBM per unit of protein is constant; disrespecting that variability may occur due to e.g. genotype, origin, processing and storage conditions [8-10]. The variability in SBMs has been nicely demonstrated by de Coca-Sinova *et al.* [10] who evaluated six SBMs from different origins and found considerable variation in the chemical composition and protein quality which translated into differences in AA digestibility in broilers, so that SBM with higher levels of CP and lower levels of trypsin inhibitor activity showed higher AA digestibility.

2.2. Carbohydrates

For broiler chickens vegetable protein sources typically have a low metabolisable energy content compared to cereals, due to a lower starch content and a higher content of indigestible NSP, which are part of the dietary fibre fraction (Table 1).

The structural features of the common vegetable protein sources are more complex than that of cereals but their cell walls still contains cellulose and hemicelluloses but in addition also high amounts of rhamnogalacturonans, assigned to the pectic polysaccharides. Quantification of the pectic fraction is not straightforward due to a more complex structure and since extraction procedures used for the analysis often causes an overestimation and the literature data may differ.

It is notable that AME table values and AME calculated on the content of CP. crude fat, starch and sucrose (Table 2) are quite similar for meals from soybean and sunflower while in meals from rape seed and cotton seed the discrepancy is 1.7 and 2.3 MJ/kg, respectively. The amount of pectin (61 g uronic acids/kg dry matter) and fibre matrix structure of rape seed most likely increases the water holding capacity of this raw material resulting in a poor nutrient availability for monogastric animals and in addition ANFs may reduce the energetic value. In cotton seed similar effects may be at play although the pectin content is quite similar to that of SBM (about 45 g/kg dry matter), while the ANFs are different and the cellulose content is considerably higher.

	Soybean	Rape seed	Sunflower	Cotton seed
Crude protein	450	337	426	426
Crude fat	10	23	29	29
Starch	5	0	0	0
Sucrose	70	58	33	16
Oligo, di and mono –saccharides, except sucrose	67	24	23	56
Cellulose	62	52	89	92
Total dietary fibre	233	354	326	340
Apparent metabolisable energy (AME) in broilers				
Table values AME (MJ/kg)[a]	9.5	5.9	6.2	6.3
Calculated AME (MJ/kg)[b]	9.3	7.6	6.9	8.6

Table 2. Average chemical composition. (g/kg fresh weight) of solvent extracted meal from soybean, rape seed, sunflower and cotton seed. Table compiled from [2,3]. [a]Table values from [3]. [b]Values based on crude protein, crude fat, starch and sucrose.

Plant cell walls are divided in primary and secondary walls and their composition will differ according to their stage of development (maturity). Primary cell walls are flexible and surround cells in growth and elongation whereas secondary walls surround cells in which growth has ceased. The secondary walls are lignified and thereby rigid. The primary cell wall is synthesized during cell expansion at the first stages of development and is composed of cellulose, hemicelluloses, pectic polysaccharides and many proteins (Table 3).

Cell walls are classified as Type I cell walls, which are generally the most common in the plant kingdom, or as Type II which is typical for grasses. The non-cellulosic polysaccharides of Type I cell walls are xyloglucans and about 35 % of the cell wall mass are pectins. Type I cell walls are found in all dicotyledons, the non-graminaceous monocotyledons, and gymnosperms. Type II walls have a low pectin and xyloglucan content and a high arabinoxylan content. Type II walls also contain mixed linked β-D-glucan and possess ester linked ferulic bridges in the xylan, which have not as yet been found in Type I walls [11].

In soybeans and other dicotelydons the pectin fraction consists of rhamnogalacturonans. The rhamnogalacturonan I consists of a main chain of galacturonic acid and rhamnose. Attached to this structure there are side chains of galactose and arabinose residues. There are also xylose and rhamnose side chains present as well as xylo-galacturonans. In addition traces of mannose polysaccharides may be found that possibly origin from an incomplete removal of the hull fraction from the meal [12].

In the cotelydons approximately 30 % of the NSP belong to the pectin fraction while in the hulls about 80-90 % of the non-starch polysaccharides are of pectic origin. The galactose content of SBM is generally higher than in other oilseed meals and is highly associated with the rhamnogalacturonans. This is not the case for rape seed meal, sunflower meal and cotton seed meal [13] where arabinans and arabinogalactans constitute the most important side chains. Since the NSP are indigestible by the endogenous enzymatic systems in the small intestine they can only be utilised through hind-gut fermentation, thereby providing short chain fatty acids that may be absorbed in the hind gut and utilised as an energy source by the animal. As a consequence the AME content of all oilseed meals is low for broiler chickens that have a limited capacity for hind gut fermentation, while it is higher for pigs [14].

The dietary fibre fraction is composed of different polysaccharide structures and their molecular structure and incorporation into the cell wall matrix determines their solubility characteristics. A high solubility favours fermentation and even poultry may to some extent ferment soluble NSP [15,16].

The principal hemicelluloses found in dicotelydons including soybeans are xyloglucans (Table 3) consisting of a glucose back bone with xylose side chains linked to the carbon 6 of the glucose residues in the back bone chain. It is a well acknowledged hypothesis that primary and secondary cell walls in dicotelydons are constructed in different ways. In the primary cell wall cellulose and xyloglucans interact in a network consisting of cellulose which is coated with a monolayer of xyloglucans. The secondary cell wall is composed of pectins, but not of the homogalacturonan type that is found in fruit and berries. In soybean the pectic polysaccharides are xylogalacturonans with a backbone of α-(1-4)-galacturonan residues and to this xylose residues are linked in β-(1-3) position. The xylogalacturonans are associated with regions that consist of rhamnogalacturonans type I and II, with a higher degree of branching in the type II [18,19]. Together with the xylogalacturonans the rhamnogalacturonan I and its side chains consisting of arabinans and arabinogalactans makes up the main part of the pectic substances [20].

Component		Approximate composition (%)	
		Primary cell walls	Secondary cell walls
Cellulose		30	45-50
Hemicellulose	Xyloglucan	25	25
	Xylans	5	
	ß-D-glucans	nd	nd
	Glucomannan	nd	nd
Pectins	Homogalacturonans	15	0.1
	Rhamnogalacturonan I	15	
	Rhamnogalacturonan II	5	
Glycoproteins	Arabinogalactanprotein	Variable	nd
Phenolics	Extensin	<5	nd
	Lignin	nd	20
	Phenolic acid	0.3	nd

Table 3. Major polymers of the growing and mature plant cell walls in dicotelydons. Table compiled from [17].

2.3. Anti nutritional factors

The occurrence and amount of ANFs and their effect on protein and energy utilisation limits the inclusion of vegetable proteins in diets for pigs and poultry. In general, ANFs among legume species are similar, however the actual amounts of ANFs varies widely between different species and cultivars [21]. The main ANFs include protease inhibitors, lectins, tannins, phytic acid and indigestible carbohydrates. In addition, lupins contain considerable amounts of alkaloids and lima beans contain increased amounts of cyanogens as their dominant ANFs [22,23]. Oilseeds, and subsequently oilseed meals, have more specific ANFs depending on the actual species. Rapeseed meal contains glucosinolates and soybean is particularly high in trypsin and chymotrypsin inhibitors while cottonseed meal contains gossypol [24]. Based on what is commonly known in the feed industry and academia it may be assumed that the content of ANFs may differ depending on growing conditions and, especially, the heat processing of the feed ingredients.

The best characterized ANFs of soybean are protease inhibitors, lectins and phytate.

The Kunitz inhibitor (KSTI) together with the Bowman-Birk inhibitor (BBI) are the most abundant protease inhibitors in soybeans, and are commonly referred to as trypsin inhibitors even though they may also inhibit e.g. chymotrypsin and other proteases belonging to the serine family. The mechanism of trypsin inhibitors is to bind to the active site of the protease and thereby cause inactivation of the enzyme, which then cannot proceed to degrade protein. When the level of protease activity in the gut is depressed the pancreas responds in a compensatory fashion by producing more of the digestive enzymes. In some species this has been shown to be related to an enlargement of the pancreas [25,26]. When animals are fed SBM with a high level of protease inhibitors the digestive proteases trypsin and chymotrypsin are inactivated leading to impaired animal performance. This has been exemplified e.g. by Sklan *et al.* [27] who showed that chicks fed raw soy had significantly reduced bodyweight gain compared to a control group fed heated soy. Furthermore, the chicks fed the raw soy had increased pancreas weight and reduced trypsin activity in the small intestine.

Lectins are carbohydrate binding proteins generally considered to have an anti nutritional effect [28,29]. This has been exemplified e.g. by Schulze *et al.* who showed that inclusion of purified soybean lectin into a pig diet increased the amount of dry matter, nitrogen (N) and AA passing into the terminal ileum [30].

Phytic acid, also commonly referred to as phytate, is a well described anti nutritional factor that under physiological pH conditions binds minerals and protein thereby preventing utilization of these nutrients by the animal [31,32].

Furthermore, the high molecular weight soy proteins glycenin and β-conglycenin act as potential antigenic factors leading to the formation of serum antibodies in particular in young animals, e.g. early-weaned piglets [33].

2.4. Impact of processing on nutritive value

As opposed to the cereal grains the composition and nutritive value of oilseed meals for animal feed not only depends on the cultivar, the climatic condition and level of fertilisation, but is also influenced by the processing conditions during the oil extraction procedure. Oilseed meals are generally obtained after a pre-press solvent extraction process. The combined effects of seed preparation, de-hulling pre-conditioning, cooking and solvent extraction will determine the nutritive value of the meal.

The processing of soybeans (Figure 1) and in particular the final heating step (toasting) is critical to the quality of the resulting SBM. The initial processing of soybeans includes cleaning, drying and cracking of the beans to remove the hulls. The dehulled soybeans are the raw material for production of full fat soybean meal which may or may not be heat treated to inactivate enzymes. When processing regular SBM the dehulled beans are conditioned at 65-70°C followed by flaking, which prepares the beans for oil extraction. The oil is usually extracted using a solvent such as hexane. Finally the resulting cake is treated in a toaster in order to remove the solvent and to heat the meal sufficiently to optimize its nutritional value. In this process control of the processing conditions such as temperature, moisture, pressure and processing time is highly important to maintain a high solubility of the SBM product. [34] After toasting a fraction of the hulls will often be transferred back into the meal to produce SBMs with different protein contents.

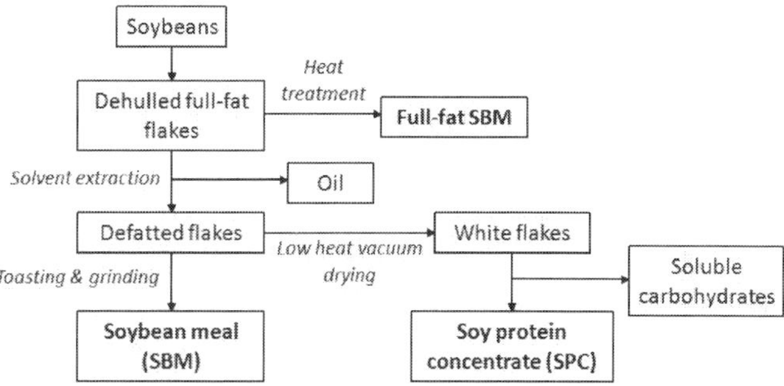

Figure 1. Overview of the processing of soybeans to obtain soybean meal (SBM), full fat SBM or Soy protein concentrate (SPC).

As mentioned earlier the nutritional value of SBM is limited by the presence of several ANFs interfering with feed intake and metabolism. ANFs are concentrated in large amounts in the hull fractions of oilseeds and de-hulling will consequently reduce the level of these substances in the meal. Furthermore, heat processing of the SBM acts to destroy the heat sensible ANFs such as protease inhibitors and lectins. However, in case SBM is heated excessively the occurrence of Maillard reaction will increase. Maillard reactions occur between the amino group of the amino acids and the reducing sugars eventually leading to a decrease in energy and amino acid digestibility [35]. Hence, the conditions applied during processing to ensure a high quality of the protein fraction are a compromise between sufficiently inactivating the ANFs and avoiding destruction of essential available nutrients.

An important problem faced by the feed industry is the lack of good techniques to correctly evaluate the quality of commercial SBMs. The available methods, of which determination of the protein dispersibility index (PDI) [36], KOH protein solubility [37,38] and urease activity [39] are the most commonly used, are based on changes in the physical and chemical properties of SBM occurring during heat treatment, and have shown not to be fully reliable [40,41]. Another means to estimate SBM quality is the determination of trypsin inhibitor content, but this method is tedious and also shows inconsistency [43].

The defatted soy flakes can also be processed to obtain a soy protein concentrate (SPC), which is a higher value protein product compared to SBM. This processing takes place by removing the solvent (e.g. hexane) by low heat vacuum drying and then removing the soluble carbohydrates yielding a final product with ~90 % protein [43]. SPC has a much lower level of ANFs than SBM and is therefore particularly well suited for young animals that do not tolerate normal SBM well, e.g. piglets. Lately the use of SPC in salmon and trout diets to replace fishmeal is also being evaluated [43].

3. Enzymes for improving the nutritive value of soybean meal

The use of exogenous microbial enzymes is today a mature concept in the animal feed industry and is used on a routine basis to improve the nutritive value of feed ingredients. The logical implication of improving the nutrient utilization is a reduced amount of nutrients in manure, which is highly beneficial for the environment especially in areas with intense animal production. The environmental benefits of using enzymes in animal diets has been exemplified in a series of published life cycle assessment studies investigating the effects of xylanase [44], phytase [45] and protease [46] when used in either pig or poultry diets. These studies together demonstrate the huge potential impacts on global warming, eutrofication and acidification that can be achieved by employing feed enzymes in animal diets to improve the utilization of nutrients.

3.1. Protein degrading enzymes

With protein being a quite expensive ingredient in animal diets, improving the nutritional value of the SBM protein fraction is an obvious target for enzyme application. The apparent ileal digestibility (AID) of CP in SBM is typically around 80-85 %, but lower values have also been reported. In an investigation of 6 different SBMs originating from South America, US and Spain the AID of CP and AA was shown to vary considerably between the batches with the US SBM having the highest digestibility value (82.3 %) followed closely by the Spanish SBM (81.8 %) and with the South American SBM's having considerably lower (75.2-76.8 %) digestibility values [10]. These results serve to demonstrate the impact of differences in SBM quality when formulating diets to achieve the necessary protein content and AA availability. Since it is both difficult and laborious to investigate SBM quality, in practice diets are often formulated to contain higher levels of nutrients than required, thereby providing a safety margin.

Early attempts to improve the nutritive value of SBM for pigs and poultry aimed at pretreating SBM in the presence of a protease to increase protein solubilisation and obtain a decrease in antigenicity. Using this approach it was demonstrated that treating SBM with either an acidic protease or an alkaline protease increased the amount of soluble α-amino N concentration and reduced the antigenic protein concentration, more so with the acidic protease compared to the alkaline protease treated SBM [47]. Furthermore, feeding studies in piglets [47] and broilers [48] showed that feeding SBM treated with the acidic protease instead of the non-treated SBM as part of a cereal-based diet led to performance improvements in both species as well as to improved N digestibility and reduced serum antisoya antibodies in broilers. In contrast feeding SBM treated with the alkaline protease reduced performance in piglets [47], and even though Ghazi *et al.* did observe reductions in chick serum antisoya antibodies upon feeding SBM treated with alkaline protease only the acid protease treated SBM resulted in a positive effect on performance and digestibility parameters [48]. Differences in digestibility may occur either directly or indirectly due to hydrolysis of ANFs interfering with the digestive process. Protease treatment in above studies did not influence the already low level of trypsin inhibitor and lectin. Hence, it was concluded that the increase in performance and N digestibility by treatment with the acidic protease was a result of general improvement in digestion of SBM protein rather than inactivation of ANFs [47,48].

More recent studies have shown that direct addition of a pure protease from *Nocardiopsis prasina* can lead to significant increases in CP and AA digestibility in broilers fed SBM or full fat SBM [49,50]. It was concluded that AA utilization was on average improved by about 5 % in SBM and 6 % in full fat SBM. Furthermore, the same protease has been demonstrated in several studies to have a positive impact on growth performance and N digestibility of broilers fed complete corn-SBM based diets [51,52,53]. The efficiency of the *Nocardiopsis prasina* protease to improve digestion of the SBM protein fraction is supported by internal laboratory studies at Novozymes A/S demonstrating the ability of the protease to improve protein hydrolysis in different SBM batches as well as in full fat SBM (Figure 2).

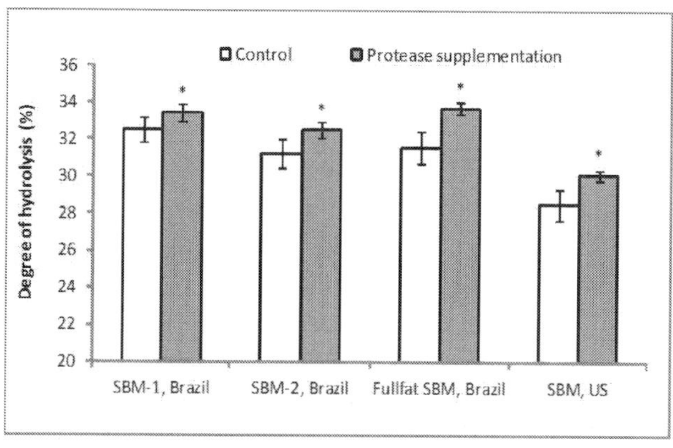

Figure 2. Increase in degree of protein hydrolysis (DH, %) by a pure protease from *Nocardiopsis prasina* dosed at 100 mg purified enzyme protein per kg of SBM (n = 5). The Various SBM batches were incubated in an *in vitro* digestion system and protease effect on top of endogenous enzymes (pepsin and pancreatic enzymes) was analysed as increase in DH (method published in [53]). Error bars indicate standard deviation and asterisks indicate a significant impact of the protease (P<0.05; Tukey HSD test). Unpublished data, Novozymes A/S.

The way by which a protease increases hydrolysis and digestion of the SBM protein fraction may be related to both general hydrolysis of the SBM proteins and to degradation of various ANFs present in SBM. A general hydrolysis of SBM protein would presumably increase the availability of the protein for further hydrolysis and absorption in the gastro intestinal tract. On the other hand, degradation of ANFs will improve the natural digestion and utilization of protein as the adverse effects of the ANFs are reduced.

In this context internal studies have shown that the *Nocardiopsis prasina* protease is capable of degrading various anti nutritional proteins from soybean. As exemplified in Figure 5 the protease efficiently degraded both the Kunitz inhibitor (KSTI) and lectin (Figure 3), leaving only 10-20 % KSTI and around 15 % lectin intact, while purified porcine trypsin and chymotrypsin could not degrade these proteins to nearly the same extent. The ability of a feed protease to degrade ANFs presents an interesting possibility to alleviate the negative impacts of including raw soy or under processed SBM in e.g. poultry or swine diets.

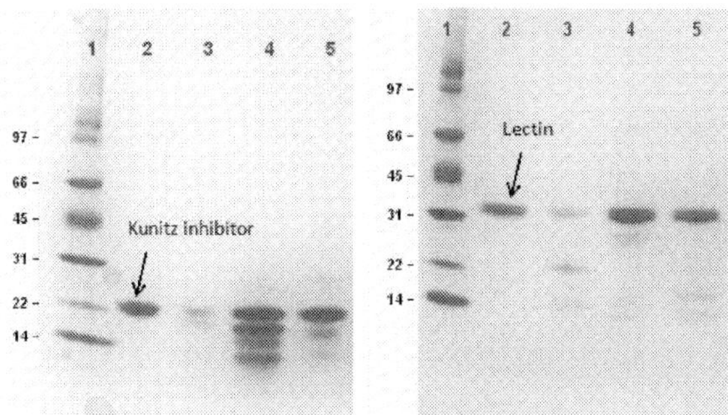

Figure 3. Degradation of purified Kunitz trypsin inhibitor (left) and lectin (right) both from *Glycine max* by purified proteases as analysed by SDS-page. Lane 1: low molecular weight marker (kDa), lane 2: no protease, lane 3: *Nocardiopsis prasina* protease, lane 4: porcine trypsin, lane 5: porcine chymotrypsin. Inhibitors (purchased from Sigma-Aldrich) and proteases were incubated in a 10:1 ratio on mg protein basis for 2 hours at 37°C, after incubation protein was precipitated with TCA, re-suspended in SDS-page sample buffer and analysed by gel electrophoresis. Unpublished data, Novozymes A/S.

3.2. Carbohydrate degrading enzymes

For degradation of the major dietary fibre constituents of importance in cereals there are many enzyme products available on the market and approved by the EU authorities based on proven efficacy in animal trials. The beneficial effects on animal performance, especially in broiler chickens, are assumed to be caused by a combination of depolymerisation of viscous arabinoxylans and a degradation of the indigestible cell wall. The resulting viscosity reduction improves nutrient absorption and the degradation of the cell walls improves the liberation of nutrients (e.g. starch and protein) enclosed by the indigestible cell walls [54,55]. This effect is commonly referred to as the cage effect. For oil seed meals the expected outcome of degradation of the cell walls is less obvious since they generally do not have an intact cell wall architecture due to extensive processing and thereby there is no cage effect. Still the water holding capacity of the material could be reduced and this could have a positive influence on nutrient absorption. In addition liberation of galactose could provide additional energy, at least from SBM. In rape seed, which has less galactan side chains associated with the pectin matrix, an anticipated energy benefit would be limited. Another possibility would be to degrade the oligosaccharides of the raffinose series and the most common attempts to improve the nutritive value of SBM have targeted these. A successful degradation of this fraction could release galactose, and also sucrose, which is the molecule from which the raffinose series is built. Raffinose, stachyose and verbascose contains (1-6)-α-galactopyranosyl units with variable chain lengths of which stachyose (two galactose units in the chain linked to sucrose) is the pre-dominant in SBM. The oligosaccharides of the raffinose series can be broken down using an α-galactosidase that liberates the galactose from the sucrose molecule. This is a straightforward enzymatic application that has been tested *in vitro* showing that α-galactosidase can degrade the oligosaccharides of the raffinose series in SBM [56,57]. Internal trials at Novozymes A/S (Figure 4) has indicated a large variability in the efficacy of α-galactosidase when exposed to the conditions prevailing in the upper gastro intestinal tract simulated in an *in vitro* digestion system. The average release of galactose from the raffinose series was about 6.7 mg/g diet and since maize only contains minor amounts of galactose the main part of the galactose originated from the SBM and about 16 g of galactose could theoretically be released from 1 kg of SBM, corresponding to a potential AME value increase of approximately 0.3 MJ/ kg SBM. The *in vitro* model does not provide information of any additional beneficial effects of degrading the raffinose series. Still the application is of interest since the oligosaccharides of the raffinose series are indigestible but readily fermented and this may cause digestive disturbances with gas production and rapid digesta passage rates in poultry [58,59]. The removal of these oligosaccharides by ethanol extraction has been shown to increase the metabolisable energy content as well as the transit time in adult roosters [60]. However, this procedure alters the general composition of the SBM by also extracting other ethanol soluble components, resulting in a meal with higher CP content and improved nutritive value. When using oligosaccharide or non-starch polysaccharide degrading enzymes the SBM composition is not altered and the effects observed can be attributed to the enzyme *per se*. Trials in pigs have generated high digestibilities in the small intestine even without α-galactosidase supplementation [61], while poultry data indicate a positive enzyme effect on the digestion of the raffinose series but not on performance [62]. Based on these and similar data it may be concluded that performance data are not strong enough to justify, from an economical point of view, the additional supplementation of this enzyme in broiler chicken diets that already may contain xylanases and phytases.

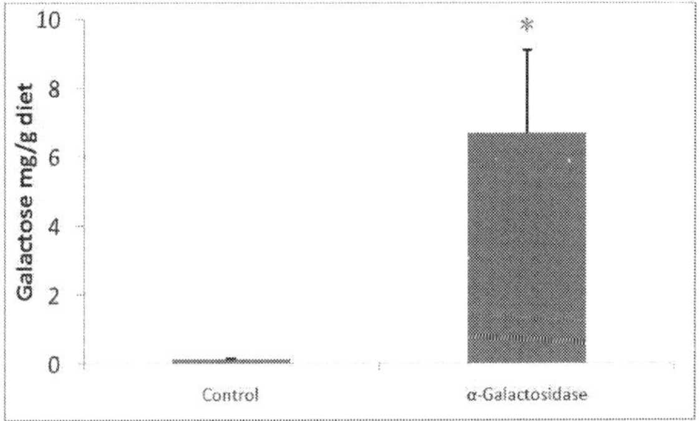

Figure 4. Release (mg/g) of galactose by α-galactosidase from 5 replicates of a model diet containing 600 g maize and 400 g soybean meal per kg and incubated at 40°C in an *in vitro* system mimicking the stomach (pH 3±0.2 and 3000 U of pepsin/g diet) and small intestine (pH 7±0.2 and 8 mg/g diet of pancreatin) for a total period of 6 h. Standard deviations are indicated by error bars and asterisks indicate a statistically significant difference (P<0.05; Tukey HSD test) compared to blank. Unpublished data, Novozymes A/S.

The use of galactanase has shown that also the galactose side chains of the rhamnogalacuron-ans may to some extent be degraded [63]. As opposed to many other feed enzyme products on the market a classical wild type fermentation product of *Aspergillus aculeatus* (RONO-ZYME® VP, DSM Nutritional Products, Basel, Switzerland), is able to significantly impact the NSP fraction of SBM. In an *in vitro* trial comparing this product with a selection of feed enzymes developed for targeting the cereal ingredients in animal feed the preparations were dosed at 100 times the commercial recommendation in a buffer system. After incubation the solubi-lised material was removed by centrifugation and the pellets containing the residual insolu-ble NSP were analysed according to Theander *et al.* [4]. The effect of the *A. aculeatus* product on NSP containing arabinose, although not statistically significant from the control, was the highest compared to all other enzyme treatments with a reduction of about 6.5 % (Figure 5). In addition the *A. aculeatus* product gave a significant reduction (P<0.05) of galactose by 9.6 % compared to the control treatment, whereas the other enzyme treatments did not provide a reduction. The residual glucose content was significantly lower for the *A. aculeatus* prod-uct and the blend product compared to the control, 22 and 13 %, respectively, and these effects were statistically significant compared to all other treatments (Figure 3). The results indi-cate that in order to degrade the complex cell wall matrix of SBM several enzymatic activi-ties are required at a high activity level and these are not generally found in single commercial products developed for targeting type II cereal cell walls. It is also obvious that only one single enzymatic activity is not enough to provide a sufficient degradation of the different polysac-charide structures of importance (Table 3).

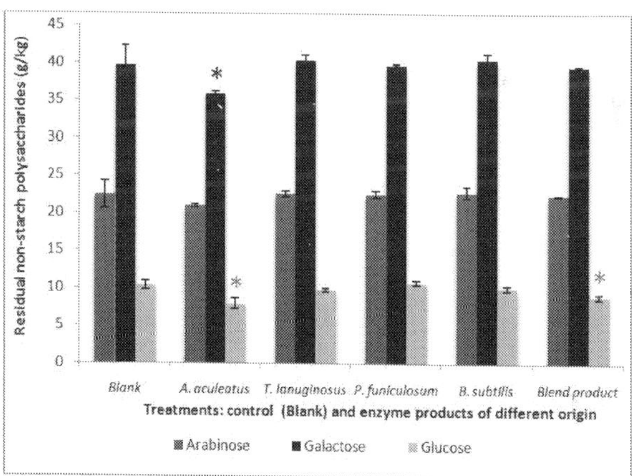

Figure 5. Residual insoluble content (g/kg) of arabinose, galactose and glucose non-starch polysaccharide residues in a soybean meal after incubation at 40°C in an acetate buffer (0.1 M pH 5.0) for 4 hours and with enzymes dosed at 100 times the commercial recommendation. The products used were the Aspergillus aculeatus wild type fermentation or different products containing only xylanase activity derived from Thermomyces lanuginosus or mainly xylanase and ß-glucanase activity in products derived from Penicillium funiculosum or Bacillus subtilis and a commercial blend prod-uct containing xylanase, ß-glucanase and α-amylase activities. Standard deviations are indicated by error bars and as-terisks indicate a statistically significant difference (P<0.05; Tukey HSD test) compared to blank. Unpublished data, Novozymes A/S.

3.3. Phytic acid degrading enzymes

The content of phytic acid in SBM is typically around 1.3-1.4 % of dry matter and constituting around 50 % of the total phosphorus pool [64-66]. Phytic acid is not absorbed in the gastro intestinal tract, but may be degraded by a phytase to render the phosphorus free and available for absorption. Phytases are enzymes that cleave of the phosphate groups from the inositol ring of phytic acid, thereby rendering free phosphorus to be utilized by the animal and also lowering the anti nutritional effect of phytic acid on mineral and protein availability. Phytase activity is present in most seeds but the activity in oilseed meals including SBM is relatively low [65], hence degradation of SBM phytic acid necessitates the presence of phytase either from cereals in the diet or from a microbial phytase source. Since a large part of the feed for poultry and swine is treated at high temperatures e.g. in a pelleting process in order to sanitize the feed from *Salmonella* infections etc., cereal phytases are often inactivated in the final feed (e.g. [67]). Hence, the use of microbial phytases in feed formulation is extensive.

Feedstuff	Residual phytic acid (%)
Wheat	21.5 ± 1.3 ab
Maize	24.5 ± 1.3 b
Barley	21.3 ± 1.3 ab
Soybean meal	47.9 ± 1.3 c
Rapeseed meal	21.3 ± 1.3 ab
Soybean meal-maize blend	17.2 ± 1.5 a

Table 4. abc: Different letters indicate significant differences (P<0.05), tested by Tukey HSD. Residual phytic acid (%) after incubation of feedstuffs with a commercial phytase (RONOZYME® HiPhos, DSM Nutritional Products, Basel, Switzerland) for 30 minutes (pH 4, 40°C). All feedstuffs were heat treated and additional calcium (6 g kg^{-1} dry matter) was added. Values are representative of the sum of inositol hexaphosphate and inositol pentaphosphate. [68]

Degradation of phytic acid in SBM has been demonstrated e.g. by Brejnholt et al. [68] showing a 50 % degradation of phytic acid upon incubation of SBM with a bacterial phytase at pH 4 (30 minutes, 40°C). Interestingly this study indicated that it was more difficult to degrade phytic acid in SBM compared to phytic acid in cereal meals and rape seed meal (Table 4). This difference might be related to the content of protein in the feedstuffs, as protein is known to form insoluble complexes with phytic acid at low pH [69]. In support of this hypothesis it has been shown that phytic acid in SBM is much less soluble than phytic acid in maize meal at low pH (Figure 6). Furthermore, internal data produced at Novozymes A/S also show that treatment of a SBM-maize mixture with pepsin to degrade the protein fraction has a positive impact on phytic acid solubility (Figure 7). In the digestive tract endogenous digestive proteases will always be present to degrade protein and thereby improve the availability of phytic acid for hydrolysis by phytase. Supporting this there are a huge amount of *in vivo* studies demonstrating that phytases effectively releases phosphate in animals fed diets containing SBM e.g. Aureli *et al.* [70] showing phytase efficacy in broilers.

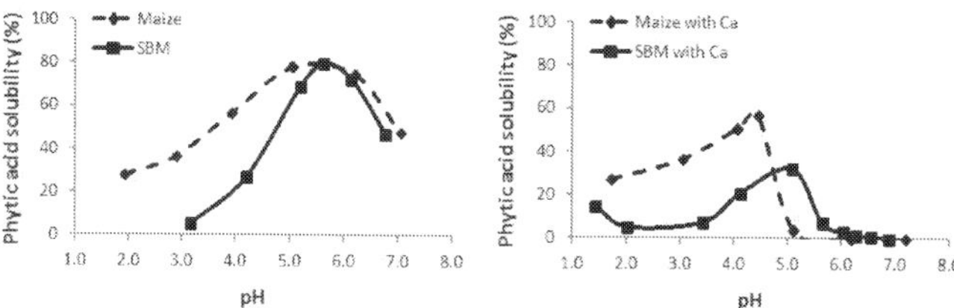

Figure 6. Phytic acid solubility (%) in maize and SBM (left) and in maize and SBM with additional 5 g calcium kg⁻¹ dry matter. Error bars represent standard deviations (2xSD) of 3 replicates. Modified from [71].

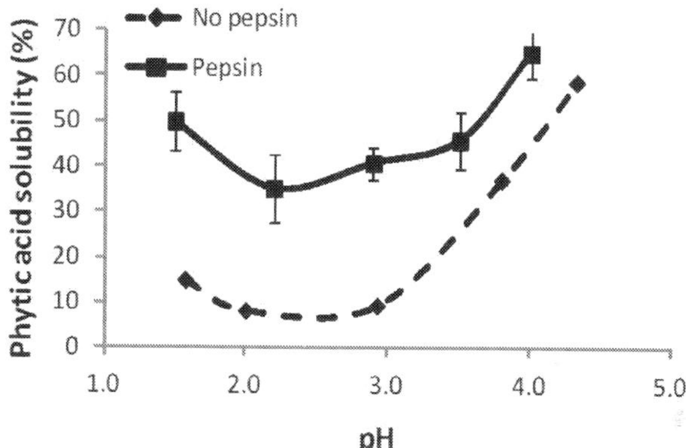

Figure 7. Phytic acid solubility (%) in a SBM-maize (30:70) mixture with additional 5 mg calcium g⁻¹ dry matter in the presence or absence of pepsin (3000 U g⁻¹ dry matter). Error bars represent standard deviations (2xSD) of 3 replicates. Unpublished data, Novozymes A/S.

4. Conclusions

SBM is the most important protein source in animal feed and it is estimated that approximately half of the SBM produced for animal feed is used in poultry diets, while another 25 % is used in pig diets. SBM is primarily added due to its high content of protein and favourable composition of AAs, while the low content of metabolisable energy and a high content of NSP may provide problems when incorporating this oil seed meal at high levels. The presence of several ANFs such as trypsin inhibitors and lectins may represent a severe problem that can restrict animal performance when feeding a meal that has not been properly processed, however there are indications that these short comings can be overcome by proper heat inactivation of the SBM as well as enzyme supplementation of the diet.

Research results indicate that the nutritive value of SBM may be enhanced by adding exogenous microbial enzymes such as carbohydrate degrading enzymes, proteases and phytases. However, there is a need for improved carbohydrate degrading enzymes to better target the oligosaccharides of the raffinose series and also to reduce the negative effects of high levels of complex NSP constituents, that evidently are not degraded by common xylanases or ß-glucanases used for improving the feeding value of cereals.

The use of protease to improve protein digestibility and reduce the presence of anti nutritional proteins represents a novel and promising application that has an environmental impact by improving protein digestibility and thereby reducing nitrogen excretion from farm animals. In a similar way phytase supplementation is already a well established environmentally friendly application that reduces phosphorus excretion when feeding SBM rich in phytic acid.

Author details

D. Pettersson* and K. Pontoppidan

*Address all correspondence to: kpon@novozymes.com

Department of Feed Applications, Novozymes A/S, Denmark

References

[1] Soystats. (2012). http://www.soystats.com/2011/accessed 6 May).

[2] Bach, Knudsen. K. E. (1997). Carbohydrate and lignin contents of plant materials used in animal feeding. *Animal Feed Science and Technology*, 67-319.

[3] Sauvant, D., Perez-M, J., & Tran, G. (2004). Tables of Composition and Nutitional Value of Feed Materials. *INRA editions, Wageningen Publishers*.

[4] Theander, O., Åman, P., Westerlund, E., Andersson, R., & Pettersson, D. (1995). Total dietary fibre determined as neutral sugar residue, uronic acid residue and Klason lignin (The Uppsala Method): Collaborative study. *Journal of AOAC International*, 74-1030.

[5] Weier, E. T., Stocking, R. C., & Barbour, M. G. (1970). Botany: An introduction to plant biology. *New York: John Wiley & Sons Inc.*

[6] Ravindran, V., Hew, L. I., & Bryden, W. L. (1998). Digestible amino acids in poultry feedstuffs. *Rural Industries Research & Development Corporation, RIRDC publication* [98].

[7] NRC. (1994). Nutrient requirements of poultry. 9th *revised edition. Washinton: National Academic Press.*

[8] Clarke, E., & Wiseman, J. (2005). Effects of variability in trypsin inhibitor content of soja bean meals on true and apparent ileal digestibility of amino acids and pancreas size in broiler chicks. *Animal Feed Science and Technology*, 121-125.

[9] Hermida, M., Valencia, D. G., Serrano, M., & Mateos, G. G. (2008). Influence of soybean meal origin on its nutritive value and quality parameters. *Poultry Science*, 87-29.

[10] De Coca-Sinova, A., Valencia, D. G., Jiménez-Moreno, E., Lázaro, R., & Mateos, G. G. (2008). Apparent ileal digestibility of energy, nitrogen, and amino acids of soybean meals of different origin in broilers. *Poultry Science*, 87-2613.

[11] Minic, Z., & Jouanin, L. (2006). Plant glycoside hydrolases involved in cell wall polysaccharide degradation. *Plant Physiology and Biochemistry*, 44-435.

[12] Huismann, M. M. H., Schols, H. A., & Voragen, A. G. J. (1998). Cell wall polysaccharides from soybean (Glycine max) meal. Isolation and characterisation. *Carbohydrate Polymers*, 37-87.

[13] Siddiqui, R. I., & Wood, P. J. (1977). Carbohydrates of rapeseed: a review. *Journal of the Science of Food and Agriculture*, 28-530.

[14] Mc Ginnis, J. (1983). Carbohydrate utilization in feedstuffs. *In: Proceedings of the Minnesota nutrition conference, St. Paul, Minnesota*, 106-107.

[15] Longstaff, M., & Mc Nab, J. M. (1989). Digestion of fibre polysaccharides of pea (pisum sativum) hulls, carrot and cabbage by adult cockerels. *British Journal of Nutrition*, 62-563.

[16] Carré, B., Derouet, L., & Leclercq, B. (1990). The digestibility of cell wall polysaccharides from wheat (bran or whole grain), soya bean meal and white lupin meal in cockerels, Muscovy ducks and rats. *Poultry Science*, 69-623.

[17] Ishii, T. (1997). Structure and functions of feruloylated polysaccharides. *Plant Science*, 127-111.

[18] Mc Cann, M. C., & Roberts, K. (1991). The cytoskeletal basis of plant growth and form. *New York, USA: Academic Press.*

[19] Carpita, N. C., & Gibeaut, D. M. (1993). Structural models of primary cell walls in flowering plants: consistency of molecular structure with the physical properties of the walls during growth. *Plant Journal*, 3-1.

[20] Whitaker, J. R., Voragen, A. G. J., & Wong, D. W. S. (2003). Handbook of food enzymology. *New York, USA: Marcel Dekker.*

[21] Warenham, C. N., Wiseman, J., & Cole, D. J. A. (1994). Processing and antinutritive factors in feedstuff. *In: Cole DJA, Wiseman J, Varley MA (eds.) Principles of Pig Science. Nottingham: Nottingham Press*, 1, 141-167.

[22] Liener, I. E. (1989). Antinutritional factors. *In: Matthews RH (ed.) Legumes: Chemistry, technology and human nutrition. New York: Marcel Dekker*, 340-366.

[23] Liener, I. E. (1989). Antinutritional factors in legume seeds: State of the art. *In: Huisman J, van der Poel TFB, Liener IE (eds.) Recent Advances of Research in Antinutritional Factors in Grain Legume seeds: proceedings of the 1st International Workshop on Antinutritional Factors in Legume Seeds, 13-25 November 1988, Wageningen*, 6-13.

[24] Dongmo, T., Pone, D. K., & Ngoupayou, J. D. N. (1989). Cottonseed cake in breeder hens diets: effects of supplementation with lysine and methionine. *Archiv für Geflügelkunde*, 53-231.

[25] Di Pietro, C. M., & Liener, I. E. (1989). Heat inactivation of the Kunitz and Bowman-Birk soybean protease inhibitors. *Journal of Agricultural and Food Chemistry*, 37, 39-44.

[26] Liener, I. E. (1994). Implications of antinutritional components in soybean foods. *Critical Reviews in Food Science and Nutrition*, 34-31.

[27] Sklan, D., Hurwitz, S., Budowski, P., & Ascarelli, I. (1975). Fat digestion and absorption in chicks fed raw soy or heated soybean meal. *Journal of Nutrition*, 105-57.

[28] Grant, G., & van Driessche, E. (1993). Legume lectins: physicochemical and nutritional properties. *In : van Driessche E, van der Poel AFB, Huisman J, Saini HS (eds.) Recent advances of research in antinutritional factors in legume seeds. Wageningen, Netherlands: Wageningen Press*, 219-233.

[29] Kennedy, J. F., Palva, P. M. G., Corella, M. T. S., Cavalcanti, M. S. M., & Coelho, L. C. B. B. (1995). Lectins, versatile proteins of recognition: a review. *Carbohydrate Polymers*, 26-219.

[30] Schulze, H., Butts, C. A., Verstegen, M. W. A., Moughan, P. J., & Huisman, J. (1995). Purified soybean lectins affect ileal nitrogen and amino acid flow in pigs. *Proceedings of the 2nd European Conference on Grain Legumes, Copenhagen*, 300-301.

[31] Kirchgessner, M., & Windisch, W. (1995). Zum einfluss von mikrobieller phytase auf zootechnische leistungen und die verdauungsquotienten von phosphor, calcium, trockenmasse und stickstoff bei abgestufter Ca-versorgung in der ferkelaufzucht. *Agribiological Research*, 48-309.

[32] Selle, P. H., Ravindran, V., & Caldwell, Bryden. W. L. (2000). Phytate and phytase: consequences for protein utilization. *Nutrition Research Reviews*, 13-255.

[33] Li, D. F., Nelssen, J. L., Reddy, P. G., Blecha, F., Klemm, R., & Goodband, R. D. (1991). Interrelationship between hypersensitivity to soybean proteins and growth performance in early-weaned pigs. *Journal of Animal Science*, 69-4062.

[34] Snyder, H. E., & Kwon, T. W. (1987). Processing of soybeans. *In: Soybean utilization. New York, USA: Avi Book.*

[35] Qin, G. X., Verstegen, M. W. A., & Van der Poel, A. F. B. (1998). Effect of temperature and time during steam treatment on the protein quality of full-fat soybean meal from different origins. *Journal of the Science of Food and Agriculture*, 77-393.

[36] Batal, A. B., Douglas, M. W., Engram, A. E., & Parsons, C. M. (2000). Protein dispersibility index as an indicator of adequately processed soybean meal. *Poultry Science*, 79-1592.

[37] Araba, M., & Dale, N. M. (1990). Evaluation of protein solubility as an indicator of overprocessing of soybean meal. *Poultry Science*, 69-76.

[38] Araba, M., & Dale, N. M. (1990). Evaluation of protein solubility as an indicator of underprocessing of soybean meal. *Poultry Science*, 69-1749.

[39] Americal Oil Chemists Society. (2000). Official Methods and Recommended Practices. *5th edition. Urbana, IL: American Oil Chemists Society.*

[40] Valencia, D. G., Serrano, M. P., Lázaro, R., Latorre, M. A., & Mateos, G. G. (2008). Influence of micronization (fine grinding) of soya bean meal and full fat soya bean meal on productive performance and digestive traits in young pigs. *Animal Feed Science and Technology*, 147-340.

[41] Valencia, D. G., Serrano, M. P., Centeno, C., Lázaro, R., & Mateos, G. G. (2008). Pea protein as a substitute of soya bean protein in diets for young pigs: Effects on productivity and digestive traits. *Livestock Science*, 188-1.

[42] Rackis, J. J., Mc Ghee, J. E., Liener, I. E., Kakade, M. L., & Puski, G. (1974). Problems encountered in measuring trypsin inhibitor activity of soy flour. *Report of a collaborative analysis. Cereal Science Today*, 19-513.

[43] Peisker, M. (2001). Manufacturing of soy protein concentrate for animal nutrition. *In: Brufau J. (ed.) Feed Manufactoring in the Mediterranean Region. Improving safety: From Feed to Food. Zaragoza, CIHEAM-IAMZ, Reus, Spain*, 1036-1107.

[44] Nielsen, P. H., Dalgaard, R., Korsbak, A., & Pettersson, D. (2008). Environmental assessment of digestibility improvement factors applied in animal production: Use of Ronozyme® WX CT xylanase in Danish pig production. *The International Journal of Life Cycle Assessment*, 13-49.

[45] Nagaraju, R. K., & Nielsen, P. H. (2011). Environmental advantages of phytase over inorganic phosphate in poultry feed. *Poultry Punch*, 27-57.

[46] Oxenböll, K. M., Pontoppidan, K., & Nji-Fru, F. (2011). Use of a protease in poultry feed offers promising environmental benefits. *International Journal of Poultry Science*, 10-842.

[47] Rooke, J. A., Slessor, M., Fraser, H., & Thomson, J. R. (1998). Growth performance and gut function of piglets weaned at four weeks of age and fed protease-treated soya-bean meal. *Animal Feed Science and Technology*, 70-175.

[48] Ghazi, S., Rooke, J. A., Galbraith, H., & Bedford, M. R. (2002). The potential for the improvement of the nutritive value of soya-bean meal by different proteases in broiler chicks and broiler cockerels. *British Poultry Science*, 43-70.

[49] Bertechini, R. L., Carvalho, J. C. C., Mesquita, F. R., Castro, S. F., Meneghetti, C., & Sorbara, J. O. B. (2009). Use of a protease to enhance the utilization of soybean meal amino acids by broilers. *Poultry Science, 88(1), 69.*

[50] Bertechini, R. L., Carvalho, J. C. C., Mesquita, F. R., Castro, S. F., Remolina, D. F., & Sorbara, J. O. B. (2009). Use of a protease to enhance the utilization of full fat soybean amino acids by broilers. *Poultry Science, 88(1), 70.*

[51] Angel, C. R., Saylor, W., Vieira, S. L., & Ward, N. (2011). Effects of a monocomponent protease on performance and protein utilization in 7- to 22-day-old broiler chickens. *Poultry Science, 290-281.*

[52] Freitas, D. M., Vieira, S. L., Angel, C. R., Favero, A., & Maiorka, A. (2011). Performance and nutrient utilization of broilers fed diets supplemented with a novel monocomponent protease. *Journal of Applied Poultry Research, 20-322.*

[53] Fru-Nji, F., Kluenter, A. M., Fischer, M., & Pontoppidan, K. (2011). A feed serine protease improves broiler performance and increases protein and energy digestibility. *Journal of Poultry Science, 48-239.*

[54] Pettersson, D., & Åman, P. (1989). Enzyme supplementation of a poultry diet containing wheat and rye. *British Journal of Nutrition, 62-139.*

[55] Steenfeldt, S., & Pettersson, D. (2001). Improvements in nutrient digestibility and performance of broiler chickens fed a wheat-and-rye based diet supplemented with enzymes. *Journal of Animal Feed Science, 10-143.*

[56] Slominski, B. A. (1994). Hydrolysis of galactooligosaccharides by commercial preparations of α-galactosidase and ß-fructofuranosidase: potential for use as dietary additives. *Journal of the Science of Food and Agriculture, 65-323.*

[57] Ao, T., Cantor, A. H., Pescatore, A. J., Pierce, J. L., & Dawson, K. A. (2010). Effects of citric acid, alpha-galactosidase and protease inclusion on in vitro nutrient release from soybean meal and trypsin inhibitor content in raw whole soybeans. *Animal Feed Science and Technology, 162-58.*

[58] Leske, K. L., Jevne, C. J., & Coon, C. N. (1993). Effect of oligosaccharide additions on nitrogen corrected true metabolizable energy of soy protein concentrate. *Poultry Science, 72-664.*

[59] Leske, K. L., & Coon, C. N. (1999). Nutrient content and energy digestibilities of ethanol extracted, low α-galactoside soybean meal as compared to intact soybean meal. *Poultry Science, 78-1177.*

[60] Coon, C. N., Leske, K. L., Akavanichan, O., & Cheng, T. K. (1990). Effect of oligosaccharide-free soybean meal on true metabolizable energy and fiber digestion in adult roosters. *Poultry Science, 69-787.*

[61] Gdala, J., Johansen, H. N., Bach-Knudsen, K. E., Knap, I. H., Wagner, P., & Joergensen, O. B. (1997). The digestibility of carbohydrates, protein, and fat in the small and

large intestine of piglets fed non-supplemented and enzyme supplemented diets. *Animal Feed Science and Technology*, 65, 15-33.

[62] Graham, K. K., Kerley, M. S., Firman, J. D., & Allee, G. L. (2002). The effect of enzyme treatment of soy-bean meal on oligosaccharide disappearance and chick growth performance. *Poultry Science*, 81, 1014-1019.

[63] Vahjen, W., Busch, T., & Simon, O. (2005). Study on the use of soya bean polysaccharide degrading enzymes in broiler nutrition. *Animal Feed Science and Technology*, 120-259.

[64] Nelson, T. S., Ferrara, L. W., & Storer, N. L. (1968). Phytate phosphorus content of feed ingredients derived from plants. *Poultry Science*, 47, 1372-1374.

[65] Eeckhout, W., & De Paepe, M. (1994). Total phosphorus, phytate phosphorus and phytase activity in plant feedstuffs. *Animal Feed Science and Technology*, 47-19.

[66] Ravindran, V., Ravindran, G., & Sivalogan, S. (1994). Total and phytate phosphorus contents of various foods and feedstuffs of plant origin. *Food Chemistry*, 50-133.

[67] Jongbloed, A. W., & Kemme, P. A. (1990). Effect of pelleting mixed feeds on phytase activity and the apparent absorbability of phosphorus and calcium in pigs. *Animal Feed Science and Technology*, 28, 233-242.

[68] Brejnholt, S. M., Dionisio, G., Glitsoe, V., Skov, L. K., & Brinch-Pedersen, H. (2011). The degradation of phytate by microbial and wheat phytases is dependent on the phytate matrix and the phytase origin. *Journal of the Science of Food and Agriculture*, 91-1398.

[69] Hill, R., & Tyler, C. (1954). The reaction between phytate and protein. *Journal of Agricultural Science*, 44-324.

[70] Aureli, R., Umar, Faruk. M., Cechova, I., Pedersen, P. B., Elvig-Joergensen, S. G., Fru, F., & Broz, J. (2011). The efficacy of a novel microbial 6-phytase expressed in Aspergillus oryzae on the performance and phosphorus utilization in broiler chickens. *International Journal of Poultry Science*, 10, 160-168.

[71] Pontoppidan, K. (2009). Factors influencing phytate degradation in piglets: feed phytate behaviour and degradation by microbial phytases. *PhD thesis. Department of Chemical and Biological Engineering Chalmers University of Technology Göteborg Sweden*.

Chapter 14

Soybean in Monogastric Nutrition: Modifications to Add Value and Disease Prevention Properties

Samuel N. Nahashon and
Agnes K. Kilonzo-Nthenge

Additional information is available at the end of the chapter

http://dx.doi.org/10.5772/52991

1. Introduction

Soybean (*Glycine max*), a leguminous oilseed and one of the world's largest and most efficient sources of plant protein, has on average crude protein content of about 37-38% and 20% fat on a dry matter basis. The crude protein content of soybean varies with geographical region and damage to the soybean crop can cause a significant decrease in the crude protein content of the soybean. On the other hand, processed soybean meal which is commonly used in monogastric feeding contains about 44-48% crude protein (NRC, 1998). This high crude protein content of soybean and soybean meal in conjunction with high energy due to significant fat content and low fiber content make soybean an ideal source of protein for humans and also ideal feed ingredient in monogastric animals feeding (Table 1). The heat processed soybean is the form primarily used for human consumption and it contains lower crude protein concentration (37%) when compared to soybean meal which is produced from solvent extracted seeds and seeds without hulls (44% and 49% CP, respectively). The soybean meal is the common form of soybean utilized in animal feeding. While other nutrients such as calcium, potassium and zinc also tend to be lower in heat treated soybean than in the soybean meals, the energy and fat content is higher in the heated soybean than the soybean meals.

Previous reports have shown that soybean and soybean meal contains a balanced amino acid profile when compared with other oilseed meals, although it is deficient in methionine and lysine (Zhou et al., 2005). The comparisons of the amino acid composition of soybean and soybean meal which are routinely utilized in human and monogastric feeding are presented in Table 2.

© 2013 Nahashon and Kilonzo-Nthenge; licensee InTech. This is a paper distributed under the terms of the Creative Commons Attribution License (http://creativecommons.org/licenses/by/3.0), which permits unrestricted use, distribution, and reproduction in any medium, provided the original work is properly cited.

Nutrient	Soybean seeds[2]	Soybean Meal[3]	Soybean Meal[4]
IFN[5]	5-04-597	5-04-604	5-04-612
Crude Protein, %	37	44	49
Energy, kcal/kg	3,300	2,230	2,440
Crude fat, %	18	0.8	1.0
Crude fiber, %	5.5	7.0	3.9
Calcium, %	0.25	0.29	0.27
Phosphorus[6], %	-	0.65	0.62
Phosphorus[7], %	0.53	0.27	0.24
Potassium, %	1.61	2.00	1.98
Iron, mg/kg	80	120	170
Zinc, mg/kg	25	40	55

[1]National Research Council 1994. [2]Heat processed. [3]Seeds, meal solvent extracted.

[4]Seeds without hulls, meal solvent extracted. [5]International feed number. [6]Total phosphorus.

[7]Non-phytate or available phosphorus.

Table 1. Comparison of selected nutrient composition of soybean and soybean meal[1]

Soybean boasts a well balanced amino acid profile with high digestibility when compared with other oilseeds. In soybean, the digestibility coefficients of lysine are estimated to be 91% (NRC, 1994) whereas that of cysteine and phenylalanine is estimated at 83-93 (Bandegan et al, 2010). Previous reports cite evidence that soya is a rich source of amino acids (Angkanaporn et al., 1996). Holle (1995) reported that soybean meal provides the best balance for amino acids when compared with other oilseeds and thus makes it a more suitable plant source protein for human and monogastric food animals. According to Kohl-Meier (1990), soybean accounts for more than 50% of the world's protein meal. The form in which soybean is utilized for human or monogastric feeding determines the nutritional value in terms of content and bioavailability of amino acids as described in the following section.

2. Anti-nutritional properties of soybean

In their natural form, soybeans contain anti-nutrients or phytochemicals which bear toxic effects when ingested by both humans and monogastric food animals. These anti-nutrients are nature's means of protection for the soybean plant from invasion by animals, bacteria, viruses and even fungi in the ecosystem. The major anti-nutrients in soybean are phytates, protease enzyme inhibitors, soyin, goitrogens, hemagglutinins or lectins, giotrogens, cyanogens, saponins, estrogens, antigens, non-starch polysaccharides and soy oligosaccharide. Although most of these anti-nutritional compounds in soybean were discussed in Nahashon and Kilonzo-Nthenge (2011), additional reviews of some of the major anti-nutritional factors are presented as follows:

Nutrient	Soybean seeds[2]	Soybean Meal[3]	Soybean Meal[4]
IFN[6]	5-04-597	5-04-604	5-04-612
	----------------------------------(%)----------------------------------		
Arginine	2.59	3.14	3.48
Lysine	2.25	2.69	2.96
Methionine	0.53	0.62	0.67
Cystine	0.54	0.66	0.72
Tryptophan	0.51	0.74	0.74
Histidine	0.90	1.17	1.28
Leucine	2.75	3.39	3.74
Isoleucine	1.56	1.96	2.12
Phenylalanine	1.78	2.16	2.34
Threonine	1.41	1.72	1.87
Valine	1.65	2.07	2.22
Glycine	1.55	1.90	2.05
Serine	1.87	2.29	2.48
Tyrosine	1.34	1.91	1.95

[1] National Research Council, 1994.

[2] Seeds, heat processed.

[3] Seeds, meal solvent extracted.

[4] Seeds without hulls, meal solvent extracted.

[5] International feed number.

Table 2. Comparison of selected amino acid composition of soybean and Soybean meals[1]

2.1. Phytates

Phytic acid (inositol hexakisphosphate), the storage form of phosphorus in seeds such as those of soybean is considered an anti-nutritional factor in monogastric nutrition. Raboy et al. (1984) cited evidence that phytic acid accounted for 67-78% of the total phosphorus in mature soybean seeds and these seed contain about 1.4-2.3% phytic acid which varies with soybean cultivars. In plants phytic acid is the principal store of phosphate and also serves as natural plant antioxidant. Earlier reports (Asada et al., 1969) suggested that phytic acid in soybean not only makes phosphorus unavailable, but also reduces the bioavailability of other trace elements such as zinc and calcium and the digestibility of amino acids (Ravindran, 1999). Ravindran et al., (1999) reported that in the presence of phytate, soybean protein forms complexes with the phytate. Heaney et al. (1991) reported that the absorption of calcium from soybean-based diets was higher in low-phytate soybean when compared with high phytate-soybean. This supports the assertion that soybean has the potential to form phytate-mineral-complex which inhibits the availability of the minerals to monogastric animals. In soybean, phytate is usually a mixture of calcium/magnesium/potassium salts of inositol hexaphosphoric acid which adversely affects mineral bioavailability and protein solubility when present in animal feeds (Liener, 1994). Reports of Vucenik and Shamsuddin (2003) point that inositol bears biological significance as antioxidant in mammalian cells. However, it interferes with mineral utilization and is the primary cause of low phosphorus utilization in soy-based poultry and swine diets. Phytin also chelates other minerals such as Calcium, Zinc, iron, Manganese and Copper, rendering them unavailable to the animals. Soybean has the highest amount of phytate when compared to all legumes and cereal grains. The phytates have been reported to be resistant to cooking temperatures.

2.2. Protease enzyme inhibitors

Proteases refer to a group of enzymes whose catalytic function is to hydrolyze, cleave or breakdown peptide bonds of proteins. They are also called proteolytic enzymes that include trypsin, chymotrypsin, elastase, carboxypeptidase, and aminopeptidase which convert protein (polypeptides, dipeptides, and tripeptides) into free amino acids which are readily absorbed through the small intestine into the blood stream. Protease or trypsin inhibitors of soybean have been reported to hinder the activity of the proteolytic enzymes trypsin and chymotrypsin in monogastric animals which in turn lowers protein digestibility (Liener and Kakade, 1980). Other reports (Liener and Kakade, 1969; Rackis, 1972) confirmed that trypsin inhibitors were key substances in soybean that adversely affected its utilization by chicks, rats and mice. Kunitz, (1946) isolated trypsin inhibitor from raw soybeans and demonstrated that it was associated with growth inhibition. These protease inhibitors were also reported to inhibit Vitamin B_{12} availability (Baliga et al., 1954). Later studies have also shown that the presence of dietary soybean trypsin inhibitors caused a significant increase in pancreatic proteases (Temler et al., 1984). To the benefit of the soybean plant, soybean protease inhibitors serve as storage proteins in seeds, regulate endogenous proteinases, and also protect the plant and seeds against insect and/or microbial proteinases (Hwang et al., 1978). These protease inhibitors contain about 20% of the sulfur-containing amino acids methionine and cysteine, which are also the most limiting essential amino acids in soybean seeds (Hwang et al., 1978).

Recent reports (Dilger et al., 2004; Opapeju et al. 2006; Coca-Sinova et al. 2008) show that the nutritional value of soybean meal for monogastric animals is significantly hindered by these protease inhibitors which interfere with feed intake and nutrient metabolism. They reported that soybeans with high content protease inhibitors, especially trypsin inhibitors adversely affect protein digestibility and amino acid availability. In earlier reports, Birth et al. (1993) cited evidence that ingestion of food containing trypsin inhibitor by pigs increased endogenous nitrogen losses hence the effect of the trypsin inhibitors affected nitrogen balance more by losses of amino acids of endogenous secretion than by losses of dietary amino acids. This may be due to compromised integrity of the gastrointestinal lining leading to reduction of absorptive surface. However, Gertler et al. (1967) attributed the depression of protein digest‐ibility to reduced proteolysis and absorption of the exogenous or dietary protein which was caused by inhibition of pancreatic proteases.

2.3. Hemagglutinins or lectins

Soybean hemagglutinins or lectins are glycoproteins that resemble some animal glycopro‐teins, such as ovalbumin and are rich in the acidic amino acids while being low in the sulfur‐containing amino acids methionine and cysteine. According to Lis *et al.* (1966), the only carbohydrates serving as constituents of soybean hemagglutinins are mannose and glucosa‐mine. Soybean hemaglutinins are a component of soybeans that were characterized by Schulze *et al.* (1995) as being anti-nutritional. Oliveira *et al.* (1989) reported that these glyco‐proteins bind to cellular surfaces via specific oligosaccharides or glycopeptides. They exhibit high binding affinity to small intestinal epithelium (Pusztai, 1991) which impairs the brush border and interfere with nutrient absorption. Hemaglutinins have also been implicated in producing structural changes in the intestinal epithelium and resisting gut proteolysis (Pusztai *et al.*, 1990), changes which in most cases result in impairment of the brush border and ulceration of villi (Oliveira *et al.*, 1989). This occurrence result in significant decrease in the absorptive surface and increase endogenous nitrogen losses as reported by Oliveira and Sgarbierri (1986) and Schulze *et al.* (1995). Consequently, Pusztai *et al.* (1990) observed that hemagglutinins depressed growth rate in young animals. Hemagglutinins are known to promote blood clotting or facilitating clumping together of red blood cells. It has however been concluded that soybean hemagglutinins play a minor role in the deleterious effect con‐tributed by anti-nutritional factors in raw soybean.

2.4. Giotrogens and estrogens

Soybean is known to produce estrogenic isoflavones which bind to the estrogen receptors. According to Doerge and Sheehan (2002), such estrogenicity was implicated in toxicity and estrogen-mediated carcinogenesis in rats. Genistein is the major soy isoflavones of great con‐cern in conferring estrogenic effect especially in women. Although the possible gitrogenic effect of the soybean isoflavones has not been researched extensively, certain soy compo‐nents may present some antithyroid actions, endocrine disruption, and carcinogenesis in an‐imals and humans as well. Messina, (2006) reported that Soybean contains flavonoids that may impair the activity of the enzymes thyroperoxidase. Earlier reports (Divi and Doerge, 1996) indicate that plant-derived foods such as soybean contain flavonoids which are widely distributed, possessing numerous biological activities that include antithyroidism in experi‐mental animals and humans. A study was conducted to evaluate inhibition of thyroid per‐oxidase (TPO), the enzyme that catalyzes thyroid hormone biosynthesis, by 13 commonly consumed flavonoids (Divi and Doerge, 1996). Consequently, most flavonoids tested includ‐ing genistein and daidzein were potent inhibitors of TPO (Figure 1). They suggested that chronic consumption of flavonoids, especially suicide substrates, could play a role in the eti‐ology of thyroid cancer. More recent reports (Messina, 2006; Xiao, 2008; Zimmermann, 2009) have also shown that use of soy-based formula without added iodine can produce goiter and hypothyroidism in infants, but in healthy adults soy-based products appear to have negligible adverse effects on thyroid function.

Other reports (Fort, 1990) have also shown that concentrations of soy isoflavones resulting from increased consumption of soy-based formulas inhibited thyroxine synthesis inducing goiter and hypothyroidism and autoimmune thyroid disease in infants. Still many questions linger on the full Impacts of soy products on thyroid function, reproduction and carcinogenesis, hence the need for further research in this context. According to Divi et al. (1997), the IC50 values for inhibition of TPO-catalyzed reactions by genistein and daidzein were ca. 1-10 microM, concentrations that approach the total isoflavone levels (ca. 1 microM). More recent finding using normal rats (Chang and Doerge, 2000) however suggest that, even though substantial amounts of TPO activity are lost concomitant to consumption of soy isoflavone, the remaining enzymatic activity is sufficient to maintain thyroid homeostasis in the absence of additional perturbations. On the other hand, additional factors other than the soybean isoflavones can also cause overt thyroid toxicity. These may include other soybean fractions, iodine deficiency, defects of hormone synthesis that may be caused by gene mutations or environmental and random factors including dietary factors that may be goitrogenic.

Environmental estrogens, on the other hand, are classified into two main categories namely phytoestrogens which are of plant origin and xenoestrogens which are synthetic (Dubey et al., 2000). Soybeans contain phytoestrogens which can cause enlargement of the reproductive tract disrupting reproductive efficiency in various species, including humans (Rosselli et al., 2000), and rats (Medlock et al., 1995). In some cases these estrogens are hydrolyzed in the digestive tract to form poisonous compounds such as hydrogen cyanide. Woclawek-potocka et al. (2004) reported that phytoestrogens acting as endocrine disruptors may induce various pathologies in the female reproductive tract. Studies have shown that soy-derived phytoestrogens and their metabolites disrupt reproductive efficiency and uterus function by modulating the ratio of PGF2a to PGE2. Because of the structural and functional similarities of phytoestrogens and endogenous estrogens, there is the likelihood that the plant-derived substances modulate prostaglandin synthesis in the bovine endometrium, impairing reproduction. Previous research has shown that phytoestrogens may act like antagonists or agonists of endogenous estrogens (Rosselli et al., 2000; Nejaty and Lacey, 2001).

2.5. Allergens and antigens

Today, food allergies are a common serious health threat and food safety concern around the world. The increasing use of soybean (*Glycine max*) products in processed foods owing to its nutritional and health promoting properties, poses a potential threat to individuals who allergic to foods and especially sinsitive to soybean. According to Cordle (2004), there are about 16 potential soy protein allergens that have been identified and the Food and Agriculture Organization of the United Nations includes soy in its list of the 8 most significant food allergens. While many of these soy allergenic proteins have not been fully characterized, their allergenicity can be mild to life-threatening anaphylaxis. Consequently, consumers who have allergies to soybean and soybean products are often at a risk of serious or life threatening allergic reaction if they consume these products.

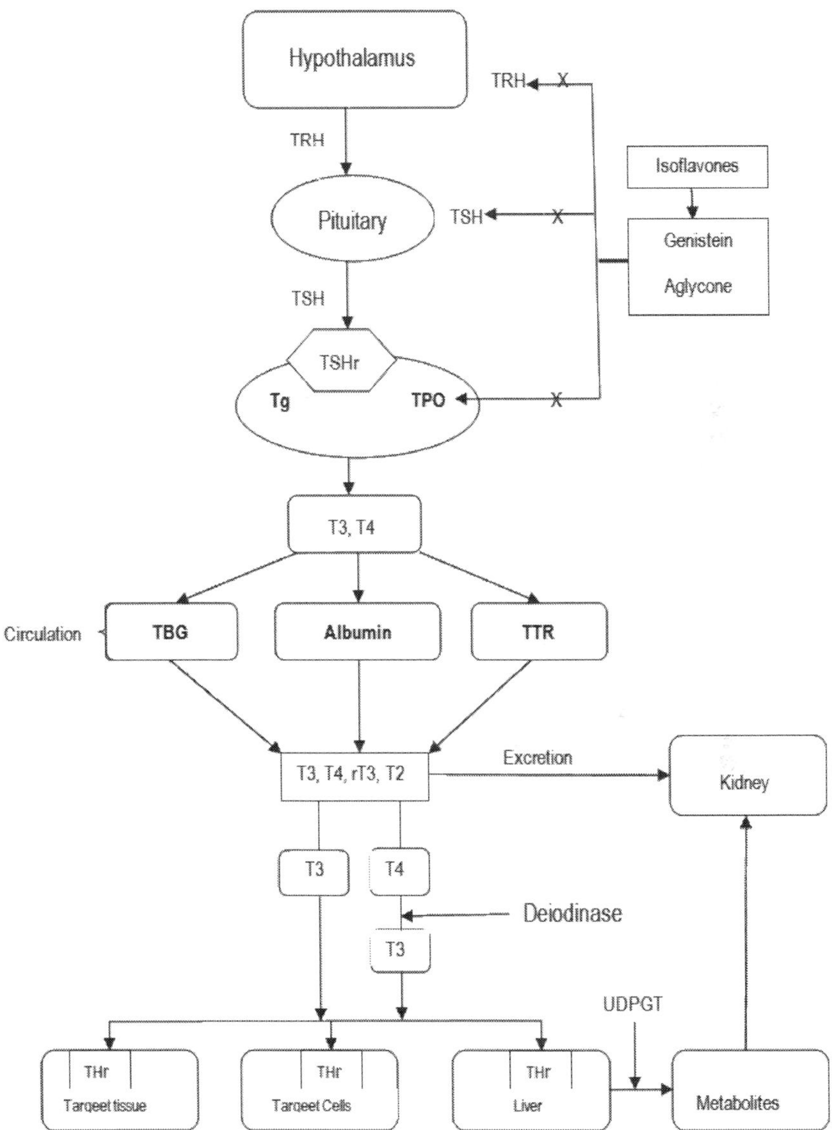

Figure 1. Schematic presentation of inhibition of TPO-catalyzed reactions by genistein and daidzein. TPO:thyroid peroxidase; UDPGT:Uridine diphosphate glucuronyltransferase; TBG:thyroxine binding globulin; TTR: Transthyretin; THr: thyroxine receptor; T:thyroxine; TSH:Thyroid stimulating hormone; TRH:Thyrotropin-releasing hormone; Tg:Thyroglobulin.

One of these low abundance proteins, Gly m Bd 30 K, also referred to as P34, is considered to be a dominant immunodominant soybean allergen. It is a member of the papain protease superfamily with a glycine in the conserved catalytic cysteine position found in all other cysteine proteases (Herman et al., 2003). The P34 protein is thought to trigger most allergic reactions to soy and it accumulates during seed maturation localizing in the protein storage vacuoles but not in the oil bodies (Kalinski et al., 1992). Several attempts have been made to understand and remove the allergenic proteins in soybean. Using a "gene silencing" technique, researchers were able to "knock out" a gene that makes Antigenic factors glycinin and β-conglycinin removal increases animal performance.

The prevalence of soybean allergy is estimated at 0.4% in children and 0.3%in adults North America (Sicherer and Sampson, 2010). A study was conducted to determine the relationship between adverse health outcomes and occupational risk factors among workers at a soy processing plant (Cummings et al, 2010). They reported that asthma and symptoms of asthma were associated with immune reactivity to soy dust. Such occurences are a occupational hazard and have led to strict regulatory oversight in soybean plants and food manufacturing plants that process food products containing soybean. There have been numerous recalls by the FDA (FDA Enforcement Reports) of several products containing soy proteins, paste, oils and flour for reasons ranging due to improper labeling. Unlisted soy protein on the product label is considered a potential hazard for people who may be allergic to soy.

2.6. Cyanogens and saponins

Soybean and other closely related legumes which are common food ingredients for human and monogastric animals have been recognized to contain cyanogenic compounds (Montgomery, 1980). The content of cyanide was reported at 0.07-0.3 pg of hydrogen cyanide/g of sample in soy protein products and 1.24 pg/g in soybean hulls when browning was kept to a minimum (Honig et al.,1983). Cyanide is considered toxic even in small amounts, hence where soy are a major constituent of a diet, there are concerns of cyanide toxicity.

On the other hand, saponins are unabsorbable glucosides of steroids, steroid alkaloids or triterpenes found in soybeans germ and cotyledons and other leguminous plants. They form lather in aqueous solutions and impart a bitter test or flavor in feed, resulting in reduction of feed consumption. In severe cases they cause haemolysis of red blood cells and diarrhea (Oakwndull, 1981). Raw soybeans have been reported to contain 2-5 g of saponins/100 g soy products. Although soybean saponins possess anti-nutritional properties, some are edible and have been reported to possess some health benefits. They have been shown to stimulate the immune system, bind to cholesterol and make it unavailable for absorption and allowing its clearance into the colon and eventual excretion (Elias et al., 1990).

2.7. Non-starch polyssacharides and soy oligosaccharides

Soybean oligosaccharides (OS) such as raffinose and stachyose are carbohydrates consisting of relatively small number of monosaccharides and they have been reported to influence ileal nutrient digestibility and fecal consistency in monogastric animals (Smiricky et al.,

2002). According to Leske et al., (1993), raffinose and stachyose represent about 4-6% of soybean dry matter. The digestion of OS in the small intestine is limited because mammals lack α-galactosidase necessary to hydrolyze the α 1,6 linkages present in OS (Slominski, 1994). Previous research has demonstrated that soy OS are responsible for increasing intestinal viscosity of digesta and as a result interfere with digestion of nutrients by decreasing their interaction with digestive enzymes (Smits and Annison, 1996). Irish and Balnave (1993) demonstrated that stachyose derived from the oligosaccharides of soyabean meals exert anti-nutritive effects in broilers fed high concentrations soyabean meal as the sole protein concentrate. The OS in soybean, raffinose and stachyose, are not eliminated by heat treatment during processing (Leske et al., 1993). In earlier reports (Coon et al.,1990) observed that removal of the OS from SBM in poultry diets increased the true metabolizable energy value of the diet by 20 percent.

The digestion of oligosaccharides in the small and large intestine is aided by beneficial microbial fermentation (Hayakawa et al., 1990). Certain oligosaccharides, however, are considered to be prebiotic compounds because they are not hydrolyzed in the upper gastrointestinal tract and are able to favorably alter the colonic microflora, conferring beneficial effects of digestion and fermentation of carbohydrates to the host. More recent studies have demonstrated that feeding a higher level of an oligosaccharide (8 g/kg) to chicks depressed metabolizable energy and amino acid digestibility (Biggs et al., 2007). Smiricky-Tjardes et al. (2003) reported the presence of significant quantities of galactooligosaccharides in soy-based swine diets. These soy oligosaccharides are partially fermented by gut microflora functioning as prebiotics which promote selective growth for beneficial bacteria. The high content of enzyme inhibitors in unfermented soybeans interferes with thecomplete digestion of carbohydrates and proteins from soybeans. When foods are not completely digested because of enzyme inhibitors, bacteria in the large intestine try to do the job, and this can cause discomfort, bloating, and embarrassment. The enzyme inhibiting properties of soybean compounds the low levels of digestive enzymes, a common phenomenon especially in elderly people.

3. Modifications of soybean to enhance nutritional value and health benefits

3.1. Genetic modifications (GMO's)

About 99 percent of the soybean that we consume is genetically modified (GMO) and referred to as GMO soybean. The genetic modifications in soybean are primarily meant to improve the yield and nutritional value of soybean, reduce allergenicity, create resistance to certain diseases or disease causing pathogens and/or confer tolerance to herbicides or adverse climatic or environmental conditions. For example, transgene-induced gene silencing was used to prevent the accumulation of Gly m Bd 30 K protein in soybean seeds. The Gly m Bd 30 K-silenced plants and their seeds lacked any compositional, developmental, structural, or ultrastructural phenotypic differences when compared with control plants (Herman et al., 2003). Current GMO crops, including soybean, have not been shown to add any additional allergenic risk beyond the intrinsic risks already present (Herman 2003).

The enhancement of the nutritional value of soybean and soybean products through genetic engineering of soybeans has been reported. Through genetic engineering completely new fatty acid biosynthetic pathways have been introduced into soybeans from exotic plants and various microorganisms (Cahoon et al., 1999; Wallis, Watts, and Browse, 2002). Soybean oil is used in many food applications and therefore alterring its composition, especially the fatty acid composition would benefit the consumer. Several fatty acids especially the omega-3 have been reported to possess health benefit to the consumer. Engineered soybean lines that are rich in oleic acid (producing stable oil that does not need to be partially hydrogenated and is thus free of trans fatty acids) and lines lower in saturated fatty acids have been produced (Kinney & Knowlton, 1998; Buhr et al., 2002). These high-oleic and low-saturated soybean oils provide the potential benefits to human health and point to the positive impact of the achievements in biotechnology that promote human health.

Bioactive polyunsaturated fatty acids are also known to confer beneficial and positive effects in humans' health (Knapp et al., 2003). These bioactive fatty acids can also be found in oils other than soybean can also be produced in soybean through genetic engineering. Thus, introducing these into soybean can enhance the existing health benefits of soybean with the complementary benefits of bioactive lipids and other compounds.These polyunsaturated fatty acids are known to mediate their heart-healthy effects by mechanisms independent of those of soy protein and they have been previously researched extensively (Kelley and Erikson, 2003; Knapp et al., 2003).

Previously attempts were made to increase the oxidative stability of soybean oil by increasing the composition in soybean of the fatty acids oleic and stearic and decrease linoleic acid content of the soybean without creating *trans* or polyunsaturated fatty acids (Clemente, 2009). According to Clemente, (2009) and Clemente and Cahoon, (2009), DuPont announced the creation of a high oleic fatty acid soybean, with levels of oleic acid greater than 80%. This product was due for release into the market in 2010. Soybean mutants with elevated and reduced palmitate have been developed (Rahman et al., 1999). While the palmitate content of commercial soybean cultivars is approximately 11%, elevated palmitate content in soybean oil may be important for the production of some food and industrial products.

Soy foods have also been reported (Sirtori and Lovati, 2001) to have the potential for reducing blood low-density lipoprotein (LDL) cholesterol concentrations in humans. According to Weggemans and Trautwein, (2003), this positive health effect appears to be directly related to the soy storage proteins rather than other components. The bulk of soy protein (more than 80%) is contributed by two major classes of storage protein, conglycinin (11S globulins) and beta-conglycinin (7S globulins). It has been possible to produce soybean transgenic lines with either 7S or no 11S protein using gene-silencing techniques (Kinney & Fader, 2002).

For many years soybean has been defined as a crop with the best amino acid composition within all cultivated protein crops and is most widely utilized for human and monogastric foods as a primary source of protein (Wenzel, 2008). Since the amino acids are directly used in the genetic formation of proteins and fatty acids, this makes the soybean invaluable in oil production and primary protein source of choice to many. There has been attempt through genetic engineering to modifiy the soybean to enhance its oxidative stability by changing the proportion of certain fatty acids, which would provide a more useful and abundant oil supply with health benefits to the consumer. The enhancement of soybean oil content, Clemente et al. (2009) achieved this goal by introducing a seed-specific transgene for a DGAT2-type enzyme from the oil-accumulating fungus *Umbelopsis ramanniana* into soybean. Without disrupting the protein content, the oil content was increased from approximately 20% of the seed weight to approximately 21.5%.

There has also been an attempt to genetically modify soybean to enhance flavors. These compounds are associated with the oxidation of the polyunsaturated fatty acids linoleic and linolenic acids (Frankel, 1987). There are hundreds of volatile compounds associated with bad flavors in soy preparations (Stephan and Steinhart, 1999), and these compounds are the predominant fatty acids in soybean oil whose oxidation during bean storage and processing results in the formation of secondary products of lipid oxidation that impart off-flavors to soy protein products.

Drought tolerant varieties of soybean have also been developed through genetic engineering. The Roundup Ready soybean, also known as soybean 40-3-2, is a transgenic soybean that has been immunized to the Roundup herbicide. Although soybean's natural trypsin inhibitors provide protection against pests, weeds still remain a major challenge in soy farming (Wenzel, 2008). An herbicide used to control weeds in soybean farming contained glyphosate which inhibited the expression of the soybean plant's ESPSP gene. According to Wenzel, (2008), the gene is involved in the maintenance of the "biosynthesis of aromatic metabolites," and killed the plant along with the weeds for which the herbicide was meant. Consequently, the soybean was genetically engineered by transferring a plasmid which provided immunity to glyphosate-containing herbicides was transferred to the soybean cells through the cauliflower mosaic virus, perfecting the Roundup Ready soybean.

Since drought stress is a major constraint to the production and yield stability of soybean, integrated approaches using molecular breeding and genetic engineering have also provided new opportunities for developing high yield and drought resistance in soybeans (Manavalan et al., 2009). Recently, Yang et al. (2010) pointed out that genetic engineering must be employed to exploit yield potential and maintaining yield stability of soybean production in water-limited environments in order to guarantee the supply of food for the growing human population and for food animals. On the other hand, new soybean varieties that are resistant to diseases and pests are being developed. Hoffman et al. (1999) observed that plants commonly respond to pathogen infection by increasing ethylene production. They suggested that soybean can be altered by genetic manipulation using mutagenesis to generate soybean lines with reduced sensitivity to ethylene. Two new genetic loci were identified, Etr1 and Etr2 and Plant lines with reduced ethylene sensitivity developed similar or less-severe disease symptoms in response to virulent Pseudomonas syringae. Other reports (Yi et al., 2004) indicate that *CaPF1*, a ERF/AP2 transcription factor in hot pepper plants may play dual roles in response to biotic and abiotic stress in plants and that through genetic engineering this factor could be modified to improve soybean disease resistance as well.

3.2. Enzyme supplementation

The diets of monogastric animals are primarily composed of feed ingredients of plant origin such as soybean. Soybean contains a variety of antinutritional factors such as phytin, non-starch polysaccharides, and protease inhibitors which limit the availability of nutrients that are essential for normal growth and performance, production or otherwise. Enzyme supplementation and soybean fermentation products have been used for a long time to improve the nutritional value and health-promoting properties of soybean (Kim et al., 1999). Recent reports (Kim et al., 2010) cite evidence that fermented soybean meal can effectively serve as an alternative protein source for nursery pigs at 3-7 weeks of age, possibly replacing the use of dried skim milk which tends to be more digestible than soybean. Feng et al. (2007) evaluated the effect of soybean meal fermented with *Aspergillus oryzae* on the activity of digestive enzymes and intestinal morphology of broilers. The fermentation had no significant influence on the activity of lipase, amylase and protease enzymes. However, they observed a significant increase in duodenal villus height and a decrease in crypts depth a sign of improved morphology of the absorptive surface. Earlier studies by Kiers et al. (2003) indicated that fermentation of soybean resulted in an increase nutrient solubility and digestibility in broilers.

The availability of phosphorus in soybean is about 30 percent; hence diets of monogastric animals must be supplemented with inorganic phosphorus or supplemented with the phytase enzyme to improve the utilization of phytate phosphorus (NRC, 1994; Richter, 1994). Phytase (myo-inositol-hexakisphosphate phosphohydrolase) is an enzyme that catalyzes the hydrolysis of phytic acid, an indigestible inorganic form of phosphorus in oil seeds such as soybean. As a result, phytases increase the digestion of phosphorus, consequently increasing its utilization and reducing its excretion by monogastric animals. The phytase enzymes commonly used in monogastric animal feeding are derived from yeast or fungi and bacteria. Nahashon et al. (1993) reported that phosphorus retention was improved in layers when the diet was supplemented with Lactobacillus bearing phytase activity. The use of phytase to hydrolyze phosphorus and possibly other mineral elements that may be bound onto phytate such as calcium, zinc, copper, manganese and ion, has been reported (Selle and Ravindran, 2007; Powell et al., 2011).

Recently, Liu et al. (2007) demonstrated that phytase supplementation in soybean-based diets significantly improved the digestibility of phosphorus and Calcium by 11.08 and 9.81%, respectively. A 2- 8% improvement of the digestibility of amino acids was also noted since phytate also binds to protein forming phytate-protein complex. This complex is less soluble resulting in decreased digestibility of soybean protein (Carnovale et al., 1988). Earlier, Singh and Kirkorian (1982) reported that phytate also inhibits trypsin and pepsin activities. These findings suggest that phytase supplementation in soybean-based diets can improve the digestibility of calcium, phosphorus and proteins and indeed amino acids. Augspurger and Baker, (2004) observed that high dietary levels of efficacious phytase enzymes can release most of the P from phytate, but they do not necessarily improve protein utilization (Boling-

Frankenbach et al., 2001). Supplemental phytase has also been reported to improving dietary phosphorus utilization by pigs (Sands et al., 2001; Traylor et al., 2001). Other reports (Pillai et al., 2006) demonstrated that addition of *E. coli* phytase to P-deficient broiler diets improved growth, bone, and carcass characteristics. Most recently Rutherford et al. (2012) demonstrated that addition of microbial phytase to diets of broiler chickens improved significantly the availability of phytate phosphorus, total phosphorus, other minerals such as calcium, zinc, manganese etc. and amino acids.

Protease enzymes on the other hand break down long protein chains into short peptides which can be readily absorbed. These proteolytic enzymes whose catalytic function is to hydrolyze or breakdown peptide bonds of proteins include enzymes such as trypsin, chymotrypsin, pepsin, papain, elastase, plasmin, thrombin, and proteinase K. These enzymes can also be supplemented in feed or indirectly by feeding microbials that have the potential to produce these enzymes in the gastrointestinal tract of the host animals. On the other hand, carbohydrases such as xylanase and amylase are enzymes that catalyze the hydrolysis of carbohydrates into sugars which are readily available or metabolizable by monogastric animals. Soybean meal contains approximately 3% of soluble non-starch polysaccharides (NSP) and 16% of insoluble NSP (Irish and Balnave, 1993). The NSP in soybean is thus of negligible amounts to yield digesta viscosity problems. Therefore, diets that comprise soybean are considered to be highly digestible, hence requiring less use of carbohydrases. Previous reports have, however, pointed out that since these cereal grains contain some soluble NSP, there is still the need to supplement soybean based diets with these enzymes to further improve their nutritional value (Maisonnier-Grenier et al., 2004).

Supplementing soybean based diets with multicarbohydrase enzymes, a preparation containing nonstarch polysaccharide-degrading enzymes, phytases and proteases reveled that these enzymes improved nutrient utilization and growth performance of broiler chickens (Woyengo et al., 2010). Cowieson and Ravindran (2008) reported that when these enzyme combinations were fed in broler diets with both adequate and reduced energy and amino acid content, a 3% and 11% increase in apparent metabolizable energy and nitrogen retention, respectively, were observed. Also feeding other multicarbohydrase combinations containing xylanase, protease, and amylase resulted in significant improvements in feed conversion and body weight gain of broilers (Cowieson, 2005). In their recent review, Adeola and Cowieson (2011) suggested that when used together with phytase, nonstarch polysaccharide-hydrolyzing enzymes may increase the accessibility of phytase to phytin encapsulated in plant cell walls.

Although the enhancement of monogastric animal performance using enzyme supplements in feed have been extensively researched and documented, the benefits of enzymes such as the phytases and multicarbohydrase in soybean-based diets of monogastric animals have not been fully explored and require further research. There is still a great deal of uncertainty regarding the mode of action of these enzymes including the phytases, carbohydrases and proteases and their combination thereof in soybean based diets of monogastric animals. It is just fair to not that the future of enzymes in nonruminant animal production is promising and will require further research to elucidate the role of enzyme supplementation in promoting health and provide an understanding of the modes of action of these enzymes in modulating gene functions and their interactions thereof.

3.3. Probiotics and prebiotics supplementation

Probiotics, also known as direct-fed microbials, are live microbial feed supplements which beneficially affect the host animal by improving its intestinal microbial balance (Fuller, 1989). They have been reported to improve feed consumption, feed efficiency, health and metabolism of the host animal (Cheeke, 1991). The total collection of these probiotics, other gut microflora, their genetic elements or genomic materials and their interactive environment or the gastrointestinal tract of the host is termed the "microbiome". Currently efforts are underway to understand the microbiome and elucidate the mode of action of both probiotics and prebiotics due to a great interest in these gut microbiota and their health promoting properties and enhancement of performance of humans and monogastric animals. The scientific basis for the modes of action of probiotics and prebiotics is, therefore, beginning to emerge. According to report of Quigley (2012) a number of human disease states may benefit from the use of probiotics; these include diarrheal illnesses, inflammatory bowel diseases, certain infectious disorders, and irritable bowel syndrome. Prebiotics promote the growth of "good" bacteria, primarily through competitive exclusion resulting in a variety of health benefits. Probiotics have also been reported to: (1) improve feed intake and digestion and production performance (Nahashon et al., 1994a, 1994b, 1994c, 1996), (2) maintain a beneficial microbial population by competitive exclusion and antagonism (Fuller, 1989), and (3) alter bacterial metabolism (Cole et al., 1987; Jin et al., 1997). Nahashon et al (1994a) evaluated the phytase activity in lactobacilli probiotics and the role in the retention of phosphorus and calcium as well as egg production performance of Single Comb White Leghorn laying chickens. They reported phytase activity in the direct-fed microbial and that supplementation of the corn-soy based diets with the probiotics to a 0.25% available phosphorus diet improved phosphorus retention and layer performance.

Prebiotics are defined as non-digestible food ingredients that beneficially affect the host, selectively stimulating their growth or activity, or both of one or a limited number of bacteria in the colon and thus improve gut health (Gibson and Roberfroid, 1995). They are short-chain-fructo-oligosaccharides (sc-FOS) which consist of glucose linked to two, three or four fructose units. They are not absorbed in the small intestine but they undergo complete fermentation in the colon by colonic flora (Gibson and Roberfroid, 1995). They benefit humans and monogastric animals by: (1) releasing volatile fatty acids which are absorbed in the large intestine and contribute to the animal's energy supply; (2) enhance intestinal absorption of nitrogen, calcium, magnesium, iron, zinc and copper in rats (Ducios et al., 2005); (3) increase the number and/or activity of bifidobacteria and lactic acid bacteria (Hedin et al., 2007); and (4) since they are non-digestible, they provide surface for attachment by pathogenic bacteria and therefore facilitate the excretion of these pathogenic microorganisms.

According to Bouhnik et al. (1994) and Gibson and Roberfroid, (1995), fructooligosaccharides such as inulin, oligofructose, and other short-chain fructooligosaccharides can be fermented by beneficial bacteria such as bifidobacteria and lactobacilli and in turn control or reduce the growth of harmful bacteria such as *Clostridium perfringens* through competitive exclusion. The bifidobacteria and lactobacilli are generally classified as beneficial bacteria (Gibson and Wang, 1994; Flickinger et al., 2003). Although the mode of action of several of these oligosaccharides are still obscure, Biggs et al. (2007, pointed out that even low concentrations (4 g/kg) of an indigestible oligosaccharide can be fed to monogastric animals with no deleterious effects on metabolizable energy and amino acid digestibility. The benefit of utilizing oligosaccharides in soy-based diets of monogastric animals are due to the ability of these oligosaccharide to pass through to the hindgut of the monogastric animals intact and to be fermented by beneficial bacteria that are stimulated to grow and produce compounds that are beneficial to the host. These beneficial bacteria are also able to prevent the growth of bacteria such as *Escherichia coli* and *Clostridium perfringens* that can be harmful to the host through competitive exclusion (Gibson and Roberfroid, 1995). Biggs and Parsons (2007) reported an increase in the digestibility of a few amino acids was by some oligosaccharides in cecectomized roosters.

3.4. Amino acids and vitamin supplementation

Soybean is an excellent source of protein and vegetable oil for human and animal nutrition due to its balanced amino acid profile. According to Berry et al. (1962), methionine is the most limiting amino acid followed by lysine, and threonine. The level of supplementation of the amino acids lysine, methionine, threonine and glycine was evaluated (Waguespack et al., 2009). Feed efficiency decreased significantly in broilers fed diets supplemented with more than 0.3% L-Lysine but not in birds fed diets containing 0.25% L-Lysine. It was also observed that up to 0.25% L-Lysine could be added to corn-soy diets of broilers supplemented with methionine, threonine and glycine. Waguespack et al. (2009) also reported that arginine and valine were equally limiting after methionine, threonine and glycine in the diets containing 0.25% L-Lysine.

Earlier reports (Douglass and Persons, 2000) demonstrated a significant improvement in feed efficiency by methionine and Lysine supplementation in broilers diets. Studies have shown that excess heating by extrusion cooking or autoclaving of soybean during oil extraction can decrease lysine availability, hence requiring supplementation of this and other amino acids in soybean meal-based diets (Persons et al., 1992). It has also been further established that supplementation of raw soybean meal with methionine is an effective way of eliminating the potential nutritional deficiencies in both the raw and heated soybean meals. Due to the fact that raw soybean contains protein of low quality, supplementation of pig raw soy-based diets with cysteine and the B-complex vitamins exhibited significant improvement in performance (Peterson et al., 1941). Evaluating the effect of supplementation of turkey diets with dL-tocopheryl acetate, Sell et al. (1997) reported that feeding soybean based diets containing 6-20 IU tocopheryl acetate/kg improved performance of male turkeys from 1-day of age to market age.

3.5. Heat treatment and autoclaving

Heat treatment is a common procedure in soybean processing during extraction of oil and the inactivation of antinutritional factors such as trypsin inhibitors. The processing inactivates these protease inhibitors, although there has to be a balance in conditions of heat inactivation since excessive heating could also destroy other essential nutrients. Kwok et al. (1993) demonstrated that excess heat in the inactivation of protease inhibitors of soybean may increases Maillard reactions between the amino group of amino acids and reducing sugars and as a result decrease the digestibility of energy and amino acids by monogastric animals. Comparing the nutritive value of different heat treated commercial soybean meals, Veltmann et al. (1986) reported that compared to the normal meal, excessively heat-treated soybean meal had lower crude protein which also reflected lower essential and non-essential amino acids, and less trypsin inhibitors. Herkelman et al. (1991) evaluated the effect of heating time and sodium metabisulfite (SMBS) on the nutritional value of full fat soybeans for chicks. They observed that chicks fed the full-fat soybean achieved maximum performance when the soybeans were heated at 121 ° C for 40 min, and the SMBS decreased by one half the heating times required inactivating trypsin inhibitors.

4. Health benefits of soybean in human nutrition

4.1. General overview

Soybeans which boast rich content of protein (38-40%) of high quality and with a balanced amino acids profile are widely grown around the world and are the most important world source of edible oil and protein. Besides it's use in livestock feeds and to some extent biofuels, soybean is processed into products that are utilized for human consumption such as soybean oil and fermented soybean products which have long been utilized to prepare healthy human foods worldwide (Kim et al., 1999). Highly purified and oil-free food grade proteins isolates containing as high as 95% crude protein are commonly utilized in human foods. In the United States, 90% of the soybean is used for food especially as soybean oil (Smith and Wolf, 1961). In Asia and other parts of the world, soybean and soybean products are routinely utilized in large quantities in various forms of foods such as mature soybean, soybean flour, soybean meal, soybean milk, and also as oriental soybean products such as tofu, natto, miso, shoyu, and sprouts. In the recent past there has been increased focus on soybean as human food because of its health benefits. As a result, considerable research effort has been directed to evaluating the health benefits and increasing the uses of soybean in human foods. Abundant supplies of high protein soybean products and the rapid development of the soybean processing industry has also contributed significantly to the increased use of soybean as human food.

4.2. Selected components of soybean that confer health benefits

Soybean and soybean products have been acclaimed to confer health benefits to consumers because they contain substances that have been confirmed to bear health conferring properties. These substances include Iron, isoflavones, high content of protein rich in balanced amino acids, the sulfur containing amino acids methionine and cysteine, saponins, phytoestrogens, and the omega-3 fatty acids present in soybean oil. Soybeans are a major source of nonheme iron in diets of humans. Although some of the iron is unavailable for it is in the form of ferritin complexed with phytate, calcium and proteins, iron in soybean is a bioavailable source for human consumption. On the other hand, the benefits of omega-3 long chain fatty acids to heart health are well established (Lemke et al., 2010) and enrichment of soybean oil with these fatty acids has been a sustainable way of increasing tissue concentration of these omega-3 fatty acids and in reducing the risk of cardiovascular disease.

Soybean also contains nonstarch polysaccharide (NSP) hydrolysis products of soybean meal. These nonstarch polysaccharide hydrolysis products of soybean meal are beneficial in maintaining fluid balance during *Enterotoxigeic Escherichia Coli* (ETEC) infection and controlling ETEC-induced diarrhea in piglets. Soybean fermented with *R. oligosporus produce* antibacterial compounds that are active against some gram-positive microorganisms. The material can be extracted with water from soybeans fermented by *R. oligosporus*. Genistein and other soybean isoflavones slow the growth of blood vessels to tumors, another action that makes it popular as a cancer fighter.

4.3. Disease prevention properties of soybean

4.3.1. Cholesterol

Soybean, a popular source of protein for both humans and other monogastric animals due to its protein content and quality, especially the balance of amino acids, is an invaluable source of oil which contains fatty acids known to be effective in prevention of cardiovascular disease. The US Food and Drug Administration (1999) indicated that soybean proteins were responsible for prevention of cardiovascular disease. According to Lovatti et al. (1996) the soybean 7S or β-conglycinin has also been implicated in the upregulation of liver high-affinity LDL receptors. This protein was also shown to reduce plasma triglycerines in humans and rats (Aoyama et al., 2001). Later studies (Duranti et al., 2004) evaluated the effect of soybean 7S globulin subunits on the upregulation of LDL receptors. They reported that it lowered the expression of β-VLDL receptors induced by soybeam subunit. The oral admoinistration of soybean 7S globulin and the α-subunit significantly reduced plasma cholesterol and tryglycerides of hypercholesterolemic rats (Duranti et al., 2004). On the other hand, feeding soybean (25 g/day) was associated with lower total cholesterol concentrations in individuals with initial cholesterol concentratons of greater than 5.7 mmol/L (Bakhit et al., 1994). Later studies (Carroll, 1991) demonstrated that soybean protein lowered blood lipids in humans and experimental animals. Sirtori et al. (1985) also demonstrated that a 50% substitution of animal protein with soybean protein significantly reduced blood cholesterol concentrations of humans with type II familial hypercholesterolemia.

Soybean fiber has also been reported to reduce blood lipids whereas comsumption of cookies containing 25 g soybean cotyledon fiber was associated with a significant reduction in total plasma LDL cholesterol in hypercholesterolemic patients (Lo et al., 1986). Various mechanisms of reduction of cholesterol in humans and other monogastric animals have been proposed. These include the direct effect of soybean peptides which may modulate the endocrine regulation of catabolism and/or reduction in cholesterol biosynthesis (Bakhit et al., 1994). Most recently, Cho et al. (2007) suggest that soy peptides can effectively stimulate LDL-R transcription in the human liver cell line and reduce blood cholesterol level. They proposed several mechanism and component of the cholesterol lowering activity of soybean which include blockage of bile acid and/or cholesterol absorption, inhibition of cholesterol synthesis, and stimulation of low-density lipoprotein receptor (LDL-R) transcription. Similar observations were reported earlier by Beynen et al. (1986) that hypocholesterolemic effect of dietary soybean protein was caused primarily by its influence on the heterohepatic circulation of bile acids and cholesterol.

4.3.2. Cancer

Soybean is a rich protein source for humans and monogastric animals and contains about 0.2-1.5 mg/g of the isoflavones daidzein and genistin, and their glycones daidzein and genistkein (Wang and Murphy, 1994). These isoflavones have been proposed to possess anticarsinogenic properties which may be associated with their ability to serve as antioxidants which prevent fat rancidity, β-carotene bleaching and glutathione peroxidase activity (Hendrich et al., 1994), antiestrogens, and inhibiting the estrogen synthetase preadipocyte aromatase in humans (Aldercreutz et al., 1993). Adlercreutz et al. (1991) suggested that the low breast cancer incidence in Japanese women may be attributed to their consumption of feeds rich in soybean, a source of isoflavones. These isoflavones in soybean such as genistein confer anticarcinogenic effect primarily by inhibiting estrogen binding to the estrogen receptors; the soy isoflavones compete for estrogen receptors.

Xu et al. (1995) also hypothesized that soybean isoflavones possess anticarsinogenic properties. They anaerobically incubated soybean isoflavones with human feces and observed that intestinal half life daidzein and genistein were as little as 7.5 and 3.3 hrs, respectively. Hence the bioavailability of these isoflavones was depended on the ability of gut microflora to degrade these compounds. They attributed the cancer protective effects of the isoflavones to also the isoflavone metabolites such as methyl p-hydroxyphenylacetate, a monophenolic compound of both exogenous flavonoids and tyrosine which are inhibitors of hormone-dependent neoplastic cell proliferation. This compound has high affinity for nuclear type II binding site which is involved in cell growth regulation by estrogenic hormones (Xu et al., 1995).

In his report, Messina (1999) stated that soybean isoflavones may reduce the risk of prostrate cancer in men and breast cancer in women. The anti-cancer properties in soybean are attributed to the isoflavone genistein which influence signal transduction and the potential role in preventing and treating cancer. McMichael-Phillips et al. (1998) observed a significant enhancement of DNA synthesis by breast cells taken from biopsies of normal breast tissue from women with benign and malignant breast disease when these women were fed soybean for about two weeks. On other studies (Jing et al., 1993) reported that daidzein, one of the two primary isoflavones in soybean exhibited anti-cancer effects by inhibiting the growth of HL-60 cells implanted into the subrenal capsules of mice. The anticancer effects of the isoflavone genistein may be attributed to its antioxidant properties and its ability to in-hibit several enzymes that are involved in signal transduction (Wei et al., 1993) including ty-rosine protein kinase (Akiyama et al., 1987), ribosomal S6 kinase (Linassier et al., 1990), Map kinase (Thoburn et al., 1994), the inhibition of the activity of DNA topoisomerase and in-creasing the concentration of transforming growth factor β (TGFβ) as reported by Benson and Colletta (1995).

The interest in soybean and soybean products has been driven by its potential health bene-fits, especially in prevention of various forms of cancer by the soybean isoflavones genistein, deidzein and glycitein. In more recent studies, Su et al. (2000) reported that isoflavones played a protective role against bladder cancer cells. They also observed that both genistein and combined isoflavones exhibited significant tumor seppressing effects. According to Messina and Barnes (1991), increased soybean consumption reduced the risk of breast, colon and breast cancer for people living in Asia as opposed to people living in the United States and Western Europe. A comprehensive review of the interelationship between diet and can-cer by The World Cancer Research Fund (1997) revealed that vegetable intake decreased the risk of colon cancer. The increased consumption of soybean and soybean products have also been reported to reduce the risk of colon cancer in some human and animal populations.

While examining the ability of dietary soybean components to inhibit the growth of prostate cancer, Zhou et al. (1999) reported that dietary soybean products inhibited experimental prostrate tumor growth through direct effects in the tumor cells and indirectly through the effect on tumor neurovesculature. Earlier reports (Herbert et al., 1998) showed that in-creased consumptio of soybean products contributed to reduction in prostrate cancer risk. Phytochemicals in soybean have been reported to posses anticarcinogenic properties (Messi-na et al., 1994). Zhou et al. (1999) further observed that soybean isoflavones and phytochem-icals inhibited LNCaP cell proliferation, blocked cell cycle progression and enhanced DNA fragmentation which is a marker for opoptosis or programmed cell death. Datta et al., (1997) reported that soybean is capable of oxidizing benzo (a) pyrene-7, 8-dihydrodiol and 2-ami-nofluorine which are known to cause developmental toxicity or transplacental carcinogenic-ity in mammals. In other studies, Wei et al., (1995) cited evidence that genistein's antioxidant properties and antiproliferative effects may be responsible for its anticarcinogenic effects. Therefore, high content of genistein in soybean and its high bioavailability increases soy-bean's potential for prevention of various forms of cancer.

Soybean saponins have also been cited as potential contributors to the health promoting properties of soybean and soybean products. Saponins are chemical structures consisting of triterpenoidal or steroidal aglycones with various carbohydrates moieties in plants. Sapo-nins are excellent emulsifyers since they bear both hydrophilic and hydrophobic regions and they tend to inhibit colon tumor cell proliferation in vitro. Various saponins have dem-onstrated antimutagenic and anticarcinogenic effecta against cell lines. More recent studies (Ellington et al., 2005) suggested that the B-group soyasaponins may be colon cancer sup-pressive component os soybean serving as potential chemopreventative phytochemical. Therefore, soybean and soybean products should be explored further for their potential in prevention and treatment of the various forms of cancer.

4.3.3. Osteoporosis

Osteoporosis is a degenerative thinning of the bones that is associated with decreasing estrogen levels which is a common problem with aging, especially in women. According to Ikenda et al. (2006), soybeans and soybean products which contain large amounts of menaquinone-7 (vitamin K2) may help prevent the development of osteoporosis. Soybeans have also generated interest in connection with osteoporosis because they contain a phytoestrogens called isoflavones, which are believed to have potential as substitute for estrogen without its adverse side effects. Intake of Natto, an ancient Japanese food of fermented soybeans, was reported to bear properties that were preventative of postmenopausal bone loss through the effects of menaquinone-7 or bioavailable isoflavones which were more abundant in natto than in other soybean products (Ikenda et al., 2006). Heaney (1996) described vitamin K functionally as a cofactor of γ-carboxylase enzyme which mediates the conversion of undercarboxylated osteocalcin to carboxylated osteocalcin by transforming the glutamyl residue of osteocalcin into carboxyglutamic acid residue. The carboxyglutamic acids have high affinity for calcium ions in hydroxyapatite and regulate the growth of these crystals in bone formation. Therefore there is sufficient evidence to suggest that fermented soybean products can effectively maintain bone stiffness (Katsuyama et al., 2002) by increasing serum levels of menaquinone-7 and γ-carboxylated osteocalcin (Kaneki et al., 2001) as well as maintaining bone mineral density.

Soybean and soybean-based diets for human contain naturally occurring bioactive compounds known as phytochemicals that have been cited to confer long-term health benefits (Setchell, 1998). These phytoestrogens primarily occur as glycosides bearing a weak estrogen-like activity which allows them to bind to the estrogen receptor (Miksicek, 1994) and therefore are of great significance as remedy where estrogen levels decline due to old age. These isoflavones can serve as alternative to estrogen therapy in the treatment of existing low bone mass or osteoporosis. They present potential naturally occuring alternative to hormone or drug therapy (estrogen) that would minimize bone loss in menopausal women.

In other studies, Picherit et al. (2001) assessed the dose-dependent effect of daily soybean isoflavones consumption in reversing bone loss in adult ovariectomized rats. They reported that in adult ovariectomized rats, daily soybean isoflavone consumption decreased bone turnover but did not reverse established osteopenia. In earlier studies using a rat model, Arjmandi et al. (1996) evaluated the potential for soybean protein isolate to prevent bone loss induced by ovarian hormone deficiency. They reported an increase in femoral and vertebral bone densities in rats that were fed soybean diets possibly due to the presence of isoflavones in soybean.

5. Nutritional benefits of soybean in other monogastric animals

5.1. Poultry

Soybean meal is the primary protein source in corn-soy based poultry rations. It is fed to poultry as soybean meal and is primarily the by-product of soybean oil extraction; it's the ground defatted flakes. Various studies have been conducted to evaluate methods of enhancing the acceptability of soybean and the enhancement of its nutritional value in poultry feeding. For instance, a study was conducted to evaluate the effect of extruding or expander processing prior to solvent extraction on the nutritional value of soybean meal (SBM) for broiler chicks. The results of this study indicate that pre-solvent processing method (expander or non-expander) had no significant effect on the nutritional value of SBM for broiler chicks. Both Methionine and Lysine supplementation increased feed efficiency (Douglas and Persons, 2000). Several other studies (Coca-Sinova et al. 2008; Dilger et al. 2004; Opapeju et al. 2006) have evaluated various methods of enhancing the digestibility of individual amino acids and protein of soybean meal.

The guinea fowl is classified as poultry and although its production is not popular as chickens, it is gradually gaining popularity and acceptance as alternative meat to chicken. It is also gradually finding its share of the global market for poultry and poultry products. Lacking however, is estimates for nutrient requirements of the guinea fowl. Recent efforts have focused on evaluating the growth pattern of the guinea fowl (Nahashon et al. 2010) and their nutrient requirements (Nahashon et al. 2009, 2010, 2011). The soybean meal has been utilized extensively as the sole protein source for the guinea fowl providing accurate estimate for the nutrient requirements for both the Pearl Grey and French varieties of the guinea fowl.

5.2. Swine

Soybean meal and soybean products have also been used extensively in swine production because of its relatively high concentration of protein (44 to 48%) and its excellent profile of highly digestible amino acids. Soy protein provides most amino acids that are deficient in most cereal grains commonly fed as energy sources in swine production. However, as opposed to feeding animal source proteins, when raw soybean is fed to young pigs as the primary protein source, dramatic slowdown in body weight gains were reported even with supplementation of amino acids such as methionine and cysteine (Peterson et al., 1942). The animal source proteins tend to exhibit higher digestibility than plant source proteins such as soybean and therefore better suited for nursery pigs (Kim et al., 2009; Gottlob et al., 2006). The low digestibility of raw soybean by young pigs is therefore attributed to the low nutritive value of the raw soybean protein. Due to the high cost of feeding and also the antinutritional factors in raw soybean, there have been attempts to minimize the amount of soybean in swine rations and also to improve its digestibility. The digestibility of the amino acids of soybean by swine has also been researched quite extensively (Smiricky-Tjardes et al. 2002; Grala et al., 1998; NRC 1998).

The supplementations of raw soybean with the amino acids threonine and the B- complex vitamins have been reported to enhance growth in young pigs. Also fermented soybean meals, soybean meals with enzyme supplements and extruded soybean meals have been used extensively in swine diets and they tend to improve performance especially of young pigs (Kim et al., 2006). Kim et al. (2009) reported an increase in crude protein concentration from 50.3 to 55.3% by fermentation of soybean meal with *A. oryzae* without affecting the balance of limiting amino acids for pigs. Bruce et al. (2006) evaluated the inclusion of soybean

(SB) processing byproducts such as gums, oil, and soapstock into soybean meal. Addition of these processing by-products significantly reduced the nutritive value of the resultant meal. Smiricky-Tjardes et al. (2003) evaluated other approaches such as the addition of galactooligosaccharides on ileal nutrient digestibility to enhance and expand the utilization of soybean in swine production.

5.3. Aquatic life

The feeding value of soybean as a rich protein source has also been extended to aquaculture. The high protein level makes soybean meal a key ingredient for aquaculture feeds since soybean meal is considerably less expensive than traditionally used marine animal meals. However, soybeans not contain complete amino acid profiles and usually are deficient in the essential amino acids lysine and methionine. Therefore, other protein sources should be used in combination with soybean to overcome the deficiencies. Soybean meal and genetically modified soybean products have also been employed in aquaculture (Hammond et al. 1995). Naylora et al. (2009) points to the importance of fish oils and fishmeal as a protein source in food animal production and also the extensive use of soybean and soybean products as protein supplements in aquaculture feeds.

5.4. Companion animals

The term "companion animals" refers to the entire spectrum of animal species which are considered as 'pets' such as cats, dogs, fish, rabbits, rodents, cage birds, and even non-indigenous species. Large animals such as horses, as well as small ruminants such as the goats and sheep have also been classified as companion animals as well because they contribute to human companionship; they have an important role to play in our society. The companion animal industry is rapidly growing sector of the global economy and so is the need for provision of adequate nutritional regimens for optimum growth, production and reproduction. Relatively few data are available on the nutrient digestibilities of plant-based protein sources by companion animals.

Plant source proteins such as soybean and soybean products are predominantly used in diets of companion animals. Soybeans are an essential part of Plant-based protein sources and are generally less variable in chemical composition than animal-based protein sources especially in nutrients such as calcium and phosphorus. The effects of including selected soybean protein sources in dog diets on nutrient digestion at the ileum and in the total tract, as well as on fecal characteristics, were evaluated (Clapper et al. 2001). Apparent amino acid digestibility at the terminal ileum, excluding methionine, threonine, alanine, and glycine, were higher (P < 0.01) for soybean protein-containing diets when compared with diets containing other sources of protein.

The effects of soybean hulls containing varying ratios of insoluble: soluble (I: S) fiber on nutrient digestibility and fecal characteristics of dogs were evaluated (Burkhalter et al., 2001). Ileal digestibility of dry matter by dogs fed the soybean hulls treatments responded quadratically (P < 0.05) to I: S fiber diets, with digestibility coefficients decreasing as the I:S approached 3.2. In other studies (Tso and Ling, 1930) reported that the blood cholesterol value of rabbits is slightly higher in animals fed the soybeans diets than in controls. However, differences in cholesterol levels between rabbits fed on cooked and raw soybeans were not statistically significant. Also the blood of rabbits fed exclusively on water-soaked raw soybeans showed an increase in uric acid, urea nitrogen, inorganic phosphorus and cholesterol. After extending this study to feeding cooked soybeans, there are no demonstrable changes in the blood composition of rabbits whether they were fed cooked or raw soybeans.

The optimum concentration of a mixture of soybean hulls and defatted grape seed meal (SHDG) for rabbits was evaluated (Necodemus et al. 2007). They observed that SHDG could be included up to 26.7% in diets for fattening rabbits and lactating does that meet ADL and particle size requirements. In another study, Angora goat doelings (average BW 22.1 kg) were used to examine the effects of dietary crude protein level and degradability on mohair fiber production (Sahlu et al. 1992). They reported that plasma glucose was elevated 2 hours after feeding in the goats fed conventional, solvent-extracted soybean meal, whereas glucagon concentrations were greater at 0 and 4 h in the group fed expelled, heat-treated soybean meal.

Author details

Samuel N. Nahashon[1*] and Agnes K. Kilonzo-Nthenge[2*]

*Address all correspondence to: snahashon@tnstate.edu

1 Department of Agricultural Sciences, Tennessee State University, Nashville, TN, USA

2 Department of Family and Consumer Sciences, Tennessee State University, Nashville, TN, USA

References

[1] Aljmandi, B. H., Elekel, L., Hollis, B. W., Amin, D., Stacewicz-Sapuntzakis, M., Guo, P., & Kukreja, S. C. (1996). Dietary soybean protein prevents bone loss in an ovarectomized rat model of osteoperosos. *J. Nutr.*, 126, 161-167.

[2] Adeola, O., & Cowieson, A. J. (2011). Board-invited review: Opportunities and challenges in using exogenous enzymes to improve nonruminant animal production. *J. Anim. Sci,*, 89, 3189-3218.

[3] Adachi, M., Kanamori, J., Masuda, T., Yagasaki, K., Kitamura, K., Mikami, B., & Utsumi, S. (2003). Crystal structure of soybean 11S globulin: Glycinin A3B4 homohexamer. *Proceedings of the National Academy of Sciences, USA,*, 100, 7395-7400.

[4] Adler-Nissen, J. (1978). Enzymatic hydrolysis of soy protein for nutritional fortification of low pH food. *Annals of Nutrition and Alimentation,*, 32, 205-216.

[5] Adlercreutz, H., Bannwart, C., Wahala, K., Makela, T., Brunow, G., Hase, T., Arosemena, P. J., Kellis, J. T., Jr , , & Vickey, E. L. (1993). Inhibition of human alromatase by mammalian lignans and isoflavonoids phytoestrogens. *J. Steroid Biochem. Mol. Biol.*, 44, 147-153.

[6] Adlercreutz, H., Honjo, S., Higashi, A., Fotsis, T., Hamalainem, E., Hasegawa, T., & Okada, H. (1991). Urinary excretion of lignans and isoflavonoid phytoestrogens in Japanese men and women consuming a traditional Japanese diet. *Am. J. Clin. Nutr.*, 54, 1093-1100.

[7] Akiyama, T., Ishida, J., Nakagawa, S., et al. (1987). . Genistein, a specific inhibitor of tyrosine specific protein kinases. J. Biol. Chem. ., 262, 5592-5595.

[8] Asada, K., Tanaka, K., & Kasai, Z. (1969). Formation of phytic acid in cereal grains. Ann NY Acad Sci , 165, 801-814.

[9] Applegate, T. J., Webel, D. M., & Lei, X. G. (2003). Efficacy of a phytase derived from Escherichia coli and expressed in yeast on phosphorus utilization and bone mineralization in turkey poults. *Poult. Sci.*, 82, 1726-1732.

[10] Angkanaporn, K., Ravindran, V., & Bryden, W. L. (1996). Additivity of apparent and true ileal amino acid digestibilities in soybean meal, sunflower meal, and meat and bone meal for broilers. *Poultry Science*, 75(9), 1098-1103.

[11] Augspurger, N. R., & Baker, D. H. (2004). High dietary phytase levels maximize phytate-phosphorus utilization but do not affect protein utilization in chicks fed phosphorusor amino acid-deficient diets. *J. Anim. Sci.*, 82, 1100-1107.

[12] Barbazan, M. (2004). Liquid swine manure as a phosphorous source for corn-soybean rotation. Thesis. Iowa State University, Ames, Iowa.

[13] Beilinson, V., Chen, Z., Shoemaker, C., Fischer, L., Goldberg, B., & Nielsen, C. (2002). Genomic organization of glycinin genes in soybean. *Theoretical and Applied Genetics*, 104, 1132-1140.

[14] Buhr, T., Sato, S., Ebrahim, F., Xing, A., Zhou, Y., Mathiesen, M., et al. (2002). Ribozyme termination of RNA transcripts down-regulate seed fatty acid genes in transgenic soybean. *The Plant Journal,*, 3, 155-63.

[15] Beynen, A. C., Van Der Meer, R., West, C. E., Cugano, M., & Kritchevsky, D. (1986). Possible mechanisms underlying he differential cholesterolemic effects of dietary casein and soy protein. In:, *Nutritional effects on Cholesterol Metabolism*, (Beynen, A. C, ed),, 29-45, Transmondial, Voorthuizen, The Netherlands.

[16] Benson, J. R., & Colletta, A. A. (1995). Transforming growth factor β. prospects for cancer prevention and treatment. *Clin. Immunother*, 4, 249-258.

[17] Biggs, P. & Pearsons, C. M. (2007). The Effects of Several Oligosaccharides on True Amino Acid Digestibility and True Metabolizable Energy in Cecectomized and Conventional Roosters. *Poultry Science*, 86, 1161-1165.

[18] Braden, C. R. (2006). Salmonella enterica serotype enteritidis and eggs: a national epidemic in the United States. *Clin. Infect. Dis.*, 43, 512-517.

[19] Barth, Christian A., Bruta Lãoehding, Martin Schmitz & Hans Hagemeister. (1993). Soybean trypsin inhibitor(s) reduce absorption of exogenous and increase loss of endogenous protein. *J. Nutr.*, 123, 2195-2200.

[20] Beynen, A. C., Winnubst, E. N. W., & West, C. E. (1983). The effect of replacement of dietary soybean protein by casein on the fecal excretion of neutral steroids in rabbits. *Z. Theraphysiol. Tierrer. naehr. Futtermittelkd.*, 49, 43-49.

[21] Berry, T. H., Becker, D. E., Rasmussen, O. G., Jensen, A. H., & Norton, H. W. (1962). The Limiting Amino Acids in Soybean Protein. *J ANIM SCI*, 21(3).

[22] Burkhalter, T. M., Merchen, N. R., Bauer, L. L., Murray, S. M., Patil, A. R., Brent, J. L., Jr , , Fahey, G. C., & Jr 200, . (2001). The Ratio of Insoluble to Soluble Fiber Components in Soybean Hulls Affects Ileal and Total-Tract Nutrient Digestibilities and Fecal Characteristics of Dogs. *J. Nutr.*, 131, 1978-1985.

[23] Biggs, P., Parsons, C. M., & Fahey, G. C. (2007). The Effects of Several Oligosaccharides on Growth Performance, Nutrient Digestibilities, and Cecal Microbial Populations in Young Chicks. *Poultry Science*, 86, 2327-2336.

[24] Bouhnik, Y., Flourie´, B., Ouarne, F., Riottot, M., Bisetti, N., Bornet, F., & Rambaud, J. (1994). Effects of prolonged ingestion of fructo-oligosaccharides on colonic bifidobacteria, fecal enzymes and bile acids in humans. *Gastroenterology*, 106, A598, (Abstr.).

[25] Balloun, S. L., & 198, . (1980). Soybean meal processing. In: K.C. Lepley (Ed) Soybean Meal in Poultry. The Ovid Bell press, Inc., Fulton, Missouri. , 6-12.

[26] Bhakit, R., Klein, B. P., Essex-Sorlie, D., Ham, J. O., Erdman, J. W., & Porter, S. M. (1994). Intake of 25 g of soybean protein with or wothout soybean fiber alters plasma lipids in men with elevated cholesterol concentrations. *J. Nutr.*, 124, 213-222.

[27] Carrol, K. K. (1991). Review of clinical studies on cholesterol lowering response to soy protein. *J. Am. Diet. Assoc.*, 91, 820-827.

[28] Cahoon, E. B., Carlson, T. J., Ripp, K. G., Schweiger, B. J., Cook, G. A., Hall, S. E., & Kinney, A. J. (1999). Biosynthetic origin of comnjugated double bonds: production of fatty acids components of high-value drying oils in transgenic soybean embryos. *Proceedings of the National Academy of Sciences, USA,*, 22, 12935-12940.

[29] Clemente, T. E., & Cahoon, E. B. (2009). Soybean Oil: Genetic Approaches for Modification of Functionality and Total Content. *Plant Physiology*, 151, 1030-1040.

[30] Cook, D. R. (1998). The effect of dietary soybean isoflavones on the rate and efficiency of growth and carcass muscle content in pigs and rats. Ph.D. Dissertation. Iowa State Univ.,.

[31] Cowieson, A. J. (2005). Factors that affect the nutritional value of maize for broilers. *Anim. Feed Sci. Technol.*, 119, 293-305.

[32] Protease inhibitors and carcinogenesis: a review. Cancer Invest , (1996). , 14, 597-608.

[33] Clapper, G. M. C. M., Grieshop, N. R., Merchen, J. C., Russett, J. L., Brent Jr, , Fahey, G. C., & Jr , . (2001). Ileal and total tract nutrient digestibilities and fecal characteristics of dogs as affected by soybean protein inclusion in dry, extruded diets. *J ANIM SCI.*, 79, 1523-1532.

[34] Clemente, Tom E., & Edgar, B. Cahoon. (2009). Soybean Oil: Genetic Approaches for Modification of Functionality and Total Content. *Plant Physiology.*, 151(3), 1030-40.

[35] Coca-Sinova, A., Valencia,†, D. G., Jimenez-Moreno,†, E., Lazaro,†, R., & Mateos, G. G. (2008). Apparent Ileal Digestibility of Energy, Nitrogen, and Amino Acids of Soybean Meals of Different Origin in Broilers. *Poultry Science*, 87, 2613-2623.

[36] Cheryan, M. (1980). Phytic acid interactions in food systems. *CRC Crit. Rev. Food Sci. Nutr.*, 13, 297-335.

[37] Cowieson, A. J., & Ravindran, V. (2008). Effect of exogenous enzymes in maize-based diets varying in nutrient density for young broilers: Growth performance and digestibility of energy, minerals and amino acids. *Br. Poult. Sci.*, 49, 37-44.

[38] Coon, C. N., Leske, K. L., Akavanichan, O., & Cheng, T. K. (1990). Effect of oligosaccharide-free soybean meal on true metabolizable energy and fiber digestion in adult roosters. *Poult. Sci.*, 69, 787-793.

[39] Cordle, C. T. . (2004). Soy Protein Allergy: Incidence and Relative Severity· 1213S EOF-1219S EOF.

[40] Crump J.A, Griffin P.M, Angulo FJ. (2002). Bacterial contamination of animal feed and itsrelationship to human foodborne illness. *Clin Infect Dis*, 35, 859-865.

[41] Cheeke, P. R. (1991). Applied Animal Nutrition. AVI Publishing Company, Inc., Wesport, CT.

[42] Cole, C. B., Fuller, R., & Newport, M. J. (1987). The effect of diluted yogurt on the gut microbiology and growth of piglets. *Food Microbiol.*, 4, 83-85.

[43] Choct, M. (2006). Enzymes for the feed industry: Past, present and future. *World's Poult. Sci. J.*, 62, 5-16.

[44] Cummings, K. J., Gaughan, D. M., Kullman, G. J., Beezhold, D. H., Green, B. J., Blachere, F. M., Bledsoe, T., Kreiss, K., & Cox-Ganser, J. (2010). Adverse respiratory outcomes associated with occupational exposures at a soy processing plant. *ERJ*, 36, 1007-1015.

[45] Datta, K., Sherblom, P. M., & Kulkarni, A. P. (1997). Co-oxidative metabolism of 4-aminobiphenyl by lipoxygenase from soybean and human term placenta. *Drug Metab. Desp.*, 25, 196-205.

[46] Dalluge, J. J., Eliason, E., & Frazer, S. (2003). Simultaneous identification of soyasaponins and isoflavones and quantification of soyasaponin Bb in soy products using liquid chromatography/electrospray ionization-mass spectrometry. *Journal of Agricultural and Food Chemistry*, 51, 3520-3524.

[47] Drewnowski, A. (2001). The science and complexity of bitter taste. *Nutrition Review,,* 59, 163-169.

[48] Duranti, M., Lovati, M. R., Dani, V., Barbiroli, A., Scarafoni, A., Castinglion, S., Ponzone, C., & Morazzoni, P. (2004). The α-subunit fro soybean 7S globulin lowers plasma lipids and upregulates liver β-VLDL receptors in rats fed a hypercholesterolemic diet. *J. Nutr.*, 134, 1334-1339.

[49] Dauglas, M. W., & Parsons, C. M. (2000). Effect of presolvent extraction processing method on the nutritional value of soybean meal for chicks. *Poult. Sci.*, 79, 1623-1626.

[50] Dilger, R. N., Sanders, J. S., Ragland, D., & Adiola, O. (2004). Digestibility of nitrogen and amino acids in soybean meal with added soyhulls. *J. Anim. Sci.*, 82, 715-724.

[51] Dixon, R. A. (2004). Phytoestrogens. Annu. Rev. Plant Biol. , 55, 225-261.

[52] Davis, Hancock. D. D., Rice, D. H., Call, D. R., Di Giacomo, R., Samadpour, M., et al. (2003). Feedstuffs as a vehicle of cattle exposure to Escherichia coli O157:H7 and Salmonella enterica. *Vet Microbiol*, 95, 199-210.

[53] De Rham, O., & Jost, T. (1979). Phytate-protein interactions in soybean extracts and low-phytate soy protein products. *J. Food Sci.*, 44, 596-600.

[54] Doerge, DR, & Sheehan, DM. (2002). Goitrogenic and estrogenic activity of soy isoflavones. *Environ Health Perspect.*, 110(3), 349-53.

[55] Divi, RL, Chang, HC, & Doerge, DR. (1997). Anti-thyroid isoflavones from soybean: isolation, characterization, and mechanisms of action. *Biochem Pharmacol.*, 54(10), 1087-96.

[56] Dubey, R. K., Rosselli, M., Imthurn, B., Keller, P. J., & Jackson, E. K. (2000). Vascular effects of environmental oestrogens: implications for reproductive and vascular health. *Hum Reprod Update*, 4, 351-363.

[57] Du, Pont. H. L. (2007). The growing threat of foodborne bacterial enteropathogens of animal origin. *Clinical Infectious Diseases*, 45, 1353-1361.

[58] Ducros, V., Arnaud, J., Tahiri, M., Coudray, C., Bornet, F., Bouteloup-Demange, C., Brouns, F., Rayssiguier, Y., & Roussel, A. M. (2005). Influence of short-chain fructo-oligosaccharides (sc-FOS) on absorption of Cu, Zn, and Se in healthy postmenopausal women. *J. American College of Nutr.*, 24(1), 30-37.

[59] DeRouchey J.M, Tokach, M.D., Nelsen J. L, Goodband, R. D., Dritz, S. S., Woodworth, J. C., & James B. W. (2002). Comparison of spray-dried blood meal and blood cells in diets for nursery pigs. *J. Anim. Sci.*, 80, 2879 EOF-86 EOF.

[60] Elias, R., De Meo, M., Vidal-Ollivier, E., et al. (1990). Antimutagenic activity of some saponins isolated from Calendula officinalis L., C arvensis L., and Hedra helix L.". *Mutagenesis*, 5, 327-331.

[61] Ellington, A. A., Barlow, M., & Singletary, K. W. (2005). Induction of macroautophagy in human colon cancer cells by soybean B-group triterpenoid saponins. *Carciniogenesis*, 26, 159-167.

[62] Eliot, M., Herman, Ricki. M., Helm, Rudolf., Jung, , & Anthony, J. Kinney. (2003). Genetic Modification Removes an Immunodominant Allergen from Soybean. (1), 1-36.

[63] Fort, P., Moses, N., & Fasano, M. (1990). Breast and soy-formula feedings in early infancy and the prevalence of autoimmune thyroid disease in children. *J Am Coll Nutr*, 9, 164-167.

[64] Frankel, E. N. (1987). Secondary products of lipid oxidation. *Chemistry and Physics of Lipids*, 44, 73-85.

[65] Friedman, M., & Brandon, D. L. (2001). Nutritional and health benefits of soy proteins. *Journal of Agricultural and Food Chemistry*, 49, 1069-1086.

[66] Flickinger, E. A., Schreijen, E. M. W. C., Patil, A. R., Hussein, H. S., Grieshop, C. M., Merchen, N. R., Fahey, G. C., & Jr , . (2003). Nutrient digestibilities, microbial populations, and protein catabolites as affected by fructan supplementation of dog diets. *J. Anim. Sci.*, 81, 2008-2018.

[67] Fuller, R. (1989). Probiotics in man and animals. A review. *J. Appl. Bacteriol.*, 66, 365-378.

[68] Gudbrandsen, O. A., Wergedahl, H., Mork, S., Liaset, B., Espe, M., & Berge, R. K. (2006). Dietary soya protein concentrate enriched with isoflavones reduced fatty liver, increased hepatic fatty acid oxidation and decreased the hepatic mRNA level of VLDL receptor in obese Zucker rats. *Br. J. Nutr.*, 96, 249-257.

[69] Green, S., Bertrand, S. L., Duron, M. J., & Maillard, R. (1987). Digestibility of amino acids in soyabean, sunflower and groundnut meals, determined with intact and caecectomised cockerels. *British Poultry Science*, 28(4), 643-652.

[70] Garcia, M. C., Torre, M., Marina, M. L., & Laborda, F. (1997). Composition and characterization of soyabean and related products. *Critical Reviews in Food Science and Nutrition*, 37, 361-391.

[71] Grala, W., Verstegen, M. A., Jansman, A. J. M., Huisman, J., & Van Leeusen, P. (1998). Ileal apparent protein and amino acid digestibilitiesand endogenous nitrogenlosses in pigs fedsoybean and rapeseed products. *J. Anim. Sci.*, 76, 557-568.

[72] Gertler, A., Birk, Y., & Bondi, A. (1967). A comparative study of the nutritional and physiological significance of pure soybean trypsin inhibitors and of ethanol-extracted soybean meals in chicks and rats. *J. Nutr.*, 91, 358-370.

[73] Goldflus, F., Ceccantini, M., & Santos, W. (2006). Amino acid content of soybean samples collected in different brazilian states-Harvest 2003/2004. *Braz. J. Poult. Sci.*, 8, 105-111.

[74] Gibson, G. R., & Wang, X. (1994). Inhibitory effects of bifidobacteria on other colonic bacteria. *J. Appl. Bacteriol.*, 77, 412-420.

[75] Gibson, G. R., & Roberfroid, M. B. (1995). Dietary modulationof the human colonic microbiota-Introducing the concept of prebiotics. *J. Nutr.*, 125, 1401-1412.

[76] Ghadge, V. N., Upase, B. T., & Patil, P. V. (2009). Effect of replacing groundnut cake by soybean meal on performance of broilers. *Veterinary World,,* 2(5), 183-184.

[77] Gaylord, B. S., Heeger, A. J., & Bazan G. C. (2003). DNA hybridization detection with water-soluble conjugated polymers and chromophore-labeled single-stranded DNA. *J. Am. Chem. Soc.*, 125, 896 EOF-900 EOF.

[78] Grieshop, C. M., Catzere, C. T., Clapper, G. M., Flickinger, E. A., Bauer, L. L., Frazier, R. L., & Fahey, G. C. Jr. (2003). Chemical and nutritional characteristics of United States soybeans and soybean meals. *J. Agric. Food Chem.*, 51, 7684-7691.

[79] Grieshop, C. M., & Fahey, G. R. (2001). Comparison of quality characteristics of soybeans from Brazil, China and the United States. *J. Agric Food Chem*, 49, 2669-2673.

[80] Halina, l., Nathan, S., & Ephrain, K. (1966). Soybean hemagglutinin, a plant glycoprotein. I. Isolation of a glycopeptide.

[81] Hong Y. H., Wang, T. C., Huang, C. J., Cheng, W. Y., & Lin, B. F. (2008). Soy isoflavones supplementation alleviates disease severity in autoimmune-prone MRL-lpr/lpr mice. *Lupus*, 17, 814-821.

[82] Heaney, R. P., Weaver, C. M., & Fitzsimmons, M. L. (1991). Soybean phytate content: effect on calciumabsorption. *Am. J. Clin Nutr*, 53, 745-747.

[83] Herman, E. M., Ricki, M., Helm, R., Anthony, J., & Kinney, J. (2003). Genetic Modification Removes an Immunodominant Allergen from Soybean. *Plant Physiology*, 132(1).

[84] Harris, L. J., Farber, J. N., Beuchat, L. R., Parish, M. E., Suslow, T. V., Garrett, E. H., & Busta, F. F. (2003). Outbreaks associated with fresh produce: incidence, growth, and survival of pathogens in fresh and fresh-cut produce. *Comprehensive Reviews in Food Science and Food Safety Supplement*, 79-141.

[85] Herman, E. M. (2003). Genetically modified soybeans and food allergies. *Journal of Experimental Botany*, 54(386), 1317-1319.

[86] Hwang, David. l., Wen-kuang, Yang., Donald, E., & Foard, . (1978). Rapid Release of Protease Inhibitors from Soybeans. *Plant Physiol.*, 61, 30-34.

[87] Hayakawa, K., Mitzutani, J., Wada, K., Masai, T., Yoshihara, I., & Mitsuoka, T. (1990). Effects of soybean oligosaccharides on human fecal flora. *Microbiol. Ecol. Health Dis.*, 3, 293-303.

[88] Harada, J. J., Barker, S. J., & Goldberg, R. B. (1989). Soybean beta-conglycinin genes are clustered in several DNA regions and are regulated by transcriptional and post-transcriptional processes. *The Plant Cell*, 1, 415-425.

[89] Herman, E. M., Helm, R. M., Jung, R., & Kinney, A. J. (2003). Genetic modification removes an immunodominant allergen from soybean. *Plant Physiology,*, 132, 36-43.

[90] Hildebrand, D. F. (1989). Lipoxygenase. *Physiologia Plantarum,*, 76, 249-253.

[91] Hitz, W. D., Carlson, T. J., Kerr, P. S., & Sebastian, S. A. (2002). Biochemical and molecular characterization of a mutation that confers a decreased raffinosaccharide and phytic acid phenotype on soybean seeds. *Plant Physiology,*, 128, 650-660.

[92] Hedin, C., et al. (2007). Evidence for the use of probiotics and prebiotics in inflammatory bowel disease: a review of clinical trials. *Proc Nutr Soc.*, 66(3), 307-15.

[93] Hendrich, S., Lee, K. W., Xu, X., Wang, H. J., & Murphy, P. A. (1994). Defining food components as new nutrients. *J. Nutr.*, 124, 1789S EOF-1792S EOF.

[94] Herkelman, K. L., Cromwel, G. L., & Stahly, T. (1991). Effect of heating time and sodium metabisulfite on the nutritional value of full-fat soybeans for chicks. *J. Anim. Sci.*, 69, 4477-4486.

[95] Hoaglund CM, VM Thomas, MK Peterson, and RW Knott. (1992). Effects of supplemental protein source and metabolizable energy intake on nutritional status in pregnant ewes. *J. Anim. Sci.*, 70, 273 EOF-80 EOF.

[96] Hammond, B. C. J. L., Vicini, C. F., Hartnell, M. W., Naylor, C. D., Knight, E. H., Robinson, R. L., Fuchs, , & Padgette, S. R. (1995). The feeding value of soybeans fed to rats, chickens, catfish and dairy cattle is not altered by genetic incorporation of glyphosate tolerance. *J. Nutr.*, 126, 3-717.

[97] Heaney, R. P. (1996). Nutrition and the risk of osteoporosis. In: Marcus R Fieldman D, Kelsey J. L., editors., *Osteoperosis*, Academic press: San Diego, CA;, 483-509.

[98] Holle, D. G. (1995). Amino Acids. *Ratite Feeds and Feeding*, 1-2, 58-59.

[99] Herkelman, K. L., Cromwell, G. L., & Stahly, T. S. (1991). Effect of heating time and sodium metabisulfite on the nutritional value of full-fat soybean for chicks. *J. Anim. Sci.*, 69, 4477-4486.

[100] Honig, D. H., Elaine, M., Hockridge, Robert. M., Gould, , Joseph, J., & Rackis, . (1983). Determination of Cyanide in Soybeans and Soybean Products. *Journal of Agricultural & Food Chemistry,*, 272-275.

[101] Irish, G. G., & Balnave, D. (1993). Non-starch polysaccharides and broiler perform‐ance on diets containing soyabean meal as the sole protein concentrate. *Australian Journal of Agricultural Research*, 44(7), 1483-1499.

[102] Ikuomola, D. S., & Eniola, K. I. T. (2010). Microbiological Quality and Public Health Risk Associated with Beske: Fried Soybean Snack, Retailed in Ikeji-Arakeji, Osun State, Nigeria. *Nigeria Journal of Microbiology,*, 24(1), 2114-2118.

[103] Irish, G. G., & Balnave, D. (1993). Non-starch polysaccharides and broiler perform‐ance on diets containing soyabean meal as the sole protein concentrate. *Aust. J. Agric. Res.*, 44, 1483-1499.

[104] Izabelawoclawek-potocka, Mamadou. M., Bah, Anna., Korzekwa, Mariusz. K., Pisku‐la, Wiesławwiczkowski., Andrzej, Depta., & Skarzynski, D. J. (2005). Soybean-De‐rived Phytoestrogens Regulate Prostaglandin Secretion in Endometrium During Cattle Estrous Cycle and Early Pregnancy. *Experimental biology and Med*, 230, 189-199.

[105] Irish, G. G., & Balnave, D. (1993). Non-starch polysaccharides and broiler perform‐ance on diets containing soyabean meal as the sole protein concentrate. *Australian Journal of Agricultural Research*, 44(7), 1483-1499.

[106] Jin, L. Z., Ho, Y. W., Abdullah, N., & , S. Jalaludin(1997). Probiotics in poultry: modes of action. World's Poultry Sci. J. , 53, 353-368.

[107] Jung, W., Yu, O., Lau, S. M., O'Keefe, D. P., Odell, J., Fader, G., & Mc Gonigle, B. (2000). Identification and expression of isoflavone synthase, the key enzyme for bio‐synthesis of isoflavones in legumes. *Nature Biotechnology*, 18, 208-212.

[108] Sherill, J. D., Morgan, Sparks., M.D., John, M. M., Kemppainen, B. W., Bartol, F. F., Morrison, E. E., & Akingbemi, B. T. (2010). Developmental Exposures of Male Rats to Soy Isoflavones Impact Leydig Cell Differentiation. *Biology of Reproduction*, 83, 488-501.

[109] Jing, Y., Nakaya, K., & Han, R. (1993). Differentiation of promyelocytic leukemia cells HL-60 induced by daidzein in vitro and in vivo. *Anti Cancer Res.*, 13, 1049S-1054S.

[110] Kaneki, M., Hedjes, S. J., Hosol, T., Fijiwara, S., Lyons, A., Crean, S. J., Ishida, N., Na‐kagawa, M., & Takechi, M. (2001). Japanese fermented soybean food as the major de‐terminant of the largegeographic difference in circulating levels of vitamin k2: possible implications of hip fracture risk. *Nutrition*, 17, 315-321.

[111] Kunihiko, U., Chieko, T. & Isao, K. (2010). Inactivation of Bacillus subtilis spores in soybean milk by radio-frequencyflash heating. *Journal of Food Engineering*, 100, 622-626.

[112] Katsuki, Y., Yasuda, K., Ueda, K, & Naoi, Y. (1978). *Annu. Rep. Tokyo Metrop. Res. Lab. Public Health*, 29, 261.

[113] Knaus, W. F, Beermann, D. H., Robinson, T. F., Fox, D. G., & Finnerty, K. D. (1998). Effects of a dietary mixture of meat and bone meal, feather meal, blood meal, and

fish meal on nitrogen utilization in finishing Holstein steers. *J. Anim. Sci.*, 76, 1481 EOF-7 EOF.

[114] Kats, L. J., Nelssen J. L., Tokach, M. D., Goodband, R. D., Hansen, J. A., & Laurin, J. L. (1994). The effect of spray-dried porcine plasma on growth performance in the early-weaned pig. *J. Anim. Sci.*, 72, 2075 EOF-81 EOF.

[115] Karr-Lilienthal, L. K., Mershem, N. R., Grieshop, C. M., Flahaven, M. A., Mahan, D. C., Watts, N. D., Fahey, G. C., & Jr , . (2004). Ileal amino acid digestibilities by pigs fed soybean meals from five major soybean producing countries. *J. Anim Sci*, 82, 3198-3209.

[116] Karr-Lilienthal, L. K., Grienshop, C. M., Spears, J. K., Fahey, G. C., & Jr , . (2005). Amino acid, carbohydrate, and fat composition of soybean meals prepared at 55 commercial U.S. soybean processing plants. *J. Agric. Food Chem.*, 53, 2146-2150.

[117] Karr-Lilienthal, L. K., Grieshop, C. N., Merchen, N. R., Mahan, D. C., Fahey, G. C., & Jr , . (2004). Chemical composition and protein quality comparisons of soybeans and soybean meals from five leading soybean-producing countries. *J. Agric. Food Chem.*, 52, 6193-6199.

[118] Katsuyama, H., Ideguchi, S., Fukunaga, M., Saijoh, K., & Sunami, S. (2002). Usual dietary intake of fermented soybeans (natto) is associated with bone mineral density in premenopausal women. *J. Nutr. Sci. Vitaminol (Tokyo).*, 48, 207-215.

[119] Kalinski, A., Melroy, D. L., Dwivedi, R. S., & Herman, E. M. (1992). A soybean vacuolar protein (p34) related to thiol proteases is synthesized as a glycoprotein precursor during seed maturation. *The Journal of Biological Chemistry*, 267, 12068-12076.

[120] Kelley, D. S., & Erikson, K. L. (2003). Modulation of body composition and immune cell functions by conjugated linoleic acid in humans and animal models:benefits vs. risks. *Lipids*, 38, 377-390.

[121] Kinney, A. J. (1996). Development of genetically engineered soybean oils for food applications. *Journal of Food Lipids*, 3, 273-292.

[122] Kinney, A. J., & Knowlton, S. (1998). Designer oils: the high oleic acid soybean. In S. Roller & S. Harlander (Eds.),, *Genetic modification in the food industry: A strategy for food quality improvement*, 193-213, London: Blackie.

[123] Kinney, A. J., & Fader, G. M. (2002). *U.S. Patent* [6], Washington, DC: US Patent and Trademark Office.

[124] Kinsella, J. E. (1979). Functional properties of soy proteins. *Journal of the American Oil Chemists Society,*, 56, 242-258.

[125] Kitamura, K. (1995). Genetic improvement of nutrional and food processing quality in soybean. *Japan Agricultural Research Quarterly,*, 29, 1-8.

[126] Kitts, D. D., & Weiler, K. (2003). Bioactive proteins and peptides from food sources. Applications of bioprocesses used in isolation and recovery. *Current Pharmaceutical Design*, 9, 1309-1323.

[127] Knapp, H. R., Salem, N., Jr , , & Cunnane, S. (2003). Dietary fats and health. *Proceedings of the 5th Congress of the International Society for the Study of Fatty Acids and Lipids,,* 38, 297-496.

[128] Kunitz, M. (1946). Crystalline soybean trypsin inhibitor. *J. Gen. Phygiol.,,* 9, 149-154.

[129] Kwok, K. C., Qin, W. H., Tsang, J. C. 1993. Heat Inactivation of Trypsin Inhibitors in Soymilk at Ultra-High Temperatures. J. Food Science 58:859–862.

[130] Liu, N., Liu, G. H., Li, F. D., Sands, J. S., Zhang, S., Zheng, A. J., & Ru, Y. J. (2007). Efficacy of Phytases on Egg Produ. ction and Nutrient Digestibility in Layers Fed Reduced Phosphorus Diets. *Poultry Science*, 86, 2337-2342.

[131] Lim, H. S., Namkung, H., & Paik, I. K. (2003). Effects of phytase supplementation on the performance, egg quality, and phosphorous excretion of laying hens fed different levels of dietary calcium and nonphytate phosphorous. *Poult. Sci.*, 82, 92-99.

[132] Lenehan, N. A., De Rouchey, J. M., Goodband, R. D., Tokach, M. D., Dritz, S. S., Nelssen, J. L., Groesbeck, C. N., & Lawrence, K. R. (2007). Evaluation of soy protein concentrates in nursery pig diets. *J. Anim. Sci.*, 85, 3013-3021.

[133] Lemke, S. L., Vicini, J. L., Su, H., Goldtein, D. A., Nemeth, M. A., Krul, E. S., & Harris, W. S. (2010). Dietary intake of steriodonic acid-enriched soybean oil increases the omega-3 index:randomized, double-blind clinical study of afficacy and safety. *J. Clin. Nutr.*, 92, 766-775.

[134] Liener, I. E. (1953). Soyin, a toxic protein from the soybean. *J. Nutrition,,* 49, 527.

[135] Liener, I. E. (1994). Implications of antinutritional components in soybean foods. *Crit. Rev. Food Sci. Nutr.*, 34, 31-67.

[136] Liu, Jiang-Gong and Lin Tser-KeShun,. (2008). Survival of Listeria monocytogenes inoculated in retail soymilk products. *Food Control,,* 19(2008), 862-867.

[137] Linassier, C., Pierr, M., Le Pacco-B, J., & Pierre, J. (1990). Mechanisms of action in NIH-3T3 cells of genistein, an inhibitor of EGF receptor tyrosine kinase activity. *Biochem Pharmacol*, 39, 187-193.

[138] Leske, K. L., Jevne, C. J., & Coon, C. N. (1993). Effect of oligosaccharide additions on nitrogen-corrected truemetabolizable energy of soy protein concentrate. *Poult. Sci.*, 72, 664-668.

[139] Lienei, I. E., & Kakade, M. L. (1969). Protease inhibitors. In I. E. Liener (Ed.), *Toxic Constituents of Plant Foodstuffs.*, Academic Press, New York.

[140] Liener, I. E. (1981). Factors affecting the nutritional quality of soy products. *J. Am. Oil. Chem. Soc.*, 58, 406-415.

[141] Liener, L. E., & Kakade, M. L. (1980). Protease Inhibitors. In: I. E. Liener (Ed), *Toxic Constituents of Plant Foodstuffs.*, 7-71, Academic Press, New York.

[142] Linz, A. L., Xiao, R. J., Ferguson, M., Badger, T. M., & Simmen, F. A. (2004). Developmental feeding of soy protein isolate to male rats inhibits formation of colonic aberrant crypt foci: Major response differences to azoxymethane (AOM) in Sprague-Dawley rats from two colonies. *FASEB J.*, 18, A127.

[143] Lovati, M. R., Manzoni, C., Corsini, A., Granata, A., Fumagalli, R., & Sirtori, C. R. (1996). 7S globulin from soybean is metabolized in human cell cultures by a specific uptake and degradation system. *Journal of Nutrition,*, 126, 2831-2842.

[144] Lovati, M. R., Manzoni, C., Corsini, A., Granata, A., Fumagalli, R., Sirtori, C., & , R. (1992). Low density lipoprotein (LDL) receptor activity.

[145] Madden, R. H., Murray, K. A., & Gilmore, A. (2004). Determination of principal points of product contamination during beef carcass dressing process in Northern Ireland. J. of. *Food Prot.*, 67, 1494-1496.

[146] Foley, S.L., White, D. G., McDermott, P.F., Walker, R.D., & Messina, M. J. (1999). Legumes and soybean: Overview of their nutritional profiles and health effect. *Am. J. Clin. Nutr.*, 70, 439S-450S.

[147] Mac, Donald. R. S., Guo, J. Y., Copeland, J., Browning, J. D., Jr , , Sleper, D., Rottinghous, G. E., & Berhow, M. A. (2005). Environmental influences on isoflavones and saponins in soybean and their role in colon cancer. *J. Nutr.*, 135, 1239-1242.

[148] Mc Andrews, G. M., Liebman, M., Cambardella, C. A., & Richard, T. L. (2006). Residual effects of composted and fresh solid swine (Sus scrofa L.) manure on soybean [Glycine max (L.) Merr.] growth and yield. *Agron. J.*, 98, 873-882.

[149] Montgomery, R. D. (1980). In, *Toxic Constituents of Plant Foodstuffs,* , 2nd ed.; Liener, I. E., Ed.; Academic Press: New York,.

[150] Maciorowski, K. G., Jones, F. T., Pillai, S. D., & Ricke, S. C. (2004). Incidence, sources, and control of foodborne Salmonella spp. in poultry feed. World's Poultry Sct. J. , 60, 446-457.

[151] Montazer-Sadegh, R., Ebrahim-Nezhad, Y., & Maheri-Sis, N. (2008). Replacement of different levels of rapeseed meal with soubean meal on broiler performance. *Asian Journal of Animal and Vet. Advances*, 3(5), 278-285.

[152] Mc Cue, P., & Shetty, K. (2004). Health benefits of soy isoflavonoids and strategies for enhancement: A review. *Crit. Rev.Food Sci. Nutr.*, 44, 361-367.

[153] Zimmermann, Michael B. (2009). *Iodine Deficiency Endocrine Reviews,*, 30(4), 376-408.

[154] Messina, M., & Redmond, G. (2006). Effects of soy protein and soybean isoflavones on thyroid function in healthy adults and hypothyroid patients: a review of the relevant literature. *Thyroid*, 16, 249-258.

[155] Manavalan, L. P., Satish, K., Guttikonda-Son, Lam., Phan, Tran., Henry, T., & Nguyen, . (2009). Physiological and Molecular Approaches to Improve Drought Resistance in Soybean. *Plant Cell Physiol*, 50(7), 1260-1276.

[156] Medlock K. L, Branham W. S, & Sheehan D. M. (1995). Effects of coumestrol and equol on the developing reproductive tract of the rat. *Proc Soc Exp Biol Med*, 208, 67-71.

[157] Moizzudin, S. (2003). Soybean meal quality in US and World markets. MS. Thesis, IA Iowa State University.

[158] Marty, B. J., Chavez, E. R., & De Lange, C. F. M. (1994). Recovery of amino acids at the distal ileum for determining apparent and true ileal amino acid digestibilities in growing pigs fed various heat-processed full-fat soybean products. *J. Anim. Sci.*, 72, 2029-2037.

[159] Martin-C, J., & Valeille, K. (2002). Conjugated linoleic acids: All the same or to everyone its own function? *Reproduction Nutrition Development,*, 42, 525-536.

[160] Maruyama, N., Adachi, M., Takahashi, K., Yagasaki, K., Kohno, M., Takenaka, et., & al, . (2001). Crystal structures of recombinant and native soybean beta-conglycinin beta homotrimers. *European Journal of Biochemistry,*, 268, 3595-3604.

[161] Maruyama, N., Mohamed, Salleh. M. R., Takahashi, K., Yagasaki, K., Goto, H., Hontani, N., et al. (2002). Structure-physicochemical function relationships of soybean beta-conglyxcinin heterotrimers. *Journal of Agricultural and Food Chemistry,*, 50, 4323-4326.

[162] Messina, M., Gardner, C., & Barnes, S. (2002). Gaining insight into the health effects of soy but a long way still to go. *Journal of Nutrition,*, 132, 547S EOF-551S EOF.

[163] Mohamed, Salleh. M. R., Maruyama, N., Adachi, M., Hontani, N., Saka, S., Kato, N., et al. (2002). Comparison of protein chemical and physicochemical properties of rapeseed cruciferin with those of soybean glycinin. *Journal of Agricultural and Food Chemistry*, 42, 525-536.

[164] Messina, M. J., Persky, V., Setchell, K. D., & Barnes, S. (1994). Soy intake and cancer risk: A review of the in-vitro and vivo data. *Nutr. Cancer*, 21, 113-131.

[165] Messina, M. J., & Loprinzi, C. L. (2001). Soy for breast cancer survivors: A critical review of the literature. *J. Nutr.*, 131, 3095S EOF-108S EOF.

[166] Messina, M. J., & Barnes, S. (1991). The role of soy products in reducing risk of cancer. *J. Nutr. Cancer Instit.*, 83, 541-546.

[167] Messina, M. J. (1999). Legumes and soybeans: Overview of their nutritional profiles and health effects. *Am. J. Clin. Nutr.*, 70, 439S EOF-450S EOF.

[168] Mc Michael-Phillips, D. F., Hardingm, C., & Morton, M. (1998). Effect of soy protein supplementation on epithelial proliferation in histologically normal human breasts. Am. J. Clin. Nutr. S., 68, 1431S EOF-1435S EOF.

[169] Milsicek, R. J. (1994). Interaction of naturally occuring nonsteroidal estrogens with expressed recombinant human estrogen receptor. *J. Steroid Biochem. Mol. Biol.*, 49, 153-160.

[170] Nahashon, S. N., & Kilonzo-Nthenge, A. (2011). Advances in soybean and soybean by-products in monogastric nutrition and health. In, *Soybean and Nutrition*, Ed. H. A. El-shemy, Intech, Rijeka, Croatia.

[171] Nahashon, S. N., Nakaue, Harry S., & Mirosh, Larry W. (1996). Performance of Single Comb White Leghorns given a diet supplemented with a live microbial during the growing and egg laying phases. *Animal Feed Science and Technology*, 57, 25-38.

[172] Nahashon, S. N., Nakaue, H. S., & Mirosh, L. W. (1994a). Phytase activity, phosphorus and calcium retention and performance of Single Comb White Leghorn layers fed diets containing two levels of available phosphorus and supplemented with direct-fed microbials. *Poultry Science*, 73, 1552-1562.

[173] Nahashon, S. N., Nakaue, H. S., & Mirosh, L. W. (1994b). Production variables and nutrient retention in Single Comb White Leghorn laying pullets fed diets supplemented with direct-fed microbials. *Poultry Science*, 73, 1699-1711.

[174] Nahashon, S. N., Nakaue, H. S., Snyder, S. P., & Mirosh, L. W. (1994c). Performance of Single Comb White Leghorn layers fed corn-soybean meal and barley-corn-soybean meal diets supplemented with a direct-fed microbial. *Poultry Science*, 73, 1712-1723.

[175] Nahashon, S. N., Adefope, N., & Wright, D. (2011). Effect of floor density on growth performance of Pearl Grey guinea fowl replacement pullets. *Poultry Science*, 90, 1371-1378.

[176] Nahashon, S. N., Nahashon, A., Akuley, , & Adefope, N. (2010). Genetic relatedness of Pearl Grey guinea fowl and Single Comb White Leghorn chickens. *Journal of Poultry Science*, 47, 280-287.

[177] Nahashon, S. N., Aggrey, S. E., Adefope, N. A., Amenyenu, A., & Wright, D. (2010). Gomperts-Laird model prediction of optimum utilization of crude protein and metabolizable energy by French guinea fowl broilers. *Poultry Science*, 89, 52-57.

[178] Nahashon, S. N., Adefope, N., Amenyenu, A., & Wright, D. (2009). The effect of floor density on growth performance and carcass characteristics of French guinea broilers. *Poultry Science*, 88, 2461-2467.

[179] Tyus, I. I. J., Nahashon, S. N., Adefope, N., & Wright, D. (2009). Production performance of single comb white leghorn chickens fed growing diets containing blood meal and supplemental blood meal. *Journal of Poultry Science*, 46, 313-321.

[180] Nagano, T., Fukuda, Y., & Akasaka, T. (1996). Dynamic viscoelastic study on the gelation properties of beta-conglycinin-rich and glycinin-rich soybean protein isolates. *Journal of Agricultural and Food Chemistry,*, 44, 3484-3488.

[181] Noordermeer, M. A., Veldink, G. A., & Vliegenthart, J. F. (2001). Fatty acid hydroperoxide lyase: a plant cytochrome p450 enzyme involved in wound healing and pest resistance. *ChemBioChem,*, 2, 494-504.

[182] Naylora, R. L., Ronald, W., Hardyb, Dominique. P., Bureauc, Alice., Chiua, Matthew., Elliottd, Anthony. P., Farrelle, Ian., Forstere, Delbert. M., Gatlinf,g, Rebecca. J., Goldburgh, Katheline., Huac, , Peter, D., & Nicholsi, . (2009). Feeding aquaculture in an era of finite resources. *PNAS*, 106, 36-15103.

[183] Nelson, A. I., Wijeratne, W. B., Yeh, S. W., Wei, T. M., & Wei, L. S. (1987). Dry extrusion as an aid to mechanically expelling of oil from soybean. *J. Am. Oil Chem. Soc.*, 64, 1341-1347.

[184] NRC. (1998). Nutrient requirements of swine. 10th ed. National Academy Press, Washington, DC.

[185] Nejaty, H., Lacey, M., & Whitehead, S. A. (2001). Differing effects of endocrine disrupting chemicals on basal and FSH-stimulated progesterone production in rat granulose-luteal cells. *Exp Biol Med*, 226, 570-576.

[186] Nicodemus, N. J., García, R., Carabaño, , & De Blas, J. C. (2007). Effect of substitution of a soybean hull and grape seed meal mixture for traditional fiber sources on digestion and performance of growing rabbits and lactating does· *J. Anim Sci.*, 85, 181-187.

[187] Opapeju, F. O., , A., Golian, A., Nyachoti, C. M., & Campbell, L. D. (2006). Amino acid digestibility in dry extruded-expelled soybean meal fed to pigs and poultry. *J. Anim. Sci.*, 84, 1130-1137.

[188] Oakwndull, O. C. (1981). Saponins in Food- A Review. *Food Chem*, 6, 19-40.

[189] Odunsi, A. A. (2003). Blend of bovine blood and rumen digesta as a replacement for fishmeal and groundnut cake in layer diets. *Inter. J. Poult. Sci.*, 2, 58.

[190] Onwudike, OC. (1981). Effect of various protein sources on egg production in a tropical environment. *Trop. Anim. Prod.*, 6, 249.

[191] Fletciier, a. P., Marks, g. S., Marshall, r. D., & Neuberger, A. (1963). *Biochem J.*, 8.

[192] Plummer, T. H., , J. R., & , C. H. W. Hirs.(1963). J. Biol. Chem., , 238, 1396.

[193] Panda, A. K., Rao, S. V. R., Raju, M. V. L. N., & Bhanja, S. K. (2005). Effect of microbial phytase on production performance of White Leghorn layers fed on a diet low in non-phytate phosphorus. *Br. Poult. Sci.*, 46, 464-469.

[194] Payne, R. L., Bidner, T. D., Southern, L. L., & Mc Millin, K. W. (2001). Dietary Effects of Soy Isoflavones on Growth and Carcass Traits of Commercial Broilers. *Poultry Science*, 80, 1201-1207.

[195] Peganova, S., & Eder, K. (2002). Studies on requirement and excess of isoleucine in laying hens. *Poult. Sci.*, 81, 1714 EOF-21 EOF.

[196] Powell, S., Bidner, T. D., & Southern, L. L. (2011). Phytase supplementation improved growth performance and bone characteristics in broilers fed varying levels of dietary calcium 1. *Poultry Science, 90*, 604-608.

[197] Potter, S. M., Baum, J. A., Teng, H., Stillman, R. J., Shay, N. F., Erdman Jr, , & , J. W. (1998). Soy protein and isoflavones: Their effects on blood lipids and bone density in postmenopausal women. *American Journal of Clinical Nutrition,, 68*, 1375S EOF-1379S EOF.

[198] Persons, C. M., Hashimoto, K., Wedekind, K. J., Han, Y., & Baker, D. H. (1992). Effect of overprocessing on availability of amino acids and energy in soybean meal. *Poult. Sci., 71*, 133-140.

[199] Paterson, W. J., Hostetler, E. H., & Shaw, A. O. (1942). Studies in feeding soybean to pigs. *J. Anim. Sci., 1*, 360.

[200] Palacios, M. F., Easter, R. A., Soltwedel, K. T., Parsons, C. M., Douglas, M. W., Hymowitz, T., & Pettigrew, J. E. (2004). Effect of soybean variety and processing on growth performance of young chicks and pigs. *J. Anim. Sci., 82*, 1108-1114.

[201] Peisker, M. (2001). Manufacturing of soy protein concentrate for animal nutrition. 103-107, *Feed Manufacturing in the Mediterranean Region,* Improving Safety: From Feed to Food. J.Brufau, ed. CIHEAM-IAMZ, Reus, Spain.

[202] Pillai, P. B., O'Connor-Dennie, T., Owens, C. M., & Emmert, J. L. (2006). Efficacy of an Escherichia coli Phytase in Broilers Fed Adequate or Reduced Phosphorus Diets and Its Effect on Carcass Characteristics. *Poultry Science, 85*, 1737-1745.

[203] Perilla, N. S., Cruz, M. P., de Belalcazar, F., & Diaz, G. J. (1997). Effect of temperature of wet extrusion on the nutritional value of full-fat soyabeans for broiler chickens. *British Poultry Science, 38*(4), 412-416.

[204] Quigley, E. M. M. (2012). Prebiotics and Probiotics Their Role in the Management of Gastrointestinal Disorders in Adults. *Nutr. Clin. Pract., 27*(2), 195-200.

[205] Rackis, J. J. (1975). Oligosaccharides of food legumes: Alpha-galactosidase activity and the flatus problem. In: A. Jeanes and J.Hodges (ed.) Physiological Effects of Food Carbohydrates., *Am. Chem. Soc.,,* Washington, DC.

[206] Ravindran, V., Cabahug, S., Ravindran, G., & Bryden, W. L. (1999). Effects of phytase supplementation, individually and in combination with glycanase, on the nutritive value of wheat and barley. *Poult. Sci., 78*, 1588-1595.

[207] Rahman, S. M. T., Kinoshita, T., Anai, , & Takagi, Y. (1999). Genetic relationships between loci for palmitate contents in soybean mutants. *J Hered, 90*(3), 423 EOF-427 EOF.

[208] Ravindran, V., Morel, P. C., Partridge, G. G., Hruby, M., & Sands, J. S. (2006). Influence of an Escherichia coli-derived phytase on nutrient utilization in broiler starters fed diets containing varying concentrations of phytic acid. *Poult. Sci., 85*, 82-89.

[209] Rhodes, P. J., Fedorka-Cray, S., Simjee, , & Zhao, S. (2006). Comparison of subtyping methods for differentiating Salmonella enterica serova Typhimurium isolates obtained from animal food sources. *J. Clin. Microbiol.*, 44, 3569-3577.

[210] Roman, S., Temler, Charles. A., Dormono, Eliane., Simon, Brigette., Morel, , & Christine, Mettraux. (1984). Proteins, Their Hydrolysates and SoybeanTrypsin Inhibitor. *J. Nutr.*, 114, 270-278.

[211] Rangngang M. D, Nelson M. L and S. M Parish. (1997). Ruminal undegradability of blood meal and effects of blood meal on ruminal and postruminal digestion in steers consuming vegetative orchard grass hay. *J. Anim. Sci.*, 75, 2788.

[212] Rosselli, M., Reinhard, K., Imthurn, B., Keller, P. J., & Dubey, R. K. (2000). Cellular and biochemical mechanisms by which environmental estrogens influence reproductive function. *Hum Reprod. Update*, 6, 332-350.

[213] Riblett, A. L., Herald, T. J., Schmidt, K. A., & Tilley, K. A. (2001). Characterization of beta-conglycinin and glycinin soy protein fractions from four selected soybean genotypes. *Journal of Agricultural and Food Chemistry*, 49, 4983-4989.

[214] Raboy, V., & Dickinson, D. B. (1984). Effect of phosphorus and zinc nutrition on soybean seed phytic acid and zinc. *Plant Physiol.*, 75, 1094-1098.

[215] Raboy, V., Dickinson, D. B., & Below, F. E. (1984). Variation in seed total phosphorus, phytic acid, zinc, calcium, magnesium, and protein among lines of Glycine Max and G. Soja. *Crop Sci.*, 24, 431-434.

[216] Rackis, J. J., Wolf, W. J., & Baker, E. C. (1986). Protease inhibitors in plant foods: Content and inactivation. In: M. Friedman (Ed), *Nutritional and Toxicological Significance of Enzyme Inhibitors in Foods*, 299-347, Plenum Publishing, New York.

[217] Rackis, J. J. (1972). Biologically active components. In Smith and Circle (Ed.), *Soybeans: Chemistry and Technology*, Vol. I. The Avi Publ. Co., Inc.,Westport, Conn.

[218] Savage, J. H., et al. (2010). The natural history of soy allergy. *J. Allergy Clin. Immunol.*, 125, 683-686.

[219] Sahlu, T. J. M., Fernandez, C. D., Lu, , & Manning, R. (1992). Dietary protein level and ruminal degradability for mohair production in Angora goats. *J ANIM SCI*, 70, 1526-1533.

[220] Sicherer, S. H., & Sampson, H. A. (2010). Food allergy. J. Allergy Clin. Immunol. S125., 125, S116.

[221] Smiricky, M. R., Grieshop, C. M., Albin, D. M., & Wubben, J. E. (2002). The influence of soy oligosaccharides on apparent and true ileal amino acid digestibilities and fecal consistency in growing pigs. *J. Anim. Sci.*, 80, 2433-2441.

[222] Sampson, H.A. (2004). Update on food allergy. *J Allergy Clin Immunol;*, 113, 805-19.

[223] Sicherer, SH. (2002). Food allergy. *Lancet*, 360, 701-10.

[224] Sell, J. L., Soto-Salanova, M., Pierre, P., & Jeffrey, M. (1997). Influence of supplementing corn-soybean diets with vitamin E on performance and selected physiological traits of male turkeys. *Poult. Sci.*, 76, 1405-1417.

[225] Setchell, K. D. (1998). Phytoestrogens: The niochemistry, physiology, and implications for human health of soy isoflavones. *Am. J. Cclin Nutr.*, 68:, S1333-S1346.

[226] Scheiber, M. D., & Reber, R. W. (1999). Isoflavones and postmenopausal bone health:A variable alternative to estrogen therapy? *Menopause*, 6, 233-241.

[227] Slominski, B. A. (1994). Hydrolysis of galactooligosaccharides by commercial preparations of α-galactosidase and β-fructofuranose:Potential for use as dietary additives. *J. Sci. Food Agric.*, 65, 323-330.

[228] Smiricky-Tjardes, M. R., Flickinger, E. A., Grieshop, C. M., Bauer, L. L., Murphy, M. R., & Fahey, G. C. Jr. (2003). In vitro fermentation characteristics of selected oligosaccharides by swine fecal microflora. *J. Anim. Sci.*, 81, 2505-2514.

[229] Sahn, K. S., Maxwell, C. V., Southern, L. L., & Buchanan, D. S. (1994). Improved soybean protein sources for early weaned pigs.:Ii. Effect on ileal amino acid digestibility. *J. Anim. Sci.*, 72, 631-637.

[230] Smiricky-Tjardes, M. R., Grieshop, C. M., Flickinger, E. A., Bauer, L. L., & Fahey, G. C. Jr. (2003). Dietary galactooligosaccharides affect ileal and total0-tract nutrient digestibility, ileal and fecal bacteria concentrations, and ileal fermentative characteristics of growing pigs. *J. Anim. Sci.*, 81, 2535-2545.

[231] Sirtori, C. R., Zucchi-Dentone, C., Sirtori, M., Gatti, E., Descovich, G. C., Gaddi, A., Cattin, L., Da, Col. P. G., Senin, U., Mannarino, E., Avellone, G., Colombo, L., Fragiacomo, C., Noseda, G., & Lenzi, S. (1985). Cholesterol-lowering and HDL-raising properties of lecithinated soy proteins in tyoe II hyperlipidemia patients. *Ann. Nutr. and Metab.*, 29, 348-357.

[232] Smiricky-Tjardes, M. R., Grieshop, C. M., Albin, D. M., Wubben, J. E., Gabert, V. M., & Fahey, G. C. Jr. (2002). The influence of soy oligosaccharides on apparent and true ileal amino acid digestibilities and fecal consistency in growing pigs. *J. Anim. Sci.*, 80, 2433-2441.

[233] Smith, A. K., & Wolf, W. J. (1961). Food uses and properties of soybean protein. I. *Food Technology.*, 15(4).

[234] Schor, A., & Gagliostro, G. A. (2001). Undegradable protein supplementation to early-lactation dairy cows in grazing conditions. *J. Dairy Sci.*, 84, 1597 EOF-606 EOF.

[235] Seong-Jun, Cho, Juillerat, Marcel A., & Lee, Cherl-Ho. (2007). Cholesterol Lowering Mechanism of Soybean Protein Hydrolysate. *J. Agric. Food Chem.*, 5, 10599-10604.

[236] Sirtori, C. R., & Lovati, M. R. (2001). Soy proteins and cardiovascular disease. *Current Atherosclerosis Reports*, 3, 47-53.

[237] Soyatech. (2003). New soyfoods market study shows Americans love their soy. Available on the World Wide Web:, http://www.soyatech.com.

[238] Stephan, A., & Steinhart, H. (1999). Identification of character impact odorants of different soybean lecithins. *Journal of Agricultural and Food Chemistry*, 47, 2854-2859.

[239] Sindt, M. H., Stock, Klop, R. A., Klopfenstein T. J. & Shain D. H. (1993). Effect of protein source and grain type on finishing calf performance and ruminal metabolism. *J. Anim. Sci.*, 71, 1047 EOF-56 EOF.

[240] Selle, P. H., & Ravindran, V. (2007). Microbial phytase in poultry nutrition. A review. *Anim. Feed Sci. Technol.*, 135, 1-41.

[241] Smits, C. H. M., & Annison, G. (1996). Non-starch plant polysaccharides in broiler nutrition-towards a physiologically valid Soy oligosaccharides and pigs 2441proach to their determination. *World's Poult. Sci. J.*, 52, 204-221.

[242] Stott, J. A., Hodgson, J. E., & Chaney, J. C. (1975). Incidence of Salmonella in animal feed and the effct of pelleting on content of Enterobacteriaceae. *J Appl. Bact.*, 39, 41-46.

[243] Sands, J. S., Ragland, D., Baxter, C., Joern, B. C., Sauber, T. E., & Adeola, O. (2001). Phosphorus bioavailability, growth performance, and nutrient balance in pigs fed high available phosphorus corn and phytase. *J. Anim. Sci.*, 79, 2134-2142.

[244] Subuh, A. M. H., Moti, M. A., Fritts, C. A., & Waldroup, P. W. (2002). Use of various ratios of extruded full fat soybean meal and dehulled solvent extracted ;soybean meal in broiler diets. *Int. J. Poult. Sci.*, 1, 9-12.

[245] Southern, L. L., Ponte, J. E., Watkins, K. L., & Coombs, D. F. (1990). amino acid-supplemented raw soybean diets for finishing swine. *J. Anim. Sci.*, 68, 2387-2393.

[246] Su, S., Yeh, T., Lei, H., & Chow, N. (2000). The potential for soybean foods as a chemoprevention approach for human urinary tract cancer. *Clin. Can. Res.*, 6, 230-236.

[247] Tyus, J., II., Nahashon S. N., Adefope, N. & Wright, D. (2008). Growth Performance of Single Comb White Leghorn Chicks Fed Diets Containing Blood Meal Supplemented with Isoleucine. *J. Poult. Sci.*, 45, 31-38.

[248] Tyus, J., II., Nahashon S. N., Adefope, N. & Wright, D. (2009). Production Performance of Single Comb White Leghorn Chickens Fed Growing Diets Containing Blood Meal and Supplemental Isoleucine. *J. Poult. Sci.*, 46, 313-321.

[249] Tharkur, M., & Hurburgh, C. R. (2007). Quality of US soybean meal compared to the quality of soybean meal from other origins. *J. Am Oil Chem Soc*, 84, 835-843.

[250] Thornburn, J., & Thornburn, T. (1994). The tyrosine kinase inhibitor, geistein, prevents α-adrenergic-induced cardiac muscle cell hypertrophy by inhibiting activation of the Ras-MAP kinase signaling pathway. *Biochem Biophys Res Commun*, 202, 1586-1591.

[251] Traylor, S. L., Cromwell, G. L., Lindermann, M. D., & Knabe, D. A. (2001). Effects of level of supplemental phytase on ileal digestibility of amino acids, calcium, and phosphorus in dehulled soybean meal for growing pigs. *J. Anim. Sci.*, 79, 2634-2642.

[252] Tso, E., & Ling, S. M. (1930). Changes in the Composition of Blood in Rabbits Fed on Raw and Cooked Soybeans. 28(3).

[253] U.S. Soybean Industry: Background Statistics and Information. (2010). United States Department of Agriculture, Economic Research Service. March 28,.

[254] U.S. Food and Drug Administration. (1999). Food labeling health claims: Soybean protein and coronary heart disease. Final rule. Fed. Regist., 645, 57699-57733.

[255] Utsumi, S., Katsube, T., Ishige, T., & Takaiwa, F. (1997). Molecular design of soybean glycinins with enhanced food qualities and development of crops producing such glycinins. *Advances in Experimental Medicine and Biology*, 415, 1-15.

[256] Vucenik, I., & Shamsuddin, A. M. (2003). Cancer Inhibition by Inositol Hexaphosphate (IP6) and Inositol:From Laboratory to Clinic. *J. Nutr.*, 133, 3778S-3784S.

[257] Veltmann, Jr., Hansen, B. C., Tanksley, T. D. Jr., Knabe, D., & Linton, S. L. (1986). Comparison of the nutritive value of different heat-treated commercial soybean meals: utilization by chicks in practical type rations. *Poult. Sci.*, 65, 1561-1570.

[258] Van Kempen, T. A. T. G., Kim, I. B., Jansman, A. J. M., Verstegen, M. W. A., Hancock, J. D., Lee, D. J., Gabert, V. M., Albin, D. M., Fahey, G. C., Jr , , Grieshop, C. M., & Mahan, D. (2002). Regional and processor variation in the ileal digestible amino acid content of soybean meals measured in growing swine. *J. Anim. Sci.*, 80, 429-439.

[259] Wang, Hwa. L., Doris, I., Ruttle, , & Hesseltine, C. W. (1969). Antibacterial Compound from a Soybean Product Fermented by Rhizopus oligosporus. *Exp Biol Med*, 131(2), 579-583.

[260] Wang, H. J., & Murphy, P. A. (1994). Isoflavon content in commercial soybean foods. *J. Agric. Food Chem.*, 42, 1666-1673.

[261] Wallis, J. G., Watts, J. L., & Browse, J. (2002). Polyunsaturated fatty acid synthesis: What will they think of next? *Trends in Biochemical Sciences*, 27, 467-473.

[262] Wansink, B., & Chan, N. (2001). Relation of soy consumption to nutritional knowledge. *Journal of Medical Food,*, 4, 145-150.

[263] Weggemans, R. M., & Trautwein, E. A. (2003). Relation between soy-associated isoflavones and LDL and HDL cholesterol concentrations in humans: A meta-analysis. *European Journal of Clinical Nutrition*, 57, 940-946.

[264] Waguespack, A. M., Powell, S., Bidner, T. D., Payne, R. L., & Southern, L. L. (2009). Effect of incremental levels of L-lysine and determination of the limiting amino acids in low crude protein corm-soybean meal diets for broilers. *Poult. Sci.*, 88, 1216-1226.

[265] Williams, J. E. (1981). Salmonella in poultry feeds a world-widereview. *World's Poult. Sci. J.*, 37, 97-105.

[266] Wu, G., Liu, Z., Bryant, M. M., Roland, D. A., & Sr , . (2006). Comparison of Natuphos and Phyzyme as phytase sources for commercial layers fed corn-soy diet. *Poult. Sci.*, 85, 64-69.

[267] Wilhelms, K. W., Scanes, C. G., & Anderson, L. L. (2006). Lack of Estrogenic or Antiestrogenic Actions of Soy Isoflavones in an Avian Model: The Japanese Quail. *Poultry Science*, 85, 1885-1889.

[268] Wenzel, G. (2008). The Experience of the ZKBS for Risk Assessment of Soybean. *Journal fuer Verbraucherschutz und Lebensmittelsicherheit/Journal of Consumer Protection and Food Safety.*, 3(2), 55-59.

[269] Wei, H., Bowen, R., Cai, Q., Barnes, S., & Wang, Y. (1995). Antioxidant and antipromotional effects of the soybean isoflavone genistein. *Exp. Biol. Med.*, 208, 124-130.

[270] Wei, H., Wei, L., Frenkel, K., Bowen, R., & Barnes, S. (1993). Inhibition of tumor promoter-induced hydrogen peroxide formation in vitro and in vivo by genistein. *Nutr. Cancer*, 20, 1-12.

[271] Woyengo, T. A., Slominski, B. A., & Jones, R. O. (2010). Growth performance and nutrient utilization of broiler chickens fed diets supplemented with phytase alone or in combination with citric acid and multicarbohydrase. *Poultry Science*, 89, 2221-2229.

[272] World production of soybean. (2009). Source: FAS/USDA, United States Department of Agriculture, Foreign Agricultural Service-, http://www.fas.usda.gov/psdonline/psdReport.aspx? hidReportRetrievalName=Table+07%3a+Soybeans%3a+World+Supply+and+Distribution&hidReportRetrievalID=706&hidReportRetrievalTemplateID=8November2009.

[273] Woodworth, J. C., Tokash, M. D., Goodband, R. D., Nelssen, J. L., O'Quinn, P. R., Knabe, D. A., & Said, N. W. (2001). Apparent ileal digestability of amino acids and the digestible and metabolizable energy content of dry extruded-expelled soybean meal and its effect on growth performance. *J. Anim. Sci.*, 79, 1280-1287.

[274] World Cancer Research Fund and American Institute for Cancer Research. (1997). Food, nutritionand the prevention of cancer. A global perspective. Banta BNook Group, Menasha, WI.

[275] Xiao, C. W. (2008). Health Effects of Soy Protein and Isoflavones in Humans. *Journal of Nutrition,*, 138, 12445-12495.

[276] Xiurong Wang, X., Xiaolong Yan and Hong Liao. (2010). Genetic improvement for phosphorus efficiency in soybean: a radical approach. *Ann Bot*, 106(1), 215-222.

[277] Xu, Xia. K. S., Harris, H., Wang, P. A., & Murphy, . (1995). Bioavailability of soybean isoflavones depends upon gut microflora in women. *J. Nutr.*, 125, 2307-2315.

[278] Wang, X., Nahashon, S. N., Tromondae, K. F., Bohannon-Stewart, A., & Adefope, A. (2010). An initial map of chromosomal segmental copy number variations in the chicken. *BMC Genomics*, 11, 351, doi:10.1186/1471-2164-11-351.

[279] Yi, S. Y., Jee-Hyub, Kim., Young-Hee, Joung., Sanghyeob, Lee., Woo-Taek, Kim., Seung, Hun., Yu, , & Doil, Choi. (2004). The Pepper Transcription Factor CaPF1 Confers Pathogen and Freezing Tolerance in Arabidopsis. *Plant Physiology*, 136, 2862-2874.

[280] Barbara V., Jiangxin W., & Huang, Y. (2010). Narrowing Down the Targets: Towards Successful Genetic Engineering of Drought-Tolerant Crops. *Mol. Plant*, 3(3), 469 EOF-490 EOF.

[281] Yagasaki, K., Takagi, T., Sakai, M., & Kitamura, K. (1997). Biochemical characterization of soybean protein consisting of different subunits of glycinin. *Journal of Agricultural and Food Chemistry*, 45, 656-660.

[282] Zimmermann, M. B. (2009). *Iodine Deficiency Endocrine Reviews*, 30(4), 376-408.

[283] Zhou, J., Gogger, E. T., Tanaka, T., Guo, Y., Blackburn, G. L., & Clinton, S. K. (1999). Soybean phytochemicals inhibit the growth of transplantable human prostrate carcinoma and tumor angiogenesis in mice. *J, Nutr.*, 129, 1628-1635.

Chapter 15

The Effects of Hydrogenation on Soybean Oil

Fred A. Kummerow

Additional information is available at the end of the chapter

http://dx.doi.org/10.5772/52610

1. Introduction

Soybeans are very versatile, both as a food product and an ingredient in many industrial products. The oil produced by soybeans is contained within many foods we eat every day. Natural soybean oil contains several essential fatty acids that our body needs to work properly, including linoleic and linolenic acids. However, much of the soybean oil consumed in many parts of the world has been partially hydrogenated; that is, it's chemical composition has been changed. This hydrogenation removes the necessary essential fatty acids contained within the original oil. Some of the partially hydrogenated soybean oil has been converted to trans fatty acids.

Trans fatty acids have been shown to increase the risk of atherosclerosis and coronary heart disease due to their in vivo effects in two ways. They effect the levels of prostacyclin and thromboxane, which increases the risk of thrombosis, and they increase sphingomyelin production by the body, which then causes calcium influx into the arterial cells to increase, leading to atherosclerosis. Consumption of partially hydrogenated soybean oil can be harmful to the body.

2. Soybeans

Soybeans have many uses. When processed, a 60-pound bushel will yield around 11 pounds of crude soybean oil and 47 pounds of meal. Soybeans are about 18% oil and 38% protein. Because soybeans are high in protein, they are a major ingredient in livestock feed. Most soybeans are processed for their oil and protein for the animal feed industry. A smaller percentage is processed for human consumption and made into products including soy milk, soy protein, tofu and many retail food products. Soybeans are also used in many non-food (industrial) products [1].

© 2013 Kummerow; licensee InTech. This is a paper distributed under the terms of the Creative Commons Attribution License (http://creativecommons.org/licenses/by/3.0), which permits unrestricted use, distribution, and reproduction in any medium, provided the original work is properly cited.

Biodiesel fuel for diesel engines can be produced from soybean oil by a process called trans-esterification. Soy biodiesel is cleaner burning than petroleum-based diesel oil. Its use re-duces particle emissions, and it is non-toxic, renewable and environmentally friendly. Soy crayons are made by replacing the petroleum used in regular crayons with soy oil, making them non-toxic and safer for children. Candles made with soybean oil burn longer but with less smoke and soot. Soy ink is superior to petroleum-based inks because soy ink is not tox-ic, renewable and environmentally friendly, and it cleans up easily. Soy-based lubricants are as good as petroleum-based lubricants, but can withstand higher heat. More importantly, they are non-toxic, renewable and environmentally friendly [1]. Soy can also be used in paint and plasticizers, and used in bread, candy, doughnut mix, frozen desserts, instant milk drinks, gruel, pancake flour, pan grease extender, pie crust, and sweet goods. Non-food items made with soybeans include anti-corrosives, anti-static agents, caulking compounds, core oils, diesel fuel, disinfectants, electrical insulation, epoxies, fungicides, herbicides, printing inks, insecticides, oiled fabrics, and waterproof cement [2].

Soybean oil is normally produced by extraction with hexane. The production consists of the following steps. The soybeans are first cleaned, dried and de-hulled prior to extraction. The soybean hulls need to be removed because they absorb oil and give a lower yield. This de-hulling is done by cracking the soybeans and a mechanical separation of the hulls and cracked soybeans. Magnets are used to separate any iron from the soybeans. The soybeans are also heated to about 75° C to coagulate the soy proteins to make the oil extraction easier. To extract the oil, first the soybeans are cut into flakes, which are put in percolation extrac-tors and emerged in hexane. Counter flow is used as extraction system because it gives the highest yield. After removing the hexane, the extracted flakes only contain about 1% of soy-bean oil and are used as livestock feed, or to produce food products such as soy protein. The hexane is recovered and returned to the extraction process. The hexane free crude soybean oil is then further purified [3].

World production of soybean oil in 2010-2011 rose 8.0% to a new record high of 41.874 million metric tons. The U.S. accounts for 20.6% of world soybean oil production, while Brazil produ-ces 15.8% and the European Union accounts for 5.8%. The consumption of soybean oil rose 9.2% worldwide in 2010-2011, with the U.S. accounting for 18.6%, Brazil accounting for 12.4%, India accounting for 6.9%, and the European Union accounting for 6.4% of demand [4].

3. Uses for soybean oil

Of the total of 18 million pounds of soybean oil consumed in 2011, approximately 9 million pounds was used for cooking and salad oil. 3.75 million pounds was used for baking, and 3.6 million pounds on industrial products. The remaining 900,000 pounds is used in various other edible products. The high smoke point of soybean oil makes it often used as a frying oil. If overused, however, it causes the formation of free radicals.

Soybean oil contains 52.5% linoleic (18:2 $\Delta^{9,12}$) acid, which is also known as 18:2n^6 or omega-6. It also contains 7.5% linolenic (18:3 $\Delta^{9,12,15}$) acid also known as 18:3n^3 or omega-3. The designation 18:2 $\Delta^{9,12}$, and 18:3 $\Delta^{9,12,15}$ means that these two fatty acids have double bonds (points of unsaturation) at position 9 and 12 or 9,12 and 15 at which hydrogen can be added. In the late 1800s, a French chemist discovered that an unsaturated fatty acid can be converted to a saturated fatty acid by bubbling hydrogen through a heated vegetable oil in a closed vessel. If completely hydrogenated, they become stearic acid. The commercial use of partially hydrogenation of soybean oil began in the early 1900s. The exact fatty acid composition of the partially hydrogenated soybean oil was essentially unknown until the development of gas chromatography (GC) by James and Martin in 1952. The Food and Drug Administration, using the American Oil Chemists Society method, labeled the isomers in partially hydrogenated fat as only one peak (elaidic acid). It is only with a GC equipped with a 200 meter column that it is possible to further separate the fatty acid isomers of partially hydrogenated fat into at least 14 separate isomeric fatty acids [5].

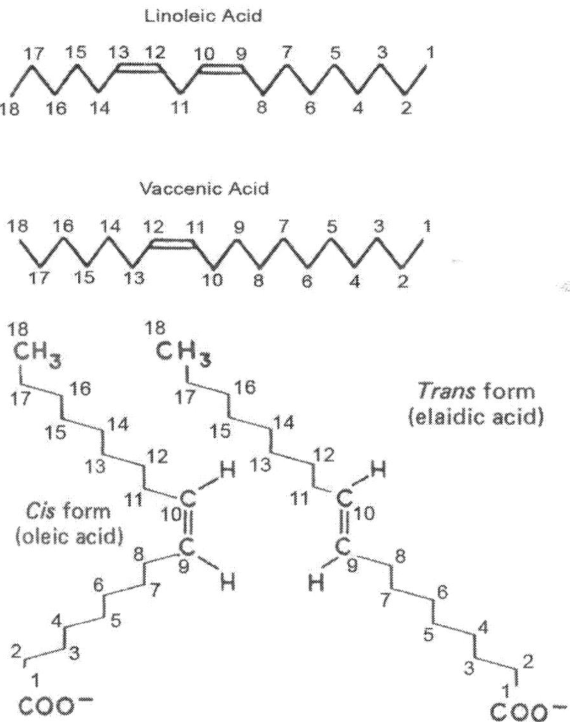

During hydrogenation, the double bond at any of these 9,12 or 9, 12, 15 positions can be shifted to form new cis and trans unsaturated fatty acid isomers not present in soybean oil.

The double bond of the cis-natural linoleic and linolenic fatty acids can also change the configuration from cis to trans, creating a geometric isomer like trans Δ^{11}-18:1 vaccenic acid in butter fat. Oleic acid, the largest percentage of the natural fatty acid in the human body, is cis Δ^9-18:1 (the number after delta indicates the position of the double bond at the 18 carbon atom chain counting from the carboxyl group).

Oleic acid goes through geometrical isomerisation during hydrogenation to trans Δ^9-18:1 acid known as elaidic acid; thus the "natural" oleic acid is turned into elaidic acid during the hydrogenation process, and becomes an "unnatural" fatty acid. It twists into a new form and can be both a cis and/or a trans fatty acid. In addition to geometrical isomerisation, the double bond of either cis or trans fatty acids can theoretically migrate along the 18 carbon chain of either oleic, linolenic, and linoleic acid, changing their position from Δ^9, $\Delta^{9,12,15}$, or $\Delta^{9,12}$, creating five monoene cis positional isomers, 6 trans monoene isomers and 3 trans diene positional isomers. Thus hydrogenated soybean oil contains 24.1% trans monoenes, 6.2% trans dienes and 9.4% cis monoene isomers or a total of 39.7% isomeric fatty acids. They were identified as cis and trans octadecenoic and octadecadienoic isomers on a GC equipped with a 200 meter column and by their mixed melting points with authentic octadecenoic and octadecadienoic acids. None of these fatty acids are present in natural soybean oil. The 14 isomers in hydrogenated fat can be used as a source of energy but they cannot substitute for EFA because they do not have the required double bond structure [5].

4. Nutrition

It was unknown until 1930 that linoleic (18:2 n^6) and linolenic (18:3 n^3) acids were essential fatty acids (EFA), and like the nine essential amino acids and the vitamins, cannot be synthesized in the human body; they must come from a diet that includes natural fats and oils. In one study, pregnant rats were fed linoleic, linolenic, and arachidonic acids by dropper. This was a sufficient amount for the mother rats to wean their young, but those pups from mothers fed only linolenic acid died before weaning. Although linolenic acid is considered an essential fatty acid, these data indicate that it may not be an essential fatty acid [6].

An increase in the sales of soy food is largely credited to the Food and Drug Administration's approval of soy as a cholesterol-lowering food [7]. A 2001 literature review argued that these health benefits were poorly supported by available evidence, and noted that data on soy's effect on cognitive function of the elderly existed [8].

The FDA issued the following claim for soy: "25 grams of soy protein a day, as part of a diet low in saturated fat and cholesterol, may reduce the risk of heart disease." [9]. Solae also submitted a petition on the grounds that soy can help prevent cancer. On February 18, 2008, Weston A. Price Foundation submitted a petition for the removal of this health claim. 25 g/day soy protein was established as the threshold intake because most trials used at least this much protein and not because less than this amount is inefficacious [10]. An American Heart Association review of a study of the benefits of soy protein casts doubt on the FDA claim for soy protein. However, AHA concludes "many soy products should be beneficial to cardiovascular and overall health because of their high content of polyunsaturated fats, fiber, vitamins, and minerals and low content of saturated fat" [11].

EFA are required to synthesize the eicosanoids that are needed to regulate blood flow in the arteries and veins. Linoleic acid (n-6) is synthesized into arachidonic acid, and linolenic acid (n-3) is synthesized into eicosapentaenoic acid. Both in turn are made into prostacyclin or thromboxane. Prostacyclins are synthesized in the endothelial cells that line the blood vessel wall. Thromboxanes are synthesized in the platelets in the blood. The balance between prostacyclin for flow and thromboxane for clotting is a very delicate one and can be changed by different diets and different drug prescriptions. Fish have already converted the linolenic acid they get from seaweed into eicosapentaenoic acid. Hence, fish oil is often recommended as a dietary supplement. Prostacyclin and thromboxane can be made from linoleic acid as well. The least expensive source of omega-3 and omega-6 is soybean oil, which is sold as vegetable oil in a supermarket [12].

However, this vegetable oil is stripped of Vitamin E, which is then sold in capsules. The removal of Vitamin E leaves the oil more susceptible to oxidation, which harms the natural fatty acids that are needed for good health.

How soybean oil is used in modern humans was developed in prehistoric humans to assure their survival. There must have been long periods of time between meals, that is fasting periods, and there were times in which they had food available, the "fed" period. During this fed period, carbohydrates were used within two hours as a quick source of energy. Extra carbohydrates were stored first as glycogen in the muscles and liver and then any excess converted to fat and stored in the adipose tissues (the fat around your middle and elsewhere). This stored fat was then available for energy during the long fasting periods. Modern humans have inherited this way of handling these fed and fasting periods. This process assured the survival of prehistoric humans but has now become one way that obesity is developing in humans today. Too much food is available all hours of the day and night, and eating it is a pleasure.

To avoid adding fat to your body, any carbohydrates you eat should be used up as a calorie source before the next meal. Any carbohydrates that have already turned into fat and any fat in your diet itself should be used for energy within the cell during the fasting period. Eating a snack between meals means adding additional carbohydrates into the system before any of the fat from the previous meal has been used for energy. It ends up adding to your adipose tissue. If you weighed yourself before a hearty meal and again the next day, you may find you have gained a pound or two, the amount depending on how much food you ate and the fat you stored. As such a meal may also contain excess salt, some of the weight gain can be due to excess water you stored. Millions of dollars are spent to try to get rid of this stored fat, and the government is planning to spend millions more dollars to solve the obesity problem. Prehistoric humans had no choice in controlling the time between fasting and fed periods because they had no refrigerators, fast food outlets, or supermarkets to run to. Modern humans do have this choice. More time between the fed periods, that is between meals, may help with the obesity problem [12].

The fat in the intestinal tract is first converted into tiny droplets of fat (chylomicrons) by the intestinal cells. The intestinal tract is not just a through highway, but is actively involved in the process of metabolizing fat so that the body can use it. The chylomicrons diffuse from the intestinal tract into the lymph system and into the veins through the thoracic duct and end up in the blood. The blood, during the fed period, carries these chylomicrons for deposit where they are resynthesized into adipose tissue and stored fat around the stomach, hips, and other locations. The fat (triglycerides) in adipose tissue is "mobilized" when the glycogen in the muscle and liver has been reduced.

The glycerin portion goes to the liver. The free fatty acids take a different route and are combined with a protein named albumin. Therefore, there must be enough albumin in the blood to carry the free fatty acids in the blood. This fatty acid albumin complex is water-soluble enough in the blood to be carried to cells of all kinds that use the fatty acid portion as an energy source. Any excess fatty acid goes to the liver and is remade into triglycerides. The cellular organelle (the endoplasmic reticulum) in the liver cells participates in coating the very small triglyceride droplets with protein and adds phospholipid and cholesterol to produce very low density lipoprotein (VLDL), which furnishes the fatty acid for the approximately 50 thousand trillion cells in the body [12].

Correction of the inhibition of lipoprotein lipase by protein binding of free fatty acids permits normal protein transport of FFA into the cellular mitochondrial oxidative phosphorylative cycle with the resultant production of high-energy phosphate which is the cellular fuel. Without this fuel, in addition to oxygen, the life process comes to a halt. Bacteria have used this method of providing energy for at least two billion years (Ratz).

5. Fried foods

Another issue with fats is the preparation of foods by frying them in fat. There are problems with deep fat fried food that affect our nutrition. These problems occur because of chemical alterations in the fat that happen as a consequence of deep fat frying food. This frying process is as follows:

1.   Food picks up oxygen from the air during frying that negatively alters the fat composition.

2.   The foods fried in these fats pick up those altered fats.

3.   These altered foods have a direct, negative influence on the nutritional value of the fat.

The changes in the fat are dependent on at least four factors:

1.   The length of time it was exposed to heat—in commercial operations, the length of time a food is fried leads to how much fat is absorbed on the cooked food item;

2.   The temperature of the fat;

3.   The exact composition of the fat used, such as corn oil, cottonseed oil, soybean oil, beef tallow, or hydrogenated fat, and

4.   What is being fried, e.g., chicken or fish.

Feeding the fats fried at varying lengths of time led to very different outcomes in the nutrition of animals. Those fed the fats fried the shortest period of time were healthier than those fed the fats fried for the longest times. Those fed fats heated at higher temperatures were not as healthy as those fed on fat heated to lower temperatures. It was interesting also that animals fed on heated margarine did not grow as well as those on fresh margarine and that their plasma cholesterol level increased. Those fed on heated butter oil grew as well as those on fresh butter oil.

Oil from commercial fat fryers was used in a set of experiments that clearly showed that poor nutrition resulted. This is important because used fat from commercial operations is typically collected and fed to animals, such as pigs, to provide energy for rapid growth. When we conducted experiments feeding the commercially used fat for frying to rats, they did not do well. When we added protein to their diets, the effect of the "bad" heated fat was countered because the added protein provided more adequate nutrition. We tried to fortify the diets with adequate vitamins, but that could not counter the growth-depressing effect of the heated oil. A few vitamins, such as riboflavin, helped a bit.

Fish contain high amounts of polyunsaturated fat that are not present in the fat of chicken or beef. Thus, when fish are fried, the polyunsaturated fat in them can leak into the frying fat, causing the fat to be changed more radically into a less healthy version. Chicken and hamburger have less of this polyunsaturated fat and thus are healthier choices to fry.

Eating excessive amounts of fried food also slows down digestion. People may get stomachaches as a result. As early as 1946, a link that heated fats may lead to cancer was shown. What we don't know yet is whether heated fats by themselves lead to cancer or whether the heated fat combined with specific foods cause cancer. Animals fed heated fat combined with a known carcinogen developed cancer, whereas those fed fresh fat combined with a known carcinogen did not. Thus the heated fat was a co-carcinogen.

Commercial frying of food has increased worldwide since our studies on heated fats. In Germany, fat fryers are required by law to test their frying fat for its freshness by a method approved by the German government. In the U.S. a test is also available, but its use is not mandatory [12].

6. Free radicals

Free radicals are produced from oxidized linoleic (n-6) and linolenic acid (n-3); they are fragments of unsaturated fatty acids. This is especially likely to happen when the essential fatty acids are heated, especially the n-3 variety. All oils change structures when they are heated, but hose high in n-3 fatty acids have more problems than those high in n-6. Free radicals provide another reason to avoid fried food. The first sign of fats becoming free radicals is that they are rancid, and they begin to smell "off" and their taste becomes bitter. Roasted peanuts, for example, can become rancid and then shouldn't be eaten.

Free radicals are "bad" since they destroy vitamins A, D, C, and E, thus preventing these vitamins from doing positive things in the body. Free radicals also destroy both the essential fatty acids and the essential amino acids. They oxidize the LDL into something called oxidized low density lipoproteins (oxLDL). These oxLDL are very powerful components in the blood that have been considered since about 1990 as involved in the development of heart disease [12].

Essential fatty acids do more than regulate the blood; they are also a key to reproduction. Since the 1930's, we've known that reproduction always fails on fat-free diets. In studies on rats, reproduction continues under low fat conditions because the rats have enough linoleic acid stored in their bodies. They manufacture arachidonic acid from the linoleic acid in their own fat, so they can reproduce healthy young even after a fat-free diet. If the rats did not have enough linoleic acid stored in their bodies (such as rats born to mothers on fat-free diets), we found they could not make enough of the arachidonic acid needed for healthy reproduction, and their young die. Women need the essential fatty acids for reproduction. The easiest way to supply them is from plant oils [5].

Data from ADM shows the composition of three different hydrogenated fats, based on a serving size of 14 grams. The first two were made of enzymatically interesterified soybean oil, and contained 0 grams of trans fat per serving. The third was made of partially hydrogenated soybean and/or cottonseed oil, and contained 4.5 grams of trans fat per serving. The take away message is that due to effective food industry lobbying, food labeling rules allow foods with up to half a gram of trans fat per serving to be labeled "0 trans fat". So look for "partially hydrogenated vegetable oil" on the label.

Several researchers have documented the effects of foods without trans fat and their positive effects on lowering CHD. Mozaffarian et al. showed that n-3 PUFAs from both seafood and plant sources may reduce CHD risk, with little apparent influence from background n-6 PUFA intake. They found lower death rates among those with high seafood and plant-based diets. Plant-based n-3 PUFAs may particularly reduce CHD risk when seafood-based n-3 PUFA intake was low, which has implications for populations with low consumption or availability of fatty fish. Kris-Etherton et al. found that nuts and peanuts routinely incorporated in a healthy diet with a composite of numerous cardioprotective nutrients reduced the risk of CHD. They also suggested that higher intake of trans fat could adversely affect endothelial function, which might partially explain why the positive relationship between trans fat and cardiovascular risk is greater than one would predict based solely on its adverse effects of plasma lipids [12].

7. Two mechanisms involved in coronary heart disease

Two mechanisms may be involved in CHD: One, the oxidation of the fatty acids and cholesterol in LDL leading to a change in sphingomyelin concentration in the arteries, which is a process that occurs over a life time; two, the deposition of trans fat in the cardiovascular system. Trans fat calcifies both the arteries and veins and causes blood clots. Trans fat leads to the reduction of prostacyclin that is needed to prevent blood clots in the coronary arteries. A blood clot in any of the coronary arteries can result in sudden death.

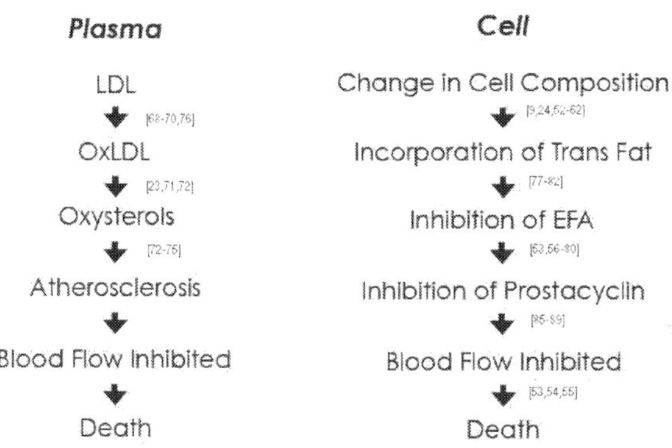

8. Mechanism one

When sufficient biological antioxidants are not present in the plasma, the LDL is oxidized to oxLDL and cholesterol is oxidized to oxysterol. Oxysterols incorporated into the endothelial layer of the arteries and veins can change the phospholipid cell membrane composition so that more sphingomyelin incorporates into the membrane which becomes "leaky" to calcium infiltration. Oxysterols were present at higher concentrations in the plasma of patients who had coronary artery bypass grafting (CABG) surgery. These patients had 40 times more calcium in their bypassed veins than normal veins in the same patient. When purchased oxysterols were added to plasma from patients who did not need CABG surgery, endothelial cells cultured in their blood and tested with radioactive calcium the incorporation of radioactive calcium did not differ from that of plasma from CABG patients. This indicates that oxysterols stimulated calcification. When endothelial cells were cultured with oxysterols in a standard culture media, the cells became calcified in a similar way to those of the CABG patient. The oxidation of cholesterol and deposition of calcium is the primary cause for the development of atherosclerosis in the arteries and veins.

In a review article entitled "The pathogenesis of atherosclerosis: Perspectives for the 1990s" Ross stated "Atherosclerosis of the extremities is most apparent at branching points of the arterial tree where blood flow is irregular with current and back currents. The cellular events that occur during the progression of lesions in hypercholesterolemic animals are almost exactly mirrored by those observed in human atherosclerotic coronary arteries in hearts removed in transplant operations" [13]. De Bakey et al. have noted similar atherosclerosis (thickening) at branching and bifurcation during coronary artery bypass grafting (CABG) surgery [14].

Keaney stated that the gene expression pattern in the arterial wall is subject to influence by modified forms of LDL [15], which altered both scavenger reception (CD36) expression and the expression of pro-inflammatory genes [16]. The disturbed laminar flow pattern of fluids occurs near branch points [17], bifurcations, at major curves and at arterial geometries [18] that are typically associated with the earliest appearance (and subsequent progression) of atherosclerotic lesions [19]. An endothelial receptor for oxLDL, a designated lectin-like oxLDL receptor (LOX-1) [20], was identified [21]. The transient application of shear stress showed that the initial stimulation of shear stress was sufficient for induced expression of LOX-1 and that sustained application of shear stress was not required [22]. The over-expression of LOX-1 receptors at the bifurcation and the higher level of modified LDL and oxysterols in the plasma of persons needing CABG surgery could lead to a higher uptake of modified LDL, resulting in a greater delivery of oxysterols to the endothelial cells at the bifurcations. The levels of sphingomyelin in plasma have been shown to be higher in patients with coronary heart disease and those with left ventricular dysfunction [23]. Furthermore, it was found that sphingomyelin levels in the blood correlate with and can be used to accurately predict coronary artery disease [24]. Sphingomyelin has long been known to accumulate in atheromas of both humans and animals, and contributes to the formation of atherosclerosis [25].

Thickening [26] was noted in the branching arteries in aging porcine on a non-cholesterol diet. It did not differ significantly in sphingomyelin composition from that of the non branching adjacent tissue of porcine at 6 months of age. By 18 and 48 months of age, however, the sphingomyelin content was significantly higher at the thickened branching areas than at the non thickened segment of the arteries. This indicated that during aging of the arteries, there was a striking increase in the amount of sphingomyelin in the membrane of the cells at the branching points of arteries [26]. Lipid extracted from both porcine and human arteries indicated that aging is a factor that increased sphingomyelin. There was more sphingomyelin in the aging arteries of both porcine and human arteries.

The non branching segment of the aorta obtained, on autopsy, from six men 21-27 years of age contained four times more sphingomyelin than in arteries isolated from human umbilical cords, indicating that the sphingomyelin content of arteries increases with age. Aging is not the only factor that increased the sphingomyelin composition of arterial cells. Women and men under 40 years of age who had been subjected to CABG surgery contained the same high percentage of sphingomyelin in their non atheromatous arterial cells as those over 40 years of age. Therefore, heart disease itself seemed to have caused an increase in non atheromatous arterial cells in sphingomyelin composition prematurely in CABG patients, pointing to a fundamental disturbance in phospholipid metabolism in their arterial cells.

The phospholipid composition of a normal arterial cell has less sphingomyelin, and this amount increases until half the artery is sphingomyelin. That is, the more sphingomyelin was in the arterial cells, the more Ca^{2+} was identified. This is because the hydroxyl group and amide group of sphingomyelin act as both donors and acceptors of hydrogen bonds [27]. Furthermore, Lehninger found that sphingomyelin's long, 18-to-26 carbon atoms chain fatty acids altered the positioning of other phospholipids. Dipalmitoylphosphatidylcholine has no amide bond [28]. As both sphingomyelin and dipalmitoylphosphatidylcholine are largely on the extracellular side of the membrane [29,30], such bilayer asymmetry would enhance binding. These in vitro results showed that sphingomyelin-Ca^{2+} binding goes beyond an isolated individual membrane binding Ca^{2+}, to lattice type matrix binding among adjacent membranes [31]. These results in vitro were simulated in vivo Ca^{2+} deposition (calcification) in arteries and veins.

9. The in vivo effect of sphingomyelin on the composition of the vascular membrane

Patients who had CABG surgery sometimes needed a second CABG surgery because the vein used in the first surgery had been occluded. During this second surgery, an unoccluded vein from the same patient was used to replace the occluded vein. The occluded veins contained, on average, significantly more sphingomyelin and Ca^{2+} than the unoccluded veins [32]. The unoccluded veins contained 24% sphingomyelin and 182 ppm of Ca^{2+} as compared to 48% of sphingomyelin and 6,345 ppm of Ca^{2+} in the occluded veins that had been used in the first CABG surgery. The increased sphingomyelin and Ca^{2+} concentrations in the occluded veins were responsible for the initial formation of atherosclerosis in these patients.

10. Oxysterols increased sphingomyelin and Ca^{2+} deposition in patients with CABG surgery

Ridgway found that 25-hydroxycholesterol stimulated sphingomyelin synthesis in Chinese hamster ovary cells [33]. Similarly in humans, an oxysterol increased sphingomyelin synthesis during the development of atherosclerosis. A significant increase in the concentrations of oxysterols, phospholipids, and Ca^{2+} were noted in patients who had CABG surgery [26, 32]. Patients who had cardiovascular disease had increased oxysterol levels in their plasma compared with the controls; that is, by comparison to cardiac catheterized patients with no stenosis [32]. The plasma from CABG patients had a higher concentration of oxysterols than was present in the controls. Human endothelial cells were cultured for 72 hours in a medium containing plasma obtained from CABG patients, or from controls patients with addition of 5 types of oxysterols (7-keto-cholesterol, cholestane-3β, 5α, 6β-triol, 7β-hydroxycholesterol, β-epoxy cholesterol, and 7α-hydroxycholesterol). These added oxysterols increased the total oxysterol level in the controls equivalent to that in the CABG plasma.

Phospholipid	Human		Porcine	
(%)	younger	older	3 weeks	2 years
Phosphatidylcholine	34.1	19.2	44.74	33.91
Phosphatidylethanolamine	8.8	2.4	25.18	24.76
Sphingomyelin	44.8	68.8	16.06	23.72
Phosphatidylinositol	+			
Phosphatidylserine	5.0	1.6	11.35	14.55
Phosphatidic acid	1.0	0.6		
Lysolecithin	3.9	8.0	trace	1.28

Table 1. Data from Kummerow F.A.. 1987. Factors which may alter the assembly of biomembranes so as to influence their structure or function *In* Membrane Biogenesis. Op den Kamp J. A. F., editor. Springer-Verlag. 95.Phospholipid composition of human and porcine arterial tissues

Oxysterols stimulated sphingomyelin synthesis and inhibited sphingomyelin metabolism [34, 23, 24]. When radioactive Ca^{2+} ($^{45}Ca^{2+}$) influx was measured, significantly higher influx of $^{45}Ca^{2+}$ was noted in the endothelial cells cultured with added oxysterols indicating that oxysterols increased Ca^{2+} influx into endothelial cells [34]. By using a radiolabeled choline, the time- and dose-dependent effects of 27-hydroxycholesterol on sphingomyelin synthesis could be observed. The increased radioactivity in sphingomyelin, which was accompanied by decreased radioactivity in phosphatidylcholine in 27-hydroxycholesterol-treated cells, was higher than that in control cells. This result indicated that 27-hydroxycholesterolincreasedthetransferofcholinefromphosphatidylcholinetosphingomyelin. An interesting finding was that the increased radioactivity in sphingomyelin by 27-hydroxycholesterol was detected first, followed by enhanced Ca^{2+} uptake and the accumulation of cytosolic free Ca^{2+}. Moreover, decreased activities of neutral and acid sphingomyelinase, which hydrolyze sphingomyelin, were also detected in 27-hydroxycholesterol treated cells [35]. Therefore, the cause for calcification was related to the structure and location of sphingomyelin in the cell membrane.

11. The concentration of cholesterol and lipid oxidation products in the plasma of cardiac catheterized patients was also determined [36]

The concentration of cholesterol, lipid oxidation products and total antioxidant capacity in the plasma of 2000 cardiac catheterized patients with 0, 10–69 and 70–100% stenosis of their arteries were analyzed. The results showed that lipid oxidation products increased with the severity of stenosis, they were 2.92 mmol/L at 0% stenosis, 3.19 mmol/L at 10–69% stenosis and 3.48 mmol/l at 70–100% stenosis. The total antioxidant capacity decreased with the severity of stenosis. The plasma cholesterol concentration, however, was not significantly different between these groups of patients. It was 201.9 mg/dL at 0% stenosis, 203.2 mg/dL at

10–69% stenosis and 207.5 mg/dL at 70–100% stenosis. Therefore, the concentration of oxidation products, rather than the concentration of cholesterol in the plasma, increased with the severity of atherosclerosis [36]. In all age groups, all of the women and men with cardiovascular atherosclerosis also had increased individual and total oxysterol levels in their plasma as compared with the controls.

The *in vivo* oxidation was enhanced by sphingomyelin. The oxidation could come from the consumption of too many polyunsaturated fatty acids in soybean oil [32, 36]. Polyunsaturated fats in vegetable oil could provide more oxidized LDL and more oxidized sterols into the plasma, which would increase the possibility of atherosclerosis. Sphingomyelin accumulates in the arterial system of humans and animals, and these increased levels mean an increased likelihood of atherosclerosis formation.

12. Mechanism two

Trans fatty acids are available on every continent. There are at least six hydrogenation plants in the United States alone; there is one in Texas, four in Illinois, and one in New Jersey. The FDA has estimated that daily intake of trans fatty acids in northern Europe to be at around 4.5g-17g/capita, and 1.34-4.9 in southern Europe. In India, 2.7-4.8g/capita/day was estimated, and only 2.7-4.8g/day in Australia and New Zealand. The least amount of trans fatty acids is consumed in Hong Kong, Japan, Korea, and China at 1.5-3g/capita/day. A large hydrogenation plant is located in a suburb of Tokyo that uses both fish and vegetable oils, as well as one in Beijing. These trans fatty acid-filled oils are liquid at room temperature, and similar to olive oil that has been used for centuries in southern Europe as an important source of fat in the diet. Butter, lard and beef tallow are saturated fats that have been used for centuries as a fat source in the diet in northern Europe [37].

The second mechanism that may be involved in CHD is trans fat. Trans fat calcifies both the arteries and veins and causes blood clots. Trans fat inhibits COX-2, an enzyme that converts arachidonic acid to prostacyclin that is needed to prevent blood clots in the coronary arteries. A blood clot in any of the coronary arteries can result in sudden death. The American Heart Association has stated that 42% of victims of a sudden heart attack do not reach a hospital still alive.

A study in 2004, with piglets from mothers fed hydrogenated soybean oil showed that their arteries contained less linoleic acid converted to arachidonic acid than the arteries of piglets from mothers fed butterfat or corn oil. This indicated that the trans fat in hydrogenated soybean oil inhibited the metabolic conversion of linoleic to arachidonic acid. Furthermore, an analysis of the fat embedded in the arteries of the piglets from mothers fed partially hydrogenated soybean oil showed that they contained 3% trans fat incorporated into their phospholipids by 48 days of age [38].

If a mother is breast-feeding her child and also eating foods containing trans fat, she would have a substantial amount of trans fat in her milk supply and pass those to her infant. Pregnant porcine fed hydrogenated fat contained 11.3% trans fat in their milk at the birth of their piglets, which decreased during lactation to 4% in 21 days. The plasma of the piglets increased from 5% trans fat three days after birth to 15.3% at six weeks of age. Transferring this result to humans, a human mother would also transfer the trans fat in her milk supply to her infant. The infant would incorporate the trans fat into his/her arterial cells inhibiting arachidonic acid synthesis and prostacyclin secretion [4].

Furthermore, calcium deposition into the endothelial cells could be enhanced. To date, the FDA has not considered the daily intake of trans fat relevant to the health of small children since they do not exhibit overt heart disease. In cases where children have died of unknown causes and had been autopsied, 99% of them showed the beginning stages of hardening (calcifications) of the arteries, which ultimately can lead to heart disease [39].

13. The effects of trans fatty acids on calcium influx into human arterial epithelial cells

The influence of trans fatty acids and magnesium on cell membrane composition and on calcium influx into arterial cells. The percentage of fatty acids incorporated into the endothelial cells was proportional to the amount added to the culture medium. Adequate magnesium was crucial in preventing calcium influx into endothelial cells. Without an adequate amount of magnesium in the culture medium, linoelaidic and elaidic acids, even at low concentrations, increased the incorporation of $^{45}Ca^{2+}$ into the cells, whereas stearic acid and oleic acid did not. A diet inadequate in magnesium combines with trans fat may increase the risk of calcification of endothelial cells [40].

Vaccenic acid in butter did not inhibit the metabolic conversion of linoleic to arachidonic acid. Epidemiological studies of intake of ruminant trans fat and risk of coronary heart disease (CHD) indicated that the intake of ruminant trans fatty acid was innocuous or even protective against CHD. Thus a study with an animal model has shown that trans-fat decreased synthesis of arachidonic acid from linoleic acid. This study was carried a step further with endothelial cells in the first layer of the artery. They were cultured in a medium that contained the fatty acids of soybean oil or in a medium that contained the fatty acids of hydrogenated soybean oil. The latter cells contained trans-fat in their membrane phospholipid and significantly less arachidonic acid and secreted less prostacyclin than endothelial cells that had been cultured with the fatty acids from unhydrogenated soybean oil [5].

We found that in the cells cultured with trans fat, the free arachidonic acid released by phospholipase action was shunted to metabolism by another pathway leaving less free arachidonic acid available as substrate for prostacyclin synthesis. Cyclooxygenase (COX) is the enzyme that is necessary to make prostacyclin to keep the blood flowing, thus lowering the potential for a heart attack. Vane et al. have shown that COX is the enzyme that converts arachidonic acid to prostaglandin H_2, is further metabolized to prostanoids. Vane et. al. stated two isoforms of COX existed, a constitutive (COX-1) and an inducible (COX-2) enzyme. COX-2 may be the enzyme that recognizes the isomers produced during hydrogenation as a foreign substrate and reacts to them by causing inflammation and reduction of prostacyclin. COX-2 is the inducible isoform of COX. COX-1 is present constitutively while COX-2 is expressed primarily after the inflammatory insult [41].

The ability to form prostacyclin from arachidonic acid was assayed using a radioimmunoassay kit. Trans-fat depressed the synthesis of prostacyclin. The addition of an excess amount of linoleic acid to this hydrogenated soybean oil fatty acids did not increase the secretion of prostacyclin in endothelial cells. The concentration of trans fatty acid rather than the concentration of linoleic acid was therefore responsible for regulating the synthesis and secretion of prostacyclin in endothelial cells. The trans fat in hydrogenated fat not only depressed the synthesis of prostacyclin that regulated the clotting of blood but also, could not serve as precursors for prostacyclin synthesis. The trans fat "incorporated" into the membrane lipids of blood vessels and muscle tissues and displaced the essential linoleic, linolenic and arachidonic acids.

In another study, rats were fed either corn oil, butter, hydrogenated vegetable oil, or coating fat for 10 weeks at 10g/100g diet. In the group fed coating fat, arachidonic acid was found to be significantly lower in the phospholipid fatty acid content of the platelets, aorta, and heart. The ratio of 20:3(n-9)/20:4(n-6) was greater than in the groups fed corn oil, butter, or hydrogenated vegetable oil, indicating that the group fed coating fat was essential fatty acid deficient. The composition of coating fat was 33% trans fat and only 0.3% linoleic acid, whereas hydrogenated oil was made up of 18% trans fat and 32.8% linoleic acid. It was then concluded that the consumption of hydrogenated fats high in trans 18:1 acids with adequate amount of linoleic acid had no effect on the amount of thromboxane or prostacyclin by platelet or aorta in vitro. The coating fat is dangerous because of its lack of linoleic acid [42].

To demonstrate the process of calcification, endothelial cells cultured with/without trans fat showed that trans fatty acid calcify arterial cells. One with a trans fatty acid added as the "unnatural" elaidic acid (t18:1 n^9) and the other with a cis fatty acid added as the "natural" oleic acid (cis 18:1 n^9) and testing with radioactive calcium. More radioactive calcium infiltration occurred into the endothelial cells cultured with elaidic acid than with oleic acid. An autopsy of 24 human specimens showed that human subjects that had died of heart disease contained up to 12.2% trans fat in their adipose tissue, 14.4% in liver, 9.3% in heart tissue, and 8.8% in aortic tissue and in atheroma.

14. The trans fatty acids in partially hydrogenated fat can cause blood clots

Partially hydrogenated soybean oil contained 14 cis and trans isomers that were formed during hydrogenation [4, 5]. They inhibited cyclooxygenase, an enzyme required for the conversion of arachidonic acid to prostacyclin, a molecule which prevents blood clots [43]. Moreover, oxidized fat enhanced thromboxane synthesis [44, 45], which caused the formation of a blood clot. Trans fatty acids in partially hydrogenated vegetable oil decreased prostacyclin synthesis by inhibiting cyclooxygenase. Oxysterols enhanced thromboxane synthesis [44, 45]. Both prostacyclin and thromboxane are involved in sudden cardiac death.

According to WebMD, "sudden cardiac death (SCD) is a sudden, unexpected death caused by loss of heart function (sudden cardiac death). It is the largest cause of natural death in the U.S., causing about 325,000 adult deaths in the United States each year. SCD is responsible for half of all heart disease deaths. SCD occurs most frequently in adults in their mid-30s to mid-40s, and affects men twice as often as it does women." [46]

Under the current Food and Drug Administration mandate [47], food items with any amount of trans fatty acids are allowed, as long as they are labeled. Products containing less than 0.5g/serving can be labeled as "trans free" or 0%. This is misleading, because it is easy to circumvent this rule by making the serving size listed on a label small enough to meet the 0.5g threshold. The food industry has taken advantage of this rule by making the serving sizes small enough to contain less than 0.5g/serving of trans fat. Fifteen foods labeled "trans fat free" were analyzed for fat content. Two contained 0% trans fatty acid, two contained higher than 0.5g/serving and the rest contained between 0.014 to 0.25g/serving. If the serving size is increased, foods would contain more than 0.5g of trans fatty acids. In 2003, the daily intake of trans fatty acids for men was estimated by the Food and Drug Administration to be nearly 7 grams per day, and almost 5 grams per day for women [47]. It is possible for people to eat the same amount of trans fatty acids today as in earlier periods, even though they have supposedly been removed from the food supply. A recent article in JAMA, "Levels of Plasma trans-fatty acids in Non-Hispanic White Adults in the United States in 2000 and 2009" listed levels in the year 2000 at 38.0, and in 2009 as 14.0µ/ml, which was considered significant [48].

15. Environmental impact of soybean use

Epidemiological data collected by the Center for Disease Control (CDC) further illustrate the potential harmful effects of trans fat. These data showed that, death from CHD in the USA increased from 265.4/100,000 in 1900 to 581/100,000 population by 1950. During this time period, both margarine and shortening had a high percentage of trans fat (ranging from 39-50%) and a low percentage of linoleic acid (ranging from 6-11%) according to the technical director of the Institute of Shortening and Edible Oils. In 1968 Dr. Campbell Moses, medical director of the AHA, appointed a five member subcommittee on fats of the AHA nutrition committee to revise the 1961 version of "Diet and Heart Disease." At the time it was known that an increase in EFA composition of a dietary fat would lower plasma cholesterol levels and there was strong evidence that trans fatty acids increased plasma cholesterol levels. The first revised version by the AHA committee stated:

"Partial hydrogenation of polyunsaturated fats results in the formation of trans forms which are less effective than cis, cis forms in lowering cholesterol concentrations. It should be noted that many currently available shortenings and margarines are partially hydrogenated and many contain little polyunsaturated fat of the natural cis, cis form." The members of the

Institute of Shortening and Edible Oils Inc objected to this version. The second revised and distributed version, omitted references to hydrogenated fat and cis fatty acids stated: "Margarines that are high in polyunsaturates usually can be identified by the listings of a liquid oil first among the ingredients. Margarines and shortenings that are heavily hydrogenated or contain coconut oil, which is quite saturated, are ineffective in lowering the serum cholesterol." The industry agreed to lower the trans fatty acids and increase the level of EFA in shortenings and margarine. Dr. R.I. Levy, director of the National Heart, Lung, and Blood Institute at the time, believed 1968 a watershed, as the incidence of CHD has steadily declined in the US since 1968. Why it decreased remained unknown in 1968.

On October 24th, 1978, ten years after the reformulation of hydrogenated fat, the National Institute of Health (NIH) held a conference in Bethesda, Maryland, on the Decline in CHD Mortality. A recent editorial in Circulation cited this symposium. Three major conclusions reached were;

1. The decrease in CHD mortality was real and not a result of artifacts or changes in death certificate coding,

2. Both primary prevention through changes in risk factor fundamentals and clinical research leading to better medical care probably have contributed to but did not fully explain the decline, and

3. A precise quantification of the causes requires further studies.

" In hindsight, the reformulation of hydrogenated fat with its lowering of the trans fatty acids and raising of linoleic acid could have also been responsible for the decline. The per capita consumption of hydrogenated fat continued to increase after 1950. However, the increase in the linoleic acid content in the reformatted 1968 fat and the increasing use of soybean oil in salad dressing and other food items could have helped to keep a decreasing death rate from CHD. The death rate from heart disease dropped substantially during the next decades even though the consumption of hydrogenated fat kept increasing and animal fat was decreasing. Lower trans fat and increased linoleic acid are possible explanations for this change.

The death rate from CHD declined after 1968 from 588.8/100,000 to 217/100,000 in 2004 in the USA. According to AHA data, 451,300 Americans died of CHD in 2004. Heart disease is still the number one cause of death. However, in a population of approximately 300 million, today the deaths would have been 1,480,000 at the 1950 rate according to the National Institute of Health (NIH). A recent study based on the autopsy of young men showed the CHD rate has been increasing since 2004. The recent reformulation of hydrogenated fat raises the trans fatty acid levels from 20% to almost 40%.

In 2003, the metabolism of the trans fat in hydrogenated oil was assumed to follow the same pathway as the natural ruminant trans fat in butterfat. The Food and Drug Administration has stated that the main reason for the trans fat in partially hydrogenated oil to remain in the diet in the USA rested on the generally held belief that trans fat is metabolized the same way as the natural trans (vaccenic acid) in butterfat. The FDA allowed the isomeric fatty acids in hydrogenated vegetable oils to remain in food products because they assumed that some of that trans fat may be from the natural vaccenic acid that has no harmful effects. Approximately 2.6% of the total daily fat intake is from trans fat and that 50% of the trans may be from vaccenic acid ($18:1n^{11}$).

16. Conclusion

The oil produced by soybeans is widely used by manufacturers of both food products and industrial manufactured goods. Crude soybean oil contains essential fatty acids that our body needs to work properly. However, much of the soybean oil consumed today has been partially hydrogenated. This hydrogenation removes the necessary essential fatty acids contained within the original oil. Additionally, some of the partially hydrogenated soybean oil has been converted to trans fatty acids.

There are two mechanisms that have been shown to lead to heart disease involving the consumption of trans fatty acids. They effect the levels of prostacyclin and thromboxane, which increases the risk of thrombosis, and they increase sphingomyelin production by the body, which then causes calcium influx into the arterial cells to increase, leading to atherosclerosis. Soybeans can be an excellent source of protein, but partially hydrogenated soybean oil can be detrimental to health.

NC Soybean Producers Assn. How soybeans are used. Retrieved from http://www.ncsoy.org/ABOUT-SOYBEANS/Uses-of-Soybeans.aspx

Author details

Fred A. Kummerow[*]

Address all correspondence to: fkummero@uiuc.edu

University of Illinois, USA

References

[1] Pedersen, P. (2007). Soy Products. http://extension.agron.iastate.edu/soybean/uses_soyproducts.htmlaccessed 1 August 2012).

[2] Soya. (2012). Information about soy and soya products. http://www.soya.be/soybean-oil-production.phpaccessed 31 July).

[3] Kummerow, F. A. (2005). Improving hydrogenated fat for the world population. *Prevention and Control*, 1, 157-164.

[4] Kummerow, F. A. (2009). The negative effects of hydrogenated trans fats and what to do about them. *Atherosclerosis*, 205, 458-465.

[5] Quackenbush, F. W., Steenbock, H., Kummerow, F. A., & Platz, B. R. (1942). Linoleic acid, pyridoxine and pantothenic acid in rat dermatitis. *Journal of Nutrition*, 24, 225-234.

[6] Wansink, B. (2003). How do front and back package labels influence beliefs about health claims? *Journal of Consumer Affairs*, 37, 305-316.

[7] Sirtori, C. R. (2001). Risks and benefits of soy phytoestrogens in cardiovascular diseases, cancer, climacteric symptoms and osteoporosis. *Drug Safety*, 24, 665-682.

[8] Henkel, J. (2000). Health claims for soy protein, questions about other components. *FDA Consumer*, 34, 13.

[9] Messina, M. (2003). Potential public health implications of the hypocholesterolemic effects of soy protein. *Nutrition*, 19, 280-281.

[10] Sacks, F. M., Lichtenstein, A., Van Horn, L., et al. (2006). Soy Protein, Isoflavones, and Cardiovascular Health: An American Heart Association Science Advisory for Professionals from the Nutrition Committee. *Circulation*, 113, 1034-1044.

[11] Kummerow, F., & Kummerow, J. (2008). Cholesterol Won't Kill You, But Trans Fat Could. *Bloomington: Trafford Publishing*.

[12] Ross, R. (1993). The pathogenesis of atherosclerosis. *New England J Med*, 297, 369-377.

[13] De Bakey, M. E., Dietrich, E. B., Garrett, H. E., & Mc Cutchen, J. J. (1967). Surgical treatment of cerebrovascular disease. *Postgrad Med J*, 42, 218-226.

[14] Keaney, J. (2000). Atherosclerosis from lesion formation to plaque activation and endothelial disfunction. *Molecular Aspects of Medicine*, 21, 118.

[15] Nicholson, A., Febbraio, M., Han, J., Silverstein, R., & Hajjar, D. (2000). CD36 in atherosclerosis, the role of class B macrophage scavenger receptor. *Ann. NY Acad. Sci*, 902, 128-131.

[16] Leschziner, M., & Dimitriadis, K. (1989). Computation of three-dimensional turbulent flow in non-orthogonal junctions by a branch-coupling method. *Computers & Fluids*, 17, 371-396.

[17] Koenig, W., & Ernst, E. (1992). The possible role of hemorheology in atherothrombogenesis. *Atherosclerosis*, 94, 93-107.

[18] Gimbrone, M., Topper, J., Nagel, T., Anderson, K., & Garcia-Cardena, G. (2000). Endothelial dysfunction, hemodynamic forces, and atherogenesis. *Ann. NY Acad. Sci*, 902, 230-240.

[19] Kataoka, H., Kume, N., Miyamoto, S., Minami, M., Moriwaki, H., Murase, T., Sawanmura, T., Masaki, T., Hashimoto, N., & Kita, T. (1999). Expression of lectin-like oxidized LDL receptor-I human atherosclerosis lesions. *Circulation*, 99, 3110-3117.

[20] Li, D., & Mehta, J. (2000). Antisense to LOX-1 inhibits oxidized LDL-mediated upregulation of monocyte chemoattractant protein-1 and monocyte adhesion to human coronary artery endothelial cells. *Circulation*, 101, 2889-2895.

[21] Murase, T., Kume, N., Korenaga, R., Ando, J., Sawamura, T., Masaki, T., & Kita, T. (1998). Fluid shear stress transcriptionally induces lectin-like oxidized low density lipoprotein receptor-1 in vascular endothelial cells. *Circ. Res*, 83, 328-333.

[22] Chen, X., Sun, A., Yunzeng, Z., et al. (2011). Impact of sphingomyelin levels on coronary heart disease and left ventricular systolic function in humans. *Nutrition & Metabolism*, 8, 25.

[23] Jiang, X., Paultre, F., Pearson, T., et al. (2000). Plasma sphingomyelin level as a risk factor for coronary artery disease. *Arteriosclerosis Thromb. & Vasc. Biol*, 20, 2614-2618.

[24] Nelson, J. C., Jiang, X. C., Tabas, I., et al. (2006). Plasma Sphingomyelin and Subclinical Atherosclerosis: Findings from the Multi-Ethnic Study of Atherosclerosis. *Am. J. Epidemiol*, 163, 903-912.

[25] Kummerow, F. A., Przybylski, R., & Wasowicz, E. (1994). Changes in arterial membrane lipid composition may precede growth factor influence in the pathogenesis of atherosclerosis. *Artery*, 21, 63-75.

[26] Bittman, R. (1988). Sterol exchange between mycoplasma membranes and vesicles. *Yeagle PL, editor. Boca Raton, FL: Biology of Cholesterol CRC Press.*

[27] Lehninger, A. L. (1975). Biochemistry. *New York: Worth Publishers Inc.*

[28] Bergelson, L. D., & Barsukov, L. I. (1977). Topological asymmetry of phospholipids in membranes. *Science*, 197, 224-230.

[29] Devaux, P. F. (1991). Static and dynamic lipid asymmetry in cell membranes. *Biochemistry*, 30, 1163-1173.

[30] Holmes, R. P., & Yoss, N. L. (1984). Hydroxysterols increase the permeability of liposomes to Ca2+ and other cations. *Biochim Biophys Acta*, 770, 15-21.

[31] Kummerow, F. A., Cook, L. S., Wasowicz, E., & Jelen, H. (2001). Changes in the phospholipid composition of the arterial cell can result in severe atherosclerotic lesions. *J Nutr Biochem*, 12, 602-607.

[32] Ridgway, N. D. (1995). Hydroxycholesterol stimulates sphingomyelin synthesis in Chinese hamster ovary cells. *J Lipid Res*, 36, 1345-1358.

[33] Zhou, Q., Wasowicz, E., Handler, B., et al. (2000). An excess concentration of oxysterols in the plasma is cytotoxic to cultured endothelial cells. *Atherosclerosis*, 149, 191-197.

[34] Zhou, Q., Band, M. R., Hernandez, A., & Kummerow, F. A. (2004). 27-Hydroxycholesterol inhibits neutral sphingomyelinase in cultured human endothelial cells. *Life Sci*, 75, 1567-1577.

[35] Kummerow, F. A., Olinescu, R., Fleischer, L., Handler, B., & Shinkareva, S. (2000). The relationship of oxidized lipids to coronary artery stenosis. *Atherosclerosis*, 149, 181-190.

[36] Kummerow, F. A. (2005). Improving hydrogenated fat for the world population. *Prevention & Control*, 1, 157-164.

[37] Kummerow, F. A., Zhou, Q., & Mahfouz, MM. (2004). Trans fatty acids in hydrogenated fat inhibited the synthesis of the polyunsaturated fatty acids in the phospholipid of arterial cells. *Life Sciences*, 74, 2707-2723.

[38] Stary, H. (1999). Atlas of atherosclerosis progression and regression. *New York: Parthenon Publishing Group*.

[39] Kummerow, F. A., Zhou, Q., & Mahfouz, MM. (1999). Effect of trans fatty acids on calcium influx into human arterial endothelial cells. *American J Clin Nutr*, 70, 832-838.

[40] Vane, J. R., & Moncada, S. (1977). The discovery of prostacyclin: a fresh insight into arachidonic acid metabolism, biochemical aspects of prostaglandins and thromboxanes. *New York: Academic Press*.

[41] Mahfouz, M. M., & Kummerow, F. A. (1999). Hydrogenated fat high in trans monoenes with an adequate level of linoleic acid has no effect on prostaglandin synthesis in rats. *Journal of Nutrition*, 129, 15-24.

[42] Kummerow, F. A., Mahfouz, MM, & Zhou, Q. (2007). Trans fatty acids in partially hydrogenated soybean oil inhibit prostacyclin release by endothelial cells in presence of high level of linoleic acid. Prostaglandins Other Lipid Mediat; , 84, 138-153.

[43] Mahfouz, MM, & Kummerow, F. A. (1998). Oxysterols and TBARS are among the LDL oxidation products which enhance thromboxane A_2 synthesis by platelets. *Prostaglandins Other Lipid Mediat*, 56, 197-217.

[44] Mahfouz, MM, & Kummerow, F. A. (2000). Oxidized low-density lipoprotein (LDL) enhances thromboxane A_2 synthesis by platelets, but lysolecithin as a product of LDL oxidation has an inhibitory effect. *Prostaglandins Other Lipid Mediat*, 62, 183-200.

[45] Maddox, T. (2012). Heart disease and sudden cardiac death. http://www.webmd.com/heart-disease/guide/sudden-cardiac-deathaccessed 31 July).

[46] FDA. (2003). Food labeling: trans fatty acids in nutrition labeling, nutrient content claims, and health claims. *Final rule. Fed Regist*, 68, 41433-41506.

[47] Vesper, H. W., Kuiper, H. C., Mirel, L. B., Johnson, C. L., & Pirkle, J. L. (2012). Levels of plasma trans-fatty acids in non-Hispanic white adults in the United States in 2000 and 2009. *Journal of American Medical Association*, 307, 562-563.

Chapter 16

Variability for Phenotype, Anthocyanin Indexes, and Flavonoids in Accessions from a Close Relative of Soybean, *Neonotonia wightii* (Wright & Arn. J.A. Lackey) in the U.S. Germplasm Collection for Potential Use as a Health Forage

J.B. Morris, M.L. Wang and B. Tonnis

Additional information is available at the end of the chapter

http://dx.doi.org/10.5772/53102

1. Introduction

The closely related soybean species, *Neonotonia wightii* Wright & Arn. J.A. Lackey is in the *Fabaceae* family and originates from several tropical countries (NPGS, 2012; Cook *et al.*, 2005). The plants produce vines with slender stems (2-3 cm in diameter) consisting of glabrous to densely pubescent trichomes and a strong taproot. The leaves are pinnately trifoliolate with elliptic, ovate, or rhombic ovate, acute to obtuse (1.5-15 cm long, and 1.3-12.5 cm wide), glabrous to densely pubescent leaflets. The stipules are lanceolate (4-6 mm long) and the petiole is 2.5-13 cm long. The inflorescence is axillary with dense or lax racemes which are 2-35 cm long on peduncles (3-12.5 cm long) and comprises 20-150 flowers. Each flower is 4.5-11 mm long with white to mauve-blue standards, however small violet streaks are noticeable on the lower part, and will change to yellow or orange at senescence. The 1.5-4 cm long by 2.5-5 mm wide, glabrous to densely pubescence with grey to reddish brown trichomes on the pods are linear, oblong, straight or slightly curved at the apex and transversely grooved with a weak septa between the seeds. Each pod contains 3-8 oblong with rounded corners, laterally compressed, olive green to reddish brown (occasionally mottled, aril white), 2-4 mm long, 1.5-3 mm wide, and 1-1.5 mm thick seeds (Cook *et al.*, 2005). *Neonotonia wightii* consists of diploid (2n = 22) and tetraploid (2n = 44) genotypes which are self-pollinated (cleistogamous) and low outcrossing (Cook *et al.*, 2005).

© 2013 Morris et al.; licensee InTech. This is a paper distributed under the terms of the Creative Commons Attribution License (http://creativecommons.org/licenses/by/3.0), which permits unrestricted use, distribution, and reproduction in any medium, provided the original work is properly cited.

Neonotonia wightii has several common names including glycine (Australia, Kenya); soja perene (Brazil); soya perenne forrajera, soya forrajera, soya perenne (Colombia, Mexico); soja perenne (French); ausdauernde soja (German); soja-perene (Portuguese); Rhodesian kudzu (Taiwan); fundo-fundo (Tanzania); and thua peelenian soybean (Thai) [Cook *et al.*, 2005]. This species also has many synonyms including *Glycine javanica* auct., *G. javanica* L. var. *paniculata* Hauman, *G. albidiflora* De Wild., *G. claessensii* De Wild., *G. javanica* sensu auct., *G. javanica* L. var. *claessensii* (De Wild.) Hauman, *G. javanica* L. var. *longicauda* (Schweinf.) Baker, *G. javanica* L. subsp. *micrantha* (A. Rich.) F.J. Herm., *G. javanica* L. var. *mearnsii* (De Wild.) Hauman, *G. longicauda* Schweinf., *G. mearnsii* De Wild., *G. micrantha* A. Rich., *G. moniliformis* A. Rich., *G. petitiana* Hermann pro parte, *G. pseudojavanica* Taub., *G. wightii* (Wight & Arn.) Verdc. var. *longicauda* (Schweinf.) Verdc., *G. wightii* (Wight & Arn.) Verdc. subsp. *petitiana* (A. Rich.) Verdc. var. *mearnsii* (De Wild.) Verdc., *G. wightii* (Wight & Arn.) Verdc. subsp. *petitiana* (A. Rich.) Verdc., *G. wightii* (Wight & Arn.) Verdc. subsp. *pseudojavanica* (Taub.) Verdc., *G. wightii* (Wight & Arn.) Verdc. subsp. *wightii*, *Johnia wightii* (Wight & Arn.) Wight & Arn., and *Notonia wightii* Wight & Arn. (Cook *et al.*, 2005).

The plant is used as pasture for grazing, hay, and silage (Cook *et al.*, 2005). However, Viswanathan *et al.*, 2001 indicated that *N. wightii* seeds are used as food by Malayali tribes in Kollihills of the Namakkal District, Tamil Nadu, India. They found several essential amino acids, fatty acids, potassium, magnesium, manganese, and copper in *N. wightii* seeds. In Kenya, *N. wightii* produced abundant organic matter contributing to soil fertility and tolerated defoliation (Macharia *et al.*, 2010). A Brazilian study showed that *N. wightii* was one of several legumes with high crude protein, low NDF, and low phenolic concentrations for use as a ruminant feed (Valarini and Possenti, 2006). Mtui *et al.* (2006) found that *N. wightii* should be a component in dairy cow diets because of its high mineral concentration. *Neonotonia wightii* plants contain low levels of tannins and alkaloids as well (Mbugua *et al.*, 2008). Tauro *et al.* (2009) found that *N. wightii* could restore productivity to soils in Zimbabwe that have been cultivated continuously and low in nutrient levels. *Neonotonia wightii* has also been found to contribute the greatest green manure effect in the absence of fertilization in Zambia (Steinmaier and Ngoliya, 2001).

Anthocyanins are chemicals responsible for natural plant colors found in leaves, stems, and flowers. An anthocyanin meter with a 520 nm LED has been used to measure the absorbance near the wavelength at which free anthocyanin aglycones, cyanidin and pelargonidin monogluscosides absorb (Macz-Pop et al., 2004). Several studies have shown potential health benefits of anthocyanins in humans. Chokeberry (*Aronia meloncarpa* E.) anthocyanins (cyanidin derivatives) have been shown to be very potent inhibitors of colon cancer cells (Zhao *et al.*, 2004). When several anthocyanins including cyanidin-3,5-diglucoside and cyanidin-3-glucoside are ingested, apoptosis effects were observed and may have potential for human hepatitis B-associated hepatoma (Shin *et al.*, 2009). Lacombe *et al.* (2010) found that cyanidin-3-glucoside caused disintegration of E. coli outer membranes. Both cyanidin-3-glucoside and pelargonidin-3-glucoside showed potential for prevention of atherosclerosis (Paixao *et al.*, 2011). Corn silage containing anthocyanins may have nutritional value as a ruminant feed (Hosoda *et al.*, 2011). *Neonotonia wightii* has also been found to be rich in crude protein content and amino acids, but low amounts of the sulfur containing amino acid, methionine (Tokita *et al.*, 2006).

Isoflavones have been associated with reducing sheep fertility (Waghorn and McNabb, 2003). The isoflavone, genistein is a secondary metabolite found in many legumes including *N. wightii* (Ingham *et al.*, 1977; Keen *et al.*, 1989). Genistein can cause reduced fertility in sheep, however after 7-10 days of adaptation, sheep rumen microbes degrade genistein and other oestrogenic compounds to non-oestrogenic metabolites. Therefore, the effects of genistein on sheep fertility is short lived (Waghorn and McNabb, 2003). However, genistein has been shown to protect against mammary and prostate cancer by regulating receptors and growth signaling pathways (Lamartiniere *et al.*, 2002).

2. Materials and methods

2.1. Phenotyping

Neonotonia wightii is a photoperiod and frost-sensitive species requiring seed regeneration in a greenhouse. Twenty-two *N. wightii* accessions from countries throughout the world were used in this study (Table 1). There was not enough room remaining in the greenhouse to accommodate the regeneration of 7 additional *N. wightii* accessions, therefore fourteen accessions were planted in 27.5 cm x 27.5 cm plastic pots containing potting soil grown in a greenhouse from August 1, 2010 – April 1, 2011 each year. Three to four seedlings per pot were maintained for plant production. Due to the vigorous growth habit, all *N. wightii* accessions were grown with trellises. The experimental design was a randomized complete block with 4 replications assigned to *N. wightii* accessions. Phenotypic descriptors including branching, foliage, plant height, plant width, and relative maturity were recorded when each accession reached 50% maturity based on visual observation, while seed numbers were counted at the end of the growing season. Branching and foliage were based on a scale of 1-5 where, 1 = > 90%, 2 = 80-89%, 3 = 70-79%, 4 = 60-69%, and 5 = 50-59%, 6 = 40-49%, 7 = 30-39%, 8 = 20-29%, and 9 = 10-19% of each plant producing branches and/or foliage based on visual observations. Relative maturity dates were based on a scale of 5 to 9 where 5 = mid-season and 9 = very late.

2.2. Anthocyanin indexes

An Opti-Sciences CCM-200 chlorophyll content meter was converted to a hand-held anthocyanin meter. The manufacturer replaced the 655 nm light emitting diode (LED) of the CCM with a 520 nm LED to measure the absorbance near the wavelength at which free anthocyanin aglycones, cyanidin and pelargonidin monogluscosides absorb (Macz-Pop et al., 2004). Anthocyanin indexes were determined by inserting each leaflet between the meter and the LED diode, followed by gently pressing the LED directly on to the leaflet and recording from each of three leaflets of 15 *Neonotonia wightii* accessions growing in the greenhouse on 14 February 2008 and 18 March 2009.

Accession (PI)	Origin
156055	Zimbabwe
189613	South Africa
213256	India
213257	India
224976	South Africa
224977	South Africa
224978	South Africa
224979	South Africa
224980 (cultivar-Tropic Verde)	Zimbabwe
224981	Zimbabwe
230324	South Africa
233148	Rhodesia
234874	Congo
235287	Zimbabwe
247677	Congo
258381	Australia
259541	Unknown
259544	South Africa
259545	Brazil
277889	Zimbabwe
314847	South Africa
612241	Taiwan

Table 1. Origin for *N. wightii* accessions used in this study.

2.3. Genistein

As plants matured in the greenhouse, most leaflets were pre-disposed to senescence and only 7 *N. wightii* accessions could be evaluated for genistein variation during 2008. Therefore, preliminary research investigating variability for leaflet weight and genistein content was conducted among these 7 *N. wightii* accessions. Leaflet tissue from each *N. wightii* accession was ground to a fine powder with liquid nitrogen, dried at room temperature, and stored at -20°C until extraction. Approximately 0.15- 0.3 g of dried tissue from each accession was placed into 5 ml test tubes, and their weights were recorded to the nearest 0.001 g. Three ml of extraction solvent consisting of 80% HPLC-grade methanol with 1.2 M HCl was added to each test tube. Each tube with extraction solvent were vortexed and incubated at 80°C for 2 hr with occasional mixing by inversion. An additional 2.0 ml of 5% methanol was added to each test tube, resulting in a final concentration of 50% methanol. A portion of the extract was filtered through a 0.45 μm membrane prior to injection. Analytes were separated and identified by high performance liquid chromatography (HPLC). The stationary phase consisted of a Zorbax Eclipse XDB-C18 column (4.6 x 150 mm, 5μM particle size) (Agilent Technologies) at 40°C with a C18 guard column. The mobile phase consisted of 20% HPLC-grade acetonitrile (B) and 80% filtered water with 0.1% formic acid, pH 2.5 (A) at 2.0 ml/min. A gradient flow was used with the following profile: 15% B for 3 min, then 15% to 40% B from 3 min to 20 min. The column was washed with 95% B for 5 min, and then equilibrated for 7 min at 15% B between injections. The injection volume was 10 µl. Flavonoid peaks were monitored at 260 and 370 nm with a diode array detector. The standards for peak identification and quantification consisted of kaempferol, quercetin, myricetin, genistein, and daidzein (Sigma-Aldrich Chemical Co.). Each standard was dissolved in extraction solvent and diluted to the following concentrations: 1, 5, 10, 25, and 50 ng/µl.

Phenotype, anthocyanin index, and isoflavonoid data were subjected to an analysis of variance using SAS (SAS Institute, 2008). Mean separations were conducted using Duncan's multiple range test ($P < 0.05$, $P < 0.01$) and correlations were accomplished using Pearson's correlation in SAS (SAS Institute, 2008). Principal component analysis using PROC PRINCOMP (SAS Institute, 2008) were then used for multivariate analysis of the data. Eigenvalues, the percentage of variances explained by each principal component, and eigenvectors were also determined. Clustering was then performed on the data by entering the similarity matrix into PROC CLUSTER for cluster analysis with the unweighted paired group method using mathematical averages (UPGMA) by specifying the AVERAGE option (SAS Institue, 2008).

3. Results and discussion

3.1. Phenotype

Significant variability for morphological, plant maturity, and seed number characteristics observed among 14 *N. wightii* accessions are reported in Table 2. Only PI 224978 from South Africa produced significantly less branching and foliage production than most of the other accessions. Many plants extended beyond the top of the trellis which allowed us to measure actual plant heights. Plant height ranged from 99.3 to 116.3 cm with both PI 224979 (South Africa) and PI 224981 (Zimbabwe) producing significantly shorter plants (averaging 99.7 cm) than the other accessions. The other twelve accessions averaged 108.4 cm tall. The accessions PI 224976 (S. Africa), PI 224977 (S. Africa), PI 230324 (S. Africa), PI 213257 (India), and PI 224981 produced the significantly narrowest plants averaging 47.6 cm while all other accessions averaged 70.0 cm wide. The accession PI 213256 from India matured the earliest while PI 224976, PI 224977, and PI 213257 averaged maturity at mid season while all others matured late or very late (Fig. 1). The significantly highest seed producing accession was PI 213256 (2214 seeds) followed by the Congonese accession, PI 234874 (725 seeds) and PI 224977 (663 seeds). The significantly lowest seed producers were PI 224979, PI 224980 (cultivar, Zimbabwe), and PI 230324 averaging 70 seeds while all other accessions averaged 307 seeds. Branching significantly correlated with foliage ($r^2 = 0.84$***) and foliage had a significant negative correlation with plant width ($r^2 = -0.28$*). Plant width was significantly correlated with maturity ($r^2 = 0.57$***) and maturity had a significant negative correlation with seed number ($r^2 = -0.39$**). Phenotypic variation for the *N. wightii* accessions can be explained by plant selection leading to the potential development of cultivated varieties or breeding material.

Accession	Branching	Foliage	Plant ht. (cm)	wd. (cm)	Maturity	Seed no.
224978	2.5a	2.3a	106.3abcd	52.5bc	7.5b	201b
224979	1.5b	2.0ab	99.3d	51.8bc	7.5b	89b
156055	1.0b	1.0b	104.0cd	70.3a	9.0a	405b
224976	1.0b	1.5ab	108.8abcd	50.0c	6.0c	304b
224977	1.0b	1.0b	110.0abcd	47.5c	6.0c	663b
213256	1.0b	1.0b	103.3cd	51.7bc	5.0c	2214a
213257	1.0b	1.0b	107.5abcd	47.5c	6.0c	461b
224980	1.0b	1.0b	116.3a	62.3ab	7.5b	78b
224981	1.0b	1.0b	100.0cd	50.5c	7.8ab	131b
230324	1.0b	1.3ab	107.5abcd	42.5c	8.3ab	42b
234874	1.0b	1.0b	105.5bcd	71.0a	9.0a	725b
235287	1.0b	1.0b	106.3abcd	69.5a	9.0a	293b
258381	1.0b	1.0b	115.0ab	71.3a	9.0a	327b
259544	1.0b	1.0b	110.3abc	66.3a	9.0a	336b

Means followed by the same letter are not significantly different.

Table 2. Phenotype and seed reproduction variability.

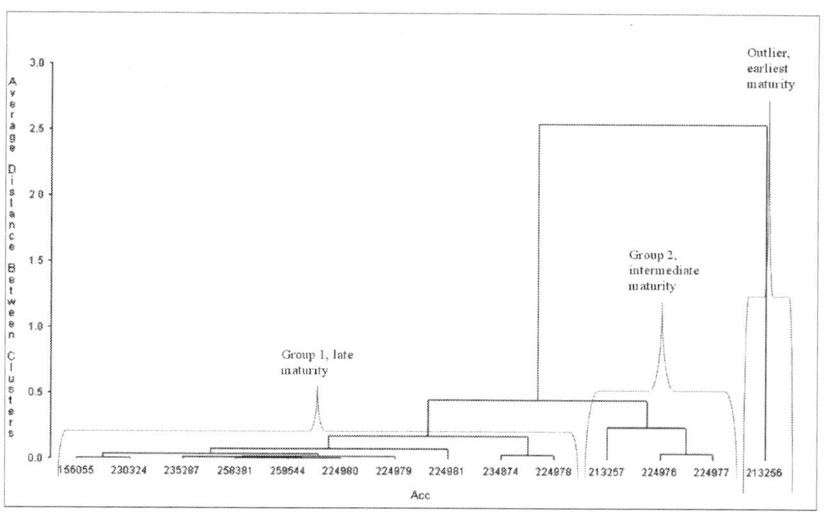

Figure 1. Dendrogram of distance between clusters based on morphological, plant maturity, and seed number differences. Accession numbers are given (Acc). Values on the baseline indicate average phenotypic distances between accessions. Two distinct clusters and one outlier (earliest maturity) for relative plant maturity can be distinguished.

3.2. Anthocyanin index, leaf weight, and genistein

Significant variation for leaf anthocyanin indexes among 15 diverse *N. wightii* accessions were observed also (Table 3). The accession, PI 213257 produced the significantly highest anthocyanin index (7.5) while all other accessions produced an average anthocyanin index of 6.1. This accession also produced significantly higher leaflet weight (0.228 g) than the other 6 accessions (averaging 0.142 g) (Table 4). However, PI 612241 from Taiwan produced significantly more genistein (90.03 μg g^{-1} of leaflet tissue) than all other accessions which averaged 51.3 μg g^{-1} (Table 4). There were no significant correlations among these traits. The flavonoids kaempferol, quercetin, myricetin, and daidzein were minutely or not detected.

Accession (PI)	N	Mean anthocyanin index
213257	4	7.50a
277889	4	7.13ab
314847	4	7.08abc
224976	4	7.00abc
247677	4	6.75abcd
259541	1	6.60abcd
189613	2	6.60abcd
224981	5	6.18abcde
234874	5	6.10bcde
259545	5	5.76cde
235287	5	5.64de
213256	5	5.46de
156055	5	5.44de
233148	5	5.14e
612241	4	5.10e

Means followed by the same letter are not significantly different.

Table 3. Preliminary leaf anthocyanin index variability among 15 diverse *N. wightii* accessions combined over 2 years (2008 and 2009).

Accession (PI)	N	Leaflet wt. (g)	Genistein ($\mu g\ g^{-1}$)
213257	4	0.228a	66.28ab
224976	4	0.175b	80.10ab
314847	4	0.162bc	59.10ab
277889	4	0.160bc	28.90b
247677	4	0.136bcd	37.95ab
189613	2	0.118cd	35.65b
612241	4	0.101d	90.03a

Table 4. Preliminary leaflet weight (g) and genistein variability among 7 *N. wightii* accessions during 2008.

3.3. Principal component analysis

Phenotypic, maturity, and seed number principal component analysis accounted for 44% of the total variation at the first principal component (Table 5). The amount of variation accounted for, cumulatively, by adding principal components 2 through 4 was 75, 88, and 96%, respectively. The first principal component was most correlated with plant width and maturity (Table 6). The second principal component accounted for 31% of the variation and was mostly due to branching and foliage while the third principal component explained 13% of the variation and was composed of primarily plant height. The fourth principal component accounted for 8% of the variation and was most correlated with seed number. Therefore, potential exists to develop cultivars with improved architecture, early or late maturity, and high or low seed yield. Anthocyanin index, leaflet weight, and genistein accounted for 63% of the total variation at the first principal component (Table 7). The cumulative amount of variation for components 2 through 3 was 98 and 100%, respectively. The first and second principal components were mostly correlated with anthocyanin index and genistein, respectively, while the third principal component correlated with both anthocyanin index and leaflet weight (Table 8). Potential exists to develop *N. wightii* cultivars with improved anthocyanin indexes, genistein content, and leaflet weight. Since all traits tested are quantitative, the variability among *N. wightii* accessions is attributed to genetic differences primarily since they were regenerated in a greenhouse.

	Principal		
component	Eigenvalue	% Variability	% Cummulative
1	2.6400	44.00	44.00
2	1.8319	30.53	74.53
3	0.8088	13.48	88.01
4	0.4535	7.56	95.57

Table 5. Eigenvalues and the proportion of total phenotypic, maturity, and seed reproduction variability among 14 *N. wightii* accessions (2010, 2011) as explained by the principal components.

Principal components	1	2	3	4	5	6
Branching	-0.26	0.58	0.06	0.49	-0.53	-0.22
Foliage	-0.35	0.52	0.23	0.01	0.72	0.15
Plant ht. (cm)	0.33	-0.16	0.88	0.27	-0.03	0.11
Plant width (cm)	0.54	0.06	-0.26	0.46	0.40	-0.49
Maturity	0.50	0.37	-0.24	0.07	-0.10	0.72
Seed no.	-0.37	-0.46	-0.19	0.67	0.11	0.37

Table 6. Eigenvectors, principal components for 6 phenotypic, maturity, and seed traits in 14 *N. wightii* accessions (2010-2011).

component	Principal Eigenvalue	% Variability	% Cummulative
1	1.8958	63.20	63.20
2	1.0540	35.13	98.33
3	0.0501	1.67	100.00

[1]Anthocyanin indexes were based on 15 *N. wightii* accessions.

[2]Leaflet weight and genistein were based on 7 *N. wightii* accessions.

Table 7. Eigenvalues and the proportion of total leaf anthocyanin index[1], leaflet weight (g)[2] (2009) and genistein[2] (2008) variability among *N. wightii* accessions as explained by the principal components.

Principal components	1	2	3
Anthocyanin index[1]	0.71	-0.03	0.69
Leaf wt. (g)[2]	0.62	0.47	-0.61
Genistein[2]	-0.30	0.87	0.36

[1]Anthocyanin index based on 15 *N. wightii* accessions.

[2]Leaflet weight and genistein based on 7 *N. wightii* accessions.

Table 8. Eigenvectors, principal components for two phytochemical traits and leaf weight in *N. wightii* accessions (2008, 2009)

4. Cluster analysis

Average distance cluster analysis grouped the original 14 *N. wightii* accessions into well de-fined phenotypes with two distinct relative plant maturity groups and one outlier (Fig. 1). Cluster or group 1 represents 10 late maturing *N. wightii* accessions and group 2 consists of three intermediate or mid-season maturing accessions. The outlier, PI 213256 represents the earliest maturing accession. The *N. wightii* accessions clustered in group 2 showed relatively closer genetic relationships than those in group 1. Using the distance values indicated in Fig. 1, the groupings at any similarity level can be identified. For example, PI 224976 and PI 224977 originate from South Africa with a phenotypic distance index of 0.0473, which indi-cates their close morphological similarities.

Average distance cluster analysis grouped 7 *N. wightii* accessions into well defined pheno-types with three distinct genistein producing accessions and one outlier (Fig. 2). Group 1 represents 2 intermediate genistein producing accessions while group 2 consists of two high genistein producing accessions. Group 3 is representative of two very high genistein pro-ducing accessions and one outlier (PI 277889) accession producing low amounts of genistein. Overall, *N. wightii* accessions showed similar genetic relationships in all groups and one out-lier. However, PI 612241 from Taiwan and PI 224976 from South Africa had a phenotypic distance index of 0.3030, which indicates their close morphological similarities.

These results show substantial variability for various phenotypic traits, maturity, seed re-production, and genistein in these *N. wightii* accessions regenerated in a greenhouse. How-ever, additional studies are warranted for investigating if similar results will occur when *N. wightii* accessions are grown in field conditions over multiple years.

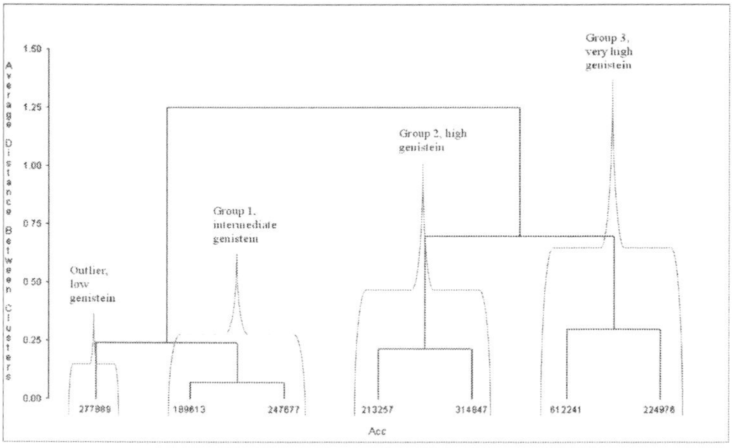

Figure 2. Dendrogram of distance between clusters based on anthocyanin indexes and genistein differences. Acces-sion numbers are given (Acc). Values on the baseline indicate average phytochemical distances between accessions. Three distinct clusters and one outlier (low) for genistein can be distinguished.

Author details

J.B. Morris, M.L. Wang and B. Tonnis

USDA, ARS, Plant Genetic Resources Conservation Unit, Griffin, GA, USA

References

[1] Cook, Bruce, Pengelly, Bruce, Brown, Stuart, Donnelly, John, Eagles, David, Franco, Arturo, Hanson, Jean, Mullen, Brendan, Partridge, Ian, Peters, Michael and Schultze-Kraft, Rainer. (2005). Tropical Forages: Neonotonia wightii. CSIRO Sustainable Eco-systems (CSIRO), Department of Primary Industries and Fisheries (DPI&F Queensland), Centro Internacional de Agricultura Tropical (CIAT) and International Livestock Research Institute (ILRI). Online version: http://www.tropicalforages.info.

[2] Hosoda, Kenji, Eruden, Bayaru, Matsuyama, Hiroki and Shioya, Shigeru. Effect of anthocyanin-rich corn silage on digestibility, milk production and plasma enzyme activities in lactating dairy cows. Animal Science Journal, 2011 doi: 10.1111/j. 1740-0929.2011.00981.x.

[3] Ingham, John L., Keen, Noel T. and Hymowitz, Theodore. A new isoflavone phytoalexin from fungus-inoculated stems of Glycine wightii. Phytochemistry, 1977 16:1943-1946.

[4] Keen, N.T., Ingham, J.L., Hymowitz, T., Sims, J.J. and Midland, S. The occurrence of glyceollins in plants related to Glycine max (L.) Merr. Biochemical Systematics and Ecology, 1989 17:395-398.

[5] Lacombe, A., Wu, V.C., Tyler, S. and Edwards, K. Antimicrobial action of the American cranberry constituents; phenolics, anthocyanins, and organic acids, against Escherichia coli O157:H7. International Journal of Food Microbiology, 2010 139:102-107.

[6] Lamartiniere, Coral A., Cotroneo, Michelle S., Fritz, Wayne A., Wang, Jun, Mentor-Marcel, Roycelynn and Elgavish, Ada. Genistein chemoprevention: Timing and mechanisms of action in murine mammary and prostate. The Journal of Nutrition, 2002 132:5525-5585.

[7] Macharia, P.N., Kinyamario, J.I., Ekaya, W.N., Gachene, C.K.K., Mureithi, J.G., and Thuranira, E.G. Evaluation of forage legumes for introduction into natural pastures of semi-arid rangelands of Kenya. Grass and Forage Science, 2010 65:456-462.

[8] Macz-Pop, G.A., Rivas-Gonzalo, J.C., Perez-Alonso, J.J. and Gonzalez-Paramas, A.M. Natural occurrence of free anthocyanin aglycones in beans (Phaseolus vulgaris L.). Food Chemistry, 2004 94:448-456.

[9] Mbugua, David M., Kiruiro, Erastus M. and Pell, Alice N. In vitro fermentation of intact and fractionated tropical herbaceous and tree legumes containing tannins and alkaloids. Animal Feed Science and Technology, 2008 146:1-20.

[10] Mtui, D.J., Lekule, F.P., Shem, M.N., Rubanza, C.D.K., Ichinohe, T., Hayashida, M. and Fujihara, T. Seasonal influence on mineral content of forages used by smallholder dairy farmers in lowlands of Mvomero District, Morogoro, Tanzania. Journal of Food, Agriculture and Environment, 2006 4:216-221.

[11] National Plant Germplasm System. (2012). Germplasm Resources Information Network (GRIN). Database Management Unit (DBMU), National Plant Germplasm System. US Department of Agriculture, Beltsville, MD.

[12] Paixao, J., Dinis, T.C. and Almeida, L.M. Dietary anthocyanins protect endothelial cells against peroxynitrite-induced mitochondrial apoptosis pathway and Bax nuclear translocation: an in vitro approach. Apoptosis, 2011 16:976-989.

[13] SAS. (2008). Cary, NC, SAS Institute.

[14] Shin, D.Y., Ryu, C.H., Lee, W.S., Kim, D.C., Kim, S.H., Hah, Y.S., Lee, S.J., Shin, S.C. Kang, H.S. and Choi Y.H. Induction of apoptosis and inhibition of invasion in human hepatoma cells by anthocyanins from meoru. Annals of the New York Academy of Science, 2009 1171:137-148.

[15] Steinmaier, N. and Ngoliya, A. Potential of pasture legumes in low-external-input and sustainable agriculture (Leisa). 1. Results from green manure research in Luapula Province, Zambia. Experimental Agriculture, 2001 37:297-307.

[16] Tauro, T.P., Nezomba, H., Mtambanengwe, F. and Mapfurro, P. Germination, field establishment patterns and nitrogen fixation of indigenous legumes on nutrient-depleted soils. Symbiosis, 2009 48:92-101.

[17] Tokita, N., Shimojo, M. and Masuda, Y. Amino acid profiles of tropical legumes, cooper (Glycine wightii), Tinaroo (Neonotonia wightii) and Siratro (Macroptilium atropurpureum), at pre-blooming and blooming stages. Asian-Australasian Journal of Animal Science, 2006 19:651-654.

[18] Valarini, M.J. and Possenti, R.A. Research note: Nutritive value of a range of tropical forage legumes. Tropical Grasslands, 2006 40:183-187.

[19] Viswanathan, M.B., Thangadurai, D. and Ramesh, N. Biochemical and nutritional evaluation of Neonotonia wightii (Wight & Arn.) Lackey (Fabaceae). Food Chemistry, 2001 75:275-279.

[20] Waghorn, Garry C. and McNabb, Warren C. Consequences of plant phenolic compounds for productivity and health of ruminants. Proceedings of the Nutrition Society, 2003 62:383-392.

[21] Zhao, Cuiwei, Giusti, M. Monica, Malik, Minnie, Moyer, Mary P. and Magnuson, Bernadene A. Effects of commercial anthocyanin-rich extracts on colonic cancer and nontumorigenic colonic cell growth. Journal of Agricultural and Food Chemistry, 2004 52:6122-6128.

Chapter 17

Soybean: Non-Nutritional Factors and Their Biological Functionality

A. Cabrera-Orozco, C. Jiménez-Martínez and
G. Dávila-Ortiz

Additional information is available at the end of the chapter

http://dx.doi.org/10.5772/51487

1. Introduction

Legumes are important for the diet of a significant part of the world's population; they are a good source of protein, carbohydrates, minerals and B-complex vitamins. In this sense, the soybean is an important legume because it has a high protein (35-48%) with a nutritionally balanced amino acid profileso their products are commonly used as a source of vegetable protein worldwide and a great proportion of high-quality oil (15-22%)[1].

The accessible price and stable supply are favourable factors for legumes to emerge as an important source of protein for human food [2]. However, the nutritional value of soybeans is lower than expected due to the presence of various non-nutritive compounds that hinder or inhibit the uptake of nutrients and produce adverse physiological and biochemical effects in humans and animals; since these could be toxic in some cases, they are referred as anti-nutritional factors [3, 4].

Recently it has been found that legumes, in the appropriate proportion, may have a beneficial role for health. It seems clear that, in many cases, the same interaction that causes legumes to be considered as anti-nutrients is responsible for its beneficial effects. Thereby,these compounds are called non-nutritional compounds or nutritionally bioactive factors, because while they lack nutritive value, are not always harmful [5]. Available data indicate that the balance between harmful and beneficial effects of these compounds is a function of concentration, exposure time and interaction with other dietary components. However, the threshold concentration at which the beneficial and harmful effects occur has not been evaluated in most cases [6].Moreover, they are compounds that do not appear

© 2013 Cabrera-Orozco et al.; licensee InTech. This is a paper distributed under the terms of the Creative Commons Attribution License (http://creativecommons.org/licenses/by/3.0), which permits unrestricted use, distribution, and reproduction in any medium, provided the original work is properly cited.

equally in all legumes, and their physiological effects in humans and other animals are different as well [7, 8].

These compounds have an important role in secondary metabolism of legumes, as reserve compounds for the biosynthesis of endogenous compounds, which accumulate in seeds and are used during the germination process, and as mechanisms of defense against bacteria, viruses, fungi, insects and animals [9].

From a biochemical point of view, these compounds have diverse nature. They may be proteins (protease inhibitors, α-amylase inhibitors and lectins), carbohydrates (α-galactosides, vicine, convicine, saponins), non-protein amino acids (L-DOPA, β-ODAP), polyphenols (condensed tannins, isoflavones), alkaloids, inositol phosphates, etc., so their extraction and quantification methodology is very specific. In soybeans, the non-nutritional factors are mainly; inositol phosphates, saponins, protease inhibitors, isoflavones, lectins, oligosaccharides and taninis[10, 11].

1.1. Phytic acid and inositol phosphates

Phytic acid (myo-Inositol hexakisphosphate or 1, 2, 3, 4, 5, 6-hexakis dihydrogen phosphate myo-Inositol), also abbreviated as $InsP_6$ or IP_6, is the main form of storage of phosphorus and inositol in seeds of cereals, legumes and oilseeds. However for humans and monogastric animals phosphorus is not available in that form, because these are not provided with sufficient activity of endogenous phosphatases (phytases) that are capable of releasing the phosphate group from phytic acid or inositol phosphates lighter phosphorylated [12].

This molecule is formed from the esterification of phosphate groups to each of the six hydroxyl groups in a molecule known as myo-Inositol (Figure 1). Usually, it represents 65 to 85% of total phosphorus in seeds while forming insoluble salts with mono and divalent cations. By releasing H+ ions from the phosphate groups, allows the molecule to interact with the ions Mn^{2+}, Fe^{2+}, Zn^{2+} and K^+ to produce the corresponding salts, which are known as phytates. The name phytin has been used to designate a mixture of salts with Ca^{2+} and Mg^{2+}. Phytates and phytins usually bind to proteins in the protein bodies, the latter are membrane-limited structures where storage proteins are deposited. Salts of phytic acid are accumulated in seeds during the maturation period and are distributed uniformly in the cotyledons and embryonic axis in legumes [13, 14].

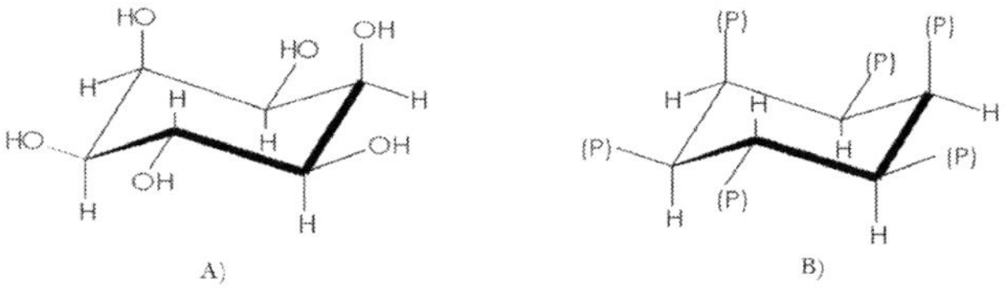

Figure 1. A) Chemical structure of myo-Inositol, B) Phytic acid structure (P) = H_2PO_4 [15]

In the soybean (*Glycine max*) phytic acid is uniformly distributed in the cotyledons, in the same way as in most legumes, probably as a soluble potassium phytate, which constitutes approximately 1.5% of the total weight of the cotyledon. One gram of soybeans contains about 9.2-16.7 mg of phytate, which represents 57% of organic phosphorus and 70% of total phosphorus [16, 17].

1.1.1. Synthesis and Function

Phytic acid is sythesized from 1D-myo-Inositol 3-phosphate (Ins_3P_1); in turn, the latter is formed from D-glucose-6-phosphate by action of synthase Ins_3P1, and from myo-Inositol by action of myo-Inositol kinase; this reaction represents the first step in the metabolism of inositol and in the phytic acid biosynthesis. Subsequently, the phosphatidylinositol kinases catalyze the gradual phosphorylation of Ins_3P_1 to produce myo-Inositol di-, tri-, tetra-, penta- and hexaphosphate (Figure 2) [14, 18, 19].

During germination, phosphorus and cations are released from phytates by the increased activity of an enzyme called phytase, then, they become available for use during the seedling growth. The enzyme phytase (myo-Inositol-hexaquisfosfatophosphohydrolase) is capable of sequentially hydrolyzing phytic acid to myo-Inositol, which produces intermediate products with a lower number of phosphate ester groups (IP_5, IP_4, IP_3, and possibly di- and mono- phosphateinositols) and inorganic phosphate [20, 21].

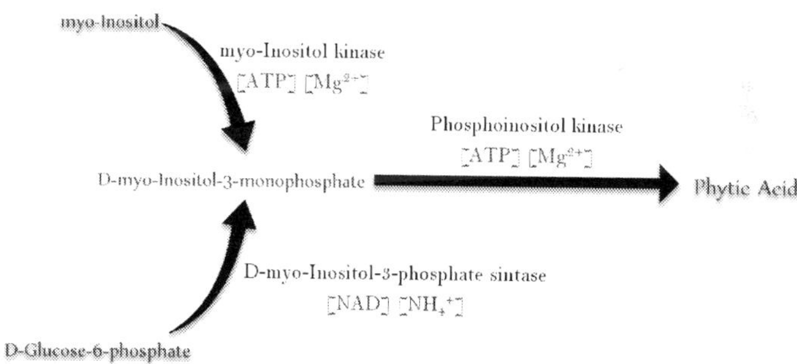

Figure 2. Biosynthesis of phytic acid [14].

A clear role of phytic acid in the seed tissues metabolism is the storage and recovery of phosphorus, minerals, and myo-Inositol during germination and growth [22]. Another physiological functions of phytic acid in plants, is the inhibition of the metabolism, since, by binding to multivalent cations required for cellular processes, the metabolism is slower, so it could be a latency-inducing molecule. Furthermore, the antioxidant capacity of phytic acid increases the time of seed latency, as it prevents lipid peroxidation [18, 23].

As well, phytates and also less-phosphorylated forms of phytic acid regulate diverse cellular functions such as DNA repair, chromatin remodelling, endocytosis, nuclear export of mRNA, and is an important hormonal marker for the development of seedlings and seeds [24-28].

Less-phosphorylated molecules of myo-Inositol are presentin free form in nature, in small amounts, as transient intermediates in biochemical reactions. The mono-, bi-, tri- myo-Inositol phosphates are important components of a group of phospholipids, known as phosphoinositides, which are present in many plants and animal tissues [18]. Raboy (2009),reported a very detailed description of the synthesis and metabolism of phytic acid and myo-Inositol phosphates in plants (Figure 3).

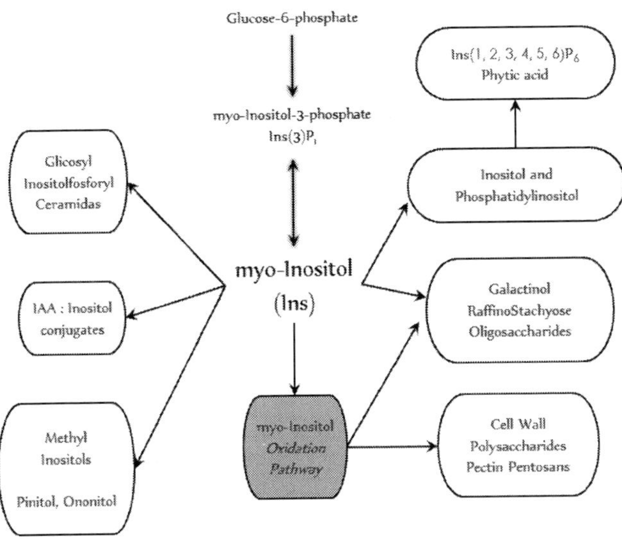

Figure 3. Pathways in plant biology that utilize myo-Inositol [29].

1.1.2. Bioavailability of minerals

Most studies on the interaction between phytic acid/inositol phosphates and minerals reveal the existence of an inverse relationship between the absorption of these micronutrients and inositol phosphates, although there are substantial differences in individual behaviour of each mineral element [16].

The interaction of phytic acid with minerals and other nutrients is pH-dependent [30], since the degree of protonation of the phosphate groups is a function of pH [31]. The molecule works in a wide region of pH as a highly negatively-charged ion, so its presence in the diet has an adverse impact on the bioavailability of mineral monovalent, divalent and trivalent ions, such as Zn^{2+}, Fe^{2+} / Fe^{3+}, Ca^{2+}, Mg^{2+}, Mn^{2+} and Cu^{2+}[32-34]; these complexes are more soluble at low or acid pH and insoluble at high or basic pH [35].

Another important aspect to be considered is that the interaction of phytic acid with minerals is due to its several phosphate groups, thereby, minerals may bind to one, two, or more phosphate groups of one, two or more phytic acid molecules [36]. Other studies have shown that the inhibitory effect of the absorption of InsP depends on the degree of phosphorylation of inositol, when it is high (5 or 6 phosphates) the absorption of Ca and Zn is significantly inhibited, however at lower levels of phosphorylation this effect is not observed [34].

The solubility of the complexes formed depends also on the InsP-mineral ratio; for instance, the solubility of the InsP-Ca complex, is extremely low in 1/8 ratios, but other ratios show higher solubility [37]. The complexes hexa-, penta-, tetra- and tri- Ca are insoluble, while complexes mono- and di- Ca are soluble [38]. Nevertheless, the absorption of Cafrom soluble complexes InsP-Ca is very low, because these complexes do not undergo passive transport in gut due to the high electric charge they have [39].

The inositol phosphates directly or indirectly interact with various minerals in the diet to reduce their bioavailability, in this context, the synergistic effect of the secondary cations (Ca^{2+}) has been widely demonstrated. Two cations may, when present at the same time, act together to increase the amount of insoluble precipitate in salt form, $i.\ e.$ a mineral has a higher affinity for certain complex (InsP-mineral) which generates more insoluble salts[40]. For instance, the phosphate inositol bound to Ca shows higher affinity to Zn, which decreases its reabsorption [41].

1.1.3. Bioavailability of proteins

The degree of interaction between phytic acid (and its phosphate inositols) and depends on the protein, net charge, conformation and interactions with minerals at a given pH.At low pH, below the isoelectric point of proteins, phytic acid phosphate esters bind to the cationic group of basic amino acids, for example, arginine, histidine and lysine, may form InsP-protein complexes.

At a pH above the isoelectric point of proteins, since the charge of proteins as well as that of the phytic acid is negative, the interaction would be impossible, however, interaction occurs through the formation of complexes with divalent such as Ca^{2+} or Mg^{2+}. This binding takes place via the formation of ionized carboxyl groups and the deprotonated imidazole group of histidine, which requires a minimum concentration of these cations to maintain these complexes. At this pH, some binary complexes may exist because lisyl and arginyl residues of the proteins are still positively charged. At high pH the interaction between proteins and phytic acid decreases, arginyl and lysyl groups lose their charge, and therefore its ability to form binary complexes [36]; as well, they may form complexes such as protein-InsPand protein-mineral-InsP(Figure 4), which reduces their bioavailability.Such complexes may affect the protein structure, which may decrease the enzyme activity, function, solubility, absorption and protein digestibility [16, 36]. Particularly, the ability to inhibit proteolytic(pepsin, trypsin, chymotrypsin), amylolytic (amylase) and lipolytic (lipase) enzymes, is responsible of their anti-nutritional properties. This inhibition may be due to the nonspecific nature of InsP-protein interactions and the chelation of calcium ions, which are essential for the activity of trypsin and α-amylase [16]. Furthermore, phytic acid can also bind to starch, either directly via hydrogen bonds or indirectly through the proteins to which it is associated [42].

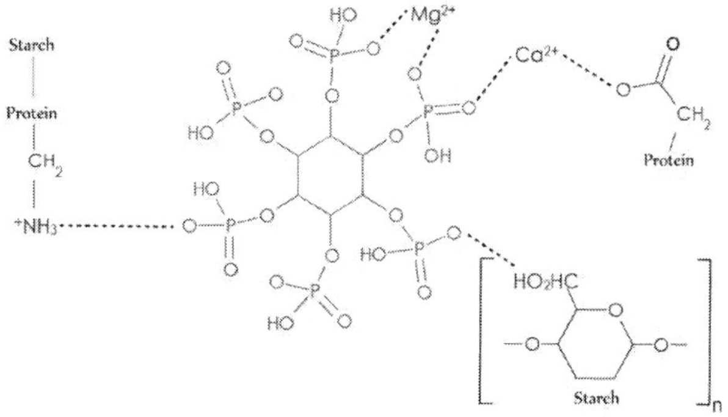

Figure 4. InsPinteractions with minerals, proteins and starch [36].

1.1.4. Pharmacological properties

In vivo and *in vitro* studies have shown that phytic acid (InsP$_6$)has prevention and therapeutic properties against cancer. Several mechanisms have been suggested to explain its anti-carcinogenic effect:

a. Experiments have shown that this compound induces apoptosis in cancer cells, causes differentiation of malignant cells and its reversion to normal phenotype, and increases the activity of natural killer cells of the immune system. In addition, IP$_3$ and IP$_4$ compounds have an important role in cellular signal transduction, regulating functions, cell growth and differentiation [43].

b. A second way in which phytic acid reduces the risk of cancer, is by chelation of Fe^{3+} and the suppression of the formation of radicals ($\bullet$OH), which also originates antioxidant properties. Fe^{3+} is an effective catalyst for many biological functions, in which this ion is reduced to Fe^{2+}. The oxidation of Fe^{2+}to Fe^{3+} leads to the formation of O$^{2-\bullet}$ that spontaneously generates O$_2$ and H$_2$O$_2$. The Fenton's reagent (Fe^{2+} + H$_2$O$_2$) quickly generates $\bullet$OH, a highly-reactive oxyradical which indiscriminately attacks most of the biomolecules. By blocking the redox cycle of Fe, which is necessary in many oxidation reactions, the lipid peroxidation and DNA damage are inhibited [37 , 44].

c. Zn is involved in DNA synthesis and cell proliferation as a cofactor for many enzymes like thymidine kinase. So, by binding to Zn^{2+}, the phytic acid indirectly reduces cell proliferation [45].

d.   Phytic acid can reduce the starch digestibility and cause low absorption, so that starch remains available in the colon to be fermented by bacteria, producing short chain fatty acids, which have protective activity against cancer [46].

Phytic acid can retard the digestion and absorption of starch in several ways: by direct binding to the polysaccharide, by its binding to the α-amylase, or by chelation of Ca^{2+} needed for activation of α-amylase. Through these mechanisms a delay occurs in the glycemic response, therefore, due to lower blood glucose, insulin is required in less amount and this reduces the risk of diabetes [47].

Respect to prevention of kidney stones and treatment of hypercalciuria, experimental evidences demonstrate that di- and tri- inositol phosphates ($InsP_2$ and $InsP_3$), are effective to prevent the formation of hydroxyapatite crystals *in vitro*, which act as the core for the formation of some kidney stones [24].

At levels from 0.2 to 9% of phytic acid in diet, the plasmatic levels of cholesterol and triglycerides are significantly reduced [48]. This seems to be related with the capability of phytic acid to be bound to Zn, which reduces the serum levels of Zn and the Zn/Cu ratio, since high values of this ratio tend to increase the risk of cardiovascular diseases, for instance, hypercholesterolemia [42].

1.2. Saponins

Saponins (Figure 5) are a big group of glycosides which are known by their surfactant properties and are widely distributed in green plants [49] .The name 'saponin' derives from the Latin word *sapo* which means soap, due to their property of generating foam in agitated aqueous solutions [50]. These substances are amphiphilic glycosides, wherein the polar constituents are sugars (pentoses, hexoses or uronic acids) that are covalently linked to a nonpolar group, which consists of an aglycone, called sapogenin, which can be either steroidal or triterpenoid. This combination of polar and nonpolar components in their molecular structure explains their surfactant property in aqueous solutions [51].

As mentioned above, the saponins are secondary metabolites that can be classified into two groups based on the nature of the aglycone skeleton. The first group consists of steroidal saponins, which are present almost exclusively in monocotyledons angiosperms. The second group is composed of triterpenoidsaponins, which occur mainly in dicototyledonous flowering plants [52]. Steroidal saponins comprise a steroidal aglycone, a spirostane skeleton of 27 carbons (C_{27}), which generally comprises a six-ring structure. In some cases, the hydroxyl group at position 26 is used to form a glycosidic bond, so that the structure of the aglycone becomes a pentacyclic structure; this structure is known as furostano skeleton. The triterpenoidsaponins have an aglycone with a backbone of 30 carbons (C_{30}), which form a pentacyclic structure (Figure 5).

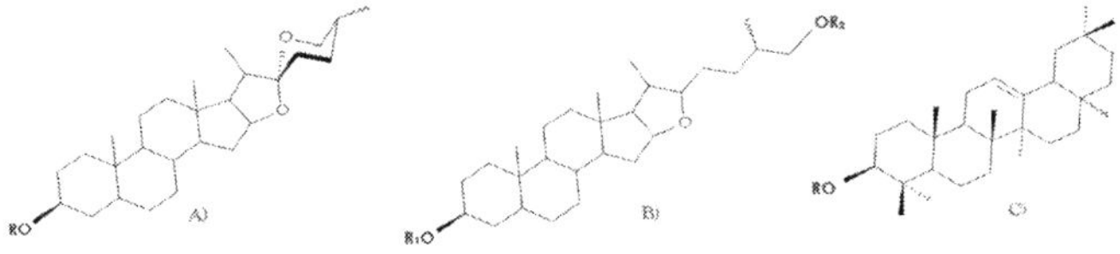

Figure 5. Skeletons ofaglycone: (A) steroidal spirostane, (B) steroidal furostane (C) triterpenoid. R = sugar residue.

It has been identified that soy contains saponins with triterpenoid-type aglycones, this kind of aglycones are subdivided into five major groups; soysapogenol A, B, C, D and E (Figure 6), and their glycosides are correspondingly called as saponins of group A, group B, and so on[53, 54].From this classification, four aglycones (soysapogenol A, B, C and E) [55]were isolated after hydrolisis of soy saponins, specifically five saponins were identified 5 with two distinct types of aglycones: soysapogenin I (the main component), soysapogenins II and III, which contain soysapogenol B, and soysaponins A1, A2 and A3, which contain soysapogenol A[55].The saponins containing soyasapogenol C and E have not been found in soybeans, so these aglycones could be formed as a product during the hydrolysis of saponins[56].Another study reported the isolation and characterization of soysaponin IV. The type of sugars attached to the aglycones found in soybeans have been identified as rhamnose, galactose, glucose, arabinose, xylose and glucuronic acid [55].

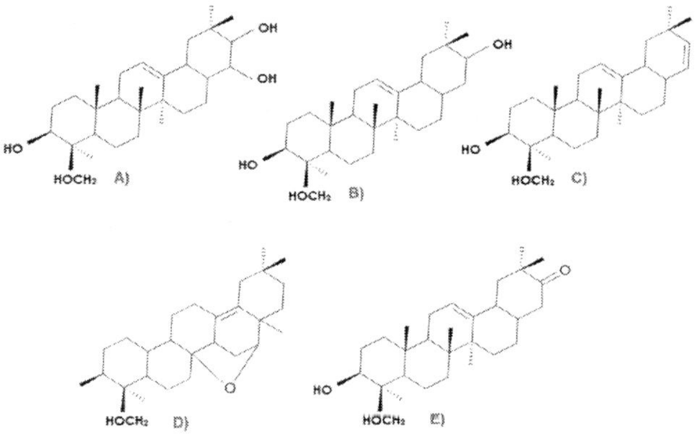

Figure 6. Groups of soyasapogenols *(Oakenfull, 1981).*

The total content of saponins in the hypocotyl fraction of soybeans, where acetyl-soyasaponins A1 and A4, mainly, are syntesized, is approximately 0.62 to 6.16%[57]. Other works have reported 5-6% saponin content in soybeans [58]. Lower values, approximately 0.6%, have been reported as well [59].

1.2.1. Synthesis and Function

The capability to synthesize saponins is widespread among plants belonging to the *Magnoliophyta* division, which includes both dicotyledons and monocotyledons. However, most of saponin-producing species are within dicotyledonous plants. The biological function of saponins is not completely understood. They are usually considered as a part of the defense system of the plant, due to their antimicrobial, fungicide, allelopathic, insecticide and molluscicide activities [60]. The synthesized saponins are accumulated during the regular growth of plants. Nonetheless, their accumulation is influenced by several environmental factors such as bioavailability of nutrients and water, solar radiation or a combination of them[61]. Some studies on soy have shown a variation in the content of saponins in soybeans with different degrees of maturity, however, the nature of this variation is not sufficient to influence on the saponindistribution in different varieties [57]. Little is known about the enzymes and biochemical pathways involved in the biosynthesis of saponins in plants [54]. However, two key aspects have been suggested for biosynthesis: the first one is the cyclization of the 2,3-oxidoscualene thought the isoprenoid pathway, which is a starting point for the biosynthesis of the sapogenin, and the second one is the glycosylation of sapogenins.

1.2.2. Membranolytic activity

Saponins have the ability to cleavege the erythrocytes. This hemolytic property is generally attributed to the interaction between saponins and sterols in the erythrocyte membrane. As a result, the membrane is broken, which causes an increase in its permeability and the consequent loss of hemoglobin. It has been investigated the effect of saponins in the membrane structure through human erythrocyte hemolysis[5, 62]. The results indicated that the fracture in the erythrocyte membrane was not closed again, so that the damage in the lipid bilayer is irreversible. However, this toxic property is difficult to occur *in vivo*, since there is evidence that no complications are detected when saponins are ingested orally, which reduces his hemolytic capability to *in vitro* studies [63]. The saponins have little anti-nutritional activity, given no damage is produced in humans when they are consumed in the amounts regularly found after food processing. However, high concentrations of saponins are also capable of breaking the membrane of other cells such as those of the intestinal mucosa, which modifies the cell membrane permeability, and then affects the active transport and the absorption of nutrients[45].

This ability to affect the cell membrane, depends on the structural characteristics of the saponins, *i. e.* the structure of the aglycone, the number of sugars in the side chains and the side chains length [64]. In Figure 7, the interaction of saponins with cell membranes is schematically shown.

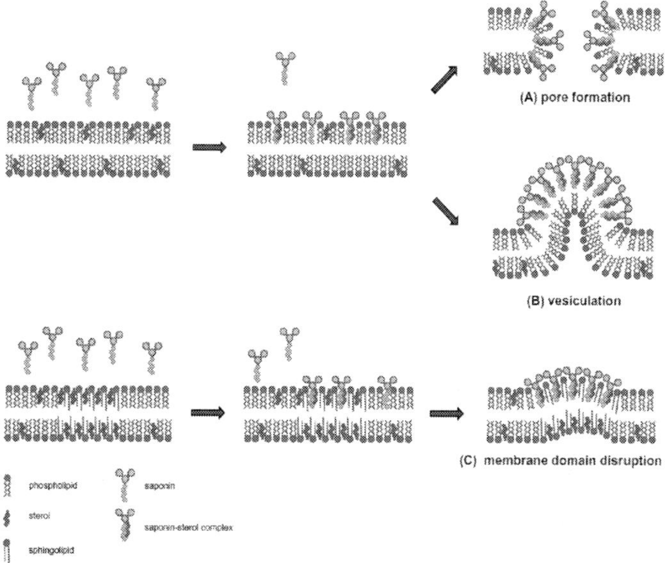

Figure 7. Schematic models of the molecular mechanisms of saponin activities towards membranes [65]Saponins integrate with their hydrophobic part (sapogenin) into the membrane. Within the membrane they form complexes with sterols, which subsequently, driven by interaction of their extra-membranous orientated saccharide residues, accumulate into plaques. Sterical interference of these saccharide moieties causes membrane curvature subsequently leading to (A) pore formation in the membrane [66] or (B) hemitubular protuberances resulting in sterol extraction via vesiculation[67]. Alternatively, after membrane integration saponins may migrate towards sphingolipid/sterol enriched membrane domains (C) prior to complex formation with the incorporated sterols, thereby interfering with specific domain functionalities [68]. Similarly to (B), accumulation of saponins in confined membrane domains has further been suggested to cause deconstructive membrane curvature in a dose-dependent manner.

1.2.3. Pharmacological properties

Many studies highlight the pharmaceutical properties of the soybean saponins, among which the anti-carcinogenic activity is mentioned, given by the membranolyticactivity these molecules have shown in human cells of colon carcinoma[69, 70]. Other studies have demonstrated hypocholesterolemic activity due to depletion of body cholesterol by preventing its reabsorption, thus increasing its excretion. Soluble fibers in legumes are known to increase the viscosity of gastric and intestinal contents, and may be one of the factors responsible for the lowering of cholesterol levels [71, 72]. Studies on health benefits of saponins suggest their hepatoprotective activity, but these studies are limited to cell culture and few animal studies. Studies on rats have shown soybean saponin to have an anabolic effect on bone components, suggesting its role as a nutritional factor in the prevention of osteoporosis [73]. Another activity that has been reported is anti-mutagenicity in breast cells [74].

1.3. Inhibitors of trypsin

Protease inhibitors are proteins widely distributed in the plant kingdom, have the ability to inhibit the proteolytic activity of digestive enzymes such as serine-proteases (trypsin and chymotrypsin) which are characteristic of the gastrointestinal tract of animals, though also may inhibit endogenous proteases and enzymes of bacteria, fungi and insects. These serine-protease inhibitors are proteins that form very stable complexes with digestive enzymes, which preventtheir catalytic activity [75].

Protease inhibitors have been classified into several families based on homology in the sequence of amino acids in the inhibitory sites. The molecular structure of the inhibitor affects both the force and the specificity of the inhibitor. The two main families of protease inhibitors found in legumes are the Kunitz inhibitor and the Bowman-Birk inhibitor, so named after its isolation [2, 51]. In the latter case, the characterization was carried out by *Birk, [76]*, so this name was added.

Both types of proteases are found in soybeans (*Glycine max*); in other legume seeds, such as beans (*Phaseolus vulgaris*) and lentil (*Lens culinaris*), protease inhibitors have been characterized as members of the Bowman-Birk family. Both inhibitors are water soluble proteins (albumin) and constitute from 0.2 to 2% of total soluble protein of legumes [75, 77], particularly soybeans havereported 50 trypsin inhibitor units / mg of dry sample [78].

1.3.1. Kunitztype inhibitor

The first protease inhibitor to be isolated and characterized was the Kunitz inhibitor. It has a MW between 18 and 24 kDa and contains between 170 and 200 amino acid residues. These inhibitors have one head, i. e., one molecule of inhibitor inactivates one molecule of trypsin. It is a competitive inhibitor, binds to the active sites of trypsin in the same way the substrate of the enzyme does, resulting in the hydrolysis of peptide bonds between amino acids of the reactive site of the inhibitor or the substrate (Figure 8).

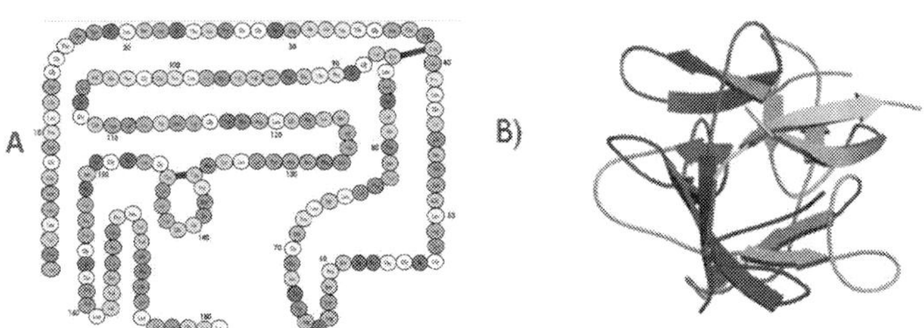

Figure 8. A) Primary structure of the Kunitz inhibitor from soybean [79]. Disulphide bonds are shown in black, B)Tridimentional structure Kunitz inhibitor from soybean [79].

Inhibitors differ from the substrate protein in the reactive site residues, which are linked via disulphide bonds. After hydrolysis, the modified inhibitor maintains the same conformation, due to the disulphide bonds. This generates a stable enzyme-inhibitor complex. This type of inhibitors are generally absent in seeds of *Phaseolus*, *Pisum*, *Vigna unguiculata* and *Glycine max* [80].

1.3.2. Bowman-Birk type inhibitor

These inhibitors are low molecular weight polypeptides (7 to 9 kDa) containing 60 to 85 amino acid residues (Figure 9). They have several disulphide bonds which make them stable to heat, acids and bases. These inhibitors have two heads (two separate sites of inhibition) and are competitive inhibitors. They can simultaneously and independently inhibit two enzymes, thus, there are trypsin/trypsin are trypsin/chymotrypsin inhibitors [74,77].

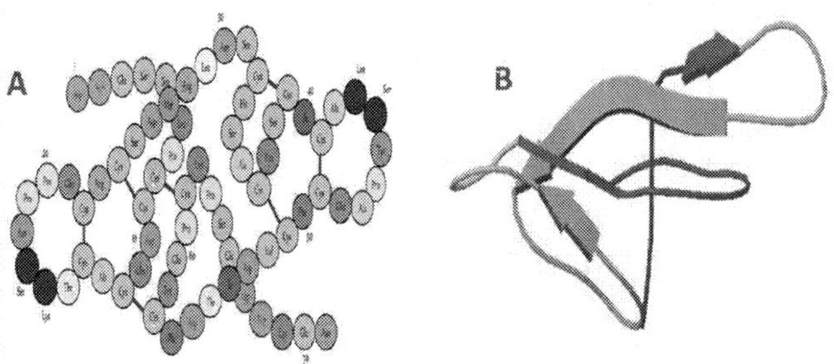

Figure 9. A) Primary structure of Bowman-Birk type inhibitor from soybean*(Odani y Ikenaka, 1973). Disulphide bondsand active sites for trypsin (Lys16-Ser17) and chymotrypsin (Leu44-Ser45) are shown in black B)* Tridimentional structure Bowman-Birk inhibitor from soybean [81]

An example of this type of inhibitor is the Bowman-Birk inhibitor from soybeans, which is constituted by a polypeptide chain of 71 amino acids, containing seven disulphide bonds. It has a MW of 8 kDa and is called dual head inhibitor because it has independent binding sites for trypsin and chymotrypsin, so that the active site for trypsin is Lys16-Ser17, whereas for chymotrypsin is Leu44-Ser45 [82, 83].

1.3.3. Synthesis and Function

Protease inhibitors have a regulatory function; they are involved the in proteolytic self-regulation process of the protein deposited in the protein bodies before and during the seed germination by inhibiting endogenous proteases. They also participate as protective agents against insects and microorganisms[76].

1.3.4. Anti-nutritional properties

Protease inhibitors ingested within legumes have adverse effects in animals. First, these compounds form inactive complexes with trypsin/chymotrypsin, so that the levels of these free digestive enzymes are reduced, thus making difficult proteolysis and amino acid absorption. In addition, these enzyme-inhibitor complexes, which are rich in sulfur amino acids, are excreted.

Finally, these inhibitors cause hypertrophy/hyperplasia of the pancreas due to chronic hypersecretion of pancreatic enzymes [trypsin and chymotrypsin), which leads to deviation of the sulfur amino acids that were used to synthesize tissue proteins to the synthesis of these enzymes [84]. All this derives in reduction of the amount of essential amino acids, which inhibits animal growth and exacerbates an already critical situation with respect to the protein of legumes which is deficient in sulfur amino acids [3, 84-87].

The mechanism by which inhibitors of proteases stimulate pancreatic secretion is not entirely clear. There is a theory about this secretion would be regulated by a negative feedback mechanism, so that, when the content of trypsin/chymotrypsin in duodenum is reduced below a certain level, the endocrine cells of the duodenal mucosa release the hormone cholecystokinin, prompting the pancreas to synthesize more serine-proteases (Figure 20). The reduced levels of trypsin and chymotrypsin are produced when the protease inhibitors ingested reach the duodenum and bind to these digestive enzymes by forming complexes. Although this does not seem to be the only mechanism by which the pancreatic secretion is activated.

Recent studies have demonstrated that both states of protease inhibitors, free and enzyme-inhibitor complexes, bind to the duodenal mucosa and stimulate the release of cholecystokinin, thus increasing the pancreatic secretion of serine proteases [84, 88]. The action of the trypsin inhibitors on the human organism is not totally understood, since human trypsin has two forms: cationic, which is the main component of pancreatic juice and is weakly inhibited; and anionic, comprising about 10 to 20% of the total trypsin, which is completely inhibited [82, 84].

1.3.5. Pharmacological properties

Since the Bowman-Birk type inhibitors are proteins with a high amount of cysteine, these inhibitors make an important contribution to the content of sulfur amino acids, thus increasing the nutritional value of legumes[85, 89]. The Bowman-Birk inhibitor from soybeans as well as their counterparts present in other legumes, are involved in the prevention and treatment of cancer (colon, breast, liver, lung, prostate, etc.) by inhibiting chymotrypsin. One mechanism through which these compounds can prevent carcinogenesis is by reducing the protein digestibility and the bioavailability of amino acids such as leucine, phenylalanine or tyrosine, which are necessary for the development of cancer cells [90-92].

1.4. Isoflavones

Isoflavones are widely distributed in the plant kingdom, mainly in plants of the legume family, being soybeans the source with the highest content of these components [93]. Isoflavones are oxygen heterocycles containing a 3-phenylchroman skeleton that is hydroxylated at 40 and 7 positions (Figura 10) [94]. Based on the substitution pattern on carbons and 6, three aglycon forms of isoflavones commonly found in soybeans are daidzein, genistein, and glycitein. These three isoflavones can also exist in conjugated forms with glucose (daidzin, geinstin, glycitin), malonylglucose (malonyldaidzin, malonylgeinstin, malonylglycitin), and acetylglucose (acetyldaidzin, acetylgeinstin, acetylglycitin) units. Thus 12 free and conjugated forms of isoflavones have been isolated from different soybean samples (Table 1) [95].

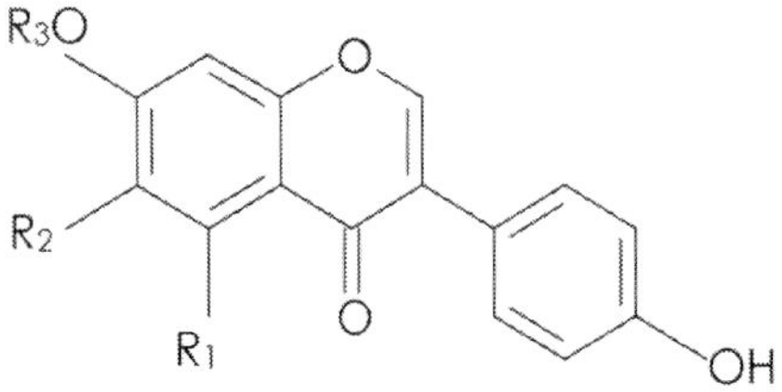

Figure 10. Chemical structureof an isoflavone [95].

Name	R_1	R_2	R_3
Daidzein	H	H	H
Glycitein	H	OCH_3	H
Genistein	OH	H	H
Daidzin	H	H	Glu
Glycitin	H	OCH_3	Glu
Genistin	OH	H	Glu
Acetyldaidzin	H	H	Glu-$COCH_3$
Acetylglycitin	H	OCH_3	Glu-$COCH_3$
Acetylgenistin	OH	H	Glu-$COCH_3$
Malonyldaidzin	H	H	Glu-$COCH_2COOH$
Malonylglycitin	H	OCH_3	Glu-$COCH_2COOH$

Table 1. Chemical structures of 12 isoflavones isolated from soybeans [95].

1.4.1. Synthesis and Function

The variation of the concentration of isoflavones in soybeans is mainly due to the soybean variety, environment, location and growing conditions, such as year, area and temperature, post-harvest storage and the methodology used to determine this concentration [95]. The content of isoflavones in soybeans ranks 1.2 to 2.4 mg of total isoflavones per gram of sample [96], distributed in different concentrations in the tissues of the seed, being higher in the embryo than in the endosperm [97].

In plants, these compounds play several roles, such as protection against UV light and phytopathogens, signal transduction during nodulation, attraction of pollinator animals and defense against insects and herbivores[98].

1.4.2. Pharmacological properties

The enriched extracts of isoflavones have been evaluated for prevention of a wide range of health problems associated with cardiovascular diseases, osteoporosis and breast cancer, prostate and colon [99]. Furthermore, soy isoflavones have a structure very similar to a phenolic estrogen known as phytoestrogen, so that these compounds have been used as a natural alternative for postmenopausal therapy [100].

1.5. Lectins

In soybeans, a class of proteins called lectins or phytohemagglutinins is present. These compounds can be defined as proteins or glycoproteins of non-immune origin, which can reversibly bind to specific sugar segments through hydrogen bonds and Van Der Waals interactions, with one or more binding sites per subunit [101]. Lectins are tetrameric proteins composed of two different types of subunits: E-type subunit (MW = 34 kDa) and L-type subunit (32 kDa). The first one has the characteristic of binding to erythrocytes, while the second one to lymphocytes[102].Therefore, it is possible to find 5 combinations of these four subunits, i. e. 5 isoforms, as follows: E_4, E_3L, E_2L_2, EL_3, and L_4. Soybean seeds show hemagglutinating activity at 2400 mg per mg of dry sample [78].

The name lectin [from Latin legere, which means to choose or to select), was adopted by Boyd for many years to emphasize the caability of some lectins to bind specifically to cells of the ABO blood groups [103]. Currently the name lectinis preferred over the haemagglutinin one and is widely used to denote all vegetable proteins that possess at least one non-catalytic domain, which binds reversibly to a specific mono- or oligosaccharide [104].

According to the overall structure of the plant lectins, these are subdivided into four main classes: Merolectins which are proteins having a single carbohydrate-binding domain; Hololectins, comprising all lectins having di- or multivalent carbohydrate-binding sites; Chimerolectins, proteins consisting of one or more carbohydrate-binding domain(s) plus an additional catalytic or another biological activity dependent on a distinct domain other than the carbohydrate- binding site; and Superlectins which also possess at least two carbohydrate-binding domains but differ from the hololectins because their sites are able to recognise structurally unrelated sugars [105].

Lectins can be divided according to the monosaccharide for which they show the highest affinity: D-mannose/D-glucose, D-galactose/N-acetyl-D-galactosamine, L-fucose and N-acetyl-glucosamineacid [106]. Thus depending on the specificity toward a given monosaccharide the lectin will selectively bind to one of these above sugars which are typical constituents of eukaryotic cell surfaces [101].

1.5.1. Synthesis and Function

The wide distribution of lectins in all tissues of plants and their ubiquitous presence in the plant kingdom suggest important roles for these proteins. One possible physiological function that has emerged is the defensive role of these carbohydrate-binding proteins against phytopathogenic microorganisms, phytophagous insects and plant-eating animals [102, 107]. Indeed it has been shown that plant lectins possess cytotoxic, fungitoxic, anti-insect and anti-nematode properties either in vitro or in vivo and are toxic to higher animals[63, 81, 104]. One of the most important features of plant lectins, compatible with the proposed defensive function, is the remarkably high resistance to proteolysis and stability over a large range of pH, even when they are out of their natural environment [103].

1.5.2. Anti-nutritional properties

Some of these were found to be toxic or antinutritional for man and animals. In general, nausea, bloating, vomiting and diarrhoea characterize the oral acute toxicity of lectins on humans exposed to them.

In experimental animals fed on diets containing plant lectins the evident symptoms are loss of appetite, decreased body weight and eventually death [84, 108].

As most lectins are not degraded during their passage through the digestive tract they are able to bind the epithelial cells which express carbohydrate moieties recognised by them. This event is undoubtedly the second one in importance for determining the toxicity of orally fed lectins. Indeed, lectins which are not bound by the mucosa usually induce little or no harmful antinutritive effect for the consumers [88]. Once bound to the digestive tract, the lectin can cause dramatic changes in the cellular morphology and metabolism of the stomach and/ or small intestine and activate a cascade of signals which alters the intermediary metabolism. Thus, lectins may induce changes in some, or all, of the digestive, absorptive, protective or secretory functions of the whole digestive system and affect cellular proliferation and turnover. In 1960, Jaffe' suggested that the toxic effects of ingested lectins were due to their ability to combine with specific receptor sites of the cells lining the small intestine and to cause a non-specific interference with absorption and nutrient utilisation [50].

Author details

A. Cabrera-Orozco, C. Jiménez-Martínez and G. Dávila-Ortiz[*]

*Address all correspondence to: crisjm_99@yahoo.com

Graduates in Food, National School of Biological Sciences.National Polytechnic Institute, Mexico

References

[1] Rex, Newkirk. P. (2010). SoyBean, Feed Industry Guide, CIGI. *Canadian International Grains Institute*, 1.

[2] Bowman, Y. (1944). Fractions derived from soybean and navy beans which retard tryptic digestion of casein. Proc Soc Exp Biol Med; , 57, 139.

[3] Grant, G. (1989). Anti-nutritional effects of soyabean: a review.

[4] E, L. I. (1989). Antinutritional factors. *In: Legumes: Chemistry, Technology, and Human Nutrition. Matthews RH, editor: M. Dekker.*

[5] Baumann, E., Stoya, G., Völkner, A., Richter, W., Lemke, C., & Linss, W. (2000). Hemolysis of human erythrocytes with saponin affects the membrane structure. *Acta Histochemica*, 102(1), 21-35.

[6] Silveira, Rodríguez. M. B., Monereo, Megías. S., & Molina, Baena. B. (2003). Alimentos funcionales y nutrición óptima: "Cerca o lejos" . *Revista Española de Salud Pública*, 77, 317-331.

[7] Khokhar, S., Frias, J., Price, K. R., Fenwick, G. R., & Hedley, C. L. (1996). Physico-Chemical Characteristics of Khesari Dhal (Lathyrus sativus): Changes in α-Galacto-sides, Monosaccharides and Disaccharides during Food Processing. *Journal of the Science of Food and Agriculture*, 70(4), 487-492.

[8] Grela, E. R., T. , S., & J., M. (2001). Antinutritional factors in seeds of Lathyrus sativus cultivated in Poland. *Lathyrus Lathyrism Newsletter*, 2, 101.

[9] Chung, T. K., Wong, T. Y., Wei, I. C., Huang , W. Y., & Lin, Y. (1998). Tannins and Human Health: A Review. *Critical Reviews in Food Science and Nutrition*, 38(6), 421-64.

[10] Liener, I. E. (1994). Implications of antinutritional components in soybean foods. *Critical Reviews in Food Science and Nutrition*, 34(1), 31-67.

[11] Osman, M. A., Reid, P. M., & Weber, C. W. (2002). Thermal inactivation of tepary bean (Phaseolus acutifolius), soybean and lima bean protease inhibitors: effect of acidic and basic pH. *Food Chemistry*, 78(4), 419-423.

[12] Holm, P. B., Kristiansen, K. N., & Pedersen, H. B. (2002). Transgenic Approaches in Commonly Consumed Cereals to Improve Iron and Zinc Content and Bioavailability. *The Journal of Nutrition*, 132(3), 514S-516S.

[13] Reddy, N. R., & Salunkhe, D. K. (1981). Interactions Between Phytate, Protein, and Minerals in Whey Fractions of Black Gram. *Journal of Food Science*, 46(2), 564-567.

[14] Loewus, F. A., & Murthy, P. P. N. (2000). myo-Inositol metabolism in plants. *Plant Science*, 150(1), 1-19.

[15] Raboy, V. (1997). Accumulation and storage of phosphate and minerals. *In: Cellular and Molecular Biology of Plant Seed Development: Kluwer Academic Publishers.*

[16] Cheryan, M., & Rackis, J. J. (1980). Phytic acid interactions in food systems. *C R C,Critical Reviews in Food Science and Nutrition*, 13(4), 297-335.

[17] Beleia, A., Thu, Thao. L. T., & Ida, E. I. (1993). Lowering Phytic Phosphorus by Hydration of Soybeans. *Journal of Food Science*, 58(2), 375-377.

[18] Cosgrove, D. J., & Irving, G. C. J. (1980). Inositol phosphates: their chemistry, biochemistry, and physiology. Elsevier Scientific Pub. Co.

[19] Honke, J., Kozłowska, H., Vidal-Valverde, C., Frias, J., & Górecki, R. (1998). Changes in quantities of inositol phosphates during maturation and germination of legume seeds. *Zeitschrift für Lebensmitteluntersuchung und-Forschung A*, 206(4), 279-83.

[20] Irving, G. C. J. (1980). Phytase. *In: Inositol phosphates: their chemistry, biochemistry, and physiology: Elsevier Scientific Pub. Co.*

[21] Nayini, N. R., & Markakis, P. (1986). Phytases. *In: Phytic acid: chemistry & applications: Pilatus Press.*

[22] Raboy, V. (2003). myo-Inositol-1,2,3,4,5,6-hexakisphosphate. *Phytochemistry*, 64(6), 1033-43.

[23] Graf, E., Empson, K. L., & Eaton, J. W. (1987). Phytic acid. A natural antioxidant. *Journal of Biological Chemistry*, 262(24), 11647-50.

[24] Zhou, J. R., & Erdman, J. W. (1995). Phytic acid in health and disease. *Critical Reviews in Food Science and Nutrition*, 35(6), 495-508.

[25] York, J. D., Odom, A. R., Murphy, R., Ives, E. B., & Wente, S. R. (1999). A Phospholipase C-Dependent Inositol Polyphosphate Kinase Pathway Required for Efficient Messenger RNA Export. *Science*, 285(5424), 96-100.

[26] Lemtiri-Chlieh, F., Mac, Robbie. E. A. C., & Brearley, CA. (2000). Inositol hexakisphosphate is a physiological signal regulating the K+-inward rectifying conductance in guard cells. *Proceedings of the National Academy of Sciences*, 97(15), 8687-92.

[27] Laussmann, T., Pikzack, C., Thiel, U., Mayr, G. W., & Vogel, G. (2000). Diphospho-myo-inositol phosphates during the life cycle of Dictyostelium and Polysphondylium. *European Journal of Biochemistry*, 267(8), 447-51.

[28] Shears, S. B. (2001). Assessing the omnipotence of inositol hexakisphosphate. *Cell Signal*, 13(3), 151-8.

[29] Raboy, V. (2009). Approaches and challenges to engineering seed phytate and total phosphorus. *Plant Science*, 177(4), 281-96.

[30] Reddy, N. R., & Sathe, S. K. (2002). Food Phytates. *CRC Press.*

[31] Nolan, K. B., Duffin, P. A., & Mc Weeny, D. J. (1987). Effects of phytate on mineral bioavailability. in vitro studies on Mg2+, Ca2+, Fe3+, Cu2+ and Zn2+ (also Cd2+) solubilities in the presence of phytate. *Journal of the Science of Food and Agriculture*, 40(1), 79-85.

[32] Fredlund, K., Isaksson, M., Rossander-Hulthén, L., Almgren, A., & Sandberg-S, A. (2006). Absorption of zinc and retention of calcium: Dose-dependent inhibition by phytate. *Journal of Trace Elements in Medicine and Biology*, 20(1), 49-57.

[33] Lopez, H. W., Leenhardt, F., Coudray, C., & Remesy, C. (2002). Minerals and phytic acid interactions: is it a real problem for human nutrition? *International Journal of Food Science & Technology*, 37(7), 727-739.

[34] Lönnerdal, B., Sandberg, A. S., Sandström, B., & Kunz, C. (1989). Inhibitory effects of phytic acid and other inositol phosphates on zinc and calcium absorption in suckling rats. *The Journal of Nutrition*, 119(2), 211-214.

[35] Torre, M., Rodriguez, A. R., & Saura-, Calixto. F. (1991). Effects of dietary fiber and phytic acid on mineral availability. *Critical Reviews in Food Science and Nutrition*, 30(1), 1-22.

[36] Thompson, L. (1987). Reduction of phytic acid concentration in protein isolates by acylation techniques. *Journal of the American Oil Chemists' Society*, 64(12), 1712-7.

[37] Graf, E., & Eaton, J. W. (1993). Suppression of colonic cancer by dietary phytic acid. *Nutrition and Cancer*, 19(1), 11-9.

[38] Gifford-Steffen, S. R. (1993). Effect of varying concentrations of phytate, calcium, and zinc on the solubility of protein, calcium, zinc, and phytate in soy protein concentrate. *Journal of food protection*, 56(1), 42.

[39] Schlemmer, U., & Müller, H. D. J. K. (1995). The degradation of phytic acid in legumes prepared by different methods. *Eur J Clin Nutr*, 49(3), S207-10.

[40] Sandberg, A. S., Larsen, T., & Sandström, B. (1993). High dietary calcium level decreases colonic phytate degradation in pigs fed a rapeseed diet. *The Journal of Nutrition*, 123(3), 559-66.

[41] Hardy, R. (1998). Phytate Aquaculture Magazine. 24(6), 77.

[42] Thompson, L. U. (1993). Potential health benefits and problems associated with antinutrients in foods. *Food Research International*, 26(2), 131-49.

[43] Shamsuddin, A. M. (2002). Anti-cancer function of phytic acid. *International Journal of Food Science & Technology*, 37(7), 769-82.

[44] Graf, E., & Eaton, J. W. (1985). Dietary suppression of colonic cancer fiber or phytate? *Cancer*, 56(4), 717-8.

[45] Gee, J. M., Price, K. R., Ridout, C. L., Wortley, G. M., Hurrell, R. F., & Johnson, I. T. (1993). Saponins of quinoa (Chenopodium quinoa): Effects of processing on their abundance in quinoa products and their biological effects on intestinal mucosal tissue. *Journal of the Science of Food and Agriculture*, 63(2), 201-9.

[46] Thompson, L. U., & Zhang, L. (1991). Phytic acid and minerals: effect on early markers of risk for mammary and colon carcinogenesis. *Carcinogenesis*, 12(11), 2041-5.

[47] Rickard, Sharon. E., & Thompson, Lilian. U. (1997). Interactions and Biological Effects of Phytic Acid. *Antinutrients and Phytochemicals in Food: American Chemical Society*, 294-312.

[48] Jariwallar, R. J. (1999). Inositol hexaphosphate (IP6) as an anti-neoplastic and lipid-lowering agent. *Anticancer research*, 19(5), 3699.

[49] Tyler, V. E., Brady, L. R., & Robbers, J. E. (1981). Pharmacognosy. *Lea & Febiger.*

[50] Jaffe', W. G. (1960). Studies on phytotoxins in beans. *Arzneimittel Forschung*, 10, 1012.

[51] Kunitz, M. (1945). Crystallization of a trypsin inhibitor from soybean. *Science*, 101(2635), 668-9.

[52] Bruneton, J. (1999). Pharmacognosy, Phytochemistry, Medicinal Plants. *Technique & Documentation.*

[53] Bondi, A., Birk, Y., & Gestetner, B. (1973). Forage saponins. *In Chemistry and biochemistry of herbage: Academic Press*, 17.

[54] Haralampidis, K., Trojanowska, M., & Osbourn, A. (2002). Biosynthesis of Triterpenoid Saponins in Plants. History and Trends in Bioprocessing and Biotransformation. *In: Dutta N, Hammar F, Haralampidis K, Karanth N, König A, Krishna S, et al., editors.: Springer Berlin / Heidelberg*, 31-49.

[55] Kitagawa, I., Yoshikawa, M., Wang, H., Saito, M., Tosirisuk, V., Fujiwara, T., et al. (1982). Revised structures of soyasapogenols A, B, and E, oleanene-sapogenols from soybean. *Structures of soyasaponins I, II, and III. Chemical & pharmaceutical bulletin*, 30(6), 2294-2297.

[56] Ireland, P. A., & Dziedzic, S. Z. (1986). Effect of hydrolysis on sapogenin release in soy. *Journal of Agricultural and Food Chemistry*, 34(6), 1037-1041.

[57] Shimoyamada, M., Kudo, S., Okubo, K., Yamauchi, F., & Harada, K. (1990). Distributions of saponin constituents in some varieties of soybean [Glycine max] plant. *Agricultural and Biological Chemistry*, 54(1), 77.

[58] Potter, B. V. L. (1990). Recent advances in the chemistry and biochemistry of inositol phosphates of biological interest. *Natural Product Reports*, 7(1), 1-24.

[59] Sodipo, O. A., & Arinze, H. U. (1985). Saponin content of some Nigerian foods. *Journal of the Science of Food and Agriculture*, 36(5), 407-408.

[60] Kerem, G. F., Makkar, Z. S. H P., & Klaus, B. (2002). The biological action of saponins in animal systems: a review. *British Journal of Nutrition*, 88, 587-605.

[61] Szakiel, A., Pączkowski, C., & Henry, M. (2011). Influence of environmental abiotic factors on the content of saponins in plants. *Phytochemistry Reviews*, 10(4), 471-491.

[62] De Geyter, E., Swevers, L., Soin, T., Geelen, D., & Smagghe, G. (2012). Saponins do not affect the ecdysteroid receptor complex but cause membrane permeation in insect culture cell lines. *Journal of Insect Physiology*, 58(1), 18-23.

[63] Committee NRCFP. (1973). Toxicants occurring naturally in foods. *National Academy of Sciences.*

[64] Kenji, O., Matsuda, H., Murakami, T., Katayama, S., Ohgitani, T., & Yoshikawa, M. (2000). Adjuvant and haemolytic activities of 47 saponins derived from medicinal and food plants. *Biological Chemistry*, 381(1), 67-74.

[65] Augustin, J. M., Kuzina, V., Andersen, S. B., & Bak, S. (2011). Molecular activities, biosynthesis and evolution of triterpenoid saponins. *Phytochemistry*, 72(6), 435-457.

[66] Armah, C. N., Mackie, A. R., Roy, C., Price, K., Osbourn, A. E., Bowyer, P., et al. (1999). The membrane-permeabilizing effect of avenacin A-1 involves the reorganization of bilayer cholesterol. *Biophys J*, 76(1), 281-290.

[67] Keukens, E. A. J., de Vrije, T., van den Boom, C., de Waard, P., Plasman, H. H., Thiel, F., et al. (1995). Molecular basis of glycoalkaloid induced membrane disruption. *Biochimica et Biophysica Acta (BBA)- Biomembranes*, 1240(2), 216-228.

[68] Lin, F., & Wang, R. (2010). Hemolytic mechanism of dioscin proposed by molecular dynamics simulations. *Journal of Molecular Modeling*, 16(1), 107-118.

[69] Rao, A. V., & Sung, M. K. (1995). Saponins as anticarcinogens. *Journal Nutrition*, 125(3), 717s.

[70] Sung, M. K., Kendall, C. W. C., & Rao, A. V. (1995). Effect of soybean saponins and gypsophila saponin on morphology of colon carcinoma cells in culture. *Food and Chemical Toxicology*, 33(5), 357-366.

[71] Amarowicz, R., & Shimoyamada, M. K. O. (1994). Hypocholesterolemic effects of saponins. *Roczniki Państwowego Zakładu Higieny*, 45(1-2), 125.

[72] Rao, A. V., & Gurfinkel, D. M. (2000). Dietary saponins and human health. *In:Saponins in Food, Feedstuffs and Medicinal Plants: Kluwer Academic Publishers*, 291.

[73] Antinutrients and Phytochemicals in Food:. (1997). *American Chemical Society*, 348.

[74] Berhow, M. A., Wagner, E. D., Vaughn, S. F., & Plewa, M. J. (2000). Characterization and antimutagenic activity of soybean saponins. *Mutation Research/Fundamental and Molecular Mechanisms of Mutagenesis*, 448(1), 11-22.

[75] Shewry, P. R., & Casey, R. (1999). Seed Proteins:. *Kluwer Academic.*

[76] Birk, Y. (1968). Chemistry and nutritional significance of proteinase inhibitors from plant sources. *Annals of the New York Academy of Sciences*, 146(2), 388-399.

[77] Savage, G. P., Morrison, S. C., & Benjamin, C. (2003). TRYPSIN INHIBITORS. *Encyclopedia of Food Sciences and Nutrition. Oxford: Academic Press*, 5878-5884.

[78] Valdebouze, P., Bergeron, E., Gaborit, T., & Delort-Laval, J. (1980). Content and distribution of trypsin inhibitors and hemagglutinins in some legume seeds. *Canadian Journal of Plant Science*, 60(2), 695-701.

[79] De Meester, P., Brick, P., Lloyd, L. F., Blow, D. M., & Onesti, S. (1998). Structure of the Kunitz type soybean trypsin inhibitor (STI): implication for the interactions between members of the STI family and tissue-plasminogen activator. *Acta Crystallogr, Sect*, 54, 589-97.

[80] Sathe, S. K. (2002). Dry Bean Protein Functionality. *Critical Reviews in Biotechnology*, 22(2), 175-223.

[81] Voss, R. H., Ermler, U., Essen, L. O., Wenzl, G., Kim, Y. M., & Flecker, P. (1996). Crystal structure of the bifunctional soybean Bowman-Birk inhibitor at 0.28-nm resolution. Structural peculiarities in a folded protein conformation. *EurJBiochem*, 242, 122.

[82] Liener, I. E., & Kakade, M. L. (1980). Protease inhibitors. *In:Toxic constituents of plant foodstuffs: Academic Press*.

[83] N., G. (1991). Proteinase inhibitors. *In:Toxic Substances in Crop Plants: Royal Society of Chemistry*.

[84] L., I. E. (1989). Antinutritional factors. *In: Legumes: Chemistry, Technology, and Human Nutrition: M. Dekker*.

[85] Hedemann, M. S., Welham, T., Boisen, S., Canibe, N., Bilham, L., & Domoney, C. (1999). Studies on the biological responses of rats to seed trypsin inhibitors using near-isogenic lines of Pisum sativum L (pea). *Journal of the Science of Food and Agriculture*, 79(12), 1647-53.

[86] Carbonaro, M., Grant, G., Cappelloni, M., & Pusztai, A. (2000). Perspectives into Factors Limiting in Vivo Digestion of Legume Proteins: Antinutritional Compounds or Storage Proteins? *Journal of Agricultural and Food Chemistry*, 48(3), 742-749.

[87] Friedman, M., & Brandon, D. L. (2001). Nutritional and Health Benefits of Soy Proteinst. *Journal of Agricultural and Food Chemistry*, 49(3), 1069-1086.

[88] P., A., G., G., B., S., & A., M.-CM. (2004). The mode of action of ANFs on the gastointestinal tract and its microflora. *In: Recent Advances of Research in Antinutritional Factors in Legume Seeds and Oilseeds: Proceedings of the Fourth International Workshop on Antinutritional Factors in Legume Seeds and Oilseeds, Toledo, Spain, 8-10March: Wageningen Academic Publishers*.

[89] Sastry, M. C. S., & Murray, D. R. (1987). The contribution of trypsin inhibitors to the nutritional value of chick pea seed protein. *Journal of the Science of Food and Agriculture*, 40(3), 253 261.

[90] Troll, W., & Frenkel, K. R. W. (1987). Protease inhibitors: possible preventive agents of various types of cancer and their mechanisms of action. *Prog Clin Biol Res*, 239, 297-315.

[91] Tamir, S., Bell, J., Finlay, T. H., Sakal, E., Smirnoff, P., Gaur, S., et al. (1996). Isolation, characterization, and properties of a trypsin-chymotrypsin inhibitor from amaranth seeds. *J Protein Chem*, 15(2), 219.

[92] Kennedy, A. R. (1998). The Bowman-Birk inhibitor from soybeans as an anticarcinogenic agent. *The American Journal of Clinical Nutrition*, 68(6), 1406S-1412S.

[93] Klejdus, B., Mikelová, R., Adam, V., Zehnálek, J., Vacek, J., Kizek, R., et al. (2004). Liquid chromatographic-mass spectrometric determination of genistin and daidzin in soybean food samples after accelerated solvent extraction with modified content of extraction cell. *Analytica Chimica Acta*, 517(1-2), 1-11.

[94] Peñalvo, J. L., Nurmi, T., & Adlercreutz, H. (2004). A simplified HPLC method for total isoflavones in soy products. *Food Chemistry*, 87(2), 297-305.

[95] Luthria, D. L., Biswas, R., & Natarajan, S. (2007). Comparison of extraction solvents and techniques used for the assay of isoflavones from soybean. *Food Chemistry*, 105(1), 325-33.

[96] Rostagno, M. A., Palma, M., & Barroso, C. G. (2004). Pressurized liquid extraction of isoflavones from soybeans. *Analytica Chimica Acta*, 522(2), 169-177.

[97] Kim, J. S., Kim, J. G., & Kim, W. J. (2004). Changes in Isoflavone and Oligosaccharides of Soybeans during Germination. *Korean Journal of Food Science and Technology*, 36(2), 294.

[98] Koes, R., Verweij, W., & Quattrocchio, F. (2005). Flavonoids: a colorful model for the regulation and evolution of biochemical pathways. *Trends in Plant Science*, 10(5), 236-242.

[99] Anderson, J. J. B. G. C. S. (1997). Phytoestrogens and Human Function. *Nutrition Today*, 32(6), 232-239.

[100] Messina, M. (1999). Soy, soy phytoestrogens (isoflavones), and breast cancer. *The American Journal of Clinical Nutrition*, 70(4), 574-5.

[101] Lis, H. , & N., S. (1998). Lectins:Carbohydrate-Specific Proteins That Mediate Cellular Recognition. *Chem Rev*, 98(2), 637.

[102] Chrispeels, M. J., & Raikhel, N. V. (1991). Lectins, Lectin Genes, and Their Role in Plant Defense. *The Plant Cell*, 3, 1.

[103] Vasconcelos, I. M., & Oliveira, J. T. A. (2004). Antinutritional properties of plant lectins. *Toxicon*, 44(4), 385-403.

[104] J., P. W., & J., V. D. E. (1995). Lectins as plant defense proteins. *Plant Physiol*, 109(2), 347.

[105] Damme, E. J. M. V., Peumans, W. J., Barre, A., & Rougé, P. (1998). Plant Lectins: A Composite of Several Distinct Families of Structurally and Evolutionary Related Proteins with Diverse Biological Roles. *Critical Reviews in Plant Sciences*, 17(6), 575-692.

[106] Goldstein, I. J., & Poretz, R. D. (2007). Isolation, physicochemical characterization, and carbohydrate-binding specificity of lectins. *In: Lectins: Springer*.

[107] Harper, S. M., Crenshaw, R. W., Mullins, M. A., & Privalle, L. S. (1995). Lectin Binding to Insect Brush Border Membranes. *Journal of Economic Entomology*, 88(5), 1197-202.

[108] Duranti, M., & Gius, C. (1997). Legume seeds: protein content and nutritional value. *Field Crops Research*, 53(1-3), 31-45.

Chapter 18

Approach for Dispersing a Hydrophilic Compound as Nanoparticles Into Soybean Oil Using Evaporation Technique

Kenjiro Koga

Additional information is available at the end of the chapter

http://dx.doi.org/10.5772/52016

1. Introduction

Most popular formulation dispersing a hydrophilic compound into oil phase such as soybean oil is emulsions. An emulsion is a dispersed system that consists of water, oil, and surfactant. In general, apparatuses of an emulsifier, a homogenizer, etc. are used for the preparation. As the pharmaceutical trial to disperse water-soluble compounds in an oil phase, the form of the emulsion is very important. Namely, for pharmaceutical preparations containing a hydrophilic drug dispersed uniformly into the oil phase, water-in-oil (w/o) and water-in-oil-in-water (w/o/w) emulsions are preferred. In these cases, hydrophilic drug molecules must retain a high-density in the dispersed water phase of the emulsion; doing so depends on the oil-to-water partition coefficient of the drug. Furthermore, decreasing the particle size in the dispersed water phase is necessary. Much pharmaceutical technical information about adjusting the size of particles is now available: for example, rotating membrane emulsification [1, 2], shirasu porous glass membrane emulsification [3, 4], electrocapillary emulsification [5, 6]. These methods adjust particle size on the basis of membrane pore size and shearing force, which depends on the flow of dispersion medium or on contact-surface dielectric constant differences between the dispersion medium and the dispersion phase. Therefore, these technologies are advantageous in that they can produce uniform particle sizes. In this chapter, a simple method of preparing w/o emulsions with a narrow range of polydispersity is described. In this method, a Polytron homogenizer and an evaporator are used as apparatuses. Namely, specific and expensive apparatuses were not used. Glycyrrhizin monoammonium (GZ) and indocyanine green (ICG) were used as a hydrophilic compound. Here, the phase behavior, stability in terms of particle size of w/o emulsions prepared using the novel method and the sustained release characteristics drug from nano-sized w/o emulsions were investigated [7, 8].

INTECH
open science | open minds

© 2013 Koga; licensee InTech. This is a paper distributed under the terms of the Creative Commons Attribution License (http://creativecommons.org/licenses/by/3.0), which permits unrestricted use, distribution, and reproduction in any medium, provided the original work is properly cited.

2. Selection of emulsifier for the preparation of stable w/o emulsion

The choice of ideal emulsifier is an important to prepare physicochemically stable w/o emulsion. Furthermore, the emulsifier must be safe in human. Therefore, mainly non-ionic surfactants added in foods or medicines were chosen. The list used in this experiment was shown in Table 1.

No.	Surfactants	Product name	HLB
1	condensed ricinoleic acid tetraglycerin ester	CR-310[1]	2.5
2	polyethyleneglycol distearate ester	CDS-400[2]	8.5
3	hexaglycerin sesquistearate ester	SS-500[1]	10.1
4	tetraglycerin monostearate ester	MS-310[1]	10.2
5	tetraglycerin monooleate ester	MO-310[1]	10.2
6	tetraglycerin monolaurate ester	ML-310[1]	10.3
7	polyethyleneglycol(10EO) monostearate ester	MYS-10[2]	11
8	hexaglycerin monostearate ester	MS-500[1]	12.2
9	hexaglycerin monooleate ester	MO-500[1]	12.2
10	polyethyleneglycol(10EO) monolaurate ester	MYL-10[2]	12.5
11	stearyl macrogol glycerides	GELUCIRE 50/13[3]	13
12	hexaglycerin monolaurate ester	ML-500[1]	13.5
13	lauroyl macrogol-32 glycerides	GELUCIRE 44/14[3]	14
14	decaglycerin monooleate ester	MO-750[1]	14.5
15	Polyethyleneglycol (25EO) monostearate ester	MYS-25[2]	15
16	decaglycerin monolaurate ester	DECAGLYN 1-L[2]	15.5
17	decaglycerin monolaurate ester	ML-750[1]	15.7
18	polyoxyethylene(20EO) oleylether	BO-20[2]	17
19	polyethyleneglycol(40EO) monostearate ester	MYS-40[2]	17.5
20	polyoxyethylene(30EO) phytosterol	BPS-30[2]	18
21	polyoxyethylene(21EO) laurylether	BL-21[2]	19
22	polyoxyethylene(18EO)nonylphenylether	NP-18TX[2]	19
23	polyoxyethylene(25EO)laurylether	BL-25[2]	19.5
24	polyoxyethylene(30EO)octylphenylether	OP-30[2]	20
25	polyoxyethylene(20EO)nonylphenylether	NP-20[2]	20

Number 1), 2), and 3) indicated in product name were gifts from Sakamoto Yakuhin Kogyo Co.Ltd., Nikko Chemicals Co. Ltd., and Gattefossé, respectively.

Table 1. Surfactants used for the preparation of w/o emulsions.

The following examination was carried out in order to estimate the stability of w/o emulsion. Each surfactant (0.75 g) and soybean oil (6.75 g) were put in a glass tube. The mixture was dissolved or dispersed uniformly at 60°C for 15 min. GZ solution (2.25 mL of 40 mg/mL in dissolved with 100 mM phosphate buffered solution, pH7.4) was added into the glass tube, then immediately, the solution with oil, water phase, and surfactant was emulsified using a Polytron homogenizer (PT-MR 3100, Kinematica AG, Littau/Luzern, Switzerland) at 20,000 rpm for 3 min. The state of the w/o emulsions was observed at 24 h after the preparation. The stability of the w/o emulsions was estimated by objective evaluation scale (OES; 0, 1, 2, 3, 4, and 5). Namely, OES 0 is a completely discrete state without emulsifying. OES 1 is a biphasic separate state with wispy cloud in bottom phase, OES 2 is a biphasic separate state with weakly white turbidity in bottom phase, OES 3 is a biphasic separate state with moderately white turbidity in bottom phase, OES 4 is a biphasic separate state with strongly white turbidity in bottom phase, and OES 5 is a stable state without phase separation.

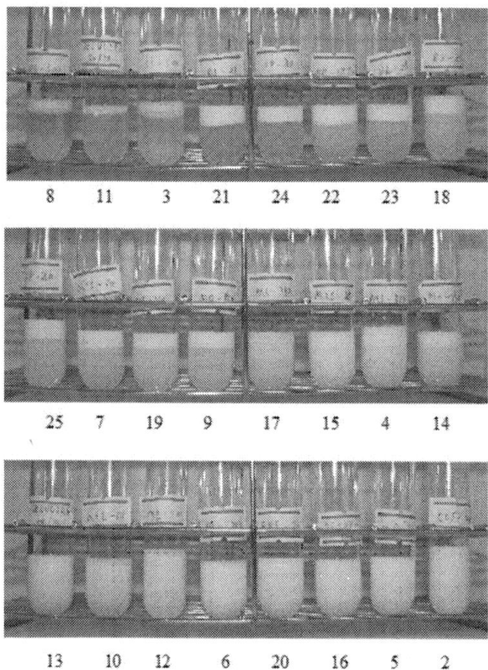

Figure 1. Observed properties of emulsions with 24 kinds of emulsifiers at 24 h after the preparation using a Polytron homogenizer. The photo of emulsion with CR-310 was referred in Figure 3A. The number under each photo is a product number shown in Table 1.

The results in the observed state after preparation of w/o emulsions are shown in Figure 1. Moreover, Figure 2 shows the relationship between HLB number of emulsifiers and OES.

CR-310 and CDS-400 among 25 kinds of surfactants were convenient for the preparation of stable w/o emulsion. The results were identified with the theory that surfactant with low HLB is suitable for the preparation of w/o emulsions. A difference of viscosity as physico-chemical properties was observed between CR-310 and CDS-400. Namely, the viscosity of the w/o emulsion with CDS-400 was high like cream, that of the emulsion with CR-310 was the same as the viscosity of soybean oil. Therefore, it was clear that CR-310 among used surfactants was most convenient emulsifier for the preparation of w/o emulsions when soybean oil was used as an oil phase.

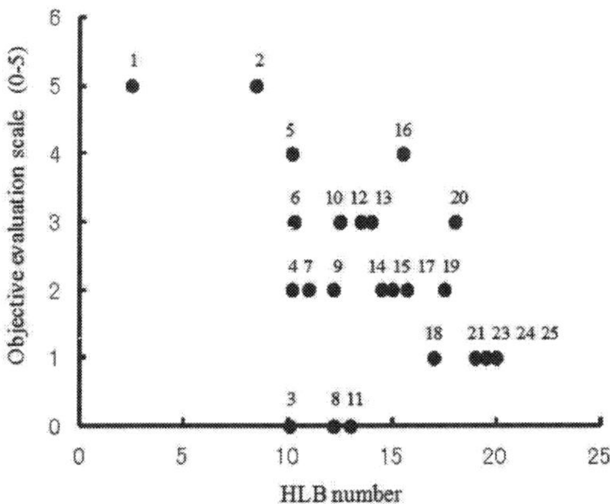

Figure 2. The relationship between HLB number of surfactants and objective evaluation scale. The number of each symbol is a product number shown in Table 1.

3. Preparation of nano-sized w/o emulsions

GZ solution (400 mg/mL) was prepared by dissolving GZ powder at 60°C in 100 mM phosphate-buffered solution (pH7.4) containing 8.0% (w/v) L-arginine. L-arginine was used to inhibit the gelation of GZ [9]. Soybean oil (4.50 g) and CR-310 (0.50 g) were mixed in a glass tube, and then the mixture was heated for 15 min at 60°C in order to blend uniformly. The GZ solution and the mixture of oil and emulsifier were cooled down at room temperature, and 400 mg/mL GZ solution (2.2, 3.3, or 4.4 g) was added to glass tubes containing the mixture. First, the w/o emulsions were prepared by agitating the mixture with a Polytron homogenizer (PT-MR 3100, Kinematica AG) at 20,000 rpm for 3 min. Second, the w/o emulsions were placed into a 50-mL round-bottom flask, which was then placed into a rotary evaporator (R-210, Buchi Labortechnik AG, Flawil, Switzerland) equipped with a vacuum controller (V-850) and a vacuum pump (V-700). The vacuum was initially set to 120 hPa, and then decreased at a rate of 10 hPa per minute until 20 hPa was reached. The mixture was then subjected to these vacuum conditions at 40°C for 90 min. The prepared emulsions were separated into either glass vials or 10-mL centrifuge tubes, depending on the analyses to be done. For phase behavior comparisons, w/o emulsions without GZ was also prepared using distilled water (3.3 g) instead of GZ solution [7].

4. Phase behavior during the preparation of w/o emulsions

The w/o emulsions prepared by adding GZ solution slowly changed in turbidity to pale white, and after 23-24, 26-27, and 31-32 min for GZ solutions of 2.2, 3.3, and 4.4 g, respectively, to clear or slightly turbidity. The samples remained clear only for approximately 2 min. When subjected to prolonged evaporation, the emulsions rapidly changed in turbidity to white as solid dispersion. Figure 3A shows photographs of the emulsions prepared with GZ before evaporation (0 min) and 10, 15, 26, and 90 min after evaporation. To confirm the relationship between evaporation time and phase behavior in w/o emulsions with or without GZ, similar experiments on w/o emulsions lacking GZ were carried out (Figure 3B). The turbidity of w/o emulsions without GZ remained white up to 15 min after evaporation, before gradually changing to pale white. The w/o emulsions finally became transparent 90 min after evaporation. These results suggest that the phase behavior may depend on the water content of the w/o emulsions.

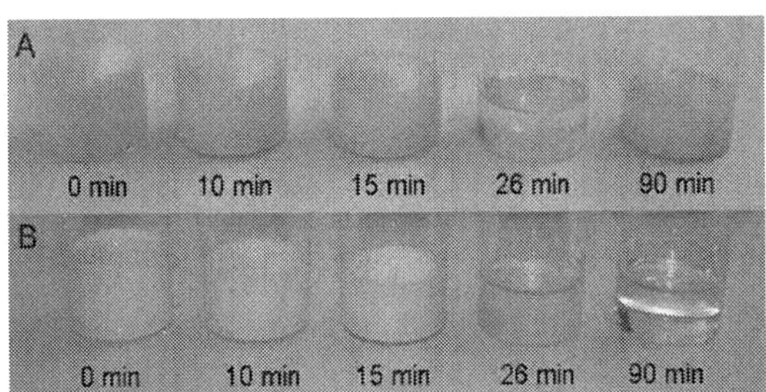

Figure 3. Photographs of w/o emulsions at the indicated times after evaporation. (A) Emulsion prepared with a GZ solution containing 3.3 g GZ. (B) Emulsion prepared without GZ (water phase is distilled water). The times shown below each vial represent the length of time samples underwent evaporation.

5. Particle size of water phase in w/o emulsions

GZ solution (400 mg/mL) was prepared by dissolving GZ powder at 60°C in 100 mM phosphate-buffered solution (pH 7.4) containing 8.0% (w/v) L-arginine and 10 μg/mL fluorescein sodium. Fluorescein sodium was used as a marker for the fluorescent observation of the dispersion state of water phase in the emulsions. After the GZ solution (3.3 g), soybean oil (4.50 g), and CR-310 (0.50 g) were mixed, w/o emulsions were prepared according to the method described above. The w/o emulsion before dehydration was analyzed with a confocal laser scanning microscope LSM510 (Carl Zeiss GmbH, Jena, Germany) and LMS Image Browser Software (Carl Zeiss GmbH). The excitation and fluorescein wavelengths were set to 405 and 488 nm, respectively.

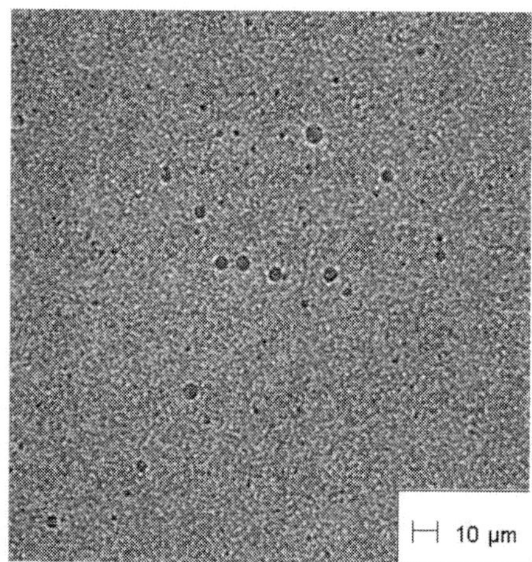

Figure 4. Fluorescence photomicrograph of a w/o emulsion following the addition of fluorescein sodium to the water phase. The dark gray areas represent the aqueous phase. The position of water phase containing GZ is indicated as green fluorescent particles.

Figure 4 shows the photo by a confocal laser scanning microscope. The water phase was uniformly distributed as small droplets (< 5 μm), indicating that before evaporation the particle size distribution in the water phase of GZ sample distributed widely after dispersal with a Polytron homogenizer.

Emulsions	Particle size in relative frequency (nm)				
	10%	25%	50%	75%	90%
2.2 g GZ sample	135	170	225	287	341
3.3 g GZ sample	219	253	299	349	394
4.4 g GZ sample	310	504	1105	1381	1610

Table 2. Particle size and size distribution of clear or slightly turbid w/o emulsions.

The particle sizes of w/o emulsions at 23, 26, and 31 min for GZ solutions of 2.2, 3.3, and 4.4 g after evaporation were analyzed using a Particle Size Analyzer (LS 13 320; Beckman Coulter, Inc., Fullerton, CA, United States). Table 2 presents the average particle sizes and size distributions of the clear or slightly turbid w/o emulsions after evaporation. The size distribution of the three kinds of emulsions was narrow, with relative frequency values of 10, 25,

75, and 90%. These results suggest that dehydration proceeded in a way that caused the dispersed phase to approach the narrow distribution. In particular, the particle size distribution of the 2.2 and 3.3 g GZ samples converged toward the nano-size range. This was consistent with our prediction that large water droplets would efficiently evaporate from the surface of the w/o emulsions as their round bottom flasks were rotated and that particle size distribution would narrow as a function of evaporation time.

During the preparation of transparent w/o emulsions containing GZ, the only component removed by evaporation was water. Thus, the components dissolved in the water phase (GZ, L-arginine, and phosphate salts) were gradually concentrated as the water content decreased. The observation that a simple evaporation process changed the turbidity of the w/o emulsions from white to clear within a short time suggests that the particles comprising the dispersed phase became extremely small in size. In fact, the particle size range of sample which prepared with 3.3 g GZ solution was 219-394 nm by dynamic light scattering assay (LS 13 320; Beckman Coulter, Inc.). The state of the nano-sized droplets was observed in transmission electron microscopy (TEM). Namely, a clear w/o emulsion was prepared by adding GZ solution (3.3 g) to the mixture of soybean oil (4.50 g) and CR-310 (0.50 g), and the resulting clear w/o emulsion was passed through a quantitative filter (No. 5B, Advantec; Toyo Roshi Kaisha, Ltd., Tokyo, Japan). The filter was then hardened with cured acryl resin. After embedding the filter, ultrathin sections were obtained by cutting the surface of the block containing the filter on an ultramicrotome equipped with a diamond knife (EM-Ultracut-s; Leica Microsystems Vertrieb GmbH, Wetzlar, Germany). The ultrathin sections were mounted onto freezing support grids and then stained with ruthenium tetrachloride. Next, the emulsified particles were observed with a transmission electron microscope (TEM; JEM-2100; Jeol Ltd., Tokyo, Japan).

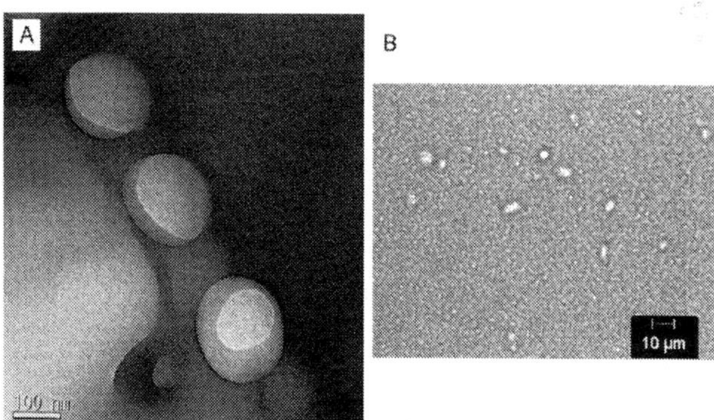

Figure 5. A) TEM photograph of dispersed particles in a clear w/o emulsion prepared by adding GZ solution. (B) Solid GZ particles after 90 min evaporation.

To determine whether the nano-sized droplets exist as a liquid state in the water phase, the shape of particles in the clear w/o emulsion (3.3 g GZ sample) was observed with TEM. As the water content of the emulsion decreases with evaporation, part of the dissolved GZ may precipitate as a solid state from the dispersed phase. If this hypothesis is accurate, then the particles may be not spherical. However, TEM analysis revealed that the particles were spheres of approximately 200 nm in diameter and were maintained a uniform globe (Figure 5A), strongly suggesting that the nanoparticles in the clear w/o emulsions existed as a liquid phase.

Following additional evaporation, the turbidity of the w/o emulsions again changed to white, indicating a change in phase behavior. This suggests that the hydrophilic components dissolved in the water phase separated as solid states. In fact, microscopic analysis demonstrated that the w/o emulsions containing GZ contained 1-10 μm-diameter solid particles >90 min after evaporation (Figure. 5B) [8].

6. Water contents in w/o emulsions

The water content (%, w/w) of the w/o emulsions was analyzed using a Karl Fischer titration apparatus (870, KF Titrino Plus; Metrohm Shibata Co., Ltd., Tokyo, Japan). The water content (%, w/w) of the 2.2, 3.3, and 4.4 g GZ samples before evaporation was 16.8, 21.9, and 25.4% (w/w), respectively. On the other hand, the water content of the clear or slightly turbid 2.2, 3.3, and 4.4 g GZ samples was 7.8 ± 0.1, 9.4 ± 0.3, and $11.9 \pm 0.3\%$ (w/w), respectively. Furthermore after 90 min of evaporation, the water content of the 2.2, 3.3, and 4.4 g GZ samples was all in the range of 1.3-1.8% (w/w). The water content of w/o emulsion lacking GZ was 9.3 ± 0.2 and $0.18 \pm 0.02\%$ (w/w), respectively, at 26 and 90 min after evaporation. These results indicate that the phase behavior of the w/o emulsions changed from white turbid to translucent or transparent when the water content reached approximately 8-12% (w/w). Since the water content of the clear or slightly turbid w/o emulsions correlated well with GZ content (0.8, 1.2, and 1.6 g for 2.2, 3.3, and 4.4 g GZ samples; correlation coefficient was 0.992), it is plausible that the precipitation of hydrophilic components, such as GZ, L-arginine and phosphate salts, was responsible for the increased turbidity resulting after prolonged evaporation.

From these results, it is concluded that the water contents of w/o emulsions changed the phase behavior in the emulsions, that is, from white turbid phase to clear phase when the water content reached to be approximately 9%, and then from clear phase to white turbid phase. The decreasing rate of the water content is affected by the setting of a vacuum controller. In the experiments, although the relationship between the decreasing rate of the water content and maintained interval with clear phase in the w/o emulsions was not investigated, the pressure condition in the evaporation process will be a major problem for the simple preparation of clear w/o emulsions.

7. Stability of w/o emulsions

The stability of the w/o emulsions were evaluated according to two criteria: i) the uniform dispersal of the water phase containing GZ in the emulsion; and ii) the size distribution at steady-state conditions. Clear or slightly turbid w/o emulsions (2.2, 3.3, and 4.4 g GZ solutions) were transferred to 10-mL centrifuge tubes (Figure 6), which remained undisturbed at 20 ± 2°C for 10 days. The w/o emulsion prepared with 3.3 g GZ solution remained under undisturbed conditions continuously for 65 days. Next, a pipette was inserted 1 cm from the top or bottom of each centrifuge tube and a small amount (50 mg) of emulsion was removed and transferred into two screw vials (one vial for the "top" sample, the other vial for the "bottom" sample). Methanol (30 mL) was added to each vial, and the vials were shaken for 15 min on a vortex mixer. The methanol-containing GZ samples were adequately diluted with 100 mM phosphate-buffered solution (pH 7.4) and were injected into an HPLC system in order to determine the GZ concentration in the samples [10]. The water content (%, w/w) in the w/o emulsions obtained from both samples was analyzed by Karl Fischer titration.

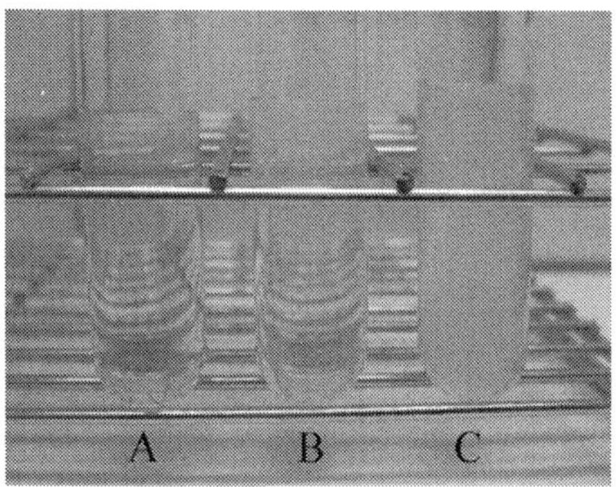

Figure 6. Photograph of clear and slightly turbid w/o emulsions. (A) 2.2 g GZ sample. (B) 3.3 g GZ sample. (C) 4.4 g GZ sample.

7.1. Uniformity of GZ concentration and water content in w/o emulsions

The GZ concentrations of 2.2, 3.3, and 4.4 g GZ samples that were clear or slightly turbid were 11.1, 14.5, and 17.0% (w/w), respectively. The GZ concentrations and water content of clear or slightly turbid w/o emulsions 10 days and 65 days after preparation are shown in Table 3. As the phase behavior changes with time, the aggregation or coalescence of the dispersed phase occurs more frequently after emulsification is achieved, because the emulsion is thermodynamically unstable. If the dispersed stability is not maintained, then the GZ concentration and water content in the emulsion will differ in the parts of the emulsion near the top and bottom of the sample. This is because the dispersed phase containing GZ moves toward the bottom due to specific gravity differences. GZ concentration in the samples were analyzed in the present study, however, the GZ concentrations in the top and bottom parts of the emulsions were identical in the 2.2 and 3.3 g GZ samples but not in the 4.4 g GZ sample. Furthermore, similar results were obtained in our comparative analysis of water content. These results suggest that the dispersed stability of the 2.2 and 3.3 g GZ w/o emulsions was extremely high. On the other hand, the difference in phase behavior in the top and bottom layers of the 4.4 g GZ sample suggests dispersed instability. Specifically, after 10 days the turbidity of the bottom layer became clearer than that of the top layer, indicating that the dispersed phase containing 4.4 g GZ gradually moved toward the bottom of the 10-mL centrifuge tube. The specific gravity of the dispersed phase in the w/o emulsions increased with increasing GZ concentration.

w/o emulsions	GZ concentration (%, w/w)		Water content (%, w/w)	
	upper position	bottom position	upper position	bottom position
2.2 g GZ sample	11.5 ± 0.7	11.2 ± 0.4	8.2 ± 0.2	8.4 ± 0.2
3.3 g GZ sample	14.9 ± 0.5	14.7 ± 0.5	10.1 ± 0.2	10.2 ± 0.3
3.3 g GZ sample**	15.0 ± 0.4	14.8 ± 0.4	9.8 ± 0.3	9.6 ± 0.2
4.4 g GZ sample	10.2 ± 0.8	17.4 ± 0.7	7.4 ± 0.1	12.0 ± 0.3

GZ concentration and water content were determined after the sample was stored undisturbed for 10 days at 20 ± 2°C. *Samples were obtained 1 cm from the top and bottom of the emulsions contained within a 10-mL centrifuge tube. **The sample was stored undisturbed for 65 days at 20 ± 2°C.

Table 3. GZ concentration and water content in different parts of the w/o emulsions*.

7.2. Uniformity of particle size of dispersed phase in w/o emulsions

The particle size distribution of the GZ sample containing 3.3 g GZ solution at 65 days after preparation was 226-421 nm. This range was similar to that measured during immediate intervals after preparation. In general, refinement of emulsion particle size lowers thermodynamic stability, because the phase behavior of an emulsion-which is a very complicated system of oil, water, and surfactant-is affected by decreases in particle size [11]. Nano-sized emulsions, in particular, are not as stable as micron-sized emulsions. Therefore, nano-sized emulsions must be stabilized with polymers and excessive amounts of surfactants [12]. In GZ sample, the concentration of surfactant in the oil phase was 10% (w/w); the emulsions did not contain other stabilizers. However, the GZ sample (nano-sized emulsions) was stable at least for 2 months. Moreover, although the particle sizes of the top and bottom layers in the 4.4 g GZ sample were not determined, the aggregation of dispersed particles suggests that the size distribution in this w/o emulsion tended to be on the large side. Taken together, these observations suggest that dispersed stability decreases with increasing GZ content.

8. Sustained release effect by nano-sized w/o emulsions

From the viewpoint of medical treatment, drug release from w/o emulsions is important for the efficiency of controlled release. The pharmacokinetics of GZ by nano-size w/o emulsion, aqueous formulation, o/w emulsion, and w/o emulsion with solid GZ was investigated in order to clarify the degree of the controlled release. Furthermore, the release characteristics of a hydrophilic compound, indocyanine green (ICG), from administered subcutaneous site in rats was observed using a near-infrared fluorescent camera (Photo Dynamic Eye, PDE).

8.1. In vivo experiments in GZ pharmacokinetics

Pharmacokinetic studies of GZ were investigated in detail in human [13, 14], in rat [15, 16], and other species. The elimination half-life of GZ in rats after the intravenous administration (20-50 mg/kg) is approximately 2-4 h in plasma [17, 18]. GZ is rapidly excreted into bile via multidrug resistance-associated protein 2 (MRP2) ATP-binding cassette transporter C2 (ABCC2) transporter [19]. Therefore, the release of GZ from w/o emulsions was estimated as GZ elimination into bile.

The protocol of this study was approved by the Committee of Animal Use of Hokuriku University. All animal experiments were conducted in accordance with the Institutional Guidelines of Care and Use of Laboratory Animals. Male Sprague-Dawley rats (180-200 g) were housed for at least 10 days in a clean room. The rats were given free access to commercial chow and water and were maintained according to the Hokuriku University Animal Guidelines. For in vivo experiments in GZ formulations, the rats (250-280 g) were randomly divided into four treatment groups as four rats per group.

A GZ stock solution (400 mg/mL) was prepared at 60°C in 100 mM phosphate-buffered solution, pH7.4, containing 8.0% (w/v) L-arginine. The GZ stock solution was stored in a refrigerator. An aqueous formulation of GZ (150 mg/mL; Rp-II) was prepared by adding 100 mM phosphate-buffered solution (pH7.4) to the GZ stock solution. Preparation of an oil-in-water (o/w) emulsion of GZ was as follows: soybean oil (1.00 g), HCO-60 (0.12 g), and egg yolk lecithin (0.12 g) were blended uniformly by heating at 90°C for 15 min on a block heater. The mixture was then cooled at room temperature. The o/w emulsion of GZ (150 mg/mL; Rp-III) was prepared by combining the soybean oil mixture (1.0 mL), GZ stock solution (1.16 mL), and 100 mM phosphate-buffered solution, pH7.4 (0.84 mL) and by using a Polytron homogenizer (PT-MR 3100) at 20000 rpm for 3 min for emulsification.

Preparation of an w/o emulsion of GZ was described above. The GZ stock solution (3.3 g) was then added to the lukewarm mixture, which was emulsified using a Polytron homogenizer (PT-MR 3100) at 20000 rpm for 3 min. The w/o emulsion was placed into a 50-mL round-bottom flask, which was then set in a rotary evaporator (R-210, Buchi Labortechnik AG) equipped with a vacuum controller (V-850). The vacuum was initially set to 120 hPa at 40°C; thereafter, the pressure was decreased at a rate of 10 hPa per min until 20 hPa was reached. To prepare Rp-I and Rp-IV, the dehydration was continued for 27 min and 120 min, respectively, and then adjusted the GZ concentration to 150 mg/mL by adding the soybean oil/CR-310 (9 : 1, w/w) mixture. Adminidtration method of four kind formulations and sampling of bile in rats were described in detail in reference [8].

As the characteristics of used formulations, the average particle sizes in the Rp-I and Rp-III formulations were 299 nm and 376 nm, respectively. The 10-90% ranges of size distribution in Rp-I and Rp-III were 208-402 nm and 255-512 nm, respectively. Two peaks in size distribution were observed in the Rp-IV formulation: 312 nm and 5000 nm. Microscopic observations revealed that large-size GZ particles were solid GZ, because they were not spherical. The water content in Rp-IV (1.5%, w/w) was very low compared with that in Rp-I (9.4%, w/w). Moreover, the small- and large-size particles were not observed after evaporation for 120 min in a w/o emulsion lacking GZ.

Almost all GZ transported to hepatocytes via the blood is eliminated into bile as unchanged GZ (i.e., not metabolized) [20, 21]. Therefore, the elimination rate of GZ in bile reflects the bioavailability of GZ in hepatocytes. Figure 7 shows the cumulative elimination (%) of GZ over time after subcutaneous administration of Rp-I, Rp-II, Rp-III, and Rp-IV in rats. After the administration of Rp-I, cumulative elimination at 8 h, 24 h, and 72 h as a function of administered GZ dose (50 mg/kg) in bile was 11%, 20%, and 47%, respectively. These results indicate that the Rp-I formulation resulted in a sustained release of apparent zero-order kinetics. The average elimination rate of GZ up to 72 h was 84.2 ± 14.2 µg/h.

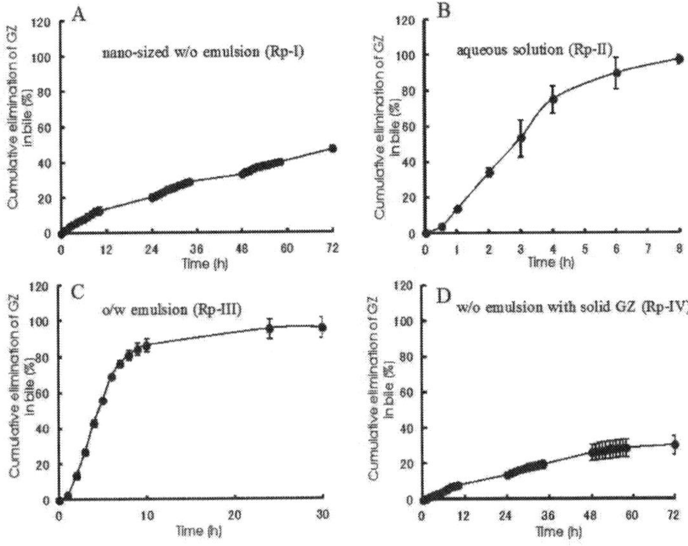

Figure 7. Cumulative elimination of GZ in bile after subcutaneous administration of GZ formulations in rats. (A) Nano-sized w/o emulsion encapsulating GZ, (B) aqueous solution of GZ, (C) o/w emulsion containing GZ, and (D) w/o emulsion with solid GZ. GZ concentrations were all adjusted to 150 mg/mL. The GZ dose administered to all rats was all 50 mg/kg. Data represent means ± S.D. of four experiments.

With the Rp-II formulation, the cumulative elimination at 4 h and 8 h as a function of administered GZ dose was 75% and 97%, respectively. In intravenous and subcutaneous administration models, no difference in the elimination rate of GZ has been observed after intravenous administration of GZ [10]. These results suggest that GZ dissolved in phosphate buffered solution rapidly diffused in the hyperdermis, before being transferred into the general circulation. With the Rp-III formulation, the cumulative elimination of GZ at 8 h and 30 h was 81% and 96%, respectively. The elimination rate of GZ in Rp-III was faster than that in Rp-I but slower than that in Rp-II, suggesting that the oil phase of the o/w emulsion inhibited the diffusion of the water phase containing GZ in subcutaneous regions, even though GZ was dissolved in the outer water phase of o/w emulsion. On the other hand, the cumulative elimination of GZ in Rp-IV was the lowest among the four formulations: The GZ elimination in bile at 8 h, 24 h, and 72 h was 7.1%, 14%, and 31%, respectively. As with the elimination kinetics of the Rp-I formulation, the elimination kinetics of GZ in Rp-IV showed that GZ was released in a sustained fashion for up to 72 h. Since Rp-IV contained solid GZ in w/o emulsions, it was speculated that the elimination rate of GZ in bile after the administration of Rp-IV would be slower than that of GZ in Rp-I. In fact, the eliminated amount of GZ in bile after the administration of Rp-IV was 0.64-fold compared to that of Rp-I. These results suggest that reduced water content in w/o emulsions delays hydration in the subcutaneous region and that much time is required to dissolve the dispersed solid GZ in Rp-IV.

To determine more precisely the characteristics of sustained GZ release from w/o emulsions, the rates of GZ elimination in bile were recalculated every 24 h after the administration of Rp-I and Rp-IV (Figure 8).

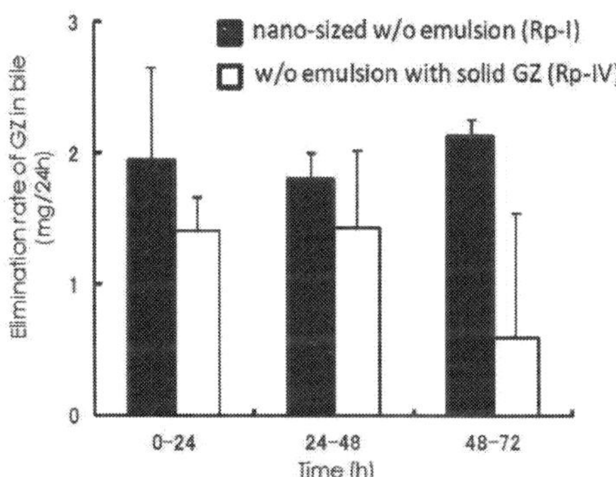

Figure 8. Elimination rate per 24 h of GZ in bile after subcutaneous administration of Rp-I and Rp-IV in Rats. Administration dose of GZ was 50 mg/kg in both formulations. Data represent means ± S.D. of four experiments.

The elimination of GZ in Rp-I occurred at a constant rate for 72 h, i.e., the rate ranged from 1.80 to 2.12 mg/day. On the other hand, elimination of GZ in Rp-IV decreased from 1.40-1.41 mg/day at 48 h to 0.60 mg/day at 48-72 h. As the reason for the decrease in GZ elimination rate at 48-72 h in Rp-IV, it was predicted that the presence of solid dispersed GZ may be involved deeply the transfer rate of GZ from subcutaneous site to liver. Actually, dissolved-state GZ and solid-state GZ exist in Rp-IV. Although it was considered that the dissolved GZ in Rp-IV was transferred to liver as similar to GZ in Rp-I, solid GZ particles must be dissolved to some extent which can be passed vascular system such as vein and lymph capillary in order to transfer GZ from subcutaneous site to liver. Therefore, after 48 h, the proportion of solid GZ for residual GZ in the subcutaneous site will increase certainly. As a result, it was guessed that GZ elimination into bile decreased based on the decrease of transfer rate from subcutaneous site to liver. These results indicate that Rp-I was a substantially superior formulation compared to Rp-IV in terms of sustained release in bile. It was hypothesized that the small and narrowrange polydispersity (208-402 nm in Rp-I) of the dispersed phase in w/o emulsions may be important for stabilizing the release rate of GZ from these emulsions.

8.2. Tissue distribution of ICG from subcutaneous site in rats

For the experiments of ICG administration, two male Sprague-Dawley rats (250, 255 g) were used. A w/o emulsion encapsulating ICG was prepared as follows: 5 mg/mL ICG solution (3.3 g) was added to the mixture of soybean oil (4.50 g) and CR-310 (0.50 g) heated at 60°C. Next procedure was the same with the preparation step of Rp-I described above. This ICG solution and w/o emulsion encapsulating ICG were used to observe the tissue distribution of drugs from the subdermal injection site. To monitor over time the delayed drug distribution and the diffusion of hydrophobic formulation in the subdermal site, a w/o emulsion encapsulating ICG instead of GZ was prepared. ICG is a hydrophilic fluorescent dye and biocompatibility marker with excitation and emission spectra in the near-infrared wavelength range of 600 to 900 nm, and the maximum emission wavelength of ICG in vivo is 845 nm [22, 23]. The kinetics of ICG in vivo was observed using a non-invasive, near-infrared fluorescent camera (Photo Dynamic Eye; PDE), because the near-infrared-wavelength monitoring barely affects biological molecules such as water and hemoglobin. For the experiments of the microscopic image using a fluorescent probe, ICG, rats were anesthetized with an intraperitoneal administration of sodium pentobarbital (50 mg/kg), and then were kept in a supine position on an operation plate. A PDE was set on the upper position of 20 cm from the rat and the photo image was monitored using a personal computer. Two samples, 5 mg/mL ICG solution (0.05 mL) and w/o emulsion encapsulating ICG (0.05 mL) were administered to right and left hind legs of rat, respectively. The diffusion of ICG was observed for 60 min.

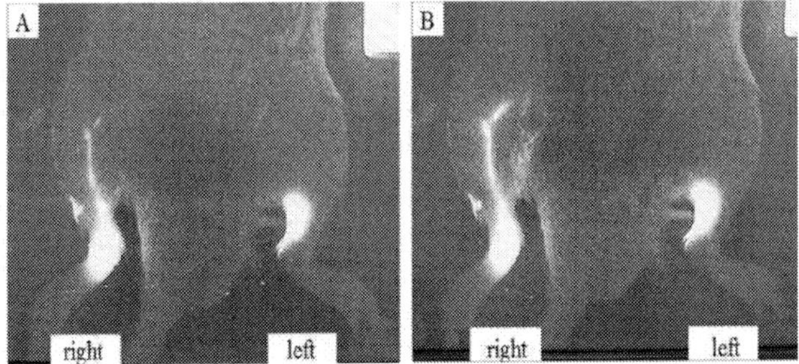

Figure 9. Photographs taken at (A) 15 min and (B) 60 min after the subcutaneous injection of ICG solution into the right leg and w/o emulsion encapsulating ICG into the left leg injection volume of ICG was 50 mL in both formulations.

The purpose of the PDE experiments was to determine whether the diffusion rate of w/o emulsions at the subdermal site is remarkably slower than that of solutions like Rp-II of GZ formulation. Figure 9 shows photographs of rats' hind legs 15 min and 60 min after injecting ICG (right leg) and after injecting w/o emulsion encapsulating ICG (left leg). The ICG solution was rapidly absorbed into capillary blood vessels within 15 min. Furthermore, ICG also reached peripheral lymphatics. On the other hand, ICG in w/o emulsions remained in the vicinity of the administration site for 60 min. These results suggested that the outer oil phase inhibited the diffusion of w/o emulsion encapsulating ICG and/or the release of ICG from the w/o emulsion. The viscosity of lipophilic formulations is generally high. Therefore, one would expect the release of ICG from a w/o emulsion to be delayed compared to that from aqueous formulations of ICG. The expectation will be correspondent with the decrease of diffusion and/or release of GZ from Rp-I of GZ formulation.

9. Conclusion

It was clarified that GZ, a hydrophilic compound, was dispersed as nanoparticles into soybean oil by using the evaporation technique. The prepared oil phase containing GZ was nano-emulsions with low polydispersity, and was stable at least 2 months. This ideal dispersion method had to make water content approximately 9%, and it was clear that the further dehydration became solid dispersion. It is concluded that a hydrophilic compound can be dispersed easily into an oil phase such as soybean oil by utilizing this method. Concretely, the w/o emulsions containing GZ (2.2 g of GZ [11.1%, w/w] and 3.3 g of GZ [14.5%, w/w]) with narrow-ranged polydispersity and high-dispersed stability were easily prepared by the measurable removal of water using a Polytron homogenizer and a rotary evaporator. The water content in 2.2 and 3.3 g GZ samples had to be 7.8 and 9.4% (w/w), respectively, because decreasing the water content beyond these levels caused the phase behavior to change (e.g., white turbid). The particle size distribution (relative frequency values ranging from 10 to 90%) of the clear w/o emulsions was in the range of 135 to 421 nm as the samples remained undisturbed for 65 days at $20 \pm 2°C$. The w/o emulsion preparation method described in the present study provides useful information on the lipophilic formulations of GZ.

A nano-sized w/o emulsion of GZ (Rp-I) showed sustained elimination of GZ in bile at a relatively constant rate for 72 h. The sustained GZ elimination in bile was strongly affected by diffusion of the w/o emulsion and by the release of GZ from the emulsion to the perimeter of the subdermal site, based on the PDE observations with ICG. Indeed, the average elimination rate of GZ in bile was 0.084 mg/h over 72 h, when Rp-I (50 mg/kg as GZ) was administered subcutaneously. If GZ release from Rp-I will be maintained as zeroorder elimination (0.084 mg/h), 6-7 day are needed until the GZ release finishes in the rats. Namely, GZ in Rp-I will slowly transfer from subcutaneous tissue to liver 20-fold periods as compared with GZ in Rp-II, the elimination of almost all GZ finished 8 h. These results indicate that the nano-sized w/o emulsion encapsulating GZ, which can be subcutaneously administered, will be useful as a new sustained-release formulation.

Acknowledgements

Author thanks Cokey Co. Ltd. for the gift of glycyrrhizin monoammonium, and thanks Sakamoto Yakuhin Kogyo Co. Ltd. and Nikko Chemicals Co. Ltd. for the gifts of several kinds of surfactants.

Author details

Kenjiro Koga*

Address all correspondence to: k-koga@hokuriku-u.ac.jp

Faculty of Pharmaceutical Sciences, Hokuriku University, Japan

References

[1] Aryanti, N., Hou, R., & Williams, R. A. (2009). Performance of a rotating membrane emulsifier for production of coarse droplets. *Journal of Membrane Science*, 326(1), 9-18.

[2] Vladisavljevic, G. T., & Williams, R. A. (2006). Manufacture of large uniform droplets using rotating membrane emulsification. *Journal of Colloid and Interface Science*, 299(1), 396-402.

[3] Vladisavljevic, G. T., Surh, J., & Mc Clements, J. D. (2006). Effect of emulsifier type on droplet disruption in repeated shirasu porous glass membrane homogenization. *Langmuir: the ACS journal of surfaces and colloids*, 22(10), 4526-4533.

[4] Hosoya, K., Bendo, M., Tanaka, N., Watabe, Y., Ikegami, T., & Minakuchi, H. (2005). An application of silica-based monolithic membrane emulsfication technique for easy and efficient preparation of uniformly sized polymer particles. *Macromolecular Materials and Engineering*, 290(8), 753-758.

[5] Sakai, H., Tanaka, K., Fukushima, H., Tsuchiya, K., Sakai, K., Kondo, T., & Abe, M. (2008). Preparation of polyurea capsules using electrocapillary emulsification. *Colloids and Surfaces, B: Biointerfaces*, 66(2), 287-290.

[6] Abe, M., Nakayama, A., Kondo, T., Morishita, H., Tsuchiya, K., Utsumi, S., Ohkubo, T., & Sakai, H. (2007). Preparation of tiny biodegradable capsules using electrocapillary emulsification. *Journal of Microencapsulation*, 24(8), 777-786.

[7] Koga, K., Liu, H., & Takada, K. (2009). A simple preparation method for dispersing glycyrrhizin solution into oil phase using an evaporation technique. *Journal of Drug Delivery Science and Technology*, 19(6), 431-435.

[8]   Koga, K., Nishimon, Y., Ueta, H., Matsuno, K., & Takada, K. (2011). Utility of Nano-sized, water-in-oil emulsion as a sustained release formulation of glycyrrhizin. *Biological & Pharmaceutical Bulletin*, 34(2), 300-305.

[9]   Koga, K., Kawashima, S., Shibata, N., & Takada, K. (2007). Novel formulations of a liver protection drug glycyrrhizin. *Yakugaku Zasshi*, 127(7), 1103-1114.

[10]   Koga, K., Tomoyama, M., Ohyanagi, K., & Takada, K. (2008). Pharmacokinetics of glycyrrhizin in normal and albumin-deficient rats. *Biopharmaceutics & Drug Disposition*, 29(7), 373-381.

[11]   Porras, M., Solans, C., Gonzalez, C., & Gutierrez, J. M. (2008). Properties of water-in-oil (W/O) nano-emulsions prepared by a low-energy emulsification method. *Colloids and Surfaces, A: Physicochemical and Engineering Aspects*, 324(1-3), 181-188.

[12]   Sood, A. (2008). Modeling of the particle size distribution in emulsion polymerization. *Journal of Applied Polymer Science*, 109(3), 1403-1419.

[13]   Yamamura, Y., Kawakami, J., Santa, T., Kotaki, H., Uchino, K., Sawada, Y., Tanaka, N., & Iga, T. (1992). Pharmacokinetic profile of glycyrrhizin in healthy volunteers by a new high-performance liquid chromatographic method. *Journal of Pharmaceutical Sciences*, 81(10), 1042-1046.

[14]   Ploeger, B., Mensinga, T., Sips, A., Seinen, W., Meulenbelt, J., & De Jongh, J. (2001). The pharmacokinetics of glycyrrhizic acid evaluated by physiologically based pharmacokinetic modeling. *Drug Metabolism Reviews*, 33(2), 125-147.

[15]   Yamamura, Y., Santa, T., Kotaki, H., Uchino, K., Sawada, Y., & Iga, T. (1995). Administration-route dependency of absorption of glycyrrhizin in rats: intraperitoneal administration dramatically enhanced bioavailability. *Biological & Pharmaceutical Bulletin*, 18(2), 337-341.

[16]   Takahashi, M., Nakano, S., Takeda, I. , Kumada, T. , Sugiyama, K., Osada, T. , Kiriyama, S., Toyoda, H. , Shimada, S. , & Samori, T. (1995). The pharmacokinetics of the glycyrrhizin and glycyrrhetic acid after intravenous administration of glycyrrhizin for the patients with chronic liver disease caused by type C hepatitis virus. *The Japanese journal of gastro-enterology*, 92(12), 1929-1936.

[17]   Ishida, S. , Sakiya, Y. , Ichikawa, T., & Taira, Z. (1992). Dose-dependent pharmacokinetics of glycyrrhizin in rats. *Chemical & Pharmaceutical Bulletin*, 40(7), 1917-1920.

[18]   Koga, K., Takekoshi, K., Kaji, A., Hata, Y., Ogura, T., & Fujishita, O. (2006). Basic study on development of glycyrrhizin formulations for self administration. *Iryo Yakugaku*, 32(5), 414-419.

[19]   Horikawa, M., Kato, Y., Tyson, C. A., & Sugiyama, Y. (2002). The potential for an interaction between MRP2 (ABCC2) and various therapeutic agents: probenecid as a candidate inhibitor of the biliary excretion of irinotecan metabolites. *Drug Metabolism and Pharmacokinetics*, 17(1), 23-33.

[20] Ishida, S., Sakiya, Y., Ichikawa, T., Taira, Z., & Awazu, S. (1990). Prediction of glycyrrhizin disposition in rat and man by a physiologically based pharmacokinetic model. *Chemical & Pharmaceutical Bulletin*, 38(1), 212-18.

[21] Ichikawa, T., Ishida, S., Sakiya, Y., Sawada, Y., & Hanano, M. (1986). Biliary excretion and enterohepatic cycling of glycyrrhizin in rats. *Journal of Pharmaceutical Sciences*, 75(7), 672-675.

[22] Baker, K. J. (1966). Binding of sulfobromophthalein sodium and indocyanine green by plasma (sub α1)-lipoproteins. *Proceedings of the Society for Experimental Biology and Medicine*, 122(4), 957-963.

[23] Benson, R. C., & Kues, H. A. (1978). Fluorescence properties of indocyanine green as related to angiography. *Physics in Medicine & Biology*, 23(1), 159-163.

Chapter 19

Recent Advances on Soybean Isoflavone Extraction and Enzymatic Modification of Soybean Oil

Masakazu Naya and Masanao Imai

Additional information is available at the end of the chapter

http://dx.doi.org/10.5772/ 52603

1. Extraction of soybean isoflavone

1.1. Soybean isoflavones and attractive potential of supercritical carbon dioxide ($SCCO_2$)

Isoflavones produced from bioresources are gaining attention as attractive components in food supplements. Isoflavones are heterocyclic phenols with a structure very similar to that of estrogens. Isoflavone displays like estrogens and has anti estrogen activity; it influences sex hormone metabolism and related biological activity [1,2] and prevents osteoporosis [3,4], arteriosclerosis [5], dementia [2], and cancer [6,7].

Soybeans contain 12 different isoflavones classified into two components, glycosides and aglycons. Glycoside isoflavone has a glucose chain in its molecular structure; aglycon isoflavone does not have a glucose structure.

Ninety-three percent of isoflavones are produced and stored as glycoside. Therefore, in practical separation processes, glycoside isoflavones were the major fraction and were recognized as the main target group rather than aglycons. This article focuses on daidzin, genistin and glycitin as typical glycosides. Their aglycons (i.e., daidzein, genistein and glycitein) were examined for comparison. The aglycons have no glycoside chain; their chemical structure is depicted in Fig. 1.

Methods of extracting isoflavones from soybean have been previously examined by using organic solvent [8], pressurized liquid [9], ultrasound [10,11], and supercritical carbon dioxide [12-16]. Supercritical carbon dioxide has been the favorite extraction medium for many food functional components, i.e. caffeine [17-20], capsaicin [21,22], carotenoids [23-26], polyphenol [27-30], aspirin [31], and coenzyme Q10 [32].

© 2013 Naya and Imai; licensee InTech. This is a paper distributed under the terms of the Creative Commons Attribution License (http://creativecommons.org/licenses/by/3.0), which permits unrestricted use, distribution, and reproduction in any medium, provided the original work is properly cited.

In general, the solubility of polar components in the $SCCO_2$-only system was very low because carbon dioxide has non-polar characteristics. The solubility of polar components has been well enhanced by adding polar components to the $SCCO_2$ system. The added component was referred to as an entrainer. Ethanol was effectively employed as an entrainer for extraction and applied to caffeine [17,19], capsaicin [21], catechin [27], epicatechin [28], aspirin[31], and coenzyme Q10 [32]. Rostagno et al. (2002) successfully extracted large amounts of isoflavones from soybean flour by using methanol aqueous solution as an entrainer [14]. Zuo et al. (2008) also extracted isoflavones from soybean meal by using methanol [16].

To design practical separation processes using $SCCO_2$, it is necessary to establish a reliable database of the entrainer's enhancement effects. This would facilitate both the choice of a suitable entrainer for an objective component and the quantitative evaluation of separation yield of a target component in actual processes.

In this chapter, we demonstrate the solubility of isoflavones in $SCCO_2$ with ethanol added. The solubility in an $SCCO_2$-only system was also measured for comparison. The effect of the entrainer on solubility is discussed with the hydrophobicity of guest components evaluated from their molecular structure. The thermodynamic relationship between the solubility and the parameter indicated a non-ideal state in $SCCO_2$ [33].

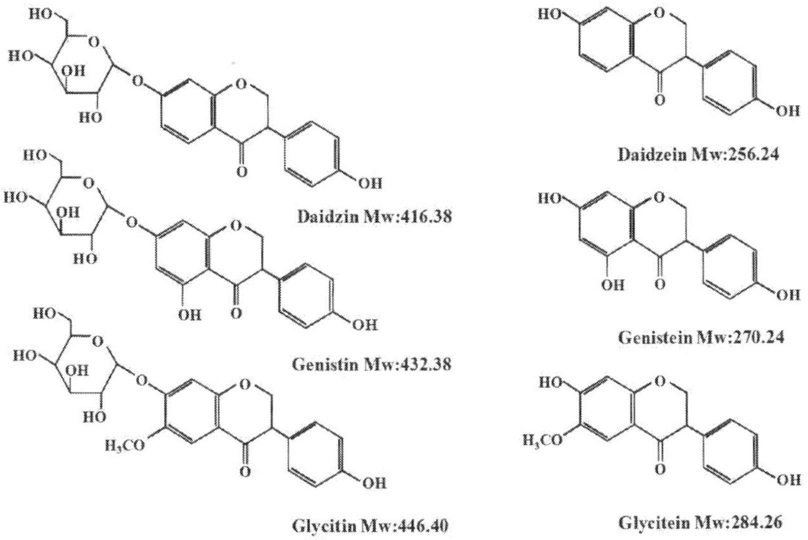

Daidzin Mw:416.38

Genistin Mw:432.38

Glycitin Mw:446.40

Daidzein Mw:256.24

Genistein Mw:270.24

Glycitein Mw:284.26

Figure 1. Chemical structure of isoflavones in soybean

1.2. Solubility of isoflavones and effect of entrainer

1.2.1. Experimental

A circulation flow of $SCCO_2$ was employed for the experimental extraction system (JASCO Co., Ltd., Tokyo) as presented in Fig. 2. The 1.0mL stainless-steel extraction vessel was instal-led in an extraction line with a total volume of 19.8mL. The extraction temperature was set at 313K. The pressure range was from 15 to 25MPa. The CO_2 volumetric flow rate in the extrac-tion line was adjusted to a constant 5mL/min at 15MPa and 25MPa.

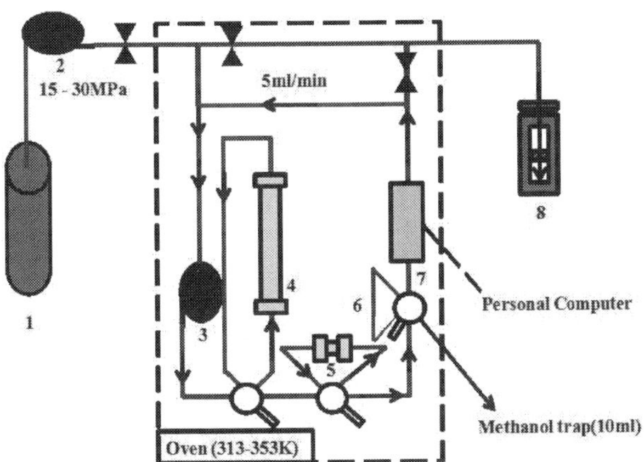

Figure 2. Schematic of experimental apparatus used in measuring the solubility in $SCCO_2$. (1) CO_2 gas cylinder, (2) com-pressor, (3) circulating pomp, (4) diffusion column, (5) extraction vessel, (6) sample loop with a methanol trap, (7) UV-detector, and (8) exhaust regulator

1.2.2. Solubility of isoflavones

Table 1 summarizes the solubility of isoflavones in the $SCCO_2$-single system. In general, the isoflavones were hardly extracted by the $SCCO_2$-single system. In particular, the solubility of glycoside isoflavones was very low it could not be detected by HPLC.

	Isoflavone	Solubility [mol-isoflavone/mol-$SCCO_2$]
Glycoside	Daidzin	not detected
	Genistin	not detected
	Glycitin	not detected
Aglycon	Daidzein	5.14×10^{-10}
	Genistein	6.38×10^{-10}
	Glycitein	not detected

Table 1. The solubility of isoflavones in pure $SCCO_2$ without ethanol at 313K, 25MPa

1.2.3. Effect of entrainer (ethanol) on solubility of isoflavones

Figure 3 presents the solubility of daidzin (as glycoside) and daidzein (as aglycon) in the $SCCO_2$ and ethanol binary system. Solubility S was increased remarkably by increasing the molar fraction of ethanol, M. This trend was also obtained at 25MPa. The solubility of genistin (as glycoside) and genistein (as aglycon) presented in Fig. 4 also exhibited the same trend. This remarkable influence of the molar fraction of ethanol also seemed to be similar between glycitin (as glycoside) and glycitein (as aglycon) (Fig. 5). The results indicated that the solubility of hydrophilic glycoside isoflavones (daidzin, genistin, and glycitin) depended more strongly on the molar fraction of ethanol.

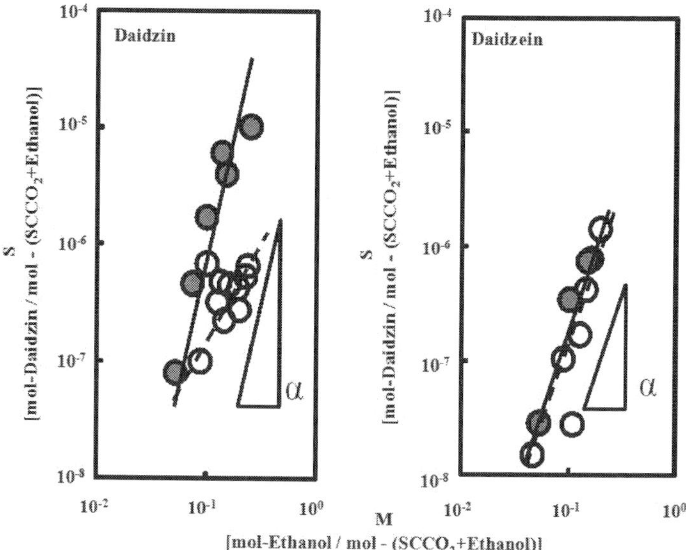

Figure 3. Solubility of daidzin and daidzein in $SCCO_2$ and ethanol binary system at 15 and 25 MPa and at 313K, o 15 MPa, ● 25 MPa.

As seen in Fig. 3, the solubility of daidzin at 25MPa was far greater than that at 15MPa. Ethanol depended more heavily on the molar fraction at 25MPa than at 15MPa. In contrast to genistin and genistein, the solubility was almost the same in spite of the increased pressure (Fig. 4). The dependency on the molar fraction of ethanol was similar for 25MPa and 15MPa.

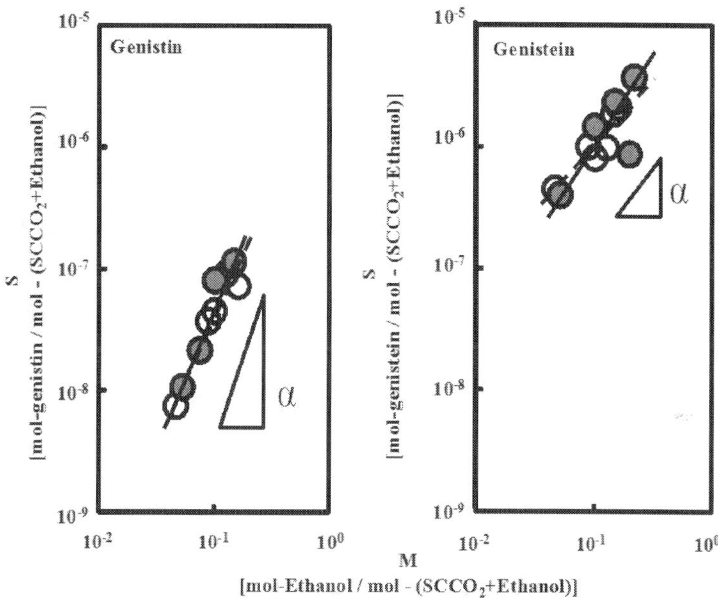

Figure 4. Solubility of genistein into $SCCO_2$ and ethanol binary system at 15 and 25 MPa and at 313K ○ 15 MPa, ● 25 MPa.

	Isoflavones	Solubility ratio
		$\equiv \dfrac{\text{The solubility } at\ 25MPa}{\text{The solubility } at\ 15MPa} [-]$
Glycoside	Daidzin	6.3
	Genistin	1.8
	Glycitin	-
Aglycon	Daidzein	1.3
	Genistein	1.8
	Glycitein	-
The solubility were referred from Fig. 3 and Fig. 4.		
The molar fraction of ethanol M was set at 0.1.		

Table 2. The solubility ratio of isoflavones in $SCCO_2$ with ethanol at 313K. The solubility were referred from Fig. 3 and Fig. 4. The molar fraction of ethanol M was set at 0.1.

The solubility ratio was defined as the solubility of 25 MPa divided by that of 15 MPa. The molar fraction of ethanol was set at 0.10, as evaluated from Figs. 3 and 4, and summarized in Table 2. In the case of daidzin, the solubility ratio was calculated as 6.3 fold. It was especially high among the tested isoflavones, i.e. 1.8 (Genistin), 1.3 (Daidzein), and 1.8 (Genistein). The solubility of daidzin was strongly affected by extraction pressure in four tested isoflavones.

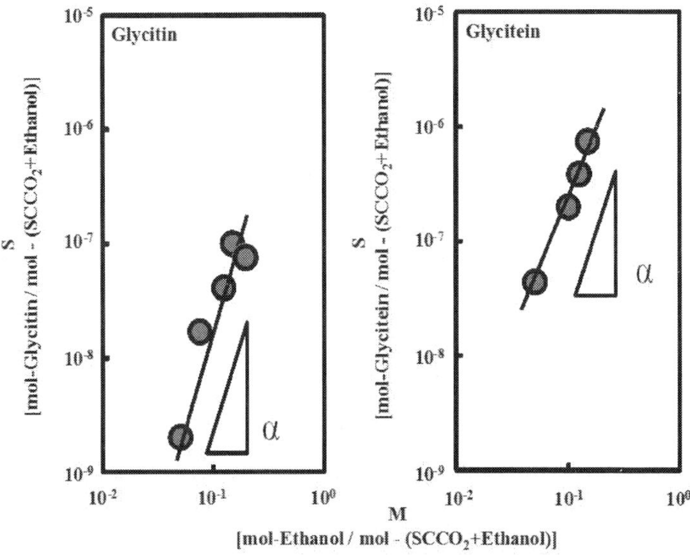

Figure 5. Solubility of glycitin and glycitein in SCCO$_2$ and ethanol binary system at 25 MPa and 313K.

The solubility of daidzin (glycoside) exceeded that of daidzein (aglycon). This trend appeared especially strong in daidzin, in contrast to that of other isoflavones. For other isoflavones, glycosides (genistin and glycitin) were less soluble than the corresponding aglycons (genistein and glycitein) due to their hydrophilic nature and the glycoside chain in their molecular structure. The detailed reasons for the special behavior of daidzin and daidzein are not clear at present.

The enhanced solubility after adding ethanol was preliminarily evaluated by the logarithmic dependency of α on the molar fraction of ethanol M. The solubility S was proportional to the $\alpha^{th.}$ power of M as indicated in empirical equation Eq. (1).

$$S \propto M^{\alpha} \tag{1}$$

The power term α is summarized in Table 3. The power term α of glycoside isoflavones (daidzin, genistin, glycitin) at both 15MPa and 25MPa often exceeded 3.0. As presented in Table 4, glycoside isoflavones are commonly more hydrophilic than their corresponding aglycons. Additive ethanol concentration in SCCO$_2$ strongly affected the solubility of hydrophilic isoflavones. The power term α of daidzein (aglycon) exceptionally exceeded 3.0 in spite of its hydrophobic nature. The detailed mechanism of solubilization must be investigated further. It may be related to a slight difference of molecular structure.

	Isoflavones	α(15MPa) [-]		α(25MPa) [-]	
Glycoside	Daidzin	3.42	R^2=0.708	3.69	R^2=0.907
	Genistin	3.08	R^2=0.996	3.41	R^2=0.932
	Glycitin	No data		3.27	R^2=0.852
Aglycon	Daidzein	3.17	R^2=0.967	3.02	R^2=0.975
	Genistein	0.728	R^2=0.985	1.72	R^2=0.999
	Glycitein	No data		1.85	R^2=0.988
	R^2 : Correlation coefficient				

Table 3. Power term α of Eq.(1) on the solubility enhancement

The dependency on the molar fraction of ethanol increased at higher pressures. The solubilities of genistin and genistein are almost the same in spite of the pressure change. The dependency on molar fraction of ethanol was also similar, suggesting that the solubility depended heavily on the amount of ethanol added. Power term α became large under $SCCO_2$ at higher pressures, except for daidzein.

	Isoflavones	Log P [-]
Glycoside	Daidzin	0.232
	Genistin	0.837
	Glycitin	0.230
Aglycon	Daidzein	1.29
	Genistein	2.09
	Glycitein	1.85

Table 4. The evaluated Log P of isoflavones

1.3. Conclusion

Solubilities of six different isoflavones were measured in an $SCCO_2$ system with ethanol added. Ethanol effectively increased the solubility of isoflavones. It served as an attractive entrainer with $SCCO_2$. The power term in the molar fraction of ethanol exceeded 3.0. The enhancement was remarkable in more hydrophilic isoflavones (daidzin, genistin, and glycitin). We experimentally determined the hydrophobicity (Log P) [34] of isoflavones from the equilibrium constant between 1-octanol and water. The hydrophobicity of daidzin was lowest among the tested isoflavones, and the enhancement due to adding ethanol was the highest.

Soybean and other natural bioresources are abundant sources of various glycoside isoflavones. Isoflavones will be successfully extracted from these sources for practical application by $SCCO_2$ with ethanol added.

2. Enzymatic modification of soybean lipid by lipase and immobilized lipase

2.1. Introduction

Soybean is beneficial in food applications and is attractive as a bioresource for functional components. Soybean contains many proteins and much oil. Furthermore, many functional components, isoflavone [35], lecithin [36], saponin [35,37], and oligosaccharide, [38,39] are desirable for promoting human health.

Soybean oil generally contains 52% linoleic acid, 22% oleic acid, 10% palmitic acid, and 8% linolenic acid. Soybean oil can be readily hydrolyzed by lipase like other vegetable oils. The produced fatty acids have several applications such as in manufacturing soaps, surfactants, and detergents, and in food.

Lipases have received attention for lipid modification [40-42]. They are used in fields such as food engineering, detergents, beverages, cosmetics, biomedical uses, and the chemical industry. They catalyze hydrolysis, alcoholysis, acidolysis, amidolysis, and esterification in the food and pharmaceutical industries [43-48]. Lipid modifications (hydrolysis, esterification, etc.) often lead to better quality products due to high specificity and selectivity of the lipase. Immobilized lipases have been applied in various hydrophobic reactions [42,49-51]. Reactivity of immobilized lipase was affected by physicochemical factors in reaction media [52,53]. A hydrophobic material is especially favorable for quick initiation of hydrophobic enzymatic reaction due to the easy diffusion of the substrate in the inner pores of the carrier. Previously, hydrophilic gels and solid porous carriers were often employed even for hydrophobic substrate reactions. Detailed technical data focused on carriers to quickly initiate hydrophobic enzymatic reactions, and high yield repeated-use immobilized enzymes are necessary in industrial design of hydrophobic enzyme reactions [54-56].

2.2. Process chemistry of soybean oil modification

Vegetable oils (olive oil [40,42]) can be hydrolyzed to produce monoglyceride, diglyceride, free fatty acids, and glycerol. Free fatty acids are value-added products because of their wide applications in surfactants, soap manufacturing, the food industry, and biomedical uses. The conventional and industrial method of oil hydrolysis has been carried out using a chemical catalyst at high temperatures and pressure. However, successful enzymatic hydrolysis reactions are possible without high temperatures and pressure.

Dalla, R. C. et al. investigated the continuous production of fatty acid ethyl esters from soybean oil in compressed fluids, namely carbon dioxide, propane, and n-butane, using immobilized Novozym 435 as a catalyst [57]. Their work evaluated the effects of some process variables on the production of fatty acid ethyl esters from soybean oil in compressed propane using Novozym 435 as a catalyst in a packed-bed reactor. In contrast to using carbon dioxide and n-butane, their results indicated that lipase-catalyzed alcoholysis was achieved in a continuous tubular reactor in compressed propane with high reaction yields at mild temperatures (70°C) and pressures (60 bar) and with short reaction times. The results demonstrated that lipase-catalyzed alcoholysis in a packed-bed reactor using compressed propane as solvent was promising as a potential alternative to conventional processes. It may be possible to manipulate process variables as well as reactor configurations to achieve acceptable yields.

Guan, F. et al. investigated the transesterification of a combination of two lipases [58]. A combination of two lipases was employed to catalyze methanolysis of soybean oil in an aqueous medium during production process. The aqueous medium was a mixture of 7 g soybean oil, methanol in various molar ratios (3:1, 4:1, 5:1, 6:1, and 9:1; methanol : oil) and 2mL (550U per mL) *P. pastoris-Rhizomucor miehei* lipase supernatant of fermentation broth (a water content of 28.6wt%, implies the total H_2O/weight of oil). The two lipase genes were cloned from fungal strains *Rhizomucor miehei* and *Penicillium cyclopium*, and each was expressed successfully in *Pichia pastoris*. Activities of the 1,3-specific lipase from *R. miehei* and the non-specific mono- and diacylglycerol lipase from *P. cyclopium* were 550U and 1545U per mL respectively. Enzymatic properties of these supernatants of fermentation broth (liquid lipase) were continuously stable at 4°C for more than 3 months. Under optimized conditions, the ratio of production conversion after 12h at 30°C, using *R. miehei* alone, was 68.5%. When *R. miehei* was assisted by adding *P. cyclopium*, the production conversion ratio increased to 95.1% under the same reaction conditions. The results suggested that combination of lipases with different specificity, for enzymatic conversion of more complex lipid substrates, is a potentially useful strategy to realize high conversion.

2.2.1. Hydrolysis

In hydrolysis, water is used to break the bonds of certain substances. In biotechnology and living organisms, these substances are often polymers. In hydrolysis involving an ester link between two amino acids in a protein, the products include the hydroxyl (OH) group, which becomes carboxylic acid with the addition of the remaining proton.

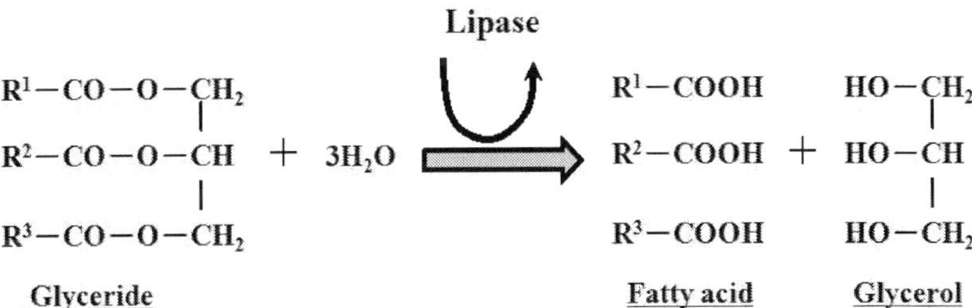

Figure 6. Hydrolysis of triglyceride catalyzed by lipase

Hydrolysis reactions in living organisms are performed with the help of catalysis by a class of enzymes known as hydrolases. The biochemical reactions that break down polymers such as proteins (peptide bonds between amino acids), nucleotides, complex sugars and starch, and fats are catalyzed by hydrolases. Within this class, lipases, amylases, and proteinases hydrolyze fats, sugars and proteins, respectively (Fig. 6).

The hydrolysis of vegetable oils is also industrially important. The complete hydrolysis of triglycerides will produce fatty acids and glycerol. These fatty acids find several applications such as in manufacturing soaps, surfactants, and detergents, and in the food industry. Since there are many kinds of natural substrates, the high specificity and selectivity of the enzymes used in the hydrolysis reaction will lead to products of better quality. Lipase has been used in the hydrolysis of different oils and fats to produce free fatty acids.

Ting, W-J. et al. investigated soybean hydrolysis by immobilized lipase in chitosan beads [59]. Their work is the culmination of their research efforts to develop an enzymatic/acid-catalyzed hybrid process for production with a view to utilizing edible and off-quality soybean oils as feedstock. They achieved a higher degree of hydrolysis. The reaction was carried out at 40°C for 12 h using binary immobilized *Candida rugosa* lipase. The conversion of free fatty acid increased rapidly from 0 to 5 h. After 5 h, the conversion of free fatty acid did not increase significantly. Almost 88% of the oil was hydrolyzed after 5 h, indicating that the feedstock for the acid-catalyzed synthesis was easily obtained by the hydrolysis of soybean oil using the binary immobilized lipase. The feedstock for acid-catalyzed production obtained after 5 h of enzymatic hydrolysis of oil contained 12% triglyceride and 88% monoglyceride, diglyceride, and free fatty acid. Problems linked to higher free fatty acid contents can be overcome by using the enzymatic/acid-catalyzed hybrid process proposed in their study. Therefore, any unrefined oil that contains different levels of free fatty acid can be used.

2.2.2. Esterification

Esterification is the chemical process of making esters, which are compounds of the chemical structure R-COOR', where R and R' are either alkyl or aryl groups (Fig. 7). The esterification process has a broad spectrum of uses from preparing highly specialized esters in chemical laboratories to producing millions of tons of commercial ester products. These commercial compounds are manufactured by either a batch or a continuous synthetic process. The batch procedure involves a single pot reactor that is filled with the acid and alcohol reactants.

Sugar fatty acid esters are widely used as non-ionic surfactants in cosmetic and food applications. Current chemical production is based on high-temperature esterification of sugars and fatty acids, using an alkaline catalyst leading to a mixture of products. Alternatively, sugar fatty acid esters can be obtained by fermentation as so-called biosurfactants. The direct esterification of sugar and fatty acid using isolated enzymes (mainly lipases) is hampered by the low solubility of sugars in most organic solvents. Good conversions can be achieved in pyridine, but this solvent is incompatible with food applications. Other solutions are based on the use of alkylglycosides or protected sugars like isopropylidene or phenylboronic acid derivatives, which require additional synthesis steps.

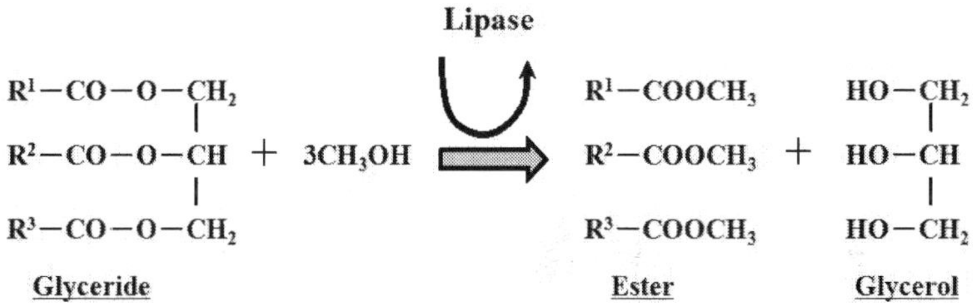

$$R^1-CO-O-CH_2 \quad\quad\quad\quad R^1-COOCH_3 \quad\quad HO-CH_2$$
$$R^2-CO-O-CH + 3CH_3OH \xrightarrow{\text{Lipase}} R^2-COOCH_3 + HO-CH$$
$$R^3-CO-O-CH_2 \quad\quad\quad\quad R^3-COOCH_3 \quad\quad HO-CH_2$$

Glyceride   **Ester**   **Glycerol**

Figure 7. Esterification of triglyceride catalyzed by lipase

Nagayama, K. et al. investigated lecithin microemulsion-based organogels as immobilization carriers for the esterification of lauric acid with butyl alcohol catalyzed by *Candida rugosa* lipase [60]. Gelatin was used as the gelling component of the microemulsion-based organogels. The maximum reaction rate was obtained at a G_{LW} (volume fraction of water in microemulsion-based organogel) of 75% v/v, a gelatin content of 18.5% w/v, and a lecithin concentration of 18 mM. The reaction proceeded under a reaction-controlled regime, and the reaction rate was influenced by microemulsion-based organogel compositional changes. The effective diffusion coefficient of lauric acid varied with the microemulsion-based organogel composition, while that of butyl alcohol remained constant. The partition coefficient of both substrates was affected by the microemulsion-based organogel composition. Immobilized lipase was reused in a batch-reaction system, and its activity was successfully maintained for 720h. During repeated batch reactions, lipase activity was enhanced, while the ester concentration at 48h was between 30 and 40 mM.

2.3. Immobilized enzymatic reaction of soybean lipid modification

Immobilization of lipase has been investigated to improve the stability and reusability of lipase in oil hydrolysis. For practical applications, a systematic strategy is necessary to select suitable support and organic solvents. Authors investigated a key factor of suitable support to improve enzyme activity and stability of immobilized lipase [61].

Author	Year	Enzyme	Carrier	Immobilization	Substrate	Solvent	Surfactant	Production	Reaction	Reference Number
Ahn, K.W. et al.	2011	*Pseudomonas cepacia* lipase *Pseudomonas fluorescens* lipase	Mesoporous silica	Stirred	Soybean oil	Methanol			Methanolysis	[63]
Cao, L. et al.	1999	*Candida antarctica* lipase	Polypropylene Silica gel PEG	Adsorption Crosslinking	Olive oil	*t*-butanol		Fatty acid	Esterification Hydrolysis	[40]
Dizge, N. et al.	2009	*Thermomyces lanuginosus* lipase	Microporous polymeric matrix		Soybean oil	Methanol			Transesterification batch reaction	[54]
Huang, D. et al.	2012	*Rhizomucor miehei* lipase			Soybean oil	Isooctane		Methyl ester	Transesterification	[64]
Khare, S.K. and Nakajima, M.	2000	*Rhizopus japonicus* lipase	Celite	Adsorption	Tripalmitin Stearic acid Soybean oil Docosahexaenoic acid *p*-nitrophenyl palmiate	Hexane		1,2-dipalmitoyl-3-stearoyl glycerol 1,3-distearoyl-2-palmitoyl glycerol *p*-nitrophenol	Transesterification Hydrolysis	[65]
Kiatsimkul, P-P. et al.	2006	*Candida rugosa* lipase *Burkholderia cepacia* lipase *Psedomonas* sp. lipase *Penicillium roquefortii* lipase *Penicillium camembertii* lipase *Aspergillus niger* lipase *Mucor javanicus* lipase *Rhizomucor miehei* lipase			Soybean oil Epoxidized soybean oil			Fatty acid Glycerides Methyl esters	Hydrolysis	[41]
Li, S-F. and Wu, W-T.	2009	*Candida rugosa* lipase	Polyacrylonitrile nanofibrous membranes	Shaking	Soybean oil				Hydrolysis batch reaction	[66]
Li, S-F. et al.	2011	*Pseudomonas cepacia* lipase	Polyacrylonitrile nanofibrous membranes	Shaking	Soybean oil			Fatty acid	Transesterification Hydrolysis	[67]
Naoe, K. et al.	2001	*Rhizopus delemar* lipase	W/O microemulsion		Oleic acid Octyl alcohol	Hexane	DK-ester	Fatty acid esters	Esterification	[52]
Naya, M. and Imai, M.	2012	*Candida rugosa* lipase	Hydrophobic porous carrier Polypropylene Accurel® MP 100	Crosslinking glutaraldehyde	Triolein	Isooctane	DK-ester	Oleic acid	Hydrolysis	[61]
Noureddini, H. et al.	2005	*Pseudomonas cepacia* lipase *Penicillium roqueforti* lipase *Psedomonas* sp. lipase *Mucor* sp. lipase *Aspergillus niger* lipase *Rhizopus oryzae* lipase *Penicillium camemberittii* lipase *Rhizopus niveus* lipase *Candida rugosa* lipase	Hydrophobic sol-gel support	Sol-gel method	Soybean oil	Methanol Ethanol		Free fatty acids Methylesters Ethylesters	Transesterification	[68]
Ozmen, E.Y. and Yilmaz, M.	2009	*Candida rugosa* lipase	β-cyclodextrin-based polymer	Crosslinking	Soybean oil			Fatty acid Glycerol	Hydrolysis batch reaction	[69]
Rodrigues, R.C. and Záchia Ayub, M.A.	2011	*Thermomyces lanuginosa* lipase *Rhizomucor miehei* lipase	Lewatil®	Multipoint-covalently immobilized	Soybean oil	Methanol		Glycerol	Transesterification Hydrolysis batch reaction	[70]
Ting, W-J. et al.	2008	*Candida rugosa* lipase	Chitosan beads	Crosslinking glutaraldehyde	Soybean oil			Free fatty acid	Hydrolysis	[59]
Uehara, A. et al.	2008	*Rhizopus delemar* lipase	W/O microemulsion		Triolein	Isooctane	DK-ester	Oleic acid	Hydrolysis	[53]
Virto, M.D. et al.	1994	*Candida rugosa* lipase	Polypropylene Accurel EP-100 (1.0-0.2 mm)	Adsorption	Beef tallow Pork lard Olive oil	Isooctane n-Heptane n-pentane Isopropanol Ethyl ether		Free fatty acid	Hydrolysis	[42]
Wang, W. et al.	2011	*Rhizomucor miehei* lipase *Thermomyces lanuginosa* lipase *Candida antarctica* lipase	Tree commercial immobilized lipase Lipozyme RM IM Lipozyme TL IM Novozym 435		Soybean oil	*t*-butanol		Diacylglycerol Fatty acid	Glycerolysis	[71]
Watanabe, Y. et al.	2002	*Candida antarctica* lipase			Soybean oil triacylglycerols	Chloroform/methanol		Fatty acid methyl esters	Methanolysis batch reaction	[47]
Xie, W. and Ma, N.	2010	*Thermomyces lanuginosa* lipase	Magnetic Fe₃O₄ nano-perticles	Mix	Soybean oil	Methanol		Fatty acid methyl esters	Transesterification batch reaction	[48]
Xie, W. and Wang, J.	2012	*Candida rugosa* lipase	Magnetic chitosan microspheres	Crosslinking glutaraldehyde	Soybean oil	Methanol		Fatty acid methyl esters	Transesterification batch reaction	[72]
Zhou, G. et al.	2009	*Candida rugosa* lipase	Mesoporous rod-like silica Mesoporous vesicle-like silica	Physical adsorption	Butyrin	Water Phosphate buffer saline			Hydrolysis batch reaction	[56]

Table 5. Previous investigations of enzymatic lipid modification by lipase immobilized

Immobilized enzymes have been examined for various industrial applications. In general, enzyme immobilization effectively enables separating the enzyme from products, thus facilitating their recovery and repeated use [40,42,62]. This is promising for industrial enzymatic production of various biomaterials. The main aspects of the currently investigated immobilized enzyme are as follows. First, the molecular structure of the enzyme is directly influenced by immobilization [50]. Second, enzyme reactivity is affected by the physicochemical characteristics of the enzyme carrier and the reaction media [40,51]. To quickly initiate hydrophobic enzymatic reactions, a water-in-oil (W/O) microemulsion system is desirable for achieving higher concentrations of hydrophobic substrate in the reaction media. Third, the diffusion of the substrate and the reaction products determines the rate-limiting condition in the reactivity of the immobilized enzyme [49]. Finally, repeated use of the immobilized enzyme in a practical process is a key factor in reducing costs in industrial applications.

Solid porous carriers are expected to resist compaction and deformation of carrier particles during practical use in bioreactors. Hydrophobic solid porous materials are preferred as immobilized enzymes for hydrophobic reactions. Table 5 summarizes previous hydrophobic substrate reactions using immobilized lipase. Hydrophobic materials, primarily a polypropylene porous commercial carrier called Accurel, have been employed for lipid hydrolysis and esterification. Lipase is adsorbed with strong multipoint interactions in Accurel [73]. Particle size plays a dominant role in determining the rate-limiting condition of the substrate [46,49,55,74]. The particle size as well as handling of particles was very important for both the practical design of the bioreactor and for determining reaction-rate-limiting conditions. In the Accurel EP-100 system, the effect of particle size on reaction rate was examined for a size range of 0.2 to 2.5 mm [49,55,62,74]. A higher reaction rate was obtained for a smaller immobilized carrier. Sabbani et al. reported that the reaction rate was increased sixfold by decreasing the particle size from 0.2 to 1.5 mm[55]. Montero et al. pointed out that cross-linking of lipase (*Candida rugosa*) by glutaraldehyde (GA) was promising for attaining higher reaction activity [62]. Naya and Imai investigated lipid hydrolysis using an immobilized lipase on Accurel MP100 [61]. It examined the effect of particle size on the apparent reaction rate. The technical data were expected to be used in designs for industrial application of Accurel MP100 for hydrophobic immobilized lipase reactions.

2.3.1. W/O microemulsion

W/O microemulsions are spontaneous aggregates composed of amphiphilic molecules in non-polar media. The properties of reverse micelles have been extensively investigated in the field of reverse micellar techniques. Reverse micelles enable hydrophilic proteins to be solubilized in organic solvent and are anticipated to be used as separation and enzymatic reaction media with hydrophobic substrates. When enzymes are micro-encapsulated, they are situated inside the water pool of the W/O microemulsion; whether or not they interact with the micellar interface depends on the enzyme species (Fig. 8). For example, an enzyme reaction involving lipase was observed on the interfacial layer between the hydrophobic phase containing substrates, and the hydrophilic phase containing dissolved lipase.

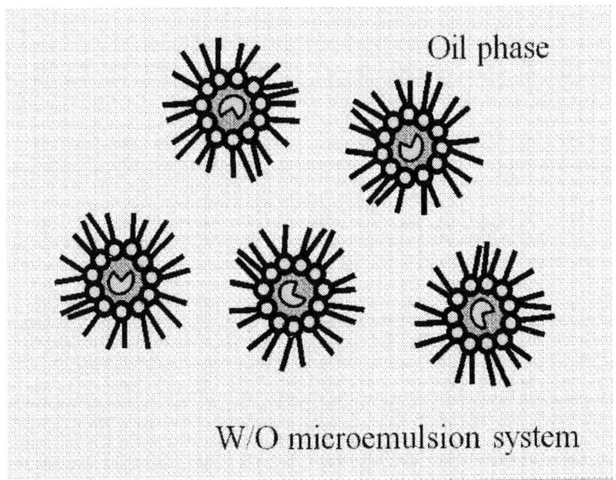

Figure 8. Schematic image of W/O microemulsion system. Micro-water pool was dispersed in bulk oil phase. ⚬⟨ : amphiphilic molecule. ♡ :enzyme

Uehara et al. defined the reaction condition producing high reactivity over a limited range of both hydrophilicity and interfacial fluidity of the microemulsion droplet [53]. Their reaction condition was identified as the most favorable condition for sugar–ester alcohol W/O microemulsion media to perform lipid hydrolysis. The critical micelle concentration depended on the concentration of 1-butanol and was found to be inversely proportional to the second power of the 1-butanol concentration. The initial reaction rate of the hydrolysis of triolein in W/O microemulsion depended on the solubilized water content, reaching a maximum in the limited range of $2 < W_{soln} < 4$. The maximum initial reaction rate increased about 2-fold following the addition of 1-butanol. The most favorable concentration of 1-butanol for hydrolysis by *Rhizopus delemar* was identified as 3.5% v/v.

Naoe et al. investigated the esterification of oleic acid with octyl alcohol catalyzed by *Rhizopus delemar* lipase in a reverse micellar system of sugar ester DK-F-110 [52]. A high initial reaction rate was obtained by preparing a micellar organic phase with extremely low water content. The initial reaction rate decreased slightly with decreasing DK-F-110 concentration. The lipase exhibited 40% of its esterification activity after 28h incubation in the DK-F-110 reverse micellar organic phase. Sodium bis(2-ethylhexyl)sulfosuccinate (AOT) is often used as an ionic amphiphilic molecule for reverse micelle formation owing to the advantages of spontaneous aggregation, thermodynamic stability, and non-additional co-surfactant. In the work of Naoe et al. the turnover number of the DK-F-110 system was larger than that of the system using AOT.

2.3.2. Gel beads carrier

The major problem that must be solved to employ a microemulsion system in industrial processes is the recovery of the products and the repeated use of enzyme. Usual techniques such as extraction and distillation lead to poor separation because of the problems of emulsion-forming and foaming caused by the presence of surfactants. One approach to simplifying the recovery of the product and the enzyme for reuse from microemulsion based-media has been to employ gelled microemulsion systems. Interestingly, many W/O microemulsions can be gelled by adding gelatin, yielding a matrix suitable for enzyme immobilization. Cooling at room temperature causes a transparent gel with reproducible physical properties to form. These enzyme-containing, gelatin-based gels are rigid and stable in various non-polar organic solvents and may therefore be used for biotransformations in organic media. Under most conditions, the gel matrix fully retains the surfactant, gelatin, water, and enzyme components, allowing the diffusion of non-polar substrates or products between a contacting non-polar phase and the gel pellets.

Natural gelling agents such as gelatin, agar and κ-carrageenan have been tested for the formation of lecithin microemulsion-based gels as well as hydrogels presented by Stamatis, H and Xenakis, A [75]. Lipase-containing microemulsions-based organogels formulated with various biopolymers have considerable potential for their application in biotransformations. Lipase immobilized in gelatin and agar organogels exhibited good stability in catalyzing esterification reactions under mild conditions with high conversion yields. High yields (80%) were obtained with agar and κ-carrageenan organogels in isooctane. The remaining lipase activity in repeated syntheses was found to depend on the nature of the biopolymer used for forming the organogels. Gelatin and agar microemulsion-based gels had the highest operational stability. Moreover, aqueous gelatin and agar gels containing only lipase, water, and biopolymer retain their integrity in organic solvents and can also be used for the synthesis of esters.

Chitosan, poly [β-(1-4)-linked-2-amino-2-deoxy-D-glucose], is non-toxic, hydrophilic, biocompatible, biodegradable, and anti-bacterial and can be used as a material for immobilized carriers since it has a variety of functional groups that can be tailored to specific applications. Xie, W. and Wang, J. investigated the effects of various transesterification parameters on the enzymatic conversion of soybean oil [72]. In their work, magnetic chitosan microspheres were prepared by the chemical co-precipitation approach using glutaraldehyde as the cross-linking reagent for lipase immobilization. Using the immobilized lipase, the conversion of soybean oil to fatty acid methyl esters reached 87% under the optimized conditions of a methanol/oil ratio of 4:1 with the three-step addition of methanol, reaction temperature 35°C, and reaction time 30h. Moreover, the immobilized lipase could be used for four times without significant decrease of activity.

2.3.3. Polypropylene carrier

The immobilized lipase (*Candida rugosa*) using polypropylene-based hydrophobic granular porous carrier Accurel MP100 was investigated in lipid hydrolysis reactions involved in the effect of particle size on the apparent reaction rate [61]. The true shape of the original Accurel was similar to a half cylinder (Fig. 9 (a)). Macro-pores existed near the particle surfaces. Inside the particle, the micro-pores formed many branched channels (Fig. 9 (b)).

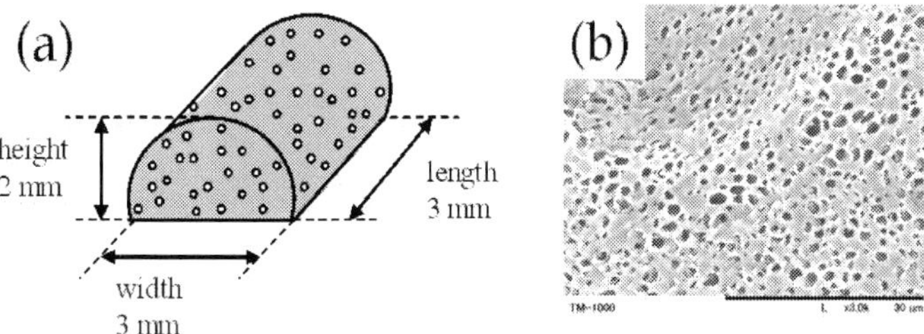

Figure 9. a). Schematic illustration of original Accurel MP100, a polypropylene-based hydrophobic granular support. The true shape of original Accurel particles seemed to be half cylinders. (b) SEM (electron microscopy) image of original Accurel particles.

The amount of immobilized lipase per unit mass of particle was increased by 19% in smaller particles (500 to 840 μm). The immobilized yield lipase based on the adsorbed amount was high (over 98%) in every class of particle size (Fig. 10). Cross-linking of lipase by glutaraldehyde (GA) holds much promise for immobilization.

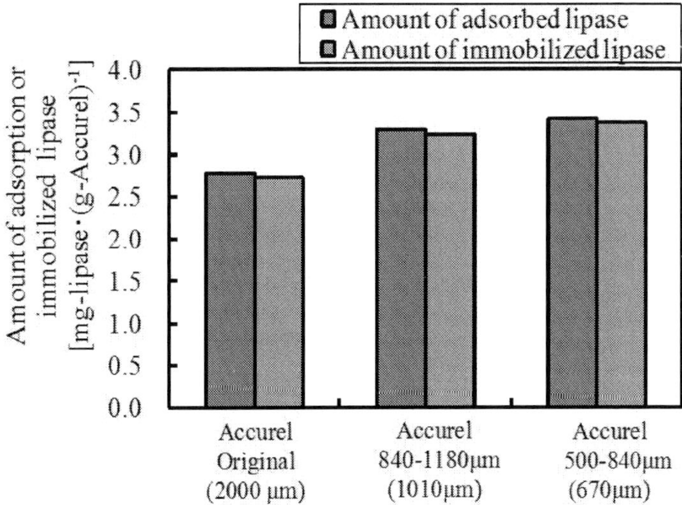

Figure 10. (a). Schematic illustration of original Accurel MP100, a polypropylene-based hydrophobic granular support. The true shape of original Accurel particles seemed to be half cylinders. (b) SEM (scanning electron microscopy) image of original Accurel particles.

The reactivity of immobilized lipase as evaluated from the oleic acid production rate strongly depended on the Accurel particle size. In particular, the 500 to 840 µm (mean diameter 670µm) particles performed significantly outstanding reactivity compared with that of 840 to 1180 µm (mean diameter 1010µm) particles and original Accurel (Fig. 11). The experimental effectiveness factor was obtained and compared with the theoretical effectiveness factor. The difference was speculated to be due to assumptions of the geometrical factor of particles and the partition equilibrium of the substrate between the carrier particle and bulk phase. Quick initiation was observed in the repeated use of immobilized lipase on the 500 to 840 µm particles. The production yield was well-preserved.

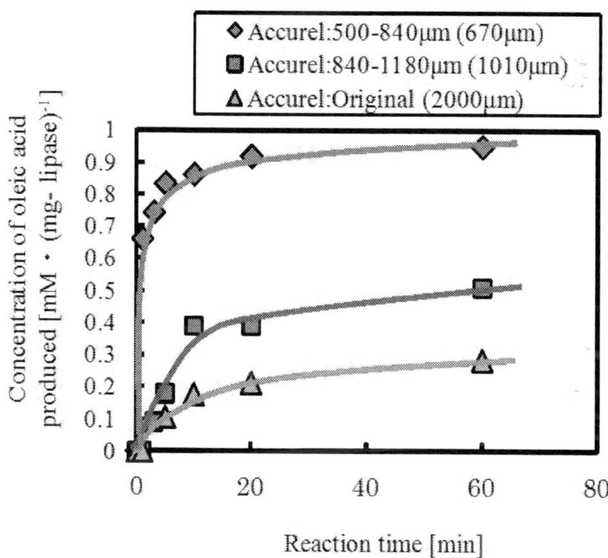

Figure 11. Comparison of reactivity of immobilized lipase for various particle sizes of Accurel.

2.3.4. Nanofiber membrane

Li, S-F. and Wu, W-T. investigated immobilized lipase activity using a nanofiber membrane [66]. The activity retention of the immobilized lipase was 87.5% of the free enzyme. Under these optimal reaction conditions, the hydrolysis conversion of soybean oil was 72% after 10min and 85% after 1.5h. In reusability, the immobilized lipase retained 65% of its initial conversion after 20 additional batch reactions. Protein loading reached 21.2mg/g material of the membrane due to the large specific surface area provided by the nanofibers. This effective enzyme immobilization method has good potential for industrial applications.

2.4. Conclusion

Soybean has been expected to be used both as a food and as a bioresource for attractive functional components. Soybean contains many proteins and much oil. Soybean oil can be hydrolyzed readily by lipase like other vegetable oils. The produced fatty acids find several applications such as in manufacturing soaps, surfactants, and detergents, and in food.

Immobilization of lipase has been investigated to improve its stability and reusability in oil hydrolysis. For practical applications, a systematic strategy is necessary to select suitable support and organic solvent. Since the novel developed method is promising, it could be used industrially for producing chemicals requiring immobilized lipases.

Nomenclature

G_{LW}: volume fraction of water in microemulsion-based organogel phase, referred from [60] (% v/v microemulsion-based organogel phase)

Log P: hydrophobicity index by Laane et al. [34]. P was defined by partition equilibrium (-)

M: molar fraction of ethanol in $SCCO_2$, referred from [33] ([mol-Ethanol]/[mol-($SCCO_2$+Ethanol)])

S: molar fraction of extracted sample in the $SCCO_2$ and ethanol binary system, referred from [33]] ([mol-extracted sample]/[mol-($SCCO_2$+Ethanol)])

W_{soln}: molar ratio of solubilized water to amphiphile, referred from [53] ([mol-H_2O_{soln}]/[mol-amphiphile])

α: the power term on the molar fraction of ethanol M, presented by Eq. (1), referred from [33]. It is summarized in Table 3 (-)

Author details

Masakazu Naya and Masanao Imai*

*Address all correspondence to: XLT05104@nifty.com

Course in Bioresource Utilization Sciences, Graduate School of Bioresource Sciences, Nihon University, Japan

References

[1] Izumi, T., Obata, A., Arii, M., Yamaguchi, H., & Matsuyama, A. (2007). Oral Intake of Soy Isoflavone Improves the Aged skin of Adult Women. *Journal of Nutritional Science and Vitaminology*, 53, 57-62.

[2] Lee, Y. B., Lee, H. J., Won, M. H., Hwang, I. K., Kang, T. C., Lee, J. Y., Nam, S. Y., Kim, K. S., Kim, E., Cheon, S. H., & Sohn, H. S. (2004). Soy isoflavones Improve Spatial Delayed Matching-to-Place Peformance and Reduce Cholinergic Neuron Loss in Elderly Male Rats. *The Journal of Nutrition*, 134, 1827-1831.

[3] Lee, Y. B., Lee, H. J., Kim, K. S., Lee, J. Y., Nam, S. Y., Cheon, S. H., & Sohn, H. S. (2004). Evaluation of the preventive Effect of Isoflavone Extract on Bone Loss in Ovariectomized Rats. *Bioscience Biotechnology Biochemistry.*, 68(5), 1040-1045.

[4] Suh, K. S., Koh, G., Park, C. Y., Woo, J. T., Kim, S. W., Kim, J. W., Park, I. K., & Kim, Y. S. (2003). Soybean isoflavones inhibit tumor facter-α-induced apoptosis and production of interleukin-6 and prostaglandin E_2 in osteoblastic cells. *Phytochemistry.*, 63, 209-215.

[5] Mahesha, H. G., Singh, S. A., & Rao, A. G. A. (2007). Inhibition of lipoxygenase by soy isoflavones: Evidence of isoflavones as redox inhibitors. *Archives of Biochemistry and Biophysics*, 461, 176-185.

[6] Chan, H. Y., Chan, H. Y., & Leung, L. K. (2003). A potential mechanism of soya isoflavoes against 7,12-dimetylbenz[a]anthracene tumour inhibition. *British Journal of Nutrition.*, 90, 457-465.

[7] Clubbs, E. A., & Bomser, J. A. (2007). Glycitein activates extracellular signal-regulated kinase via vascular endothelial growth factor receptor signaling in nontumorigenic (RWPE-1) prostate epithelial cells. *Journal of Nutritional Biochemistry.*, 18, 525-532.

[8] Murphy, P. A., Barua, K., & Hauck, C. C. (2002). Solvent extraction selection in the determination of isoflavones in soy foods. *Journal of Chromatography B.*, 777, 129-138.

[9] Luthria, D. L., Biswas, R., & Natarajan, S. (2007). Comparison of extraction solvents and techniques used for the assay of isoflavones from soybean. *Food Chemistry*, 105, 325-333.

[10] Rostagno, M. A., Palma, M., & Barroso, C. G. (2003). Ultrasound-assisted extraction of soy isoflavones. *Journal of Chromatography A.*, 1012, 119-128.

[11] Rostagno, M. A., Palma, M., & Barroso, C. G. (2007). Ultrasound-assisted extraction of isoflavones from beverages blended with fruit juices. *Analytica Chimica Acta*, 597, 265-272.

[12] Araújo, J. M. A., Silva, M. V., & Chaves, J. B. P. (2007). Supercritical fluid extraction of daidzein and genistein isoflavones from soybean hypocotyls after hydrolysis with endogenous β-glucosidases. *Food Chemistry.*, 105, 266-272.

[13] Kao, T. H., Chien, J. T., & Chen, B. H. (2008). Extraction yield of isoflavones from soybeans cake as affected by solvent and supercritical carbon dioxide. *Food Chemistry*, 107, 1728-1736.

[14] Rostagno, M. A., Araújo, J. M. A., & Sandi, D. (2002). Supercritical fluid extraction of isoflavones from soybean flour. *Food Chemistry*, 78, 111-117.

[15] Yu, J., Liu, Y. F., Qiu, A. Y., & Wang, X. G. (2007). Preparation of isoflavones enriched soy protein isolate from defatted soy hypocotyls by supercritical CO_2. *LWT-Food Science and Technology*, 40, 800-806.

[16] Zuo, Y. B., Zeng, A. W., Yuan, X. G., & Yu, K. T. (2008). Extraction of soybean isoflavones from soybean meal with aqueous methanol modified supercritical carbon dioxide. *Journal of Food Engineering*, 89, 384-389.

[17] Iwai, Y., Nagano, H., Lee, G. S., Uno, M., & Arai, Y. (2006). Measurement of entrainer effect of water and ethanol on solubility in supercritical carbon dioxide by FT-IR spectroscopy. *Journal of Supercritical Fluids.*, 38, 312-318.

[18] Johannsen, M., & Brunner, G. (1994). Solubility of the xanthenes caffeine, theophylline and theobromine in supercritical carbon dioxide. *Fluid Phase Equilibria.*, 95, 215-226.

[19] Kopcak, U., & Mohamed, R. S. (2005). Caffeine solubilities in supercritical carbon dioxide/co-solvent mixtures. *Journal of Supercritical Fluids.*, 34, 209-214.

[20] Li, S., Varadarajan, G. S., & Hartland, S. (1991). Solubility of theobromine and caffeine in supercritical carbon dioxide: correlation with density-based models. *Fluid Phase Equilibria.*, 68, 263-280.

[21] Duarte, C. M. M., Crew, M., Casimiro, T., Aguiar-Ricardo, A., & Ponte, M. N. (2002). Phase equilibrium for capsaicin + water + ethanol + supercritical carbon dioxide. *Journal of Supercritical Fluids*, 22, 87-92.

[22] de la Fuente, J. C., Valderrama, J. O., Bottini, S. B., & del Valle, J. M. (2005). Measurement and modeling of solubilities of capsaicin in high-pressure CO_2. *Journal of Supercritical Fluids*, 34, 195-201.

[23] Cygnarowicz, M. L., Maxwell, R. J., & Selder, W. D. (1990). Equilibrium solubility of β-carotene in supercritical carbon dioxide. *Fluid Phase Equilibria*, 59, 57-71.

[24] de la Fuente, J. C., Oyarzún, B., Quezada, N., & del Valle, J. M. (2006). Solubility of carotenoid pigment (lycopene and astaxanthin) in supercritical carbon dioxide. *Fluid Phase Equilibria*, 247, 90-95.

[25] Škerget, M., Knez, Ž., & Habulin, M. (1995). Solubility of β-carotene and oleic acid in dense CO_2 and data correlation by a density based model. Fluid Phase Equilibria. , 109, 131-138.

[26] Subra, P., Castellani, S., Ksibi, H., & Garrabos, Y. (1997). Contribution to the determination of the solubility of β-carotene in supercritical carbon dioxide and nitrous oxide: experimental data and modeling. *Fluid Phase Equilibria*, 131, 269-286.

[27] Berna, A., Cháfer, A., Montón, J. B., & Subirats, S. (2001). High-pressure solubility data of system ethanol (1) + catechin (2) + CO_2 (3). *Journal of Supercritical Fluids*, 20, 157-162.

[28] Cháfer, A., Berna, A., Montón, J. B., & Nuñoz, R. (2002). High-pressure solubility data of system ethanol (1) + epicatechin (2) + CO_2 (3). *Journal of Supercritical Fluids.*, 24, 103-109.

[29] Nunes, A. V. M., Matias, A. A., da Ponte, M. N., & Duarte, C. M. (2007). Quaternary Phase Equilibria for scCO$_2$ + Biophenolic Compound + Water + Ethanol. *Journal of Chemical & Engineering Data*, 52, 244-247.

[30] Wang, L. H., & Cheng, Y. Y. (2005). Solubility of Puerarin in Ethanol + Supercritical Carbon Dioxide. *Journal of Chemical Engineering Data.*, 50, 1747-1749.

[31] Huang, Z., Chiew, Y. C., Lu, W. D., & Kawi, S. (2005). Solubility of aspirin in supercritical carbon dioxide/alcohol mixture. *Fluid Phase Equilibria*, 237, 9-15.

[32] Matias, A. A., Nunes, A. V. M., Casimiro, T., & Duarte, C. M. M. (2004). Solubility of coenzyme Q10 in supercritical carbon dioxide. *J. of Supercritical Fluids.*, 28, 201-206.

[33] Nakada, M., Imai, M., & Suzuki, I. (2009). Impact of ethanol addition on the solubility of various soybean isoflavones in supercritical carbon dioxide and the effect of glycoside chain in isoflavones. *Journal of Food Engineering*, 95, 564-571.

[34] Laane, C., Boeren, S., Vos, K., & Veeger, C. (1987). Rules for optimization of biocatalysis in organic solvents. *Biotechnology and Bioengineering*, 30, 81-87.

[35] Paucar-Menacho, L. M., Amaya-Farfan, J., Berhow, M. A., Mandarino, J. M. G., Mejia, E. G., & Chang, Y. K. (2010). A high-protein soybean cultivar contains lower isoflavones and saponins but higher minerals and bioactive peptides than a low-protein cultivar. *Food Chemistry*, 120, 15-21.

[36] Comas, D. I., Wagner, J. R., & Tomas, M. C. (2006). Creaming stability of oil in water (O/W) emulsions: Influence of pH on soybean protein-lecithin interaction. *Food Hydrocolloids*, 20, 990-996.

[37] Berhow, M. A., Wagner, E. D., Vaughn, S. F., & Plewa, M. J. (2000). Characterization and antimutagenic activity of soybean saponins. *Mutation Research*, 448, 11-22.

[38] Viana, P. A., Rezende, S. T., Falkoski, D. L., Leite, T. A., Jose, I. C., Moreira, M. A., & Guimaraes, V. M. (2007). Hydrolysis of oligosaccharides in soybean products by Debaryomyces hansenii UFV-1 α-galactosidases. *Food Chemistry.*, 103, 331-337.

[39] Wang, Q., Ying, T., Jahangir, M. M., & Jiang, T. (2012). Study on removal of coloured impurity in soybean oligosaccharides extracted from sweet slurry by adsorption resins. *Journal of Food Engineering, in press.*

[40] Cao, L., Bornscheuer, U. T., & Schmid, R. D. (1999). Lipase-catalyzed solid-phase synthesis of sugar esters. Influence of immobilization on productivity and stability of the enzyme. *Journal of Molecular Catalysis B: Enzymatic*, 6, 279-285.

[41] Kiatsimkul-P, P., Sutterlin, W. R., & Suppes, G. J. (2006). Selective hydrolysis of epoxidized soybean oil by commercially available lipases: Effects of epoxy group on the enzymatic hydrolysis. *Journal of Molecular Catalysis B: Enzymatic.*, 41, 55-60.

[42] Virto, M. D., Agud, I., Montero, S., Blanco, A., Solozabal, R., Lascaray, J. M., Llama, M. J., Serra, J. L., Landeta, L. C., & Renobales, M. (1994). Hydrolysis of animal fats by immobilized Candida rugosa lipase. *Enzyme and Microbial Technology.*, 16, 61-65.

[43] Hita, E., Robles, A., Camacho, B., Gonzalez, P. A., Esteban, L., Jimenez, M. J., Munio, M. M., & Molina, E. (2009). Production of structured triacylglycerols by acidolysis catalyzed by lipases immobilized in a packed bed reactor. *Biochemical Engineering Journal*, 46, 257-264.

[44] Jimenez, M. J., Esteban, L., Robles, A., Hita, E., Gonzalez, P. A., Munio, M. M., & Molina, E. (2010). Production of triacylglycerols rich in palmitic acid at sn-2 position by lipase-catalyzed acidolysis. *Biochemical Engineering Journal*, 51, 172-179.

[45] Pilarek, M., & Szewczyk, K. W. (2007). Kinetic model of 1,3-specific triacylglycerols alcoholysis catalyzed by lipases. *Journal of Biotechnology*, 127, 736-744.

[46] Salis, A., Sanjust, E., Solinas, V., & Monduzzi, M. (2003). Characterisation of Accurel MP1004 polypropylene powder and its use as a support for lipase immobilization. *Journal of Molecular Catalysis B: Enzymatic.*, 24-25, 75-82.

[47] Watanabe, Y., Shimada, Y., Sugihara, A., & Tominaga, Y. (2002). Conversion of degummed soybean oil to biodiesel fuel with immobilized Candida antarctica lipase. *Journal of Molecular Catalysis B: Enzymatic.*, 17, 151-155.

[48] Xie, W., & Ma, N. (2010). Enzymatic transesterification of soybean oil by using immobilized lipase on magnetic nano-particles. *Biomass and Bioenergy*, 34, 890-896.

[49] Salis, A., Svensson, I., Monduzzi, M., Solinas, V., & Adlercreutz, P. (2003). The atypical lipase B from Candida antarctica is better adapted for organic media than the typical lipase from Thermomyces lanuginose. *Biochimica et Biophysica Acta.*, 1646, 145-151.

[50] Palomo, J. M., Fernandez-Lorente, G., Mateo, C., Ortiz, C., Fernandez-Lafuente, R., & Guisan, J. M. (2002). Modulation of the enantioselectivity of lipases via controlled immobilization and medium engineering: hydrolytic resolution of mandelic acid esters. *Enzyme and Microbial Technology*, 31, 775-783.

[51] Persson, M., Mladenoska, I., Wehtje, E., & Adlercreutz, P. (2002). Preparation of lipases for use in organic solvents. *Enzyme and Microbial Technology*, 31, 833-841.

[52] Naoe, K., Ohsa, T., Kawagoe, M., & Imai, M. (2001). Esterification by Rhizopus delemar lipase in organic solvent using sugar ester reverse micelles. *Biochemical Engineering Journal*, 9, 67-72.

[53] Uehara, A., Imai, M., & Suzuki, I. (2008). The most favorable condition for lipid hydrolysis by Rhizopus delemar lipase in combination with a suger-ester and alcohol W/O microemulsion system. *Colloids and Surfaces A: Physicochemical and Engineering Aspects*, 324, 79-85.

[54] Dizge, N., Aydiner, C., Imer, D. Y., Bayramoglu, M., Tanriseven, A., & Keskinler, B. (2009). Biodiesel production from sunflower, soybean, and waste cooking oils by

transesterification using lipase immobilized onto a novel microporous polymer. *Bioresource Technology.*, 100, 1983-1991.

[55] Sabbani, S., Hedenstrom, E., & Nordin, O. (2006). The enantioselectivity of Candida rugosa lipase is influenced by the particle size of the immobilising support material Accurel. *Journal of Molecular Catalysis B: Enzymatic.*, 42, 1-9.

[56] Zhou, G., Chen, Y., & Yang, S. (2009). Comparative studies on catalytic properties of immobilized Candida rugosa lipase in ordered mesoporous rod-like silica and vesicle-like silica. *Microporous and Mesoporous Materials*, 119, 223-229.

[57] Dalla, Rosa. C., Morandim, M. B., Ninow, J. L., Oliveira, D., Treichel, H., & Vladimir, Oliveira. J. (2009). Continuous lipase-catalyzed production of fatty acid ethyl esters from soybean oil in compressed fluids. *Bioresource Technology*, 100, 5818-5826.

[58] Guan, F., Peng, P., Wang, G., Yin, T., Peng, Q., Huang, J., Guan, G., & Li, Y. (2010). Combination of two lipases more efficiently catalyzes methanolysis of soybean oil for biodiesel production in aqueous medium. *Process Biochemistry*, 45, 1677-1682.

[59] Ting , W. J., Huang, C. M., Giridhar, N., & Wu, W. T. (2008). An enzymatic/acid-catalyzed hybrid process for biodiesel production from soybean oil. *Journal of the Chinese Institute of Chemical Engineering*, 39, 203-210.

[60] Nagayama, K., Yamasaki, N., & Imai, M. (2002). Fatty acid Esterification catalyzed by Candida rugosa lipase in lecithin microemulsion-based organogels. *Biochemical Engineering Journal*, 12, 231-236.

[61] Naya, M., & Imai, M. (2012). Regulation of the hydrolysis reactivity of immobilized Candida rugosa lipase with the aid of a hydrophobic porous carrier. *Asia-Pacific Journal of Chemical Engineering*, 7(S1), S157-S165.

[62] Montero, S., Blanco, A., Virto, M. D., Landeta, L. C., Agud, I., Solozabal, R., Lascaray, J. M., Renobales, M., de Llama, M. J., & Serra, J. L. (1993). Immobilization of Candida rugosa lipase and some properties of the immobilized enzyme. *Enzyme and Microbial Technol.*, 15, 239-247.

[63] Ahn, K. W., Ye, S. H., Chun, W. H., Rah, H., & Kim, S. G. (2011). Yield and component distribution of biodiesel by methanolysis of soybean oil with lipase-immobilized mesoporous silica. *Microporous and Mesoporous Materials*, 142, 37-44.

[64] Huang, D., Han, S., Han, Z., & Lin, Y. (2012). Biodiesel production catalyzed by Rhizomucor miehei lipase-displaying Pichia pastoris whole cells in an isooctane system. *Biochemical Engineering Journal*, 63, 10-14.

[65] Khare, S. K., & Nakajima, M. (2000). Immobilization of Rhizopus japonicas lipase on celite and its application for enrichment of docosahexaenoic acid in soybean oil. *Food Chemistry.*, 68, 153-157.

[66] Li, S. F., & Wu, W. T. (2009). Lipase-immobilized electrospun PAN nanofibrous membranes for soybean oil hydrolysis. *Biochemical Engineering Journal*, 45, 48-53.

[67] Li, S. F., Fan, Y. H., Hu, R. F., & Wu, W. T. (2011). Pseudomonas cepacia lipase immobilized onto the electrospun PAN nanofibrous membranes for biodiesel production from soybean oil. *Journal of Molecular Catalysis B: Enzymatic.*, 72, 40-45.

[68] Noureddini, H., Gao, X., & Philkana, R. S. (2005). Immobilized Pseudomonas cepacia lipase for biodiesel fuel production from soybean oil. *Bioresource Technology*, 96, 769-777.

[69] Ozmen, E. Y., & Yilmaz, M. (2009). Pretreatment of Candida rugosa lipase with soybean oil before immobilization on β-cyclodextrin-based polymer. *Colloids and Surfaces B: Biointerfaces.*, 69, 58-62.

[70] Rodrigues, R. C., & Záchia, Ayub. M. A. (2011). Effects of the combined use of Thermomyces lanuginosus and Rhizomucor miehei lipases for the transesterification and hydrolysis of soybean oil. *Process Biochemistry*, 46, 682-688.

[71] Wang, W., Li, T., Ning, Z., Wang, Y., Yang, B., & Yang, X. (2011). Production of extremely pure diacylglycerol from soybean oil by lipase-catalyzed glycerolysis. *Enzyme and Microbial Technology*, 49, 192-196.

[72] Xie, W., & Wang, J. (2012). Immobilized lipase on magnetic chitosan microspheres for transesterification of soybean oil. *Biomass and Bioenergy*, 36, 373-380.

[73] Gitlesen, T., & Bauer, M. (1997). Adlercreutz, P., Adsorption of lipase on polypropylene powder. *Biochimica et Biophysica Acta.*, 1345, 188-196.

[74] Al-Duri, B., & Yong, Y. P. (1997). Characterisation of the equilibrium behavior of lipase PS (from Pseudomonas) and lipolase 100L (from Humicola) onto Accurel EP100. *Journal of Molecular Catalysis B: Enzymatic.*, 3, 177-188.

[75] Stamatis, H., & Xenakis, A. (1999). Biocatalysis using microemulsion-based polymer gels containing lipase. *Journal of Molecular Catalysis B: Enzymatic*, 6, 399-406.

Chapter 20

Soybean, Nutrition and Health

Sherif M. Hassan

Additional information is available at the end of the chapter

http://dx.doi.org/10.5772/54545

1. Introduction

Soybean (*Glycine max* L.) is a species of legume native to East Asia, widely grown for its edible bean which has several uses. This chapter will focuses on soybean nutrition and soy food products, and describe the main bioactive compounds in the soybean and their effects on human and animal health.

2. Soybean and nutrition

Soybean is recognized as an oil seed containing several useful nutrients including protein, carbohydrate, vitamins, and minerals. Dry soybean contain 36% protein, 19% oil, 35% carbohydrate (17% of which dietary fiber), 5% minerals and several other components including vitamins [1]. Tables 1 and 2 show the different nutrients content of soybean and its by-products [2]

Soybean protein is one of the least expensive sources of dietary protein [3]. Soybean protein is considered to be a good substituent for animal protein [4], and their nutritional profile except sulfur amino acids (methionine and cysteine) is almost similar to that of animal protein because soybean proteins contain most of the essential amino acids required for animal and human nutrition. Researches on rats indicated that the biological value of soy protein is similar to many animal proteins such as casein if enriched with the sulfur-containing amino acid methionine [5]. According to the standard for measuring protein quality, Protein Digestibility Corrected Amino Acid Score, soybean protein has a biological value of 74, whole soybeans 96, soybean milk 91, and eggs 97[6]. Soybeans contain two small storage proteins known as glycinin and beta-conglycinin.

© 2013 Hassan; licensee InTech. This is a paper distributed under the terms of the Creative Commons Attribution License (http://creativecommons.org/licenses/by/3.0), which permits unrestricted use, distribution, and reproduction in any medium, provided the original work is properly cited.

Nutrient	Soybean				
	Flour	Protein concentrate	Seed heat processed	Meal solvent extracted	Seed without hulls, meal solvent extracted
Protein%	13.3	84.1	37.0	44.0	48.5
Fat%	1.6	0.4	18.0	0.8	1.0
Linoleic acid %	-	-	8.46	0.40	0.40
Crude fiber%	33.0	0.2	5.5	7.0	3.9
Calcium%	0.37	0.0	0.25	0.29	0.27
Total phosphorus%	0.19	080	058	065	0.62
Non phytate phosphorus %	-	0.32	-	0.27	0.22
Potassium %	1.50	0.18	1.61	2.00	1.98
Chlorine%	0.02	0.02	0.03	0.05	0.05
Iron (mg/kg)	-	130	80	120	170
Magnesium %	0.12	0.01	0.28	0.27	0.30
Manganese(mg/kg)	29	1	30	29	43
Sodium %	0.25	0.07	0.03	0.01	0.02
Sulfur %	0.06	0.71	0.22	0.43	0.44
Copper (mg/kg)	-	7	16	22	15
Selenium (mg/kg)	-	0.10	0.11	0.10	0.10
Zinc (mg/kg)	-	23	25	40	55
Biotein(mg/kg)	0.22	0.3	0.27	0.32	0.32
Choline(mg/kg)	640	2	2.860	2794	2731
Folacin (mg/kg)	0.30	2.5	4.2	1.3	1.3
Niacin(mg/kg)	24	6	22	29	22
Pantothenic acid (mg/kg)	13.0	4.2	11.0	16.0	15.0
Pyridoxine(mg/kg)	2.2	5.4	10.8	6.0	5.0
Riboflavin(mg/kg)	3.5	1.2	2.6	2.9	2.9
Thiamin(mg/kg)	2.2	0.2	11.0	4.5	3.2
Vitamin B12 (µg/kg)	-	-	-	-	-
Vitamin E (mg/kg)	-	-	40	2	3

Table 1. Shows composition of soybean and some soybean by-product.

Nutrient	Soybean				
	flour	Protein concentrate	Seed heat processed	Meal solvent extracted	Seed without hulls, meal solvent extracted
Arginine%	0.94	6.70	2.59	3.14	3.48
Glycine %	0.40	3.30	1.55	1.90	2.05
Serine%	-	5.30	1.87	2.29	2.48
Histidine%	0.18	2.10	0.99	1.17	1.28
Isoleucine%	0.40	4.60	1.56	1.96	2.12
Leucine %	0.57	6.60	2.75	3 39	3.74
Lysine%	0.48	5.50	2.25	2.69	2.96
Methionine%	0.10	0.81	0.53	0.62	0.67
Cystine%	0.21	0.49	0.54	0.66	0.72
Phenylalanine%	0.37	4.30	1.78	2.16	2.34
Tyrosine%	0.23	3.10	1.34	1.91	1.95
Threonine%	0.30	3.30	1.41	1.72	1.87
Tryptophan%	0.10	0.81	0.51	0.74	0.74
Valine%	0.37	4.40	1.65	2.07	2.22

Table 2. Shows amino acids contain of soybean and some soybean by-product.

On the other hand, Soy vegetable oil is another product of processing the soybean crop used in many industrial applications. Soybean oil contains about 15.65% saturated fatty acids, 22.78% monounsaturated fatty acids, and 57.74% polyunsaturated fatty acids (7% linolenic acid and 54% linoleic acid) [7]. Furthermore, soybeans contain several bioactive compounds such as isoflavones among other, which possess many beneficial effects on animal and human health [8].

Soybean is very important for vegetarians and vegans because of its rich in several beneficial nutrients. In addition, it can be prepared into a different type of fermented and non-fermented soy foods. Asians consume about 20–80 g daily of customary soy foods in many forms including soybean sprouts, toasted soy protein flours, soy milk, tofu and many more. Also fermented soy food products consumed include tempeh, miso, natto, soybean paste and soy sauce among other [9, 10]. This quantity intake of soy foods is equivalent daily to 25 and 100 mg total isoflavones [11] and between 8 and 50 g soy protein [12]. On the other hand, western people consume only about 1–3 g daily soy foods mostly as soy drinks, breakfast cereals, and soy burgers among other processed soy food forms [10].

Soybean is used as the raw material for oil milling, and the residue (soybean meal) can be mainly used as source of protein feedstuff for domestic animals including pig, chicken, cattle, horse, sheep, and fish feed and many prepackaged meals as well [1]. It is widely used as a filler and source of protein in animal diets, including pig, chicken, cattle, horse, sheep, and fish feed [13]. In general, soybean meal is a great source of protein ranged from 44-49%, but methionine is usually the only limiting amino acid and contains some anti-nutritional factors such as trypsin inhibitor and hemaglutinins (lectins) which can be destroyed by heating and fermenting the soybean meal before use. Textured vegetable protein (TVP) is another soybean byproduct has been used for more than 50 years as inexpensively and safely extending ground beef up to 30% for hamburgers or veggie burgers, without reducing its nutritional value and in many poultry and dairy products (soy milk, margarine, soy ice cream, soy yogurt, soy cheese, and soy cream cheese). as well [1, 13, 14, 15]. The total estimates of feed consumed for broilers, turkeys, layers and associated breeders production over the world in 2006 was about 452 million tones [16]. This estimated value is calculated depending on poultry feeds containing about 30% soybean meal on average. Therefore, 136 million tones of soybean meal are used annually in poultry feeds. As a generalization, the numbers shown can be multiplied by 0.3 for an estimate of the needs of soybean meal. Soy-based infant formula (SBIF) is another soybean product that can be used for infants who are allergic to pasteurized cow milk proteins. It is sold in powdered, ready-to-feed, and concentrated liquid forms without side effects on human growth, development, or reproduction [17, 18, 19].

There are several types commercially available of non fermented soy foods, including soy milk, infant formulas, tofu (soybean curd), soy sauce, soybean cake, tempeh, su-jae, and many more. However, fermented foods include soy sauce, fermented bean paste, natto, and tempeh, among others. Fermented soybean paste is native to the East and Southeast Asia countries such as Korea, China, Japan, Indonesia, and Vietnam [20]. Korean soy foods including kochujang (fermented red pepper paste with soybean flour) and long-term fermented soybean pastes (doenjang, chungkukjang, and chungkookjang) are now internationally accepted foods [20]. Furthermore, natto and miso are originally Japanese soy food types of chungkukjang and doenjang, respectively. China also has different fermented soybean products including doubanjiang, douche (sweet noodle sauce), tauchu (yellow soybean paste), and dajiang. Chungkukjang is a short-term fermented soy food similar to Japanese natto, whereas doenjang, kochujang, and kanjang (fermented soy sauce) undergo long term fermentation as do Chinese tauchu and Japanese miso.

In general, this fermentation of soy foods changes the physical and chemical properties of soy food products including the color, flavor and bioactive compounds content. These changes differ according to different production methods such as the conditions of fermentation, the additives, and the organisms used such as bacteria or yeasts during their manufacture. These changes differ as well as whether the soybeans are roasted as in chunjang or aged as in tauchu before being ground. In addition to physicochemical properties, the fermentation of these soybean products changes the bioactive components, such as isoflavonoids and peptides, in ways which may alter their nutritional and health effects.

Also, the nutritional value of cooked soybean depends on the pre-processing and the method of cooking such as boiling, frying, roasting, baking, and many more. The quality and quantity of soybean components is considerably changed by physical and chemical or enzymatic processes during the producing of soy-based foods [21, 22, 23, 24, 25, 26]. Fermentation is a great processing method for improving nutritional and functional properties of soybeans due to the increased content of many bioactive compounds. On the other hand, the conformation of soy protein (glycinin) is easily altered by heat (steaming) and salt [27]. Many large molecules in raw soybean are broken down by enzymatic hydrolysis during fermentation to small molecules, which are responsible for producing new functional properties for the final products. For example, isoflavones, which are mostly present as 6-O-malonylglucoside and β-glycoside conjugates and associated with soybean proteins, are broken down by heat treatment and fermentation [28]. In general, the chemical profiles of various minor components related to health benefits and nutritional quality of products are also affected by fermentation [29]. It is usual to heat-treat legume components to denature the high levels of trypsin inhibitors soybean [30]. The digestibility of some soy foods are as follows: steamed soybeans 65.3%, tofu 92.7%, soy milk 92.6%, and soy protein isolate 93–97% [1].

3. Bioactive compounds of soybean

Many bioactive compounds are isolated from soybean and soy food products including isoflavones, peptides, flavonoids, phytic acid, soy lipids, soy phytoalexins, soyasaponins, lectins, hemagglutinin, soy toxins, and vitamins and more [31]. Flavonoids are low-molecular-weight polyphenolic compounds classified according to their chemical structure into flavonols, flavones, flavanones, isoflavones, catechins, anthocyanidins and chalcones [32]. Typical flavonoids are kaempferol, quercetin and rutin (the common glycoside of quercetin), belonging to the class of flavonols. Isoflavones (soy phytoestrogens) is a subgroup of flavonoids. The major isoflavones in soybean are genistein, daidzein, and glycitein, representing about 50, 40, and 10% of total isoflavone profiles, respectively. Soy isoflavones, daidzein and genistein, are present at high concentrations as a glycoside in many soybeans and soy food products such as miso, tofu, and soy milk. Soybeans contain 0.1 to 5 mg total isoflavones per gram, primarily genistein, daidzein, and glycitein, the three major isoflavonoids found in soybean and soy products [33]. These compounds are naturally present as the β-glucosides genistin, daidzin, and glycitin, representing 50% to 55%, 40% to 45%, and 5% to 10% of the total isoflavone content, respectively depending on the soy products [8]. Formononetin is another form of isoflavone found in soybeans and can be converted in the rumen (in sheep and cow) into a potent phytoestrogen called equol [34].

Recently, there has been increased interest in the potential health benefits of other bioactive polypeptides and proteins from soybean, including lectins (soy lectins are glycoprotein) and lunasin. Lunasin is a novel peptide originally isolated from soybean foods [35]. Lunasin concentration is ranged from 0.1 to 1.3 g/100 g flour [36, 37], and from 3.3 to 16.7 ng/mg seed [38]. Soybean phytosterols usually include four major or types: β-sitosterol, stigmasterol, campesterol, and brassicasterol, all of which make good raw materials for the production of steroid hormones. Triterpenoid saponins in the mature soybean are divided into two groups; group A soy saponins have undesirable astringent taste, and group B soy saponins have the health promoting properties [39, 40]. Group A soy saponins are found only in soybean hypocotyls, while group B soy saponins are widely distributed in legume seeds in both hypocotyls (germ) and cotyledons [39]. Saponin concentrations in soybean seed are ranged from 0.5 to 6.5% [41, 42].

Soybeans also contain isoflavones called genistein and daidzein, which are one source of phytoestrogens in the human diet. Soybeans are a significant source of mammalian lignan precursor secoisolariciresinol containing 13–273 μg/100 g dry weight [43]. Another phytoestrogen in the human diet with estrogen activity is coumestans, which are found in soybean sprouts. Coumestrol, an isoflavone coumarin derivative is the only coumestan in foods [44, 45]. Soybeans and processed soy foods are among the richest foods in total phytoestrogens present primarily in the form of the isoflavones daidzein and genistein [46].

4. Soybean and health

4.1. Beneficial effects of soybean

Recent research of the health effects of soy foods and soybean containing several bioactive compounds received significant attention to support the health improvements or health risks observed clinically or *in vitro* experiments in animal and human.

4.1.1. Effects on cancer

Recent studies suggested that soy food (soy milk) and soybean protein containing flavonoid genistein, Biochanin A, phytoestrogens (isoflavones) consumption is associated with lowered risks for several cancers including breast [11,47,48,49,50,51,52], prostate [53,54], endometrial [52,55], lung [56], colon [57], liver [58], and bladder [59] cancers.

Isoflavones (genistein) use both hormonal and non-hormonal action in the prevention of cancer, the hormonal action of isoflavones has been postulated to be through a number of pathways, which include the ability to inhibit many tyrosine kinases involved in regulation of cell growth, to enhance transformation growth factor-β which inhibits the cell cycle progression, as well as to influence the transcription factors that are involved in the expression of stress response-related genes involved in programmed cell death [60,61]. Other nonhormonal mechanisms by which isoflavones are believed to increase their anticarcinogenic effects are via their anti-oxidant, anti-proliferative, anti-angiogenic and anti-inflammatory properties [62].

On the other hand, soy proteins and peptides showed potential results in preventing the different stages of cancer including initiation, promotion, and progression [63]. They noted that Kunitz trypsin inhibitor (KTI), a protease inhibitor originally isolated from soybean, inhibited carcinogenesis due to its ability to suppress invasion and metastasis of cancer cells. Also, [64] found that soybean lectins and lunasin were able to possess cancer chemopreventive activity *in vitro, in vivo* (in human).

Cell culture experiments have demonstrated that a novel soybean seed peptide (lunasin) prevented mammalian cells transformation induced by chemical carcinogens without affecting morphology and proliferation of normal cells [65]. Lunasin purified from defatted soybean flour showed potent activity against human metastatic colon cancer cells. Lunasin caused cytotoxicity in four different human colon cancer cell lines [66]. It has been also demonstrated that lunasin causes a dose-dependent inhibition of the growth of estrogen independent for human breast cancer [67].

4.1.2. Effect on hypercholesterolemia and cardiovascular diseases

Soy food and soybean protein containing isoflavones consumption lowered hypercholesterolemia [68, 69, 70, 71]. Many studies reported that soybean protein consumption lowered incidence of cardiovascular diseases [68]. Soy isoflavone suppress excessive stress-induced hyperactivity of the sympatho-adrenal system and thereby protect the cardiovascular system [72].

Several studies reported a relation between soybean protein consumption and the reduction in cardiovascular risk in laboratory animal's models by reducing plasma cholesterol levels [68, 69]. Reduction in the incidence of hypercholesterolemia and cardiovascular diseases in Asian countries depending on their diets rich in soy protein was reported [73]. Another study found that the substitution of the animal protein with soybean protein resulted in a significantly decrease in plasma cholesterol levels, mainly LDL (low-density lipoprotein) cholesterol [74]. In the same way, [69] showed that after replacing animal protein with soybean protein consumption for hypercholesterolemia persons resulted in a significant decrease of 9.3% of total plasma cholesterol, mainly 12.9% of LDL cholesterol level and 10.5% of triglycerides. The health beneficial effect for replacing animal protein with soy protein consumption showed the most effective in the highest hypercholesterolemic depend on the initial plasma cholesterol levels [70, 71] without or with the lowest effects in normocholesterolemic persons.

Several research attentions have been paid to the high dietary intake of isoflavones because of their potentially beneficial effects associated with a reduction in the risk of developing cardiovascular diseases. On the other hand, other studies conducted out to establish whether soybean protein and/or isoflavones could be responsible for the hypocholesterolemic effects of soybean diets and therefore their beneficial effects on cardiovascular disease. By studying the effect of soy bean protein and isoflavones, [75] reported that these major components of soybean flour (soybean proteins and soybean isoflavones) independently decreased serum cholesterol. Recent study reported that soybean protein containing isoflavones significantly reduced serum total cholesterol, LDL cholesterol, and triacylglycerol and significantly increased HDL (high-density lipoprotein) cholesterol, but the changes were related to the level and duration of intake, and gender and initial serum lipid concentrations of the persons [76].

Some studies have shown that soybean oil effective in lowering the serum cholesterol and LDL levels, and likely can be used as potential hypocholesterolemic agent if used as a dietary fat and ultimately help prevent atherosclerosis and heart diseases [77]. Soybean oil is a rich source of vitamin E, which is essential to protect the body fat from oxidation and to scavenge the free radicals and therefore helps to prevent their potential effect upon chronic diseases such as coronary heart diseases and cancer [78]. The FDA granted the following health claim for soy: "25 grams of soy protein a day, as part of a diet low in saturated fat and cholesterol, may reduce the risk of heart disease [79].

4.1.3. Effect on osteoporosis and menopause

Soy food and soybean isoflavones consumption lowered osteoporosis, improved bone health and other bone health problems [80, 81, 82]. In addition, consumption of soy foods may reduce the risk of osteoporosis and help alleviate hot flashes associated with meno-pausal symptoms which are major health concerns for women [83].

4.1.4. Hypotensive activity

Soy food kochujang extract consumption lowered hypertension [84]. The angiotensin I converting enzyme inhibitory peptide isolated from soybean hydrolysate and Korean soybean paste enhanced anti-hypertensive activity *in vivo* [85], causing a fall in blood pressure compared with thiazide diuretics or beta-blockers for mild essential hypertension [86].

4.1.5. Effect on insulin secretion and energy metabolism

Increasing the insulin secretion followed by glucose challenge was recorded when male monkeys fed soybean protein and isoflavones [87, 88]. Flavonoid genistein, tyrosine kinase inhibitor, inhibited insulin signaling pathways [60]. Dietary isoflavones induced alteration in energy metabolism in human [89]. They also noted an inhibition of glycolysis and a general shift in energy metabolism from carbohydrate to lipid metabolism due to isoflavone interference.

4.1.6. Effect on blood pressure and endothelial function

Reduction in the blood pressure via renin-angiotensin system activity (one of the most important blood pressure control systems in mammals) was recorded by feeding rats on diet containing commercial purified soybean saponin [90]. They found that soybean saponin inhibited renin activity *in vitro* and that oral administration of soybean saponin at 80 mg/kg of body weight daily to spontaneously hypertensive rats for 8 weeks significantly reduced blood pressure. In addition, [91] studied the effects of dietary intake of soybean protein and isoflavones on cardiovascular disease risk factors in high risk, 61 middle aged men in Scotland. For five weeks, half the men fed diets containing at least 20 g of soybean protein and 80 mg of isoflavones daily. The effects of isoflavones on blood pressure, cholesterol levels, and urinary excretion were measured, and then compared to those of the remaining men who were fed placebo diet containing olive oil. Men that fed soybean in their diet showed significant decrease in both diastolic and systolic blood pressure. In addition, [92] found that feeding soy nut significantly decreased systolic and diastolic pressure in hypertensive post-menopausal women. On the other hand, [93] found no effect of soybean protein with isoflavones on blood pressure in hypertensive persons. Soy protein and soy isoflavones intake improved endothelial function and the flow-induced dilatation in postmenopausal hyper-cholesterolemia women by raising the levels of endothelial nitric oxide synthase (eNOS), a regulator of the cardiovascular function [94, 95, 96, 97]. Furthermore, chronic administration of genistein increased the levels of NOS in spontaneously hypertensive rats [98, 99].

4.1.7. Effects on platelet aggregation and fibrinolytic activity

The effect of genistein, a protein tyrosine kinase inhibitor on platelet aggregation was exhib-ited [100,101]. Nattokinase, a strong fibrinolytic enzyme, in the vegetable cheese natto (a popular soybean fermented Japanese food) showed approximately fourtimes stronger activ-ity than plasmin in the clot lysis assay [102]. However, intraduodenal administration natto-kinase decreased fibrinogen plasma levels in rats [103,104] and in humans [105]. In addition, soybean protein and peptides exhibited anti-fatigue activity helping in performing exercise and delaying fatigue [106], antioxidant [107,108], anti-aging, skin moisturizing, anti-solar, cleansing, and hair-promoting agent [109].

The beneficial effect of Soy isoflavonne (daidzein) on human health extends to the preven-tion of cancer [110], cardiovascular disease [111]. Also, soybean isoflavones (genistein, daid-zein, and their beta glycoside conjugates) showed antitumor [112], estrogenic [113], antifungal activities [114]. Soy isoflavonne (daidzein) stimulates catecholamine synthesis at low concentrations [115]. However, daidzein at high concentrations (1-100 μM) inhibited catecholamine synthesis and secretion induced by stress or emotional excitation. Recent studies recoded an improvement in cognitive function, particularly verbal memory [116] and in frontal lobe function [117] with the use of soy supplements. Glyceollins molecules are also found in the soybean and exhibited an antifungal activity against *Aspergillus sojae*, the fungal ferment used to produce soy sauce [118]. They are phytoalexins with an antiestrogen-ic activity [119].

4.2. Harmful effects

Despite the several beneficial effects documented of soybean consumption, there are some controversial effects claimed in recent studies on animal and human health. Soybean con-tains several naturally occurring compounds that are toxic to humans and animals such as the trypsin (a serine protease found in the digestive system) inhibitors, phytic acid, toxic components such as lectins and hemagglutinins, some metalloprotein such as soyatoxin and many more other biological of soyatoxin. Some studies reported high levels of protease or trypsin inhibitors (1-5% of total protein) in legume seeds such as soybean [120]. *In vivo* stud-ies using rat, high levels of exposure to trypsin inhibitors isolated from raw soy flour cause pancreatic cancer whereas moderate levels cause the rat pancreas to be more susceptible to cancer-causing agents. However, the US FDA concluded that low levels of soybean protease (trypsin) inhibitors cause no threat to human health. For human consumption, soybeans must be cooked with "wet" heat to destroy the trypsin inhibitors (serine protease inhibitors). Raw soybeans, including the immature green form, are toxic to humans, swine, chickens, and in fact, all monogastric animals [121]. Tofu intake was associated with worse memory, but tempeh (a fermented soy product) intake was associated with better memory [122]. Isoflavones might increase breast cancer risk in healthy women or worsen the prognosis of breast cancer patients [123].

Soy compounds	Biological properties	Selected references
Genistein, daidzein, lectins and lunasin	Anticancer	[11, 47, 48,49,50, 51,52, 53,54, 55,56,57,58,59,62,63, 65,66,67,110,113]
Isoflavones and oil	Hypercholesterolemia	[68,69,70,71,73,74,75,76,77]
Daidzein and oil	Cardiovascular diseases	[68,72,73,77,78,79,91, 111]
Isoflavones	Osteoporosis and menopause	[80,81,82,83]
Genistein	Hypertensive	[84,85,86,98,99]
Genistein	Insulin secretion and energy metabolism	[60,87,88,89]
Saponin and genistein	Blood pressure and endothelial function	[90,91,92,94,95,96,97,98,99]
Genistein	Plate aggregation and fibrinolyic activity	[100,101,102,103,104,105]
Genistein	Antioxidant	[107,108]
Protein and peptides	Anti aging	[109]
Protein and peptides	Anti-fatigue	[106]
Genistein	Anti-flammation	[62]
Genistein, daidzein and Glyceollins	Antifungal	[114,118]
Genistein, daidzein and Glyceollins	Estrogenic activity	[113,119]
Daidzein	Catecholamine synthesis	[115]
Genistein	Anti-angiogenic	[62]

Table 3. Summarizes some beneficial effects of some soybean compounds on animal and human health

Phytic acid is also criticized for reducing vital minerals due to its chelating effect, especially for diets already low in minerals [124]. Phytic acid present in soybean seeds binds to minerals and metals to form phytate (chelated forms of phytic acid with magnesium, calcium, iron, and zinc). Phytate is not digestible and impermeable molecules through cell membranes for humans or nonruminant animals. In addition, phytic acid prevents the body to use many essential minerals such as magnesium, calcium, iron and especially zinc. Unfermented soy products contain high levels of lectins/hemagglutinins. Hemagglutinin makes red blood cells unable to absorb oxygen. However, the soybean fermentation process deactivates soybean hemagglutinins and makes the amounts of lectins present in soybeans inconsiderable. However, some dried soybean products may still contain a large amount of active or toxic lectins. These lectins are believed to cause allergic in a human body. Recently, a metalloprotein named soyatoxin exhibiting toxicity to mice (LD_{50} 7-8 mg/kg mouse upon intraperitoneal injection) was identified. Regardless of the beneficial effect of genistein, there are some controversies about safety and harmful effect of soybean food supplementation rich in genistein. Some studies reported that genistein is not safe and has harmful effects on human health. Consumption of genistein-rich soy food and supplements during pregnancy has been suggested to raise the risk of infant leukemias [125]. In addition, some researches showing stimulatory effect of genistein on proliferation of some breast cancer cells lines increase the concerning problem about the safety of genistein intake for breast cancer women [126]. Recent study reported that administration 56g soy protein powder daily caused a reduction in serum testosterone up to 4% in four weeks in a test group of 12 healthy males [127]. Finally, allergy to soy is common, and the food is listed with other foods. Only a few reported studies have attempted to confirm allergy to soy by direct challenge with the food under controlled conditions [127]. Table (3) shows several beneficial effects reported of some soybean compounds on animal and human health.

Acknowledgements

This work was supported by a grant from the Deanship of Scientific Research, King Faisal University, Kingdom of Arabia Saudi. The authors have no conflict of interest for the information presented in this review.

Author details

Sherif M. Hassan

College of Agricultural and Food Sciences, King Faisal University, Kingdom of Saudi Arabia

References

[1] Liu K.S. Chemistry and Nutritional Value of Soybean Components. In: Soybean: Chemistry, Technology, and Utilization. New York: Chapman & Hall; 1997.p25-113.

[2] Composition of Feedstuffs Used in Poultry Diets. In: National Research Council (NRC). Washington, DC: National Academy of Sciences; 1994.p61-68.

[3] Derbyshire, E, Wright DJ, Boulter D. Review: Legumin and vicilin, storage proteins of legume seeds. Phytochemistry 1976:15:3.

[4] Sacks FM, Lichtenstein A, Van Horn L, Harris W, Kris-Etherton P, Winston M. "Soy Protein, Isoflavones, and Cardiovascular Health. An American Heart Association Sci-

ence Advisory for Professionals from the Nutrition Committee". Circulation 2006;113(7):1034–1044.

[5] Hajos G, Gelencsér E, Grant G, Bardocz S, Sakhri M, Duguid TJ, Newman AM, Pusztai A. Effects of Proteolytic Modification and Methionine Enrichment On the Nutritional Value of Soya Albumins For Rats. The Journal of Nutritional Biochemistry 1996;7:481-487.

[6] FAO/WHO (1989). Protein Quality Evaluation: Report of the Joint FAO/WHO Expert Consultation. Bethesda, MD (USA): Food and Agriculture Organization of the United Nations (Food and Nutrition Paper) 1989:51.

[7] Wolke RL"Where There's Smoke, There's a Fryer". The Washington Post.2007.

[8] Young VR. Soy Protein in Relation to Human Protein and Amino Acid Nutrition. Journal of American Dietetic Association 1991;91:828-835.

[9] Wang H, Murphy, P. Isoflavone content in commercial soybean foods. Journal of Agricultural and Food Chemistry 1994;42(8):1666-1673.

[10] Fournier DB, Erdman JW, Gordon GB. Soy, its components, and cancer prevention: a review of the in vitro, animal, and human data. Cancer Epidemiology, Biomarkers and Prevention 1998;7(11):1055–1065.

[11] Messina M, McCaskill-Stevens W, Lampe JW. Addressing the soy and breast cancer relationship: review, commentary, and workshop proceedings. Journal of the National Cancer Institute 2006;98:1275-1284.

[12] Erdman JrJ, Jadger T, Lampe J, Setchell KDR, Messina M. Not all soy products are created equal: caution needed in interpretation of research results. The Journal of Nutrition 2004;134(5):S1229–S1233.

[13] Riaz, MN. Soy Applications in Food. Boca Raton. Florida: CRC Press;2006.

[14] Hoogenkamp HW. Soy Protein and Formulated Meat Products. Wallingford. Oxon, UK: CABI Publishing; 2005:14.

[15] Joseph GE. Soy Protein Products. AOCS Publishing. 2001.

[16] Leesons S., Summers J. Commercial Poultry Nutrition. Guelph, Ontario, Canada: Nottingham University Press; 2005.

[17] Giampietro PG, Bruno G, Furcolo G, Casati A, Brunetti E, Spadoni GL, Galli E. Soy Protein Formulas in Children: No Hormonal Effects in Long-term Feeding. Journal of Pediatric Endocrinology and Metabolism 2004;17(2):191–196.

[18] Strom BL. "Exposure to Soy-Based Formula in Infancy and Endocrinological and Reproductive Outcomes in Young Adulthood". The Journal of the American Medical Association 2001;286(7):807–814.

[19] Merritt RJ, Jenks BH. Safety of Soy-Based Infant Formulas Containing Isoflavones: The Clinical Evidence. The Journal of Nutrition2004;134(5):1220S–1224S.

[20] Kwon DY, Daily JW, Kim HJ, Park S. Antidiabetic Effects of Fermented Soybean Products on Type 2 Diabetes. Nutrition Research 2010;30:1-13.

[21] Park JS, Lee MY, Kim JS, Lee TS. Compositions of Nitrogen Compound and Amino acid in Soybean Paste (Doenjang) Prepared with Different Microbial Sources. Korean Journal of Food Science and Technology 1994; 26:609-615.

[22] Garcia MC, Torre M, Marina ML, Laborda F. Composition and Characterization of Soybean and Related Products. Critical Reviews in Food Science and Nutrition 1997;37:361-391

[23] Nakajima N, Nozaki N, Ishihara K, Ishikawa A, Tsuji H. Analysis of Isoflavone Content in Tempeh, a Fermented Soybean, and Preparation of a New Isoflavone enriched Tempeh. Journal of Bioscience and Bioengineering 2005;100: 685-687.

[24] Yamabe, S, Kobayashi-Hattori, K, Kaneko, K, Endo, K, Takita, T. Effect of Soybean Varieties on the Content and Composition of Isoflavone in Rice-koji Miso. Food Chemistry 2007;100:369-374.

[25] Jang CH, Park CS, Lim JK, Kim JH, Kwon DY, Kim YS, Shin DH, Kim JS. Metabolism of Isoflavone Derivatives during Manufacturing of Traditional Meju and Doenjang. Food Science and Biotechnology 2008; 17:442-445.

[26] Baek LM, Park LY, Park KS, Lee SH. Effect of Starter Cultures on the Fermentative Characteristics of Cheonggukjang. Korean Journal of Food Science and Technology 2008;.40:400-405.

[27] Kim KS, Kim S, Yang HJ, Kwon DY. Changes of Glycinin Conformation due to pH, Heat and Salt Determined by Differential Scanning Calorimetry and Circular Dichroism. International Journal of Food Science and Technology 2004;39:385-393.

[28] Choi HK, Yoon JH, Kim YS, Kwon DY. Metabolomic Profiling of Cheonggukjang during Fermentation by 1H NMR Spectrometry and Principal Components Analysis. Process Biochemistry 2007;42:263-266

[29] Kim, NY, Song EJ, Kwon DY, Kim HP, Heo MY. Antioxidant and Antigenotoxic Activities of Korean Fermented Soybean. Food and Chemical Toxicology 2008; 46:1184-1189.

[30] Teakle R., Jensen J. Heliothis punctiger. In: Singh R., Moore R. (ed.) Handbook of Insect Rearing. Amsterdam: Elsevier; 1985.p312 – 322.

[31] Davis J, Iqbal MJ, Steinle J, Oitker J, Higginbotham DA, Peterson RG. Soy Protein Influences the Development of the Metabolic Syndrome in Male Obese ZDFxSHHF Rats. Hormone and Metabolic Research 2007;37:316-325.

[32] Rice-Evans C. Flavonoid antioxidants. Current Medicinal Chemistry 2001; 8(7): 797-807.

[33]   Park OJ, Surh Y-H. Chemopreventive potential of epigallocatechin gallate and genistein: evidence from epidemiological and laboratory studies. Toxicology Letters 2004; 150(1):43-56.

[34]   Tolleson WH, Doerge DR, Churchwell MI, Marques MM, Roberts DW. Metabolism of Biochanin A and Formononetin by Human Liver Microsomes in Vitro. Journal of Agricultural and Food Chemistry 2002;50(17):4783-4790.

[35]   Galvez AF, Revilleza MJR, de Lumen BO. A novel methionine-rich protein from soybean cotyledon: cloning and characterization of cDNA (accession No. AF005030). Plant Register #PGR97-103. Plant Physiology 1997;114:1567-1569.

[36]   de Mejia EG, Vasconez M, de Lumen BO, Nelson R. Lunasin concentration in different soybean genotypes, commercial soy protein, and isoflavone products. Journal of Agricultural and Food Chemistry 2004;52(19):5882-5887.

[37]   Wang W, Dia VP, Vasconez M, Nelson RL, de Mejia EG. Analysis of soybean protein-derived peptides and the effect of cultivar, environmental conditions, and processing on lunasin concentration in soybean and soy products. Journal of AOAC International 2008;91(4):936-946.

[38]   de Mejia EG., Dia VP. Chemistry and biological properties of soybean peptides and proteins. In: Cadwallader K. et al. (ed.) Chemistry, Texture and Flavor of Soy. USA:American Chemical Society; 2010b.p133-154. Available from http://pubs.acs.org.

[39]   Shiraiwa M, Harada K, Okubo K. Composition and structure of group B saponin in soybean seed. Agricultural and biological chemistry 1991;55:911-917.

[40]   Kuduo S, Tonomura M, Tsukamoto C. Isolation and structure elucidation of DDMP-conjugated soyasaponins as genuine saponins from soybean seeds. Bioscience, Biotechnology, and Biochemistry. 1993;57:546-550.

[41]   Ireland, PA, Dziedzic SZ, Kearsley MW. Saponin content of soya and some commercial soya products by means of high-performance liquid chromatography of the sapogenins. Journal of the Science of Food and Agriculture 1986;37:694-698.

[42]   Berhow MA, Kong SB, Vermillion KE, Duval SM. Complete quantification of group A and group B soyasaponins in soybeans. Journal of Agricultural and Food Chemistry 2006;54:2035-2044.

[43]   Adlercreutz H, Mazur W, Bartels P, Elomaa V, Watanabe S, Wähälä K, Landström M, Lundin E, Bergh A, Damber JE, Aman P, Widmark A, Johansson A, Zhang JX, Hallmans G. "Phytoestrogens and Prostate Disease". The Journal of Nutrition 2000;130(3): 658S–659S.

[44]   De Kleijn MJ, Van Der Schouw YT, Wilson PW, Grobbee DE, Jacques PF. Dietary Intake of Phytoestrogens is Associated With a Favorable Metabolic Cardiovascular Risk Profile in Postmenopausal U.S. Women: The Framingham Study. The Journal of Nutrition 2002;132(2):276–282.

[45] Valsta LM, Kilkkinen A, Mazur W, Nurmi T, Lampi A-M, Ovaskainen M-L, Korho-nen T, Adlercreutz H, Pietinen P. Phyto-oestrogen Database of Foods and Average Intake in Finland. British Journal of Nutrition 2003;89 (5):S31–S38.

[46] Thompson, Lilian U.; Boucher, Beatrice A.; Liu, Zhen; Cotterchio, Michelle; Kreiger N. Phytoestrogen Content of Foods Consumed in Canada, Including Isoflavones, Lignans, and Coumestan". Nutrition and Cancer 2006;54 (2): 184–201.

[47] Peterson G, Barnes S Genistein and Biochanin A. Inhibit the Growth of Human Prostate Cancer Cells but not Epidermal Growth Factor Receptor Autophosphorylation. Prostate 1993;22:335-345.

[48] Wu AH, Ziegler RG, Horn-Ross PL, Nomura AMY, West DW, Kolonel LN, Rosenthal JF, Hoover RN, Pike MC. Tofu and risk of breast cancer in Asian-Americans. Cancer Epidemiological Biomarkers Preview 1996;5(11):901–906.

[49] Wu AH, Ziegler RG, Nomura AM, West DW, Kolonel LN, Horn-Ross PL, Hoover RN, Pike MC. Soy intake and risk of breast cancer in Asians and Asian Americans. The American Journal of Clinical Nutrition 1998;68:1437S-1443S.

[50] Zheng W, Dai Q, Custer LJ, Shu XO, Wen WQ, Jin F, Franke AA. Urinary excretion of isoflavonoids and the risk of breast cancer. Cancer Epidemiology Biomarkers & Prevention 1999;8(1):35-40.

[51] Boyapati SM, Shu XO, Ruan ZX, Dai Q, Cai Q, Gao YT, Zheng W. Soyfood intake and breast cancer survival: a followup of the Shanghai Breast Cancer Study. Breast Cancer Res Treat 2005;92:11-17.

[52] Lof M, Weiderpass E. Epidemiological evidence suggests that dietary phytoestrogens intake is associated with reduced risk of breast, endometrial and prostate cancers. Nutrition Research 2006; 26(12):609-619.

[53] Peterson G, Barnes S. Genistein Inhibition of the Growth of Human Breast Cancer Cells: Independence from Estrogen Receptors and the Multi-drug Resistance Gene. Biochemical and Biophysical Research Communications 1991; 179: 661-667.

[54] Jacobsen BK, Knutsen SF, Fraser GE. Does high soy milk intake reduce prostate cancer incidence? The Adventist health study (United States). Cancer Causes Control 1998; 9(6):553-557.

[55] Goodman MT, Wilkens LR, Hankin JH, Lyu LC, Wu AH, Kolonel LN. Association of soy and fiber consumption with the risk of endometrial cancer. American Journal of Epidemiology 1997;146(4):294-306.

[56] Swanson CA, Mao BL, Li JY, Lubin JH, Yao SX, Wang JZ, Cai SK, Hou Y, Luo QS, Blot WJ. Dietary determinants of lung-cancer risk - results from a case-control study in Yunnan province, China. International Journal of Cancer 1992;50(6):876-880.

[57] Azuma N, Machida K, Saeki T, Kanamoto R, Iwami K. Preventive effect of soybean resistant proteins against experimental tumorigenesis in rat colon. Journal of Nutritional Science and Vitaminology 2000;46(1):23–29.

[58] Kanamoto R, Azuma N, Miyamoto T, Saeki T, Tsuchihashi Y, Iwami K. Soybean resistant proteins interrupt an enterohepatic circulation of bile acids and suppress liver tumorigenesis induced by azoxymethane and dietary deoxycholate in rats. Bioscience, Biotechnology, and Biochemistry 2001; 65(4):999–1002.

[59] Sun CL, Yuan JM, Arakawa K, Low SH, Lee HP, Yu MC. Dietary soy and increased risk of bladder cancer: The Singapore Chinese health study. Cancer Epidemiological Biomarkers Preview 2002;11(12):1674-1677.

[60] Akiyama T, Ishida J, Nakagawa S, Ogawara H, Watanabe S, Itoh N, Shibuya M, Fukami Y. Genistein a specific inhibitor of tyrosine-specific protein kinases. The Journal of Biological Chemistry 1987;262:5592-5595.

[61] Zhou Y, Lee AS. Mechanism for the suppression of the mammalian stress response by genistein, an anticancer phytoestrogen from soy. Journal of National Cancer Institute 1998:9(5):381–388.

[62] Gilani G.S., Anderson J.J.B., editor. Phytoestrogens and Health. IL, USA: AOCS Press; 2002.

[63] de Mejia EG, Dia VP. The role of nutraceutical proteins and peptides in apoptosis, angiogenesis, and metastasis of cancer cells. Cancer Metastasis Review 2010a; 29(3): 511-528.

[64] de Mejia EG. Bradford T, Hasler C The anticarcinogenic potential of soybean lectin and lunasin. Nutrition Reviews 2003;61(7):239-246.

[65] Galvez A.F, Chen N, Macasieb J, de Lumen BO. Chemopreventive property of a soybean peptide (Lunasin) that binds to deacetylated histones and inhibit acetylation. Cancer Research 2001; 61(20):7473-7478.

[66] Dia VP, and de Mejia EG. Lunasin induces apoptosis and modifies the expression of genes associated with extracellular matrix and cell adhesion in human metastatic colon cancer cells. Molecular Nutrition and Food Research 2011;55(4): 623-634.

[67] Hsieh C-C, Hernandez-Ledesma B, de Lumen BO. Lunasin, a novel seed peptide, sensitizes human breast cancer MDA-MB-231 cells to aspirin-arrested cell cycle and induced-apoptosis. Chemico-Biological Interactions 2010;186(2):127-134.

[68] Carroll KK. Hypercholesterolemia and atherosclerosis: effects of dietary protein. Fed Proc 1982;41:2792-2796.

[69] Anderson JW, Johnstone BM, Cook-Newell ME. Meta-analysis of the effects of soy protein intake on serum lipids. The New England Journal of Medicine 1995;333:276-282.

[70] Crouse JR III, Morgan T, Terry JG, Ellis J, Vitolins M, Burke GL. A randomized trial comparing the effect of casein with that of soy protein containing varying amounts of isoflavones on plasma concentrations of lipids and lipoproteins. Archives of Internal Medicine 1999;159:2070-2076.

[71] Jenkins DJ, Kendall CW, Jackson CJ, Connelly PW, Parker T, Faulkner D, Vidgen E, Cunnane SC, Leiter LA, Josse RG. Effects of high- and low isoflavone soyfoods on blood lipids, oxidized LDL, homocysteine, and blood pressure in hyperlipidemic men and women. The American Journal of Clinical Nutrition 2002; 76:365-372.

[72] Yanagihara N, Toyohira Y, Shinohara Y. Insights into the pharmacological potentials of estrogens and phytoestrogens on catecholamine signaling. Annals of the New York Academy of Sciences 2008;1129: 96-104.

[73] Keys A. Seven Countries: A Multivariate Analysis of Death and Coronary Heart Disease. Cambridge, Massachusetts:. Harvard University Press;1980.

[74] Descovich GC, Ceredi C, Gaddi A, Benassi MS, Mannino G, Colombo L, Cattin L, Fontana G, Senin U, Mannarino E, Caruzzo C, Bertelli E, Fragiacomo C, Noseda G, Sirtori M, Sirtori CR. Multicentre study of soybean protein diet for outpatient hypercholesterolaemic patients. Lancet 1980;2:709-712.

[75] Potter SM, Baum JA, Teng H, Stillman RJ, Shay NF, Erdman Jr JW. Soy protein and isoflavones: their effects on blood lipids and bone density in postmenopausal women, The American Journal of Clinical Nutrition 1998;68 (Suppl):1375S-1379S.

[76] Zhan S, Ho SC. Meta-analysis of the effects of soy protein containing isoflavones on the lipid profile. The American Journal of Clinical Nutrition 2005;81:397-408.

[77] Kummerow FA, Mahfouz MM, Zhou Q. Trans fatty acids in partially hydrogenated soybean oil inhibit prostacyclin release by endothelial cells in presence of high level of linoleic acid, Prostaglandins & other Lipid Mediators 2007;84(3-4):138–153.

[78] Lu C, Liu Y. Interaction of lipoic acid radical cations with vitamins C and E analogue and hydroxycinnamic acid derivatives. Archives of Biochemistry and Biophysics 2002;406:78–84.

[79] Henkel J. Soy:Health Claims for Soy Protein, Question About Other Components. FDA Consumer (Food and Drug Administration) 2000;34 (3): 18–20.

[80] Toda, T, Uesugi, T, Hirai, K, Nukaya, H, Tsuji, K, Ishida, H. New 6-O-acyl isoflavone Glycosides from Soybeans Fermented with Bacillus subtilis (natto). I. 6-OSuccinylated Isoflavone Glycosides and Their Preventive Effects on Bone Loss in Ovariectomized Rats Fed a Calcium-Deficient Diet. Biological & Pharmaceutical Bulletin 1999a; 22(11):1193-1201.

[81] Toda T, Uesugi T, Hirai K, Nukaya H, Tsuji K, Ishida H. New 6-O-acyl isoflavone glycosides from soybeans fermented with Bacillus subtilis (natto). I. 6-Osuccinylated isoflavone glycosides and their preventive effects on bone loss in ovariectomized rats fed a calcium-deficient diet, Biological & Pharmaceutical Bulletin 1999b22:1193-1201.

[82]   Messina M, Messina V. Soyfoods, soybean isoflavones, and bone health: a brief overview. Journal of Renal Nutrition 2000;10:63-68.

[83]   Persky VW, Turyk ME, Wang L, Freels S, Chatterton RJ, Barnes S, Erdman JJ, Sepkovic DW, Bradlow HL, Potter S. Effect of soy protein on endogenous hormones in postmenopausal women. American Society for Clinical Nutrition 2002;75(1):145–153.

[84]   Kim SJ, Jung KO, Park KY. Inhibitory Effect of Kochujang Extracts on Chemically Induced Mutagenesis. Journal of Food Science and Nutrition 1999; 4:38-42.

[85]   Ahn SW, Kim KM, Yu KW, Noh DO, Suh HJ. Isolation of angiotensin I converting enzyme inhibitory peptide from soybean hydrolysate. Food Science Biotechnology 2000;9(3):378–381.

[86]   Pool JL, Smith SG, Nelson EB, Taylor AA, Gomez HJ. Angiotensin converting enzyme inhibitors compared with thiazide diuretics or beta-blockers as monotherapy for treatment of mild essential hypertension. Current Opinion Cardiology 1989;4:ll-15.

[87]   Sites CK, Cooper BC, Toth MJ, Gastaldelli A, Arabshahi A, Barnes S. Effect of Daily Supplement of Soy Protein on Body Composition and Insulin Secretion in Postmenopausal Women. Fertility and Sterility 2007; 88:1609-1617.

[88]   Wagner JD, Zhang L, Shadoan MK, Kavanagh K, Chen H, Tresnasari K, Kaplan JR, Adams MR. Effects of soy protein and isoflavones on insulin resistance and adiponectin in male monkeys. Metabolism 2008;57:S24-S31.

[89]   Solanky KS, Bailey NJ, Beckwith-Hall BM, Bingham S, Davis A, Holmes E, Nicholson JK, Cassidy A. Biofluid 1H NMR-based metabonomic techniques in nutrition research - metabolic effects of dietary isoflavones in humans. The Journal of Nutritional Biochemistry 2005;16:236-244.

[90]   Hiwatashi K, Shirakawa H, Hori K, Yoshiki Y, Suzuki N, Hokari M, Komai M, Takahashi S. Reduction of blood pressure by soybean saponins, rennin inhibitors from soybean, in spontaneously hypertensive rats. Bioscience, Biotechnology, and Biochemistry 2010;74:2310-2312.

[91]   Sagara M, Kanda T, NJelekera M, Teramoto T, Armitage L, Birt N, Birt C, Yamori Y. Effects of dietary intake of soy protein and isoflavones on cardiovascular disease risk factors in high risk, middle-aged men in Scotland. Journal of the American College of Nutrition 2004; 23:85-91.

[92]   Welty, FK, Lee, KS, Lew, NS, Zhou, JR. Effect of soy nuts on blood pressure and lipid levels in hypertensive, prehypertensive, and normotensive postmenopausal women. Archives of Internal Medicine 2007;167:1060-1067.

[93]   Teede HJ, Giannopoulos D, Dalais, FS, Hodgson, J, McGrath, BP. Randomised, controlled, cross-over trial of soy protein with isoflavones on blood pressure and arterial function in hypertensive subjects. Journal of the American College of Nutrition 2006;25:533-540.

[94] Cuevas AM, Irribarra VL, Castillo OA, Yanez MD, Germain AM. Isolated soy protein improves endothelial function in postmenopausal hypercholesterolemic women. European Journal of Clinical Nutrition 2003;57:889-894.

[95] Lissin LW, Oka R, Lakshmi S, Cooke JP. Isoflavones improve vascular reactivity in post-menopausal women with hypercholesterolemia. Vascular Medicine 2004;9:26-30.

[96] Colacurci N, Chiantera A, Fornaro F, de NV, Manzella D, Arciello A, Chiantera V, Improta L, Paolisso G. Effects of soy isoflavones on endothelial function in healthy postmenopausal women. Menopause 2005;12:299-307.

[97] Hall WL, Formanuik NL, Harnpanich D, Cheung M, Talbot D, Chowienczyk PJ, Sanders TA. A meal enriched with soy isoflavones increases nitric oxidemediated vasodilation in healthy postmenopausal women. Journal of Nutrition 2008;138:1288-1292.

[98] Vera R, Sanchez M, Galisteo M, Villar IC, Jimenez R, Zarzuelo A, Perez-Vizcaino F, Duarte J. Chronic administration of genistein improves endothelial dysfunction in spontaneously hypertensive rats: involvement of eNOS, caveolin and calmodulin expression and NADPH oxidase activity. Clinical Science 2007;112:183-191.

[99] Si H, Liu, D. Genistein, a soy phytoestrogen, upregulates the expression of human endothelial nitric oxide synthase and lowers blood pressure in spontaneously hypertensive rats. Journal of Nutrition 2008;138:297-304.

[100] Nakashima S, Koike T, Nozawa Y. Genistein, a protein tyrosine kinase inhibitor, inhibits thromboxane A2-mediated human platelet responses. Molecular Pharmacology 1991;39:475-480.

[101] McNicol A. The effects of genistein on platelet function are due to thromboxane receptor antagonism rather than inhibition of tyrosine kinase. Prostaglandins Leukot Essent Fatty Acids 1993;48:379-384.

[102] Fujita M, Nomura K, Hong K, Ito Y, Asada A, Nishimuro S. Purification and characterization of a strong fibrinolytic enzyme (nattokinase) in the vegetable cheese natto, a popular soybean fermented food in Japan. Biochemical and Biophysical Research Communications 1993;197:1340-1347.

[103] Fujita M, Hong K, Ito Y, Fujii R, Kariya K. Nishimuro Shrombolytic effect of nattokinase on a chemically induced thrombosis model in rat. Biological & Pharmaceutical Bulletin 1995;18:1387-1391.

[104] Suzuki Y, Kondo K, Matsumoto Y, Zhao BQ, Otsuguro K, Maeda T, Tsukamoto Y, Urano T, Umemura K. Dietary supplementation of fermented soybean, natto, suppresses intimal thickening and modulates the lysis of mural thrombi after endothelial injury in rat femoral artery. Life Sciences 2003;73:1289-1298.

[105] Hsia CH, Shen MC, Lin JS, Wen YK, Hwang KL, Cham TM, Yang NC. Nattokinase decreases plasma levels of fibrinogen, factor VII, and factor VIII in human subjects. Nutrition Research 2009;29:190-196.

[106] Marquezi ML, Roschel HA, Costa ADS, Sawada LA, Lancha AH. Effect of aspartate and asparagine supplementation on fatigue determinants in intense exercise. International Journal of Sport Nutrition and Exercise Metabolism 2003;13(1):65–67.

[107] Chen HM, Muramoto K, Yamauchi F, Fujimoto K, Nokihara K. Antioxidative properties of histidine-containing peptides designed from peptide fragments found in the digests of a soybean protein. Journal of Agricultural and Food Chemistry 1998; 46(1): 49–53.

[108] Tsai P, Huang P. Effects of isoflavones containing soy protein isolate compared with fish protein on serum lipids and susceptibility of low density lipoprotein and liver lipids to in vitro oxidation in hamsters. The Journal of Nutritional Biochemistry 1999;10(11):631–637.

[109] Sudel KM, Venzke K, Mielke H, Breitenbach U, Mundt C, Jaspers S, Koop U, Sauermann K, Knussman-Hartig E, Moll I, Gercken G, Young AR, Stab F, Wenck H, Gallinat S. Novel aspects of intrinsic and extrinsic aging of human skin: beneficial effects of soy extract. Photochemistry and Photobiology 2005;81(3), 581–587.

[110] Birt DF, Hendrich S, Wang W. Dietary agents in cancer prevention: flavonoids and isoflavonoids. Pharmacology & therapeutics 2001;.90(2-3):157-177.

[111] Rimbach G, Boesch-Saadatmandi C, Frank J, Fuchs D, Wenzel U, Daniel H, Hall WL, Weinberg PD. Dietary isoflavones in the prevention of cardiovascular disease – A molecular perspective. Food and Chemical Toxicology 2008;46(4):1308–1319.

[112] Coward L, Barnes N, Setchell K, Barnes S. Genistein, daidzein, and their beta glycoside conjugates: antitumor isoflavones in soybean food from American and Asian diets. Journal of Agricultural Food Chemistry 1993; 41(11):1961-1967.

[113] Doerge DR, Sheehan DM. Goitrogenic and Estrogenic Activity of Soy Isoflavones, Environmental Health Perspectives Supplements 2002;110(3):349-353.

[114] Weidenboerner, M, Hindorf, H, Jha, HC, Tsotsonos, P, Egge, H. Antifungal Activity of Isoflavonoids in Different Reduced Stages on Rhizoctonia solani and Sclerotium rolfsii. Phytochemistry 1990;29(3):801-803.

[115] Liu M, Yanagihara N, Toyohira Y, Tsutsui M, Ueno, S, Shinohara Y. Dual effects of daidzein, a soy isoflavone, on catecholamine synthesis and secretion in cultured bovine adrenal medullary cells. Endocrinol 2007;148:5348-5354.

[116] Kritz-Silverstein D, Von Mühlen D, Barrett-Connor E, Bressel MA. Isoflavones and Cognitive Function in Older Women: The Soy and Postmenopausal Health in Aging (SOPHIA) Study. Menopause 2003;10(3):196–202.

[117] File SE, Hartley DE, Elsabagh S, Duffy R, Wiseman H. Cognitive Improvement After 6 Weeks of Soy Supplements in Postmenopausal Women is Limited to Frontal Lobe Function. Menopause 2005;12(2):193–201.

[118] Kim HJ, Suh H-J, Lee CH, Kim JH, Kang SC, Park S, Kim J-S. Antifungal Activity of Glyceollins Isolated From Soybean Elicited with Aspergillus Sojae. Journal of Agricultural and Food Chemistry 2010;58 (17): 9483–9487.

[119] Tilghman SL, Boué SM, Burow ME. Glyceollins, a Novel Class of Antiestrogenic Phytoalexins. Molecular and Cellular Pharmacology 2010;2(4):155–160.

[120] Sastry M, Murray D. The contribution of trypsin inhibitors to the nutritional value of chickpea seed protein. Journal of Science Food and Agriculture 1987;40: 253 – 261.

[121] Circle SJ, Smith AH. Soybeans: chemistry and technology. Westport, Conn: Avi Pub. Co.; 1972.

[122] Hogervorst E, Sadjimim T, Yesufu A, Kreager P, Rahardjo TB. High Tofu Intake is Associated with Worse Memory in Elderly Indonesian Men and Women". Dementia and Geriatric Cognitive Disorders 2008;26 (1): 50–57.

[123] Messina MJ, Loprinzi CL. Soy for breast cancer survivors: a critical review of the literature. The Journal of Nutrition 2001;131(11):3095–3108.

[124] National Research Council (NRC), Committee on Food Protection, Food and Nutrition Board "Phytates". Toxicants Occurring Naturally in Foods. Washington, DC: National Academy of Sciences; 1973.p363–371.

[125] Hengstler JG, Heimerdinger CK, Schiffer IB, Gebhard S, Sagemuller J, Tanner B, Bolt HM, Oesch F Dietary topoisomerase II-poisons: contribution of soy products to infant leukemia? Experimental and Clinical Sciences International online journal for advances in science Journal 2002;1:8-14.

[126] Lavigne JA, Takahashi Y, Chandramouli GVR, Liu H, Perkins SN, Hursting SD, Wang TTY Concentration-dependent effects of genistein on global gene expression in MCF-7 breast cancer cells: an oligo microarray study. Breast Cancer Research and Treatment 2008;110(1):85-98.

[127] Goodin S, Shen F, Shih WJ, Dave N, Kane MP, Medina P, Lambert GH, Aisner J, Gallo M, DiPaola RS Clinical and Biological Activity of Soy Protein Powder Supplementation in Healthy Male Volunteers. Cancer Epidemiology Biomarkers & Prevention 2007;16 (4): 829–833.

[128] Cantani A, Lucenti P Natural History of Soy Allergy and/or Intolerance in Children, and Clinical Use of Soy-protein Formulas. Pediatric Journal of Allergy and Clinical Immunology 1997;8 (2):59–74.

Chapter 21

Brazilian Soybean Varieties for Human Use

Neusa Fátima Seibel, Fernanda Périco Alves,
Marcelo Álvares de Oliveira and
Rodrigo Santos Leite

Additional information is available at the end of the chapter

http://dx.doi.org/10.5772/52602

1. Introduction

In the present days, the export trade in soybean and its derivatives has a major impact on the Brazilian agro-industrial system and economy. Brazil is the second largest producer, behind only the United States, and three states represent 63% of national production: Mato Grosso, Paraná and Rio Grande do Sul. The 2010/2011 crop has maintained its growth momentum, with higher volume than the previous one, with the climatic factor as primarily responsible for these results.

Soybean has a high nutritional and functional value, is source of quality protein and some essential nutrients to human diet. Due to this nutritional quality, high production, low cost and variety of derivate products, the soybean grain is an alternative for feed [10]. The benefits of soybean have increased its consumption both *in natura* and processed. Studies have shown the association between soy consumption and reduced incidence of esophageal, lung, prostate, breast and colorectal cancer, cardiovascular disease, osteoporosis, diabetes, Alzheimer's disease and menopausal symptoms. The joint action of high-quality protein and polyunsaturated and saturated fats present in soybean helps reduce LDL [37, 28].

In Brazil, the consumption of soybean and its products is still not widespread, due to the few options, exotic flavor to the Brazilian palate and presence of antinutritional factors in the grain. Some of these factors, as the protease inhibitors and lipoxygenase enzymes, can be reduced by suitable thermal processing. Coupled with this, the genetic breeding is responsible by eliminate lipoxygenase enzymes, reducing the flavor which limits the acceptability of the soy products [10, 28].

The soymilk is nutritive, lactose-free, contains no cholesterol and is highly digestible. Can be sold in liquid or powder, pasteurized or sterilized, and commonly flavored, such as juices

© 2013 Seibel et al.; licensee InTech. This is a paper distributed under the terms of the Creative Commons Attribution License (http://creativecommons.org/licenses/by/3.0), which permits unrestricted use, distribution, and reproduction in any medium, provided the original work is properly cited.

and vitamins. The extract can also be incorporated as ingredient in breads, cakes, biscuits, chocolates and more.

The results of production, yield, chemical composition and nutritional value of soymilk depend directly on the soybean cultivar, and the quality of soymilk may also be interfered by the water proportion and initial conditions of the grains. There are 316 soybean cultivars currently available in Brazil, with different characteristics of productivity, production cycle, grain size, adaptation to regional climate and lipoxygenase presence. Some are considered commodities, and other cultivars have special purpose.

Cultivars specially developed for human consumption can contribute to the sensory quality of the extract, which directly increases the acceptability of soy as a food, since the sensory quality is decisive in the buying process. Even with important nutritional characteristics, products with undesirable sensory aspects normally lose market to other similar foods. Therefore, sensory evaluation is important to determine the consumer preference, in order to provide support for research, manufacturing, marketing and quality control in new product development [15].

2. Brazilian soybean cultivars

Soybean is currently the most important source of edible oil and high-quality plant protein for feeding both human and animals worldwide [43, 20, 40]. Originated from mid latitude regions, these species are expanding in tropical areas as a result of the development of new genotypes tolerant to the environmental adversities of these localities [9, 40]. One of the largest soybean producers of the world is Brazil, a tropical country that comprises an extensive ecological region with wide variation in the environmental conditions. In Brazil, soybean was firstly grown in the South (in mid latitude areas) and more recently next to Equator line, in the Northeastern region, owing to the development of genotypes with high productivity, well adapted to photoperiod effect and resistant to local pathogens and pests [1, 40]. Presently, in these places, soybean cultivation has great economic and social importance [40]

Water is the main factor changing soybean productivity in time and space [32, 19]. Water use by soybeans varies with climatic conditions, management practices and the life cycle of the cultivar. This crop's response to photoperiod and temperature defines the areas to which it is adapted. Water use by soybean crop increases as the crop grows and is maximal during flowering and pod-fill [19].

Most soybean cultivars respond to photoperiod as quantitative short-day plants and are adapted in a narrow band of latitudes. The soybean has a juvenile stage after emergence when it is especially sensitive to temperature and insensitive to day length [23, 19]. Cultivars with the genetically controlled long juvenile trait have wider adaptability and can be utilized over a wider range of latitudes and planting dates than cultivars without these characteristics [19].

Soybean develops well under a wide range of temperatures, although regions in which the warmest mean monthly temperature is below 20ºC are considered inappropriate for soybean [7, 19]. Brown (1960) affirmed that vegetative growth is slow or nil at temperature 10ºC or less and optimum at 30ºC, decreasing thereafter. Temperatures above 40ºC are known to have adverse effects on growth rate, flower initiation and pod-set [19].

Nearly all soybean cultivars exhibit one of two possible growth habits. Cultivars with determinate growth habit have rather distinct vegetative and reproductive development periods. In the other side, indeterminate cultivars have overlapping vegetative and reproductive growth periods.

In recent years, the early planting date and harvest of soybeans, this ensures a lower use of pesticides and makes possible the cultivation of winter maize, resulted in a growth of cultivars of indeterminate habit, principal in South of Brazil. Now, they are dominating the market. Therefore, all breeding programs in Brazil have been working with the introduction of the specific characteristics on indeterminate cultivars.

Embrapa Soybeans has a specific breeding program that develops cultivars with special characteristics for human consumption. However there is still no one cultivar of indeterminate growth habit, but will be released in the near future. It is noteworthy that all cultivars and genotypes that are part of the active Germplasm Bank that give rise to these are conventional. The main cultivars released to date by this program are:

Embrapa 48 – cultivar with more than 15 years on the market. It is knew to processing soymilk with superior flavor when compared with other cultivars. However, due the market need for early cultivars, the cycle has become very long. Regarding the productivity also produces about 20% less than the current more productive cultivars.

BRS 213 – cultivar triple-null for lipoxigenase enzyme, which is responsible for a taste of the "beany flavor" in the extract. This cultivar has light hilum, but almost no more seed on the market, due to some fitossanitary problems and productivity. The cycle is also too long for the demands of today's market.

BRS 216 – cultivar with very small seeds and high protein value but the productivity is at least 30% less compared with the current cultivars. Mainly because of this very small size, a higher loss in the harvest occurs. It is indicated to produced soybean sprouts, especially because the high protein value and the small seed size.

BRS 257 – cultivar triple-null for lipoxigenase enzyme, with similar productivity with current cultivars. The soymilk and soybean flour industries are very interested in this cultivar.

BRS 258 – cultivar originated from an old Embrapa Soybean cultivar called BR 36. It also has a long cycle for the current market requirements and a lower productivity, however the soymilk and flour of this cultivar is well accepted.

BRS 267 – cultivar with very large seeds, sweet flavor and ideal for prepare soymilk and tofu. Also ideal to be consumed as a vegetable soybeans. However the cycle is long and the productivity at least 20% lower when compared with the current cultivars.

BRS 282 – cultivar originated from Embrapa 48 and was launched three years ago. This cultivar does have a cycle consistent with what the market wants today, but studies of the special characteristics of this cultivar are still scarce. The productivity is similar with current cultivars. The soymilk has excellent acceptance and is a cultivar that should be encouraged to be cultivated.

Among the cultivars released by Embrapa, there is a cultivar that did not originate in the program of special cultivars for human consumption but is suitable for this purpose, the BRS 232 cultivar. It has a size large seed and light hilum, ideal characteristics for this purpose. It is always recommended for human consumption when there isn't a special cultivar. This cultivar has well accepted soymilk and flour when compared with current cultivars.

3. Characterization of eight brazilian soybean cultivars for human use

3.1. Chemical composition

The grains of cultivars EMBRAPA 48, BRS 213, BRS 216, BRS 232, BRS 257, BRS 258, BRS 267 and BRS 282, planted in various locations in the state of Paraná, Brazil, during the 2009/10 crop were characterized, and the average results of the composition are shown in Table 1.

The calculated values were similar to those reported by other authors [6, 31]. The highest protein content was found for BRS 258 (44.37%), and differed significantly ($p > 0.05$) from other grains.[36] have analyzed the same variety, in organic cultivation, and reported lower levels (42.84%). [16] also reported lower contents, with values of 41.70%.

Cultivar	Moisture	Protein	Lipids	Ash	Carbohydrate
Embrapa 48	6.14 ± 0.95^a	40.11 ± 0.58^{bc}	22.45 ± 1.31^a	4.97 ± 0.10^{de}	32.47
BRS 213	5.35 ± 0.19^a	39.50 ± 0.26^c	21.86 ± 0.65^{ab}	4.90 ± 0.30^e	33.74
BRS 216	5.61 ± 0.23^a	41.08 ± 0.54^{bc}	19.19 ± 1.32^{cd}	4.45 ± 0.15^e	35.28
BRS 232	5.69 ± 0.07^a	40.99 ± 0.51^{bc}	20.72 ± 0.71^{abcd}	5.47 ± 0.16^{cd}	32.82
BRS 257	5.67 ± 1.11^a	41.66 ± 1.38^b	21.17 ± 0.70^{abc}	6.60 ± 0.12^a	30.57
BRS 258	6.63 ± 0.18^a	44.37 ± 0.06^a	18.76 ± 0.62^d	5.86 ± 0.21^{bc}	31.01
BRS 267	6.02 ± 0.16^a	39.41 ± 1.08^c	20.03 ± 0.39^{bcd}	6.45 ± 0.30^a	34.11
BRS 282	6.16 ± 0.38^a	39.96 ± 0.27^{bc}	20.70 ± 0.90^{abcd}	6.35 ± 0.13^{ab}	32.99

Table 1. Centesimal composition of eight soybeans cultivars (g.100g-1). Means followed by same letters in columns do not differ by Tukey test ($p \leq 0.05$). Means from three replicates on a dry basis. * Calculated by difference.

As reported by [11], the soybean features a unique high quality protein source. In general, the industry focus is the production of soybean meal and soybean oil. Therefore, the cultivar BRS 258, due to higher protein content, can be an interesting alternative to the industry, which seeks yield and for high protein content in soybean meal.

According with Embrapa Soja results, the average levels of protein from cultivars Embrapa 48, BRS 213, BRS 232, BRS 257, BRS 267 and BRS 282 are very similar to those found in this study [16]. An exception was found for BRS 216, which presented values of 41.08%, lower than those reported in the literature (43.06%) [17]. [25] has determined the composition of different soybeans cultivars, and found a mean value of 38% in protein. [12] reports values between 33% and 42%. In the present study, the cultivar which surpassed this variation was BRS 258. BRS 216, BRS 232, BRS 257, BRS 258 and BRS 267 were analyzed by [6], and the protein content varied between 38.47% and 39.61%.

Regarding lipids, Embrapa 48 had the highest content (22.45%), but did not differ significantly from BRS 213, BRS 232, BRS 257 and BRS 282. BRS 258 presented the lowest lipid content (18.76%) and did not differ significantly ($p \leq 0.05$) from BRS 216, BRS 232, BRS 267 and BRS 282.

In general, literature reports levels between 13 and 25% for lipids in soybean [6]. According to [10], the oil content in soybeans (20%) provides enough calories, so the consumed protein is metabolized for the synthesis of new tissues, and not converted into energy, as commonly seen in diets with low caloric content. However, since industry has as main objective the production of soy oil, cultivars Embrapa 48, BRS 213, BRS 232, BRS 257 and BRS 282 are the best choice for this market.

Some authors report an inverse relation between lipids and protein in soybean [29, 41]. This relation is confirmed by the results for BRS 258, with higher protein, and consequently, lower lipids contents. The average results found in this study were lower compared to lipids and higher for the protein, when compared to that reported in the literature [16, 36].

According to [3], increasing of the planting site temperatures directly affects the oil content in the grains, increasing it. Woodrow and [31] studies about the harvest in 1999, a hot and dry year, showed smaller grains with a reduction in protein concentration, when compared to the previous crop grains (1998). However, there was an increase in lipid content of the grains. For the protein content, the temperature directly influences the composition of amino acids. At higher temperatures, the proteins are rich in methionine, desirable for human consumption.

The higher ash content was the BRS 257 (6.60%), but this did not differ significantly from cultivars BRS 267 and BRS 282. The lowest content was the BRS 216 (4.45%), with no significant differences from the levels of BRS 213 and Embrapa 48. The mineral composition of soybean has quantities that normally exceed the recommended daily dose, when consumed 100 grams of grain, with calcium as the less useful in the consuming of the whole grain [10, 37]. The highest content for total carbohydrates was found in BRS 216 (35.28%), and the lowest in grains of BRS 257 (30.57%).

It is noteworthy that the variations in results between the cultivars, and comparison with literature data using the same varieties, are normal, since the planting site, year and climatic conditions affect these values [29, 33, 31, 36].

3.2. Trypsin inhibitor of soybean grains

Trypsin inhibitor is normally present in the soybean fresh grains, and considered an antinutritional factor. The average values found in literature goes up to 18 milligrams of inhibitor per gram of soybean (HAFEZ, 1983 apud [2]). In this work, however, the average value for BRS 232 (13.82) was lower than usually reported in the literature, and did not differ statistically from BRS 216 (Table 2).

Cultivar	Trypsin Inhibitor(mg TI/g)
Embrapa 48	20.28 ± 0.35^a
BRS 213	22.97 ± 2.42^a
BRS 216	18.12 ± 1.63^{ab}
BRS 232	13.82 ± 0.73^b
BRS 257	21.02 ± 2.18^a
BRS 258	19.61 ± 0.90^a
BRS 267	23.18 ± 1.64^a
BRS 282	22.76 ± 1.92^a

Table 2. Trypsin inhibitor in soybean grains for eight different cultivars. Means followed by same letters in columns do not differ by Tukey test ($p \leq 0.05$). Means from three replicates.

Although the soybean provides high quality protein, these biochemical agents (protease inhibitors) cause a limitation in the biological utilization of the amino acids present in the grains, and may reduce protein digestibility [28, 21], by the blockade of some proteases, including human digestive enzymes. Trypsin is an enzyme secreted by the pancreas, responsible for digestion of proteins by peptide bonds break, and the presence of trypsin inhibitor causes metabolic changes in the pancreas, since the inhibitor binds with the trypsin and inhibits the digestion of proteins. With the protein concentration increasing, the pancreas is stimulated to produce more trypsin, causing pancreatic hypertrophy. Most of these proteases inhibitors are inactivated or inhibited when suitable thermal treatments are applied [11, 21, 30, 35].

3.3. Soybean Isoflavones

The Isoflavones, present in soybean with greater concentration than in the other legumes, belong to the class of phytoestrogens, and have the capacity to assist in the effects of menopause. Besides, the isoflavones are known as having anticancer properties, and antioxidant action that neutralizes free radicals, contributing to reduce LDL (bad cholesterol). The main isoflavones determined in soybean are genistein, daidzein and glycitein, which can be found in the form of aglycones (unconjugated) and glycosylated (conjugated) [4, 18].

In the present study, there was a large variation in the total isoflavones content for the studied cultivars, (Tables 3 and 4), with the highest levels in the BRS 213 (386.60 mg.100g-1) and BRS 282 (364.56 mg.100 g-1) and the lowest in BRS 258 (54.06 mg.100g-1).

The soybean grain naturally presents the isoflavones in the aglycone and glycoside form. The aglycones are absorbed directly, since they are not linked to a sugar, while the other conjugated forms require a hydrolysis for their absorption [27, 18].

The isoflavones profiles for the studied cultivars were very similar, with higher levels of the M-genistein form. However, daidzein, genistein and glycitein forms have been receiving most of attention from researchers. According to [42], genistein has the potential effect of inhibiting the growth of cancer cells at physiological concentrations, and daidzein has effect only if combined with genistein. In the present study, BRS 213 presented the highest level of genistein. Only BRS 267 and BRS 282 showed levels of glycitein, while, BRS 232 showed no levels for the isoflavones highlighted.

The acetyl form was not found in any samples, proving that the soybean did not suffer thermical treatment. According to [2], and [26], in the heat treated products the malonyl form is unstable and may be transformed into the acetyl form. [28] also points out that the processing parameters, the varieties and planting condition affect the composition and / or the isoflavones profile in soy products.

Isoflavones	EMBRAPA 48	BRS 213	BRS 216	BRS 232
G-Daidzin	34.03 ± 1.44	78.26 ± 8.82	75.64 ± 3.38	13.06 ± 0.76
G-Glycitin	8.19 ± 1.05	11.35 ± 1.61	16.84 ± 1.17	4.65 ± 0.53
G-Genistin	22.94 ± 0.59	63.71 ± 5.28	50.03 ± 1.76	10.14 ± 0.20
M-Daidzin	88.91 ± 4.83	75.12 ± 7.81	73.08 ± 3.02	33.84 ± 1.42
M-Glycitin	21.69 ± 3.35	13.94 ± 1.33	20.74 ± 1.63	12.43 ± 1.20
M-Genistin	107.45 ± 2.70	111.46 ± 9.00	89.11 ± 3.34	48.89 ± 0.97
A-Daidzin	0.00 ± 0.00	0.00 ± 0.00	0.00 ± 0.00	0.00 ± 0.00
A-Glycitin	0.00 ± 0.00	0.00 ± 0.00	0.00 ± 0.00	0.00 ± 0.00
A-Genistin	0.00 ± 0.00	0.00 ± 0.00	0.00 ± 0.00	0.00 ± 0.00
Daidzein	2.47 ± 0.21	19.03 ± 0.81	5.79 ± 0.43	0.00 ± 0.00
Glycitein	0.00 ± 0.00	0.00 ± 0.00	0.00 ± 0.00	0.00 ± 0.00
Genistein	1.87 ± 0.03	13.71 ± 0.65	3.63 ± 0.15	0.00 ± 0.00
TOTAL	**287.57 ± 14.04**	**386.60 ± 33.66**	**334.86 ± 12.80**	**123.01 ± 2.74**

Table 3. Isoflavones profile in soybean grains from the cultivars EMBRAPA 48, BRS 213, BRS 216 and BRS 232 (mg. 100g-1).

Isoflavones	BRS 257	BRS 258	BRS 267	BRS 282
G-Daidzin	39.34 ± 1.19	7.23 ± 0.32	29.82 ± 5.89	29.53 ± 2.96
G-Glycitin	10.38 ± 0.44	2.70 ± 0.32	10.55 ± 3.70	17.19 ± 1.62
G-Genistin	33.98 ± 0.94	4.09 ± 0.01	23.61 ± 1.47	35.43 ± 2.12
M-Daidzin	88.89 ± 2.35	18.31 ± 0.42	25.94 ± 6.95	69.78 ± 4.59
M-Glycitin	24.47 ± 1.41	5.97 ± 0.51	11.13 ± 4.66	31.91 ± 2.68
M-Genistin	134.59 ± 3.96	15.04 ± 0.35	35.82 ± 2.10	150.94 ± 4.97
A-Daidzin	0.00 ± 0.00	0.00 ± 0.00	0.00 ± 0.00	0.00 ± 0.00
A-Glycitin	0.00 ± 0.00	0.00 ± 0.00	0.00 ± 0.00	0.00 ± 0.00
A-Genistin	0.00 ± 0.00	0.00 ± 0.00	0.00 ± 0.00	0.00 ± 0.00
Daidzein	2.05 ± 0.25	0.43 ± 0.12	4.77 ± 0.68	8.89 ± 2.08
Glycitein	0.00 ± 0.00	0.00 ± 0.00	3.08 ± 0.66	10.96 ± 2.30
Genistein	2.59 ± 0.09	0.30 ± 0.09	4.02 ± 0.13	9.93 ± 2.57
TOTAL	329.29 ± 6.24	54.06 ± 1.05	148.74 ± 25.69	364.56 ± 12.87

Table 4. Isoflavones profile in soybean grains from the cultivars BRS 257, BRS 258, BRS 267 and BRS 282 (mg.100g^{-1}).

4. Soymilk production

The soymilk was produced at a 1:6 ratio [soybean (g): water volume (mL)], with the eight characterized cultivars (Embrapa 48, BRS 213, BRS 216, BRS 232, BRS 257, BRS 258, BRS 267 and BRS 282). Initially, the beans were submitted to soaking for five minutes at 95°C in 1:3 ratio with boiling water, and then water was discarded. After this, the soybean was submitted to heat treatment at 95°C for ten minutes, at the proportion of 1:6 with water, and seeds were ground for three minutes. The soymilk was separated from the wet okara by filtration, in which it was applied a heat treatment for two minutes under boiling.

4.1. Efficiency of the soymilk process

The yield is an important processing variable for the food industry, and should be calculated by the ratio between the mass of raw materials and final volume of extract. Thus, from the volume of processed grain, the greater the volume obtained, the better the utilization of production. According to [17], from 500 grams of grain and 4.5 liters of water, 1.5 liters of soymilk are produced. Following the same method, the extraction was performed with 250g of grains and 2.25 liters of water, splitted in 750 mL for maceration (which was discarded) and 1500 mL for grinding. The extraction yields are calculated on the weight of macerated grains, which absorbs water during this process, and the water used in grinding (Table 5).

Cultivar	Weight after soaking (g)	Soymilk(mL)	Yield (%)
Embrapa 48	385	800	42.44
BRS 213	370	820	43.85
BRS 216	390	670	35.44
BRS 232	405	720	37.79
BRS 257	365	820	43.96
BRS 258	385	760	40.32
BRS 267	390	605	32.01
BRS 282	405	840	44.09

Table 5. Yield of the process for soymilk of eight different soybean cultivars.

In this study, it was found that the different soybean cultivars resulted in different yield for the soymilk. The cultivar that showed the best results was BRS 282 (44.09%), higher than that reported by [17]. BRS 213 and BRS 257 showed very similar yields, 43.85% and 43.96% respectively. And BRS 232 (37.79%), BRS 216 (35.44%) and BRS 267 (32.01%) had the lowest yields.

4.2. Soymilk characterization

4.2.1. Freeze-dried

In accordance with the results of fresh grains, the freeze-dried soymilk with the highest protein content was the BRS 258 (42.25%) (Table 6). The lowest level was the extract of BRS 267 (35.34%), but did not differ significantly from extracts of BRS 282 and BRS 213.

Cultivar	Moisture	Protein	Lipids	Ash	Carbohydrates*
Embrapa 48	3.52 ± 0.08^d	36.25 ± 0.40^d	18.34 ± 0.14^a	8.37 ± 0.33^c	33.52
BRS 213	7.78 ± 0.16^a	36.02 ± 0.10^{de}	18.13 ± 0.22^a	9.08 ± 0.24^b	28.99
BRS 216	7.95 ± 0.26^a	33.48 ± 0.54^f	16.95 ± 0.89^{abc}	9.19 ± 0.27^b	32.43
BRS 232	7.49 ± 0.40^a	38.60 ± 0.39^c	14.99 ± 0.89^{cd}	8.69 ± 0.15^{bc}	30.23
BRS 257	7.88 ± 0.09^a	40.44 ± 0.12^b	17.73 ± 0.25^{ab}	8.52 ± 0.02^{bc}	25.43
BRS 258	4.74 ± 0.02^c	42.45 ± 0.23^a	13.57 ± 0.24^{de}	8.52 ± 0.30^{bc}	30.72
BRS 267	5.31 ± 0.27^{bc}	35.34 ± 0.22^e	12.24 ± 0.48^e	9.91 ± 0.30^{bc}	37.20
BRS 282	5.91 ± 0.16^b	35.63 ± 0.19^{de}	15.52 ± 1.20^{bcd}	10.07 ± 0.20^a	32.87

Table 6. Composition of freeze-dried soymilk from eight different soybeans cultivars $(g.100g^{-1})$. Means followed by same letters in columns do not differ by Tukey test $(p \leq 0.05)$. Means from three replicates on a dry basis. * Calculated by difference.

The higher lipid content was found in the soymilk of Embrapa 48 (18.35%), and did not differ significantly from extracts of BRS 213, BRS 216 and BRS 257. The lowest level was the soymilk of BRS 267 (12.24%), which did not differ from the extract of BRS 258. The soymilk of BRS 267 showed the highest content of carbohydrate, 37.20%.

4.2.2. In natura (Liquid)

The results for chemical composition of the fresh soymilk were determined indirectly, through the results of total solids (Tables 7 and 8).

Cultivar	Moisture	Solids
Embrapa 48	92.16 ± 0.06^{cd}	7.84 ± 0.06^{cd}
BRS 213	92.35 ± 0.06^{c}	7.65 ± 0.06^{d}
BRS 216	92.97 ± 0.06^{b}	7.03 ± 0.06^{e}
BRS 232	89.00 ± 0.06^{f}	11.00 ± 0.06^{a}
BRS 257	91.85 ± 0.06^{de}	8.15 ± 0.06^{bc}
BRS 258	93.71 ± 0.06^{a}	6.29 ± 0.06^{f}
BRS 267	93.68 ± 0.06^{a}	6.32 ± 0.06^{f}
BRS 282	91.67 ± 0.06^{f}	8.33 ± 0.06^{b}

Table 7. Moisture and solids of the soymilk from eight different soybean cultivars (%). Means followed by same letters in columns do not differ by Tukey test (p ≤ 0.05). Means from three replicates.

Cultivar	Protein	Lipids	Ash	Carbohydrates*
Embrapa 48	5.49 ± 0.06^{c}	2.78 ± 0.02^{ab}	1.27 ± 0.05^{cde}	6.14
BRS 213	5.08 ± 0.01^{d}	2.56 ± 0.21^{b}	1.28 ± 0.03^{c}	6.38
BRS 216	4.33 ± 0.07^{e}	2.19 ± 0.12^{c}	1.19 ± 0.04^{de}	6.35
BRS 232	7.86 ± 0.08^{a}	3.05 ± 0.18^{a}	1.77 ± 0.03^{a}	9.32
BRS 257	6.07 ± 0.02^{b}	2.66 ± 0.04^{b}	1.28 ± 0.00^{cd}	6.29
BRS 258	5.09 ± 0.03^{d}	1.63 ± 0.03^{d}	1.02 ± 0.04^{f}	4.84
BRS 267	4.23 ± 0.03^{e}	1.46 ± 0.06^{d}	1.19 ± 0.04^{e}	5.76
BRS 282	5.59 ± 0.03^{c}	2.43 ± 0.19^{bc}	1.58 ± 0.03^{b}	7.06

Table 8. Centesimal composition of soymilk from eight different soybeans cultivars (g.200mL^{-1}). Means followed by same letters in columns do not differ by Tukey test (p ≤ 0.05). Means from three replicates, wet basis. * Calculated by difference.

A simple comparison between the composition of soybeans and their respective soymilk allows observing that the solubilization rate of compounds in aqueous solutions is critical to the final results. The extract of BRS 232 showed the highest levels of soluble compounds (11.01%) obtained during the processing, being superior to others in protein levels (7.86 g. 200mL^{-1}). BRS 267 has presented high protein content in the grains (39.41%), but after processing the content was reduced and the soymilk showed the lowest level (4.23 g.200mL^{-1}), when compared to the other extracts. This reduction indicates that the cultivar has a low content of soluble proteins.

4.2.3. Trypsin Inhibitor and Isoflavones in the soymilk

The results for soymilk confirmed that the heat treatment for 15 minutes at 100°C, performed during the processing, was enough for complete inactivation of the inhibitor, with final values equal to zero. According to [35] and [28] when foods are submitted to appropriate heat treatment, the inhibitor is inactivated.

Regarding to the isoflavones in the freeze-dried soymilk, all cultivars were significantly different, and the extract of BRS 213 had the highest average (421.61 mg.100g-1). The lowest level was found in the extract of BRS 258. For the liquid soymilk, the liquor obtained from BRS 213 maintained the highest isoflavones content (64.50 mg.200mL-1). However, for the equivalent amount to one cup of drink, this did not differ from cultivar BRS 257 (62.01 mg.200mL-1). There was no significant difference between the BRS 216 (53.02 mg.200mL-1) and BRS 282 (53.65 mg.200mL-1) (Table 9).

Cultivar	Freeze-dried soymilk (mg.100g^{-1})	Liquid soymilk (mg.200mL^{-1})*
Embrapa 48	370.65 ± 0.88^b	58.12 ± 0.91^b
BRS 213	421.61 ± 2.55^a	64.50 ± 1.31^a
BRS 216	377.18 ± 3.57^b	53.02 ± 0.50^c
BRS 232	143.45 ± 3.70^e	31.55 ± 0.81^e
BRS 257	380.44 ± 3.51^b	62.01 ± 1.25^a
BRS 258	79.59 ± 0.55^f	10.01 ± 0.07^f
BRS 267	279.91 ± 8.61^d	39.58 ± 2.20^d
BRS 282	322.03 ± 4.41^c	53.65 ± 0.74^c

Table 9. Total isoflavones content of the soymilk produced from eight different cultivars. Means followed by same letters in columns do not differ by Tukey test (p ≤ 0.05). Means values from three replicates. * Equivalent to a glass of drink.

The isoflavones concentration reported in the literature, for soy beverages with original or chocolate flavor, varies between 4 and 13 mg.200mL-1 [8]. [13] reported 12.2 mg.200mL-1, and [22] found a content of 16.6 mg in 200mL. The isoflavones content and profile are also affected by the processing, environment, and the soybean varieties [8]. In the present study, the soymilk prepared with cultivars Embrapa 48, BRS 213, BRS 216, BRS 257 and BRS 282 showed values over 50 mg.200mL-1, which surpass the literature values in more than three times.

The isoflavones profile of soymilk was also determined (Tables 10 and 11), in order to verify alterations after the processing. However, it was observed that the pasteurization did not cause the appearance of acetyl form, as observed by [2] and [26]. According to them, malonyl form in products that suffered heat treatment is unstable, and can be converted in the acetyl form.

M-genistein maintained its high concentration, and after processing the glycitein form was found in all cultivars. The appearance of glycitein after soybean processing has already been reported by [24] who noted the absence of this form in raw soybean, with subsequent detection in soy-based beverage.

Isoflavones	EMBRAPA 48	BRS 213	BRS 216	BRS 232
G-Daidzin	47.16 ± 2.61	89.45 ± 1.72	86.38 ± 1.26	15.28 ± 0.41
G-Glycitin	10.99 ± 0.86	12.22 ± 1.13	19.56 ± 0.68	5.63 ± 0.16
G-Genistin	29.47 ± 0.70	63.73 ± 1.43	49.32 ± 1.11	11.99 ± 0.31
M-Daidzin	116.17 ± 0.52	91.39 ± 2.87	90.35 ± 1.47	38.12 ± 0.89
M-Glycitin	27.55 ± 0.47	16.62 ± 0.63	26.11 ± 0.65	13.50 ± 0.45
M-Genistin	127.95 ± 2.14	121.76 ± 2.62	94.01 ± 1.43	50.38 ± 0.81
A-Daidzin	0.00 ± 0.00	0.00 ± 0.00	0.00 ± 0.00	0.00 ± 0.00
A-Glycitin	0.00 ± 0.00	0.00 ± 0.00	0.00 ± 0.00	0.00 ± 0.00
A-Genistin	0.00 ± 0.00	0.00 ± 0.00	0.00 ± 0.00	0.00 ± 0.00
Daidzein	1.98 ± 0.09	9.78 ± 0.26	3.02 ± 0.26	1.13 ± 0.50
Glycitein	8.27 ± 0.89	9.77 ± 1.04	6.35 ± 0.96	7.07 ± 1.94
Genistein	1.12 ± 0.07	6.89 ± 0.13	2.03 ± 0.10	0.37 ± 0.07
TOTAL	**370.65 ± 5.77**	**421.61 ± 8.59**	**377.14 ± 3.57**	**143.45 ± 3.69**

Table 10. Isoflavones profile in the freeze-dried soymilk obtained from soybean cultivars EMBRAPA 48, BRS 213, BRS 216 and BRS 232 (mg.100g-^{1}).

Isoflavones	BRS 257	BRS 258	BRS 267	BRS 282
G-Daidzin	43.46 ± 0.41	8.33 ± 0.63	37.88 ± 1.27	38.99 ± 1.11
G-Glycitin	12.84 ± 0.84	3.35 ± 0.19	16.40 ± 0.95	18.30 ± 0.41
G-Genistin	36.41 ± 1.06	6.15 ± 0.47	32.63 ± 1.92	35.74 ± 1.01
M-Daidzin	100.23 ± 1.75	23.84 ± 0.23	60.07 ± 3.94	70.42 ± 0.89
M-Glycitin	29.21 ± 1.35	9.02 ± 0.38	25.78 ± 1.77	30.11 ± 0.58
M-Genistin	144.84 ± 6.87	18.43 ± 0.20	95.25 ± 7.09	116.37 ± 1.75
A-Daidzin	0.00 ± 0.00	0.00 ± 0.00	0.00 ± 0.00	0.00 ± 0.00
A-Glycitin	0.00 ± 0.00	0.00 ± 0.00	0.00 ± 0.00	0.00 ± 0.00
A-Genistin	0.00 ± 0.00	0.00 ± 0.00	0.00 ± 0.00	0.00 ± 0.00
Daidzein	1.62 ± 0.53	0.44 ± 0.17	1.75 ± 0.03	1.66 ± 0.05
Glycitein	9.42 ± 1.63	9.63 ± 0.31	7.92 ± 0.56	8.69 ± 0.21
Genistein	1.76 ± 0.04	0.40 ± 0.12	1.84 ± 0.06	1.76 ± 0.09
TOTAL	**380.44 ± 7.64**	**79.59 ± 0.54**	**279.91 ± 17.44**	**322.03 ± 4.41**

Table 11. Isoflavones profile in the freeze-dried soymilk obtained from soybean cultivars BRS 257, BRS 258, BRS 267 and BRS 282 (mg.100g-^{1}).

4.2.4. Sensory analysis of soymilk

Sensory analysis was performed in order to differentiate the studied cultivars, and to discuss the best features of each one in the food industry. The panel consisted of 59 judges, comprising 40% women and 60% men, aged between 16 and 54 years, and with good educational level (82.24%), ranging from Superior Incomplete (32.25%), Superior (20.96%) and Postgraduate (29.03%).

When asked about their consumption habits of soybeans "milk", 55% of the judges affirmed to consume the commercial soymilk regularly. Of these, 68% consumed with the addition of flavor, 11% consumed the original extract, and 21% affirmed to consume both (Figure 1).

Judges evaluated the samples, applying scores from 1 (dislike very much) to 10 (like very much), with 5 as an intermediary, in the scale. The mean scores given varied between 4.14 and 6.75, close to "did not like, nor dislike."

Averages were very close between the attributes and cultivars, and this turned into one of the difficulties of implementing the analysis. Judges accustomed to the consumption of commercial extract may have been hindered due to their lack of consumption habit of original extract, with no sugar added (Table 12).

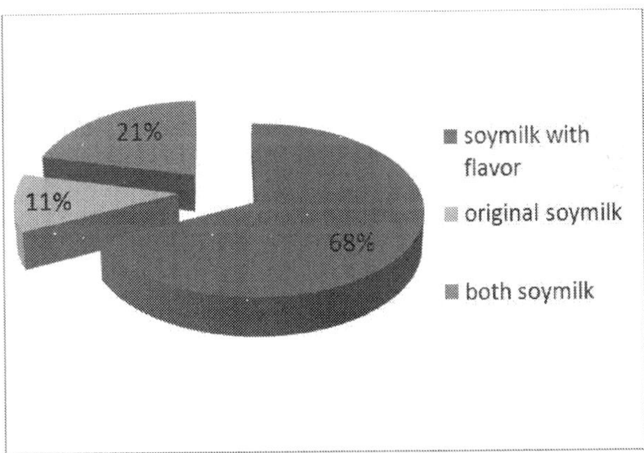

Figure 1. Type of soymilk normally consumed by the judges.

The evaluated attributes showed significant differences ($p \leq 0.05$) only for the flavor and aftertaste, with the extract of BRS 232 receiving the highest average in flavor. The highest average for aftertaste was found in BRS 213, considered the preferred for this attribute. The differences in the averages are directly linked to the composition of the grains and their extracts, and it is important to notice the relationship between the presence of lipoxygenase and the aftertaste of the extract. The soybean grain from BRS 213 has none of the lipoxygenases, which certainly contributed to achieving the highest score in the aftertaste attribute.

Cultivar	Taste	Flavor	Aftertaste	Overall appearance
Embrapa 48	4.27 ± 2.52^a	5.16 ± 2.50^b	4.14 ± 2.28^b	4.75 ± 2.39^a
BRS 213	5.24 ± 2.44^a	6.44 ± 2.05^a	5.61 ± 2.08^a	5.76 ± 2.11^a
BRS 216	4.31 ± 2.36^a	5.75 ± 2.00^{ab}	4.71 ± 2.32^{ab}	5.12 ± 2.22^a
BRS 232	5.12 ± 2.84^a	6.50 ± 2.33^a	4.73 ± 2.78^{ab}	5.82 ± 2.65^a
BRS 257	5.26 ± 2.26^a	6.05 ± 2.18^{ab}	5.14 ± 2.27^{ab}	5.90 ± 2.11^a
BRS 258	5.17 ± 2.35^a	5.87 ± 2.04^{ab}	5.37 ± 2.37^{ab}	5.78 ± 2.26^a
BRS 267	4.74 ± 2.23^a	5.80 ± 2.17^{ab}	4.84 ± 2.11^{ab}	5.34 ± 2.16^a
BRS 282	4.97 ± 2.53^a	5.98 ± 2.17^{ab}	4.98 ± 2.51^{ab}	5.51 ± 2.63^a

Table 12. Points attributed to soymilk of eight different soybean cultivars. Means followed by same letters in columns do not differ by Tukey test ($p \leq 0.05$).

The lipoxygenase enzymes (L1, L2 and L3) can be considered the primarily responsible for the undesirable taste of soybean in Brazil. The beany flavor is result of the three isoenzymes present in the grain, which catalyze the lipids oxidation. The enzyme action only begins with the breakdown and hydration of the grain, since the reaction substrate does not remain exposed in the intact grain. N-hexanal is the volatile compound produced in greater quantity, responsible for the characteristic flavor and taste [11, 29].

Although there was no significant difference between samples for the overall appearance and taste, the extract of BRS 257 deserves special attention for the highest average in both attributes. In other hand, the extract obtained from Embrapa 48 had the lowest average scores for all attributes (flavor, taste, aftertaste and overall appearance). It is interesting to note the need to produce an extract with sensory characteristics similar to the usual habits of consumption, such as the addition of flavor and aroma, allowing a better sensory evaluation.

5. Soybean products application in the food industry

Knowing the consumers profile is critical to the food industry. During the research and development of a new product, the industry focuses on knowing the market and its potential consumers, and many industries apply the sensory analysis as a tool to start or even innovate their activities. The changing of habits related to consumption of soy products has been essential for the growth of the sector [34].

[5] conducted a survey on consumer attitudes in relation to soybeans and their derivatives. They interviewed 100 individuals, 50 men and 50 women, aged 18-40 years, and mostly between 18 and 25, featuring a younger audience. When asked about soy products, tofu and "milk" were the most remembered products, and 40% of the interviewed reported never having consumed these products. A very small portion (8%) reported the consumption of soybean "milk" at least once a week. The soymilk consumption has gradually increased over

the years, by the addition of flavors capable to create a product of good flavor, which little resembles with soybean flavor.

The link between chemical composition and sensory analysis must be directly connected to yield, for the studied cultivar to become an industrial alternative. The extracts from eight soybean cultivars differed significantly in their chemical composition, and the highest protein content was found in BRS 232 (7.86 g.200mL^{-1}). In the sensorial analysis, the samples differed only in aroma and aftertaste, with the extract from BRS 232 achieving satisfactory mean. In the aroma attribute, this same cultivar had the highest average, 6.50.

The highest yield was found in BRS 282 (44.90%), followed by the cultivars Embrapa 48 (42.44%), BRS 213 (43.85%), BRS 257 (43.96%) and BRS 258 (40.32%). These values allow the use of all the studied cultivars, even those with lower yields, when considering the results of chemical composition and sensory analysis. An example is BRS 232, which showed higher levels for all compounds, but had a yield of 37.79%.

6. Final considerations

The soybean cultivars currently available in Brazil have different characteristics of productivity, production cycle, grain size, climate adaptation, lipoxygenase activity, and others. However, the Brazilian consumption of soybean as a food is still small, due to its exotic flavor to the palate, since it is an Asian grain and its development was based on the habits and customs of the orientals. These exotic flavors can be assigned to the presence of lipoxygenase enzymes, saponins and phenolic compounds, responsible for rancid or beany flavors, bitter and astringent, respectively.

A lot of products can be obtained from the soybean. However, the Brazilian food industry had to adapt them to the consumers habits, like the soymilk applied in soy beverages, which for a better acceptance is developed and commercialized with the addition of flavors or fruit juice. This is the most popular and consumed soy derivative in Brazil.

Fortunately, with more information published about the benefits of soybean consumption, its nutritional value and functional properties, this scenario is changing. After recognizing the great importance of this legume, several studies on the development of cultivars with better acceptability started, with the goal to insert soybean as an essential part of the human food. To achieve this, however, in addition to the cultivar adaptation, it is also necessary to check if this cultivar is interesting for industrialization.

The yield is a very important variable in the food industry, but cannot be considered an exclusion factor, since other important variables, such as composition, functional characteristics (isoflavones) and antinutritional compounds (trypsin inhibitor) should be observed for the choice of a soybean cultivar. Therefore, the use of cultivars specially developed for soybean based products, directed to human consumption, may contribute to improve the sensory quality and increase the soybean acceptability as a food.

Author details

Neusa Fátima Seibel[1*], Fernanda Périco Alves[2], Marcelo Álvares de Oliveira[3] and
Rodrigo Santos Leite[4]

*Address all correspondence to: neusaseibel@utfpr.edu.br

1 Mestrado Profissional em Tecnologia de Alimentos. Universidade Tecnológica Federal do
Paraná (UTFPR). Londrina/PR, Brasil

2 Tecnóloga em Alimentos (UTFPR). Londrina/PR, Brasil

3 Embrapa Soja. Londrina/PR, Brasil

4 Mestrando em Tecnologia de Alimentos (UTFPR). Embrapa Soja. Londrina/PR, Brasil

References

[1] Almeida, L. A., Kiihl, R. A. S., Miranda, M. A. C., & Campelo, G. J. A. (1999). Melhoramento da soja para regiões de baixas latitudes. In: Recursos genéticos e melhoramento de plantas para o nordeste brasileiro Available in http://www.cpatsa.embrapa.br/catalogo/livrorg/temas.html , 15.

[2] Anderson, R. L., & Wolf, W. J. (1995). Compositional Changes in Trypsin Inhibitors, Phytic Acid, Saponins and Isoftavones Related to Soybean Processing. *The Journal of Nutrition*, 125, 581S-588S.

[3] Barros, H. B., & Sediyama, T. (2009). Luz, umidade e temperatura. *In: SEDIYAMA, T. Tecnologias de produção e usos da soja. Londrina: Mecenas*, 17-27.

[4] Bedani, R., & Rossi, E. A. (2005, jul.-dez.) Isoflavonas; Bioquímica, fisiologia e implicações para a saúde. *Boletim CEPPA. Curitiba*, 23(2), 231-264.

[5] Behrens, BJ. H., & Silva, M. A. A. P. (2004, jul.-set..) Atitude do consumidor em relação à soja e produtos derivados. *Ciênc. Tecnol. Aliment., Campinas*, 24(3), 431-439.

[6] Benassi, V. T., Benassi, M. T., & Prudêncio, S. H. (2011). Cultivares brasileiras de soja: características para a produção de tofu e aceitação pelo mercado consumidor. *Semina: Ciências Agrárias, Londrina*, 32(1), 1901-1914.

[7] Berlato, M. A. (1981). Bioclimatologia da soja. *In: Miyasaka, S.; MEDINA, J.C. A soja no Brasil, Institute of Food Technology (ITAL), Campinas*, 175-184.

[8] Callou, K. L. A. (2009). Teor de isoflavonas e capacidade antioxidante de bebidas à base de soja. *124 f. Dissertação (Mestrado- Ciência dos Alimentos)- Faculdade de Ciências Farmacêuticas, Universidade de São Paulo, São Paulo*.

[9] Campelo, G. J. A., Kiihl, R. A. S., & Almeida, L. A. (1999). Características agronômicas e morfológicas das cultivares de soja desenvolvidas para as regiões de baixas latitudes. *In: Recursos genéticos e melhoramento de plantas para o nordeste brasileiro.* Available in http://www.cpatsa.embrapa.br/catalogo/livrorg/temas.html , 15.

[10] Carrão-Panizzi, M.C., & Mandarino, J.M.G. (1998). Soja: potencial de uso na dieta brasileira. *In: EMBRAPA SOJA. Documento 113. Londrina: Embrapa Soja.*

[11] Carrão-Panizzi, M.C., Mandarino, J.M.G., Bordingnon, J. R., & Kikuchi, A. (2000). Alternativa alimentar na dieta humana. *In: A cultura da soja no Brasil. Londrina: Embrapa Soja,. CD-ROM.*

[12] Cecchi, H.M. (2003). Fundamentos teóricos e práticos em análise de alimentos. 2 ed.., 122-133.

[13] Chan, S., Ho, S. C., Kreiger, N., Darlington, G., So, K. F., & Chong, P. Y. Y. (2007). Dietary sources and determinants of soy isoflavone intake among midlife chinese women in Hong Kong. *Journal of Nutrition.*, 137, 2451-2455.

[14] Chang, Y. K. (2001). Alimentos Funcionais e AplicaçãoTecnológica: Padaria de Saúde e Centro de Pesquisas em Tecnologia de Extrusão. *In: EMBRAPA Soja. (Org.). Anais do I Simpósio Brasileiro sobre os benefícios da soja para a Saúde Humana. Londrina: Embrapa da soja,* 41-45.

[15] Dutcosky, S. D. (2007). Análise Sensorial de alimentos. *2 ed., Curitiba: Champagnat.*

[16] Embrapa soja. (2010). Cultivares de soja 2010/2011 região centro-sul. *Londrina: Embrapa Soja: Fundação Meridional,* 60.

[17] Embrapa soja. (2003). Manual de receitas com soja. Londrina: Embrapa Soja Documentos, 206 , 60.

[18] Esteves, E. A., & Monteiro, J. B. R. (2001, jan-abr.) Efeitos benéficos das isoflavonas de soja em doenças crônicas. *Revista de Nutrição,Campinas,* 14(1), 43-52.

[19] Farias, J. R. B. (1994). Climatic requirements. *In: EMBRAPA-CNPSo. (Ed.) Tropical soybean: improvement and production. Rome: FAO,* 13-17.

[20] Friedman, N. M., & Brandon, D. L. (2001). Nutritional and health benefits of soy proteins. *Journal of Agricultural and Food Chemistry,* 49, 1069-1086.

[21] Genovese, M. I., & Lajolo, F.M. (2006, jan./fev.) Fatores antinutricionais da soja. *Informe Agropecuário, Belo Horizonte,* 27(230), 18-33.

[22] Genovese, M. I., & Lajolo, F. M. (2002). Isoflavones in soy based foods consumed in Brazil: levels, distribution and estimated intake. *Journal of Agricultural and Food Chemistry,* 50(21), 5987-5993.

[23] Hodges, T., & FRENCH, V. (1985). Soyphen: soybean growth stages modeled from temperature, day length and water availability. *Agronomic Journal,* 77, 500-505.

[24] Jackson, J. C., Dini, J. P., Lavandier, C., Rupasinghe, H. P. V., Faulkner, H., Poysa, V., Buzzell, D., & DeGrandies, S. (2002). Effects of processing on the content and composition of isoflavones during manufacturing of soy beverage and tofu. *Process Biochemistry*, 37, 1117-1123.

[25] Kagawa, A. (1995). Standard table of food composition in Japan. *Tokyo University of Nutrition for women*, 104-105.

[26] Kurzer, M. S., & Xu, X. (1997). Dietaryphytoestrogens. *Annual Review Nutrition*, 17, 353-381.

[27] Laudanna, E. (2006, jan.-fev.) Propriedades funcionais da soja. *Informe Agropecuário, Belo Horizonte*, 27(230), 15-18.

[28] Mandarino, J. M. G. (2010). Compostos antinutricionais da soja: caracterização e propriedades funcionais. *In: COSTA, N.M.B.; ROSA, C.O.B. (ed.). Alimentos funcionais: componentes bioativos e efeitos. Rio de Janeiro: Rubio*, 177-192.

[29] Morais, A. A. C., & Silva, A. L. (1996). Complicações e resistência ao consumo. *In: MORAIS, A.A.C.; SILVA, A.L.. Soja: suas aplicações. Rio de Janeiro: Medsi*, 151-155.

[30] Morais, A. A. C., & Silva, A. L. (2000). Valor nutritivo e funcional da soja. *Rev. Bras. Nutr. Clin*, 14, 306-315.

[31] Poysa, V., & Woodrow, L. (2002). Stability of soybean seed composition and its effect on soymilk and tofu yield and quality. *Food Research International, Barking*, 35(4), 337-345.

[32] Ravelo, A. C., & Decker, W. L. (1979). Soybean weather analysis models. *In: 14ᵗʰConf. Agric. Forest Meteorol*, 72-74.

[33] Rocha, V. S. (1996). Cultura. *In: MORAIS, A.A.C.; SILVA, A.L. Soja: suas aplicações. Rio de Janeiro: Medsi*, 29-66.

[34] Sediyama, T. (2009). Tecnologias de produção e usos da soja. *Londrina: Mecenas*, 306.

[35] Silva, M. R., & Silva, M. A. P. (2000). Fatores antinutricionais: inibidores de proteases e lectinas. *Rev. Nutr.*, 13(1), 3-9.

[36] Santos, H. M. C., Oliveira, M. A., Oliveira, A. F., & Oliveira, G. B. A. (2010, jul-dez). Composição centesimal das cultivares de soja BRS 232, BRS 257 e BRS 258 cultivadas cm sistema orgânico. *Revista Brasileira de Pesquisa em Alimentos, Campo Mourão*, 1(2), 07-10.

[37] Teixeira, R. C., Sediyama, H. A., & Sediyama, T. (2009). Composição, valor nutricional e propriedades funcionais. *In: SEDIYAMA, T. Tecnologias de produção e usos da soja. Londrina: Mecenas*, 247-259.

[38] Vasconcelos, I.M., Siebra, E. A., Maia, A. A. B., Moreira, R. A., Neto, A. F., Campelo, G. J. A., & Oliveira, J. T. A. (1997). Composition, toxic and antinutritional factors of

newly developed cultivars of Brazilian soybean (Glycine max). *Journal of the Science of Food and Agriculture*, 75, 419-426.

[39] Vasconcelos, VI. M., Maia, A. A. B., Siebra, E. A., Oliveira, J. T. A., Carvalho, A. F. F. U., Melo, V. M. M., Carlini, C. R., & Castelar, L. I. M. (2001). Nutritional study of two Brazilian soybean (Glycine max) cultivars differing in the contents of antinutritional and toxic proteins. *Journal of Nutritional Biochemistry*, 12, 1-8.

[40] Vasconcelos, I. M., Campello, C. C., Oliveira, J. T. A., Carvalho, A. F. U., Souza, D. O. B., & Maia, F. M. M. (2006). Brazilian soybean Glycine max (L.) Merr.cultivars adapted to low latitude regions: seed composition and content of bioactive proteins. *Rev. Bras. Bot., São Paulo*, 29(4).

[41] Wilcox, J. R., & Shibles, R. M. (2001). Interrelationships among seed quality attributes in soybean. *Crop Science*, 41, 11-14.

[42] Zava, D. T., & Duwe, G. (1997). Estrogenic and antiproliferative properties of genistein and other flavonoids in human breast cancer cells in vitro. *Nutrition and Cancer, Hillsdale*, 27(1), 31-40.

[43] Zeller, F. J. (1999). Soybean (Glycine max (L.) Merril): utilization, genetics, biotechnology. *Bodenkultur*, 50, 191-202.

Chapter 22

Effects of Soybean Trypsin Inhibitor on Hemostasis

Eugene A. Borodin, Igor E. Pamirsky,
Mikhail A. Shtarberg, Vladimir A. Dorovskikh,
Alexander V. Korotkikh, Chie Tarumizu,
Kiyoharu Takamatsu and Shigeru Yamamoto

Additional information is available at the end of the chapter

http://dx.doi.org/10.5772/52600

1. Introduction

Soy bean trypsin inhibitor (SBTI) belongs to the family of serpins – serine protease inhibitors widely distributed in the nature [Silverman G.A. et al., 2004; 21]. Serpins participate in the regulation of proteopytic reactions underling very important physiological and pathological processes such as digestion [16], blood clotting [3, 14, 17], immunity [44] apoptosis [36, 42], inflammation [10], dystrophy [31], carcinogenesis [22, 11] and so on. Despite the apparent importance of the correction of imbalances of the proteolysis in pathology, in fact only the pancreatic trypsin inhibitor (aprotinin) is used more or less widely as a protease inhibitor drug (Trasisol, Contrical, Gordox etc) in the treatment of some diseases [24, 13, 40]. Being the animal protein aprotinin possesses substantial disadvantages [23, 12] and attempts to develop new drug on the base of plant or recombinant proteins or peptidomimetics are undertaken [35, 27]. SBTI is one of the candidates for such a role [34].

Among proteolytic processes hemostasis is of special importance because of the high frequency of it's disturbances, accompanied many pathological processes and diseases [20, 18]. At first sight it looks strange to expect that plant protease inhibitor SBTI should influence hemostasis in humans because blood clotting represents the cascade of proteolytic reactions catalyzed by highly specific proteases [43]. However, results of the very early studies support such possibility [26, 41]. The present study was aimed to investigate the influence of SBTI and aprotinin on hemostasis using modern bioinformatics approach and classical in vitro methods. The concrete purposes of the study included:

© 2013 Borodin et al.; licensee InTech. This is a paper distributed under the terms of the Creative Commons Attribution License (http://creativecommons.org/licenses/by/3.0), which permits unrestricted use, distribution, and reproduction in any medium, provided the original work is properly cited.

1. To establish the extent of structural homology, compare functional activities and evaluate potential targets of SBTI and aprotinin among the human proteases by bioinformatics (*in silico*) methods.

2. To investigate the influence of trypsin, SBTI and aprotinin on some indexes of blood clotting, fibrilolysis and platelet aggregation as well as on the hemolytic activity of the complement *in vitro*.

3. To elucidate the possibility of the regulation of the total proteolytic and tripsin-inhibiting activity of blood plasma *in vivo* by the consumption of soy foods enriched with soy protein isolate (SPI) possessing thermo-stable fraction of SBTI.

2. The study of aprotinin and SBTI *in silico*

In silico methods (bioinformatics) entails the creation and advancement of databases, algorithms, computational techniques to solve formal and practical problems arising from the management and analysis of biological data. One of the potential areas for exploiting these methods is computational drug design. We used *in silico* methods in our study for the comparison of structural homology, functional activities and evaluation of the potential targets of SBTI and aprotinin among the human proteases.

From fifty databases we had browsed in the INTRNET nine contained information on aprotinin and SBTI (Table 1). From these PDB (Protein Data Bank) and UniProt/Swiss-Prot were the most informative and we used these databases in our study.

Database	ID in the Database	
	Aprotinin	**SBTI**
UniProt-Swiss-Prot http://www.expasy.org	BPT1_BOVIN (P00974)	ITRA_SOYBN (P01070)
Blocks - most highly conserved regions of proteins http://www.ebi.ac.uk	P00974	IPR002160
COG - the database of Clusters of Ortologous Groups of proteins http://www.ncbi.nlm.nih.gov	Precursor P00974; NP_001001554;	-
GTOP - Genomes TO Protein structures and functions http://spock.genes.nig.ac.jp	btau0:ENSBTAGU0000017328	?atha0:At1g17860.1
iProClass - an integrated, comprehensive and annotated Protein Classification database http://pir.georgetown.edu	P00974/BPT1_BOVIN; PIRSF001621	-
LIGAND - LIGAND chemical database for enzyme reactions	50059016; 3809839; bta:616039; 100156830; Bt.32343;BPT1_BOVIN	-

Database	ID in the Database	
	Aprotinin	SBTI
http://www.pasteur.fr		
MMDB - Molecular Modeling Data Base http://www.ncbi.nlm.nih.gov	P00974 (Precursor); 1QLQ0	1AVU
PDB - Protein Data Bank http://www.rcsb.org	1OA6	1AVU; 1BA7
MEROPS (peptidases) http://merops.sanger.ac.uk	I02.001	I03.001

Table 1. Databases containing the information on aprotinin and SBTI.

2.1. The homology of the primary structures of SBTI and aprotinin

To investigate the homology of the primary structures of SBTI and aprotinin we used BLAST (Basic Local Alignment Search Tool) algorithm, sequence alignment editor Bio Edit 5.0.9 and got information on these proteins from Protein Data Bank and Uni Prot using FAS-TA (Table 2), MOL and PDB format-files. The extent of homology of the total sequences of SBTI and aprotinin, calculated by the method of the multiple paired alignment, makes up 10% (Figures 1 and 2). The low homology of the total sequences of should be attributed to the different lengths of polypeptide chains these proteins – 58 amino acids in aprotinin [45] and 181 in SBTI [39].

Traditional abbreviation	Ala	Asn	Cys	Asp	Glu	Phe	Gly	His	Ile	Lis	Met
FASTA fromat	A	B	C	D	E	F	G	H	I	K	M
Traditional abbreviation	Asn	Pro	Gln	Arg	Ser	Tre	Val	Trp	Tyr	Glu	*
FASTA fromat	N	P	Q	R	S	T	V	W	Y	Z	-

Table 2. Designation of aminoacids in FASTA format. *- symbol in FASTA format corresponds to the gap of unlimited length

Comparison of the separate domains of SBTI and aprotinin reveals greater sequence similarity. Thus, the extent of homology of the N-terminal region of SBTI (5-60 amino acid residues) and the sequence of aprotonin without last two amino acids makes up 21%. The central region of SBTI (85-116 amino acid residues) and the region of aprotinin within 25-56 amino acids show 24% similarity. The highest extent of homology up to 35% is characteristic for C-terminal region of SBTI (132-181 amino acid residues) and aprotenin sequence without six C-terminal amino acids.

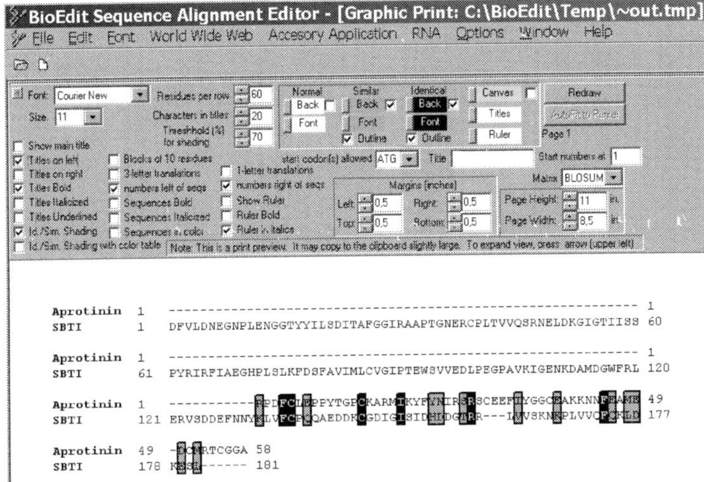

Figure 1. The screen of the Bio Edit 5.0.9 with the results of calculation of homology of Aprotinin and SBTI. The identical amino acids are marked with black and similar with gray. Amino acid sequence are represented in FASTA format.

Figure 2. Homological regions of Aprotinin and SBTI (according to Bio Edit 5.0.9). The identical amino acids are marked with black and similar with gray.

2.2. Tertiary structures of SBTI and aprotinin

For the visualization of the 3D-structures of SBTI and aprotinin we used Chem Office 5.0 and Yasara 6.2.5 software. The files in MOL format, containing the information on the sequences of these protease inhibitors, were imputed into Chem Office 5.0 software and converted in PDB format. The obtained PDB files were imputed into Yasara 6.2.5 software. The results of the generation of the electronic 3D-structures of aprotinin and SBTI are presented in Figure 3. Apparently there there is similarity of 3D-structures of these proteins.

We failed to find molecular targets of SBTI and aprotinin because there was no necessary software in free excess.

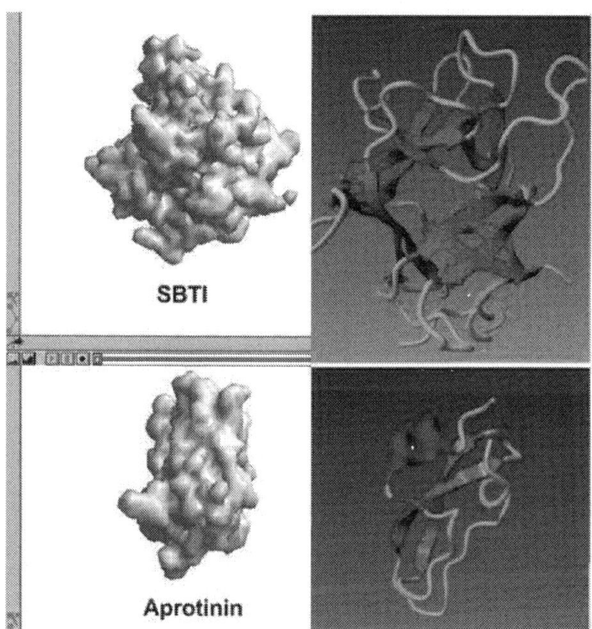

Figure 3. Electronic 3D-structures of SBTI and aprotinin. Left – model with the single surface (Chem Office 5.0). Right-ribbon diagram (Yasara 6.2.5).

2.3. Spectra of biological activity of SBTI and aprotinin

The exploiting the PASS software shows four potential activities of aprotinin and three for SBTI. Both SBTI and aprotinin may inhibit rennin as well as angiotensin- and endothelein-converting enzymes according to the revealing of the possible biological activities of these protease inhibitors by ISIS/Draw 2.4 and PASS Professional chemical structure drawing programs (Figure 4). The possibility of exerting of such effects (drug-likeness) is nearly the same in both protease inhibitors. Aprotinin may inhibit neutral endopeptidase, also. For all the abovementioned activities probability of the activity (P_a) is higher than the probability of the lack of activity (P_i). SBTI and aprotinin should not reveal serious toxicity, mutagenic, carcinogenic and teratogenic effects according to in silico, in vitro and in vivo studies.

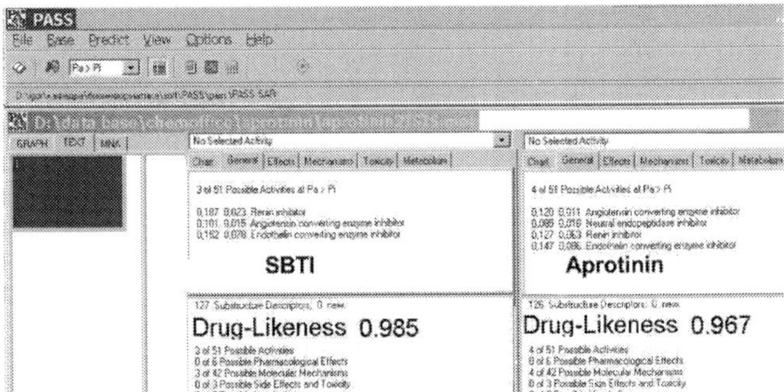

Figure 4. Spectra of the biological activities of aprotinin and SBTI according to PASS software. Drug-Likeness – probability of exerting the effect. Probability of the activity (P_a) and the lack of activity (P_i). The average accuracy of the prognosis makes up more than 85%.

3. The study of aprotinin and SBTI *in vitro*

In silico methods testify the common features in the primary structure, 3D-structures, functional activities of t SBTI and aprotinin and allow to propose that SBTI should influence processes of hemostasis similar to aprotinin. To prove this assumption we studied the influence of trypsin, SBTI and aprotinin on blood clottting, fibrilolysis and platelets aggregation *in vitro*.

3.1. Trypsin-inhibitory activity of SBTI and aprotinin

We measured tyipsin-inhibitory activity of aprotinin, SBTI and low molecular weight protease inhibitors (ε-aminocapronic acid and D-lysine) by their ability to inhibit BAEE-esterase activity of the trypsin [33]. The results are presented in the table 3.

Trypsin inhibitor	Aprotinin	SBTI	ε-aminocapronic acid	D-lysine
Trypsin-inhibitory activity (IU/mg)	141±13	61±0,9	1,36±0,04	1,31±0,02
Trypsin-inhibitory activity (IU/nmole)	0,922±0,085	1,23±0,02	-	-

Table 3. Trypsin-inhibitory activity of aprotinin, SBTI, ε-aminocapronic acid and D-lysine.

Trypsin-inhibitory activity aprotinin, SBTI, ε-aminocapronic acid and D-lysine was measured as their ability to inhibit BAEE-esterase activity of trypsin.

Tyipsin-inhibitory activity, calculated per mg of inhibitor, decrease down in the order: aprotinin > SBTI > aminocapronic acid > D-lysine. Thus, tyipsin-inhibitory activity of aprotinin is

2-times higher than similar activity of SBTI. However, taking into account that molecular weight of SBTI (20100) is 3-times higher than molecular weight of aprotinin (6514), it should be concluded that the trypsine-inhibitory activity of SBTI, calculated per mole of the inhibitor, should be nearly 1,5-times higher than trypsine-inhibitory activity of aptotinin. Trypsine-inhibitory activity of ε-aminocapronic acid and D-lysine when calculated per mole of the inhibitor is negligibly low because of the low molecular weight.

3.2. The influence of SBTI and aprotinin on coagulation hemostasis

The following indexes, characterizing mainly the first and the second phases of blood clotting, were measured: prothrombin time (reflects activity of coagulation factors YII, V, X and II), activated partial thromboplastin time (reflects activity of coagulation factors YII and VIII), activated clotting time (a measure of the anticoagulation affects of heparin) and thrombin time (time of the formation of fibrin clot). Results are summarized in the table 4.

Indexes	Time,sec			
	Plasma (control) (1)	Plasma + trypsin (2)	Plasma + aprotinin (3)	Plasma + SBTI (4)
Prothrombin time	$20\pm0,9$	$13\pm0,9$ $P_{1-2}<0,0001$	$21\pm0,9$ $P_{1-3}"/0,05$	$31\pm0,9$ $P_{1-4}<0,0001$
Activated partial thromboplastin time	$36\pm1,7$	$3\pm0,9$ $P_{1-2}<0,0001$	$113\pm1,8$ $P_{1-3}<0,0001$	$105\pm2,7$ $P_{1-4}<0,0001$
Thrombine time	$16\pm0,9$	$2,5\pm0,5$ $P_{1-2}<0,0001$	$24\pm0,9$ $P_{1-3}<0,0001$	∞
Ativated clotting time	$88\pm9,7$	$58\pm4,7$ $P_{1-2}<0.002$	157 ± 12 $P_{1-4}<0.0001$	258 ± 17 $P_{1-4}<0.001$
Euglobulin lysis time	$400\pm34,7$	182 ± 7 $P_{1-2}<0,0001$	No lysis of blood clot	No lysis of blood clot

Table 4. The influence of trypsin, SBTI and aprotinin on the indexes of hemostasis *in vitro*. The final concentration of trypsin in the incubation medium consisted of 0,01%, SBTI and aprotinin - 0,1%..

Prothrombin time decreases 33-35% in the presence of trypsin from $20\pm0,9$ to $13\pm0,9$ sec, increases insignificantly in the presence of aprotinin and increases 1,5-times up to $31\pm0,9$ sec in the presence of SBTI. Activated partial thromboplastin time sharply decreases by trypsin from $36\pm1,7$ sec in the control samples to $3,0+0,9$ sec and increases 3-times by aprotinin ($113\pm1,8$ sec) and SBTI ($105\pm2,7$ sec). The changes of the activated clotting time under the influence of trypsin, aprotinin and SBTI are similar. Thrombin time decreases nearly 7-fold in the presence of trypsin from $16\pm0,9$ to $2,5\pm0.5$ sec and increases on 50% by aprotinin to $24\pm0,9$ sec. SBTI completely blocked the formation of fibrin clot at least within 10 min. Similar effects trypsin and it's inhibitors exerts on fibrinolysis. Euglobulin lysis time (factor XII–

callicrein-dependent fibrinolysis) decreases 60% by trypsin from 400±35 to 182±7 sec. Aptotinin and SBTI in the abovementioned concentrations entirely inhibit the lysis of fibrin clot (Table 4).

3.3 The influence of SBTI and aprotinin on platelet aggregation

Results of the study of the influence of SBTI and aprotinin on platelets aggregation are presented in the table 5 and Figures 5-8.

	Plasma (control) (1)	Plasma + aprotinin (2)	Plasma + SBTI (3)	Plasma + trypsin (4)
Reversible (ADP)	22±1,8	14±0,9 $P_{1-2}<0,01$	15±0,9 $P_{1-3}<0,01$	34±0,9 $P_{1-4}<0,01$
T_{MA}	55±4,7	55±4,7	55±4,7	65±15
Two-phase (ADP):				
Fist phase				
MA	48±1,8	35±0,9 $P_{1-2}<0,01$	34±1 $P_{1-3}<0,01$	57±3 $P_{1-4}<0,01$
TMA	85±4,7	68±10,5	70±9	80±10
Second phase				
MA	70±2	56±2 $P_{1-2}<0,02$	58±1 $P_{1-3}<0,05$	99±1 $P_{1-4}<0,02$
TMA	350±9,2	325±9,2	320±9,3	310±9,2
Irreversible (ADP)				
MA	49±0,9	33±0,9 $P_{1-2}<0,02$	29±0,9 $P_{1-3}<0,02$	100 $P_{1-4}<0,001$
T_{MA}	400±18	410±9,3	410±9,3	210±26,4of
Two-phase (adrenalin):				
Fist phase				
MA	23±0,9	16±1,8 $P_{1-2}<0,02$	16±1,8 $P_{1-3}<0,02$	32±0,9 $P_{1-4}<0,01$
TMA	110±9,3	125±13,1	125±13,1	130±9,3
Second phase				
MA	47±2,4	44±2 $P_{1-2}"/0,05$	44±1,8 $P_{1-3}"/0,05$	55±0,9 $P_{1-4}<0,02$
TMA	390±9,3	430±18	430±18	435±16,2

Table 5. The effects of aprotinin, SBTI and trypsin on the aggregation of platelets initiated by ADP (reversible, two-phase and irreversible aggregation) and adrenalin (two-phase aggregation). MA – maximum aggregation (transmittance, %), T_{MA} – time of accomplishment of maximum aggregation (sec).

The effects of trypsin, SBTI and aprotinin on the aggregation of platelets initiated by ADP (reversible, two-phase and irreversible aggregation) and adrenalin (two-phase aggregation) are similar to their effects on blood coagulation and fibrinolysis. Trypsin increases platelet aggregation (the increase of the maximum of all types of aggregation and decrease of maximum aggregation time of irreversible ADP-initiated aggregation). SBTI and aprotinin reveals anti-aggregation effect manifested by the decrease of maximum aggregation.

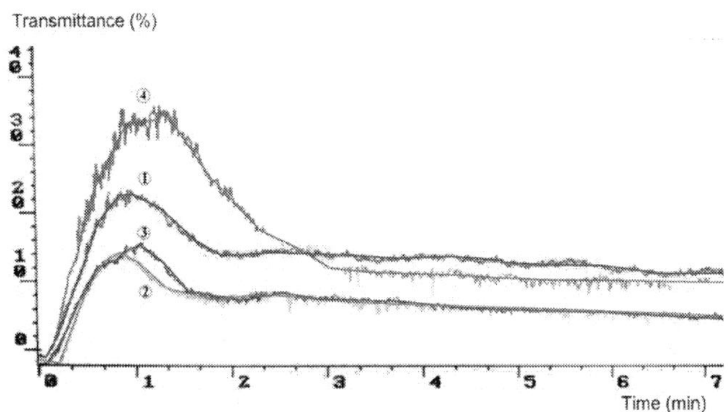

Figure 5. ADP-initiated reversible aggregation of platelets in the presence of aptotinin, SBTI and trypsin in vitro. The platelet aggregation was initiated by the addition of the 5 mkM ADP. 1 – controle; 2 – aprotinin - 1%; 3 – SBTI - 1%; 4 – trypsin – 0,1%.

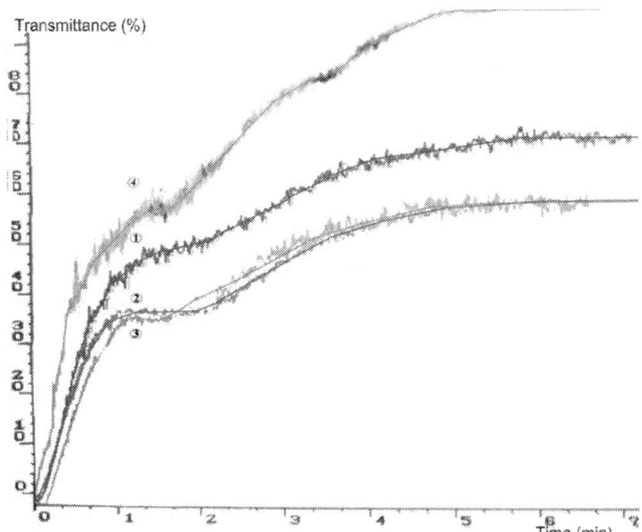

Figure 6. ADP-initiated two-phase aggregation of platelets in the presence of aptotinin, SBTI and trypsin in vitro. The platelet aggregation was initiated by the addition of the 5 mkM ADP. 1 – controle; 2 – aprotinin - 1%; 3 – SBTI - 1%; 4 – trypsin – 0,1%.

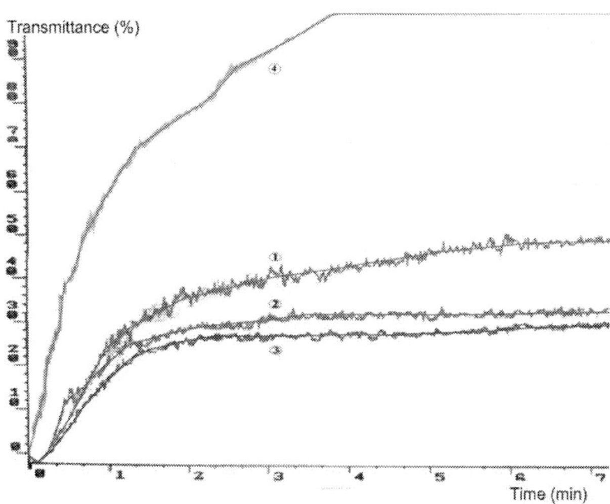

Figure 7. ADP-initiated irreversible aggregation of platelets in the presence of aptotinin, SBTI and trypsin in vitro. The platelet aggregation was initiated by the addition of the 25 mkM ADP. 1 – controle; 2 – aprotinin - 1%; 3 – SBTI - 1%; 4 – trypsin – 0,1%.

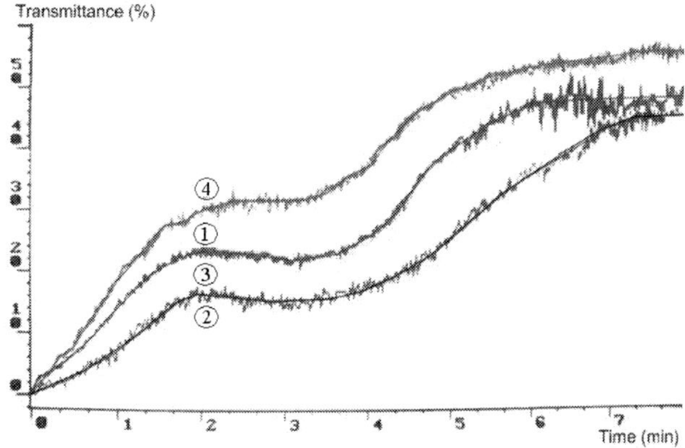

Figure 8. Adrenalin-initiated two-phase aggregation of platelets in the presence of trypsin, aptotinin and SBTI in vitro. The platelet aggregation was initiated by the addition of the 25 mkM adrenalin. 1 – controle; 2 – trypsin - 0,1%; 3 – aprotinin - 1%; 4 – SBTI - 1%.

So, the results of *in vitro* experiments proves the ability of SBTI to influence the hemostatsis predicted by the *in silico* studies.

3.4. The influence of SBTI and aprotinin on the hemolytic activity of the complement

Limited cascade proteolysis underlies the activation of the complement system, which includes nearly 20 individual proteins [30, 37]. Majority of these proteins belongs to the serine protease family (subunits C1r and C1s, C3/C5-cinvertase of the classical way of activation, factors I and D) [15]. While complement system does not have direct relation to hemostasis and targeted on alien cells and microorganisms, in some conditions complement system may play a role in the disruption of intrinsic cells of the organism and among them blood cells [15, 37]. Because of this we investigated the influence of trypsin, SBTI and aprotinin on the hemolytic activity of the complement assuming that SBTI and aprotinin may retard hemolysis of red blood cells by inhibiting activation of some components of the complement system. Results are presented in the table 6 and Figures 9-10.

	Lag-phase (min)	Total time of hemolysis, including lag-phase (min)
Control (1)	3,5±0,5	8,5±1,5
Aprotinin (1%) (2)	3,2±0,15 $P_{1\text{-}2}$"/0,05	7,25±0,25 $P_{1\text{-}2}$"/0,05
SBNI (1%) (3)	3,2±0,15 $P_{1\text{-}3}$"/0,05	7,25±0,25 $P_{1\text{-}3}$"/0,05
Trypsin(0,01%) (4)	4,25±0,25 $P_{1\text{-}4}$<0,02	15,5±0,5 $P_{1\text{-}4}$<0,02

Table 6. The influence of aprotinin, SBTI and trypsin on the hemolytic activity of complement *in vitro*

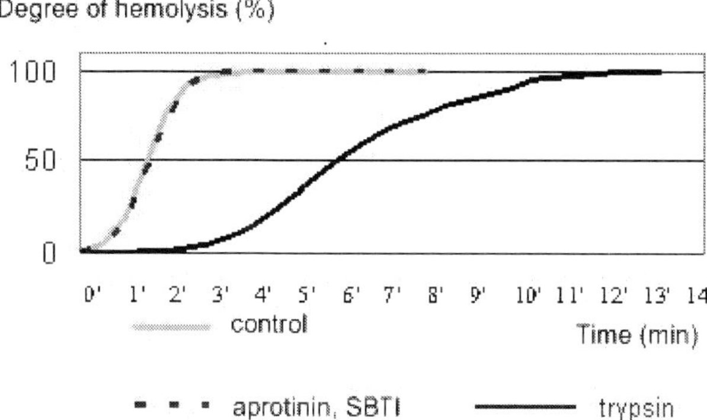

Figure 9. The hemolytic activity of the complement in the presence of aprotinin, SBTI and trypsin in vitro. The lag-phase is not shown. 1 – controle; 2 – aprotinin - 1%; 3 – SBTI - 1%; 4 – trypsin – 0,1%.

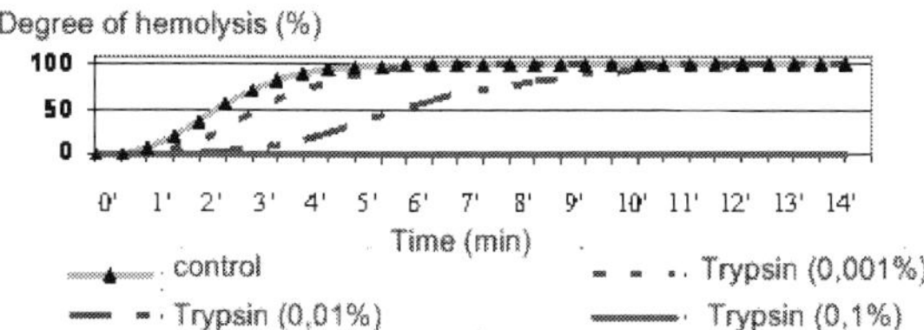

Figure 10. The hemolytic activity of the complement in the presence of different concentrations of trypsin in vitro. The lag-phase is not shown.

SBTI and aprotinin in the final concentrations 0,001-0,1% do not influence hemolysis time and the only statistically insignificant effect consists of small decrease of lag-phase from 3,5±0,3 to 3,2±0,15 min (p>0,05) and total time of hemolysis from 8,5±1,5 to 7,25±0,15 min (p>0,05). Trypsin suppresses the hemolytic activity of the complement in the concentration-dependent manner. In the presence of 0.0001% and 0,001% trypsin the lag-phase of the hemolysis increases to 4,25±0,25 min (p<0,02) and total time of hemolysis to 11,5±0,5 min (p<0,02) and 15,5±1,6 min (p<0,01), respectively. 0,01% trypsin entirely blocks the hemolysis. The inhibiting effect of trypsin should be attributed to the disruption of C3 component of the complement system.

Thus, the results obtained show that aprotinin and SBTI in the abovementiioned concentrations do not influence the hemolytic activity of the complement.

4. The influence of the consumption of soy priotein isolate by healthy persons on the indexes of proteolysis in the blood serum

Soya is cultivated as a foodstuff, having favorable effect on human health, more than 3000 years [28]. Within the last decades soy foods have started to be used in Europe, USA and in Russia as dietotherapy means at a number of diseases [6, 29]. In particular, we have shown the cholesterol-lowering effect of soy protein, resulting from the consumption of SPI enriched cookies by persons with modern hypelipidemia [7, 9]. The consumption of soy foods is followed by the antioxidant effect [Santana et al., 2008, 8, 9] due to the high antioxidant content in these foods [2, 9]. Antioxidant effect is important for the treatment of diseases followed by the development of the oxidative stress [1, 38].

The other possible use of soy foods in dietotherapy may be the correction of disturbances of proteolytic processes. Activation of the tissue proteolysis accompanies different diseases and pathological processes [34]. Soy protein foods contain Bouman-Birk type trypsin inhibitor – SBTI [4]. This protease inhibitor has been officially recognized as a component of foodstuff [25]. There are data showing the possibility of absorbtion of SBTI in the intestin after the consumption of soy protein foods [22]. So, it seams reasonable to assume that the consumption of soy foods will follow anti-proteolytic effect. Because of this it was interesting to elucidate whether prolonged consumption of the cookies enriched with soy protein isolate (SPI) will influence total proteolytic and trypsin-inhibiting activity of the blood plasma in humans or not?

4.1. Tripsin-inhibiting activity of SPI

According to the characteristics, specified by manufacturer, protein content of SPI consists of 90%. From this 10% belongs to protease inhibitors [32]. Within the processing of soy beans into SPI the former are expoused to high temperatures, which inactivates the major part of soy proteins. However, about 5-20 % of SBTI are represented by the thermostable fraction [19, 22]. The measurement of the trypsin-inhibiting activity of SPI shows that trypsin-inhibiting activity of SPI consists of 1,4±0,1 IU/mg of SPI. For example 1 mg of pure SBTI possesses activity equal nearby 60 IU. Thus, the active SBTI makes up 2,7% from SPI mass.

4.2. The influence of the consumption of SPI by healthy persons on the total proteolytic and trypsin inhibiting activity of blood serum

30 adult people aged 35-67 years without the expressed signs of chronic diseases consumed cookies, enriched with SPI (30% protein content), for two months [7]. Fasting blood samples were drawn before and after the dietary treatment. Serum samples were frozen and analyzed for total proteolytic (BAEE-esterase activity) and tripsin-inhibiting activity [33]. Daily intake of 100g of cookies corresponds to consumption of 30g of SPI and about 0,85g of the active SBTI (nearly 52 00 IU). The total consumption of SPI for two months consists of 1,8 kg or 50 g of an active SBTI. Twenty-eight participants (19 females and 9 males) could complete the trial. Results are presented in the table 7.

	Total proteolytic activity (relative units)	Trypsin-inhibitory activity (IU/ml)
Before the consumption	0,343±0,010	113±3,6
After the consumption	0,282±0,008 p<0,05	137±5,3 p<0,01

Table 7. Total proteolytic and trypsin-inhibitory activity of blood serum of healthy persons before and after two months consumption of soy protein isolate.

Total proteolytic activity of blood serum was measures as serum BAEE-esterase activity and trypsin-inhibitory activity by inhibiting by serum of BAEE-esterase activity of trypsin.

Two months long SPI consumption was followed by 18% decrease of total proteolytic activity of blood serum from 0,343±0,010 to 0,282±0,008 relative units (p<0,05) and by 21% increase of trypsin-inhibiting activity from 113±3,6 to 137±5,3 IU/ml (p<0,01). The results obtained show possibility of the regulation of proteolysis *in vivo* by the consumption of soy foods.

5. Discussion

The disturbances of the hemostasis accompany many pathological processes and diseases [20, 18]. Both blood clotting and fibrinolysis represents the cascade of proteolytic reactions catalyzed by highly specific proteases [43]. To correct disturbances of the fibrinolysis as well as imbalances of the other proteolytic processes the animal trypsin inhibitor aprotinin is used as a drug (Trasisol, Contrical, Gordox etc) [24, 13, 40]. The effectiveness of the aprotinin in the treatment of some diseases has been questioned recently because of it's substantial disadvantages [23, 12]. Following consultation with the German Federal Institute for Drugs and Medical Devices, the U.S. Food and Drug Administration, Health Canada, and other health authorities, the producer of Trasilol - Bayer announced in 2007 that it has elected to temporarily suspend worldwide marketing of Trasylol® (aprotinin injection) until final results from the Canadian BARTtrial can be compiled, received and evaluated. Information regarding the decision has been posted to Bayer's websites [5]. Because of this attempts to develop new protease inhibitor drug are undertaken [35, 27]. One of the candidates for such a role is plant protease inhibitor - SBTI.

In the present study we exploited the advantages of the modern bioinformatics for establishing the extent of structural homology, comparing functional activities and evaluating potential targets of SBTI and aprotinin among the human proteases. The results of *in silico* study testifies apparent homology of SBTI and aprotinin manifested by the common features in the primary structure, 3D-structures, functional activities of these proteins and allow us to propose that SBTI should influence processes of hemostasis similar to aprotinin. The investigation of the influence of SBTI and aprotinin on coagulation and thrombocyte hemostasis by *in vitro* methods prove this assumption and show that both proteins inhibit blood clotting, fibrinolusis and platelet aggregation which is evident from the increase of prothrombin time, activated partial thromboplastin time, activated clotting time, thrombin time and inhibition of fibrinolysis. We investigated the influence of SBTI and aprotinin on the hemolytic activity of the complement assuming that SBTI and aprotinin may retard hemolysis of red blood cells by inhibiting activation of some components of the complement system. However, both inhibitors do not influence hemolysis time.

While the major part of SBTI is disrupted by heating, nearly 20% of inhibitor are thermostable and remains active in soy foods [19]. Part of the consumed SBTI are absorbed in the intestine after the consumption of soy foods [22]. Because of this we elucidated whether prolonged consumption of the cookies, enriched with SPI, will influence total proteolytic and trypsin-inhibiting activity of the blood plasma in humans or not? Daily intake of 100g of cookies corresponds to consumption of about 0,85g of the active SBTI. The total consumption of SBTI for the study period consists of 50 g of an active inhibitor. The consumption of 100 g of cookies daily for two months was followed by 18% decrease of total proteolytic activity of blood serum and by 21% increase of trypsin-inhibiting activity. The results obtained testify possibility of the soft regulation of proteolysis *in vivo* by the consumption of soy foods.

At first, the ability of SBTI to inhibit blood clotting was shown more than half a centaury ago [26, 41]. However, these results had no any practical consequences and in fact were forgotten for a long time. Our study represent new attempt to revive interest to SBTI as possible protease inhibiting drug.

Author details

Eugene A. Borodin[1*], Igor E. Pamirsky[2], Mikhail A. Shtarberg[1], Vladimir A. Dorovskikh[1], Alexander V. Korotkikh[1], Chie Tarumizu[3], Kiyoharu Takamatsu[4] and Shigeru Yamamoto[3]

*Address all correspondence to: borodin@amur.ru

1 Amur State Medical Academy, Blagoveshchensk, Russian Federation

2 Institute of Geology and Natural Management of the Far Easteren Branch of Russia Academy of Sciences, Russian Federation

3 Asian Nutrition and Food Culture Research Center, Jumonji University, Saitama, Japan

4 Fuji Oil Co., Ltd Osaka, Japan

References

[1] Aivatidi, C., Vourliotakis, G., Georgopoulos, S., Sigala, F., Bastounis, E., & Papalambros, E. (2011, March). Oxidative stress during abdominal aortic aneurysm repair--biomarkers and antioxidant's protective effect: a review. *Eur. Rev. Med. Pharmacol. Sci.*, 15(3), 245-252, 1128-3602.

[2] Anderson, R. L., & Wolf, W. J. (1995, March). Compositional changes in trypsin inhibitors, phytic acid, saponins and isoflavones related to soybean processing. *J Nutr.*, 125(3), 581S-588S, 0022-3166.

[3] Bernstein, J. A. (2008). Hereditary angioedema: a current state-of-the-art review, VIII: current status of emerging therapies. *Ann. Allergy Asthma Immunol* [1, 2], S41-S46, 1081-1206.

[4] Birk, Y. (1985). The Bowman-Birk inhibitor. Trypsin- and chymotrypsin-inhibitor from soybeans. *Int J Pept Protein Res.*, 25(2), 113-31, 0367-8377.

[5] Bayer (2007). Available from http://www.trasylol.com http://www.pharma.bayer.com http://www.bayerscheringpharma.de/trasylol/en http://www.bayerhealthcare.com/trasylol/en

[6] Borodin, E. A., Aksyonova, T. V., & Anishchenko, N. I. (2000). Soy foods. New role. *Vestnik DVO RAN* [3], 72-85.

[7] Borodin, E. A., Menshikova, I. G., Dorovskikh, V. A., Feoktistova, N. A., Shtarberg, M. A., Yamamoto, T., Takamatsu, K., Mori, H., & Yamamoto, S. (2009, December). Ef-

fects of soy protein Isolate and casein on blood lipids and glucose in Russian adults with moderate hyperlipidemia. *J. Nutr. Sci. Vit*, 50(6), 492-497, 0301-4800.

[8] Borodin, E.A., Dorovskikh, V. A., Aksyonova, T. V., & Shtarberg, M. A. (2001, October-December). Lipid composition and the antioxidant properties of soy milk in vitro and in vivo. *Far Eastern Medical Journal (Russian)* [4], 26-30, 1994-5191.

[9] Borodin, E.A., Menshikova, I. G., Dorovskikh, V. A., Feoktistova, N. A., Shtarberg, M. A., Aksenova, T. V., Yamamoto, T., Takamatsu, K., Mori, H, & Yamamoto, S. (2011). Antioxidant and hypocholesterolemic effects of soy foods and cardiovascular disease. *In: Soybean and Health Ed. by Hany A. El-Shemy*. INTECH, 407-424, 978-9-53307-535-8.

[10] Correale, M., Brunetti, N. D., De Gennaro, L., & Di Biase, M. (2008). Acute phase proteins in atherosclerosis (acute coronary syndrome) Cardiovasc. *Hematol. Agents Med. Chem.*, 6(4), 272-277, 1871-5257.

[11] Clemente, A., Moreno, F. J., Marнn-Manzano, Mdel. C., Jimĭnez, E., & Domoney, C. (2010). The cytotoxic effect of Bowman-Birk isoinhibitors, IBB1 and IBBD2, from soybean (Glycine max) on HT29 human colorectal cancer cells is related to their intrinsic ability to inhibit serine proteases. *Mol. Nutr. Food. Res*, 54(3), 396-405, 1613-4125.

[12] Deanda, A. Jr, & Spiess, B. D. (2012, July). Aprotinin revisited. *J Thorac Cardiovasc Surg*, 13, 1344-4964.

[13] Dietrich, W. (2009, Feb). Aprotinin: 1 year on. *Curr Opin Anaesthesiol*, 22(1), 121-127, 0952-7907.

[14] Dupont, D. M., Madsen, J. B., Kristensen, T., Bodker, J. S., Blouse, G. E., Wind, T., & Andreasen, P. A. (2009). Biochemical properties of plasminogen activator inhibitor-1. *Front Biosci*, 1(14), 1337-1361, 1093-4715.

[15] Ehrnthaller, C., Ignatius, A., Gebhard, F., & Huber-Lang, M. (2011, March-April). New insights of an old defense system: structure, function, and clinical relevance of the complement system. *Mol Med.*, 17(3-4), 317-329, 1076-1551.

[16] Erickson, R. H., & Kim, Y. S. (1990). Digestion and absorption of dietary protein. *Annu Rev Med*, 41, 133-139, 0066-4219.

[17] Gray, E., Hogwood, J., & Mulloy, B. (2012). The anticoagulant and antithrombotic mechanisms of heparin. *Handb. Exp. Pharmacol.*, 207, 43-61, 0171-2004.

[18] Grabowski, E. F., Yam, K., & Gerace, M. (2012). Evaluation of hemostasis in flowing blood. *Am. J. Hematol.*, 87(1), 51-55, 0361-8609.

[19] Hathcock, J. N. (1991). Residue trypsin inhibitor: data needs for risk assessment. *Adv Exp Med Biol.*, 289, 273-279, 0065-2598.

[20] Hook, K. M., & Abrams, C. S. (2012, February). The loss of homeostasis in hemostasis: new approaches in treating and understanding acute disseminated intravascular coagulation in critically ill patients. *Clin Transl Sci.*, 5(1), 85-92, 1579-2242.

[21] Huntington, J. A. (2011, July). Serpin structure, function and dysfunction. *J Thromb Haemost.*, 9, 26-34, 1538-7933.

[22] Kennedy, A. R. (1998). Bowman-Birk inhibitor from soybeans as anticarcinogenic agent. *J. Clin. Nutr.*, 68(6), 1406-1412, 0002-9165.

[23] Kristeller, J. L., Roslund, B. P., & Stahl, R. F. (2008). Benefits and risks of aprotinin use during cardiac surgery. *Pharmacotherapy.*, 28(1), 112-124, 0277-0008, 1875-9114.

[24] Levy, J. H., & Sypniewski, E. (2004). Aprotinin: a pharmacologic overview. *Orthopedics.*, 27(6), 653-658, 0147-7447.

[25] Losso, J. N. (2008). The biochemical and functional food properties of the bowman-birkinhibitor. *Crit. Rev. Food Sci. Nutr.*, 48(1), 94-118, 1040-8398.

[26] Macfarlane, R.G., & Pilling, J. (1946). Anticoagulant action of soya-bean trypsin inhibitor. *Lancet* [2], 888, 0140-6736.

[27] Mazza, F., Tronconi, E., Valerio, A., Groettrup, M., Kremer, M., Tossi, A., Benedetti, F., Cargnel, A., & Atzori, C. (2006). The non-peptidic HIV protease inhibitor tipranavir and two synthetic peptidomimetics (TS98 and TS102) modulate Pneumocystis carinii growth and proteasome activity of HEL299 cell line. *J. Eukaryot. Microbiol.*, 53(1), S144-S146, 1066-5234.

[28] Messina, M., & Messina, V. (2003, May-June). Provisional Recommended Soy Protein and Isoflavone Intakes for Healthy Adults: Rationale. *Nutr Today*, 38(3), 100-109, 0002-9666X.

[29] Messina, M. (2010, July). A brief historical overview of the past two decades of soy and isoflavone research. *J. Nutr.*, 140(7), 1350S-1354S, 0022-3166.

[30] Mizuno, M., & Matsuo, S. (2010, January). Recent knowledge of complement system and the protective roles. *Nihon Rinsho.*, 68(6), 49-52, 0047-1852.

[31] Morris, C. A., Selsby, J. T., Morris, L. D., Pendrak, K., & Sweeney, H. L. (2010). Bowman-Birkinhibitor attenuates dystrophic pathology in mdx mice. *J. Appl. Physiol*, 109(5), 1492-1499, 8750-7587.

[32] Mosse, D., & Pernolle, D. (1986). *In: Chemisry and Biochemistry of legumes. Moscow, "Agropromizdat", Russian*, 248, 0369-8629.

[33] Nartikova, V. F., & Paskhina, T. S. (1977). Determination of anti-trypsine activity in human blood plasma. *In: Modern Methods in Biochemistry. Ed. By V.N. Orekhovich. Moskow, "Medicine" (Russian)*, 188-191, 0042-8787.

[34] Pamirsky, I., Borodin, E., & Shraberg, M. (2012). Regulation of proteolysis by plant and animal inhibitors. *LAP Lambert Academic Publishing.*, 105, 0000-9783.

[35] Pogue, G. P., Vojdani, F., Palmer, K. E., Hiatt, E., Hume, S., Phelps, J., Long, L., Bohorova, N., Kim, D., Pauly, M., Velasco, J., Whaley, K., Zeitlin, L., Garger, S. J., White, E., Bai, Y., Haydon, H., & Bratcher, B. (2010). Production of pharmaceutical-grade re-

combinant aprotinin and a monoclonal antibody product using plant-based transient expression systems. *Plant Biotechnol J.*, 8(5), 638-654, 1467-7644.

[36] Radović, N., Cucić, S., & Altarac, S. (2008). Molecular aspects of apoptosis. *Acta Med. Croatica.*, 62(3), 249-256, 1330-0164.

[37] Sarma, J. V., & Ward, P. A. (2011, January). The complement system. *Cell Tissue Res.*, 343(1), 227-235, 0030-2766X.

[38] Shi, Y. C., & Pan, T. M. (2012, April). Red mold, diabetes, and oxidative stress: a review. *Appl Microbiol Biotechnol.*, 94(1), 47-55, 0175-7598.

[39] Sigma. (2006). Catalog reagents ang labware. Sigma-Aldrich, Inc. 0013-7227 , 1440.

[40] Smith, M., Kocher, H. M., & Hunt, B. J. (2010, January). Aprotinin in severe acute pancreatitis. *J. Clin. Pract.*, 4(1), 84-92, 1368-5031.

[41] Tagnon, H. J., & Soulier, J. P. (1946). Anticoagulant activity of the trypsin inhibitor from soy bean flour. *Proc. Soc. Exper. Biol. & Med*, 61, 440, 0003-4819.

[42] Tang, M., Asamoto, M., Ogawa, K., Naiki-Ito, A., Sato, S., Takahashi, S., & Shirai, T. (2009). Induction of apoptosis in the LNCaP human prostate carcinoma cell line and prostate adenocarcinomas of SV40T antigen transgenic rats by the Bowman-Birk inhibitor. *Pathol Int.*, 59(11), 790-796, 13205-4630.

[43] Walsh, P. N., & Ahmad, S. S. (2002). Proteases in blood clotting. *Essays Biochem.*, 38, 95-111, 0071-1365.

[44] Wouters, D., Wagenaar-Bos, I., van Ham, M., & Zeerleder, S. (2008). C1 inhibitor: just a serine protease inhibitor? New and old considerations on therapeutic applications of C1 inhibitor. *Expert. Opin. Biol. Ther*, 8(8), 1225-1240, 1744-7682.

[45] Zhirnov, O. P., Klenk, H. D., & Wright, P. F. (2011, July). Aprotinin and similar protease inhibitors as drugs against influenza. *Antiviral Res.*, 92(1), 27-36, 16635429-9999.

Chapter 23

Potential Use of Soybean Flour (*Glycine max*) in Food Fortification

O. E. Adelakun, K. G. Duodu, E. Buys and
B. F. Olanipekun

Additional information is available at the end of the chapter

http://dx.doi.org/10.5772/52599

1. Introduction

Protein-energy malnutrition (PEM) results from prolonged deprivation of essential amino acids and total nitrogen and/or energy substrates. Growth which is a continuous process results from the complex interaction between inheritance and environment. Protein-energy malnutrition (PEM) also results from food insufficiency as well as from poor social and economic conditions [6, 20]. Malnutrition originates from a cellular imbalance between nutrient/energy supply and the body's demand to ensure growth and maintenance [22]. Dietary energy and protein deficiencies usually occur together, although one sometimes predominates the other and if severe enough, may lead to the clinical syndrome of kwashiorkor (predominant protein deficiency) or marasmus (mainly energy deficiency).

The origin of PEM can be primary, when it is the result of inadequate food intake, or secondary, when it is the result of other diseases that lead to low food ingestion, inadequate nutrient absorption or utilization, increased nutritional requirements and/or increased nutrient losses [13]. In as much as protein-energy malnutrition applies to a group of related disorders that develop in children and adults whose consumption of protein and energy is insufficient to satisfy the bodies nutritional needs, about one third of children worldwide suffer from malnutrition [8]. This continues to be an important problem mostly in developing countries. One way to curb the global menace of PEM is through food fortification of plant origin. Food fortification is broadly aimed to allow all people to obtain from their diet all the energy, macro- and micronutrients they need to enjoy a healthy and productive life [3].

Legumes are one of the world's most important sources of food supply especially in the developing countries in terms of food, energy as well as nutrients. It has been recognized as

© 2013 Adelakun et al.; licensee InTech. This is a paper distributed under the terms of the Creative Commons Attribution License (http://creativecommons.org/licenses/by/3.0), which permits unrestricted use, distribution, and reproduction in any medium, provided the original work is properly cited.

an important source of protein and in some cases oil. As a legume, soybeans are an important global crop that provides oil and protein for users. It is the richest sources of protein among the plant foods. Features that motivate researchers to explore its utility in a wide range include the good quality and functionality of its proteins, surplus availability and low cost. Soybeans in the form of full fat flour, concentrate, isolate and texturised proteins have been used in a wide range of food products. These attributes made soybeans to be considered as an ideal food for meeting the protein needs of the population.

The objective of this chapter is to briefly describe the quality attributes of soybean and the potential use of its flour in food fortification.

2. Soybeans

Soybeans *(Glycine max)* belonging to the family *leguminosae* constitute one of the oldest cultivated crops of the tropics and sub-tropical regions, and one of the world's most important sources of protein and oil. Soybeans are probably the most important oil seed legume which has its origin in Eastern Asia, mainly China. The cultivar *Glycine max* is thought to be derived from *Glycine ussuriensis* and *Glycine tomentosa* which grow wild in China, and can be found in great quantities in Asian countries such as Japan and Indonesia [16].

The seeds vary in shape and colour depending on the cultivar. In shape, they can be spherical to flatten while the colour varies from white, yellow and brown to black. Also, the chemical composition of each variety of soybeans differs from each other. Due to the long and tedious processing technique of soybeans, Japan which is one of the largest suppliers of soybean has developed highly advanced processing technologies in the processing and manufacturing of highly acceptable and palatable soya products [14]. As a result of high protein content in soybean, it can be used as a substitute for expensive meat and meat products [5].

Soybean is particularly very unique for different reason and hence classify as a valuable and economical agricultural commodity. In the first instance, it possesses agronomic characteristics with its ability to adapt to a wide range of soil and climate; and its nitrogen fixing ability. This makes it to be a good rotational crop for use with high nitrogen – consuming crops such as corn and rice. Secondly, soybean unique chemical composition on an average dry matter basis is about 40% of protein and 20% of oil. This composition makes it to rank highest in terms of protein content among all food crops and second in terms of oil content after peanut (48%) among all food legumes. Furthermore, soybean is a very nutritious food crop.

3. Nutritional content of soybean

Soybean (*Glycine max*) first emerged as a domesticated crop in the eastern half of North China around the 11th century B.C of Zhou Dynasty. It is easy to grow and has adaptability to a wide range of soils and climate. Because it contains high amount of protein and oil, the soybean was considered one of the five sacred grains along with rice, wheat, barley and millet [4]. The protein and oil component of soybean are not only high in terms of quantity but also in quality. For instance, soy oil has a highest proportion of unsaturated fatty acids such as linoleic and linolenic acid making it a healthy oil to use. Soybeans are known to be typical of such crops that contain all three of the macro-nutrients required particularly for human nutrition. They also contain protein which provides all the essential amino acids in the amounts needed for human health. Most of the essential amino acid present in soybean is available in an amount that is close to those required by animals and humans. The protein – digestibility – correlated amino acid score is close to 1, a rating that is the same for animal proteins such as an egg white and casein. The profile of various nutrients in raw matured seeds of soybeans as highlighted in USDA Nutrient database is as indicated in Table 1 below. Additionally, soybean contains phytochemicals which have been shown to offer unique health benefit. Soybean also has versatile end uses which include human food, animal feed and industrial materials [10].

4. Utilization of soybeans

Soybean can be processed to give soy milk, a valuable protein supplement in infant feeding, soycurds and cheese [23]. It is also used to produce soysauce used extensively in cooking and as a sauce. Soybeans are also used for making candies and ice cream and soybean flour which could be mixed with wheat flour to produce a wide variety of baked goods such as bread and biscuits [16]. Soybean oil is used for edible purposes, particularly as a cooking, and salad oil and, for manufacture of margarine [16]. The oil can also be used industrially in the processing of paints, soap, oil, cloth and printing inks. The meal and soybean proteins are used in the manufacture of synthetic fibre (artificial wool) adhesives and textile [14]. Soybeans could be made into such products as tempeh, miso and natto which may include other sub-products [16].

Soybean protein fibre has been reportedly produced from bean dregs that are produced when extracting oil [14]. From these, globular protein is extracted, made into a spinning solution of a consistent concentration with the addition of a functional auxiliary, and spun into yarn by the wet method [21]. Effect of fermentation on soybean has the tendency of altering the features of the arising dregs when oil is so extracted from soybean. This thus has tendency of either skewing up or otherwise the various arrays of benefits known to accrue from the development of soybean fiber blends with other fibers [21]. [7] has also reported on the chemical composition and total digestible nutrients (TDN) of fermented soybean paste residue. This is usually exploited for the utilization of such residues in livestock rations. Furthermore, [9] had suggested the likelihood of the use of fermented soybean paste residue for livestock feed in the near future as a form of turning waste to wealth and thus serving as anchor for many other accruing benefits.

Nutrient	value per 100g
Energy	1,866kJ (446kcal)
Carbohydrates	30.16g
Sugars	7.33g
Dietary fiber	9.3g
Fat	19.94g
Saturated	2.884g
Monounsaturated	4.404g
Polyunsaturated	11.255g
Protein	36.49g
Tryptophan	0.591g
Threonine	1.766g
Isoleucine	1.971g
Leucine	3.309g
Lysine	2.706g
Methionine	0.547g
Phenylalanine	2.122g
Tyrosine	1.539g
Valine	2.029g
Arginine	3.153g
Histidine	1.097g
Alanine	1.915g
Aspartic acid	5.112g
Glutamic acid	7.874g
Glycine	1.880g
Proline	2.379g
Serine	2.357g
Water	8.54g
Vitamin A equiv.	1µg (0%)
Vitamin B6	0.377mg (29%)
Vitamin B12	0µg (0%)
Vitamin C	6.0 mg (10%)
Vitamin K	47 µg (45%)
Calcium	277mg (28%)
Iron	15.70 mg (126%)
Magnesium	280 mg (76%)
Phosphorus	704 mg (101%)
Potasium	1797 mg (38%)
Sodium	2 mg (0%)
Zinc	4.89 mg (49%)

Table 1. Source: USDA Nutrient database, Percentage relative to US recommendations for adultsNutritional content of raw mature seeds of soybean

5. Antinutritional factors in soybeans

Flatulence is characterized by stomach cramps, nausea, diarrhoea, intestinal and gastric discomfort resulting from the production of large amounts of gas in the gastrointestinal tract. Although all the causative factors in flatulence formation are still unknown, it has been suggested and generally accepted that low-molecular weight oligosaccharides, primarily raffinose, stachyose and verbascose present in most legume seeds are linked to flatulence [2]. Researchers have suggested practical procedures for the reduction of flatulence in cooked and processed soybeans. These include fermentation, removal of seed coat prior to cooking or processing, soaking in water, germination and cooking with a mixture of sodium carbonate and bicarbonate [11]. Oligosaccharides can also be reduced through fermentation thereby leading to elimination of flatulence.

6. Soybean in food fortification

In many developing nations, cereal based foods are widely utilized as food and as dietary staples for adults and weaning foods for infants including Africa where it accounts for up to 77% of total caloric consumption. The major cereal based foods in these regions are derived mainly from maize, sorghum, millet, rice, or wheat. Oilseeds are the largest single source for production of protein concentrates. Of these, soybeans, in terms of tonnage produced, are the most important source. Properly defatted soybean flour will contain 50% or more of protein. By removing soluble carbohydrates and minerals, concentrates containing up to 70% protein can be prepared, and dispersible isolates containing 90% or more of protein are being made. The isolates are of interest as highly concentrated fortification media and also as bases for a variety of high-protein beverages, desserts, and similar products. Soybean concentrates have the virtue of low cost and good nutritive value. Fortification of staple cereals with soybean can help improves their nutritive value and may aid in alleviating malnutrition in developing countries [17].

[12] determined the effect of germination and drying on the functional and nutritional properties of common red bean flour and evaluated the effect of incorporating different levels of cowpea and a constant level of soybean into red bean flour on the functional properties of the composite. Incorporation of soybean and cowpea flour into germinated bean flour at levels of 10 and 30%, respectively, produced a composite with higher functional properties. [15] studied the effect of fermentation with *Rhizopus oligosporus* on some physico-chemical properties of starch extracts from soybean flour. Their study revealed that the length of fermentation with *R. oligosporus* within the period of 0–72 h on soybean (G. max) affected many physico-chemical properties of starch extract as well as pasting characteristics of extracted starch. The physico-chemical attributes of starch 'extract' from fermented soybean flour revealed that water binding capacity, water absorption capacity, swelling capacity and solubility power decreased slightly with increases in fermentation period. Also, while the starch yield and amylose content decreased and amylopectin contents increased with fermentation

period. The pasting characteristics of starch 'extract' from the fermented soybean flour revealed the potential use of the flours in weaning food formulation due to reduce viscosity trend as fermentation period increased.

[1] also worked on the effect of soybean substitution on some physical, compositional and sensory properties of kokoro (a local maize snack). Kokoro which is a finger-shaped snack made from maize is widely acceptable and consumed by children and adults, especially in the southwestern part of Nigeria in the Oyo and Ogun states. Soybean (*Glycine max*), which has high quality protein with high contents of sulphur-containing amino acids, is a good supplement in maize products. Because Snacks provide an avenue for introducing plant proteins such as soybeans to people who normally resist trying any unfamiliar food, kokoro was prepared from maize–soybean flour mixtures in ratios of 100:0, 90:10, 80:20, 70:30, 60:40 and 50:50. The physical, compositional, and sensory characteristics of kokoro were evaluated. Protein and fat contents increased, while carbohydrate content decreased as the soy flour proportion of the flour mixture used in the kokoro was increased.

The bulk density and water-holding capacity increased with increasing proportion of soybean flour, while the swelling capacity was found to decrease. High soy-substitution significantly reduced the sensory acceptance of kokoro. Sensory evaluation indicated that maize:soybean

flour mixture ratios of 100:0 and 90:0 were the most acceptable to the panellists. Several studies have been conducted to improve the protein quality of food products by fortification with plant proteins such as soybean, which is less expensive. Amino acid fortification was suggested by [18]. [19] also determined the acceptability of fermented maize meal fortified with defatted soy-flour in traditional Ghanaian foods.

7. Conclusion

Soybean flour has huge potentials of being used to enrich foods in order to provide adequate nutrients for individuals not meeting daily needs. Based on the available information on the nutrients profile of soybean including the amino profiles, human consumption of soybean flour can be promoted because of its positive effect on nutritional enhancement on different fortified food products. However, more efforts need to be directed at addressing associated technological issue which is flatulence to further increase effective utilization of the food product in food fortification.

Author details

O. E. Adelakun[1*], K. G. Duodu[1], E. Buys[1] and B. F. Olanipekun[2]

*Address all correspondence to: olu_yemisi2000@yahoo.com

1 Department of Food Science, University of Pretoria, Pretoria 0002, South Africa

2 Department of Food Science and Engineering, Ladoke Akintola University of Technology, Nigeria

References

[1] Adelakun, O. E., Adejuyitan, J. A., Olajide, J. O., & Alabi, B. K. (2005). Effect of Soybean Substitution on some Physical, Compositional and Sensory Properties of Kokoro (A Local Maize Snack). *European Food Research and Technology.*, 220, 79-82.

[2] Akpapunan, M. A., & Markakis, P. (1979). Oligosaccharides of 13 Cultivars of Cowpea (Vigna Sinensis). *J. Food Science*, 44, 317-318.

[3] Allan, L., Benoist, B., Dary, O., & Hurrell, R. (2006). Guidelines on food fortification with micronutrients. *A Report: Word Health Organization and Food and Agriculture Organization of the United States.*

[4] Ang, C. Y. W., Liu, K., & Huang, Y. W. (1999). Asian Foods Science and Technology. *Technomic Publishing Company Inc*, 139.

[5] Charles, A., & Guy, L. (1999). Food Biochemistry. *Aspen Publishers Inc. Gaitheasburg, Maryland*, 39-41.

[6] Dulger, H., Arik, M., Sekeroglu, M. R., Tarakcioglu, M., Noyan, T., & Cesur, Y. (2002). Proinflammatory cytokines in Turkish children with protein energy malnutrition. *Mediators Inflamm*, 11, 363-5.

[7] FFTC. (2002). Food and Fertilizer Technology. *An International Center for Farmers in the Asia Pacific Region. Taiwan.*

[8] Gray, V. B., Cossman, J. S., & Powers, E. L. (2006). Stunted growth is associated with physical indicators of malnutrition but not food insecurity among rural school children in Honduras. *Nutr Res*, 26(11), 549-55.

[9] Jong-Kyu, Ha., Kim, S. W., & Kim, W. Y. (1996). Use of Agro-Industrial. By-products a Animal Feeds in Korea. *A publication of FFTC. An International Centre for Farmers in the Asia Pacific Region.*

[10] Liu, K. (2000). Expanding soybean food utilization. *Food Technol*, 54(7), 46-47.

[11] Ndubuaku, V. O., Nwaegbule, A. C., & Nnanyeluyo, D. O. (1989). Flatulence and Other Discomfort Associated with Consumption of Cowpea (Vigna uguiculata). *Appetite*, 13, 171-181.

[12] Njintang, N. Y., Mbofung, C. M. F., & Waldron, K. W. (2001). In Vitro Protein Digestibility and Physicochemical Properties of Dry Red Bean (Phaseolus vulgaris) Flour:

Effect of Processing and Incorporation of Soybean and Cowpea Flour. *J. Agric. Food Chem.*, 49, 2465-2471.

[13] Nnakwe, N. (1995). The effect and causes of protein-energy malnutrition in Nigerian children. *Nutrition Research*, 15(6), 785-794.

[14] NSRL. (2002). A Publication of China National Soybean Research Laoratory on Soybean Procesing:. *From Field to Consumer*, http://www/NSRL.

[15] Olanipekun, B. F., Otunola, E. T., Adelakun, O. E., & Oyelade, O. J. (2009). Effect of fermentation with Rhizopus oligosporus on some physico-chemical properties of starch extracts from soybean flour. *Food and Chemical Toxicology*, 1401-1405.

[16] Onwueme, I. C., & Sinha, T. D. (1999). Field crop production in tropical Africa. *England, Michael Health Ltd.*, 402-414.

[17] Parman, G. K. (1968). Fortification of Cereals and Cereal Products with Proteins and Amino Acids. *J. Agric Food Chem.*, 16.

[18] Potter, N. N. (1978). Food Science. *3rd Edition, Ilhiala, New York*, 356.

[19] Plahar, W. A., & Leung, H. K. (1983). Composition of Ghanian fermented maize meal and the effect of soya fortification on sensory properties. *J Sci Food Agric*, 34, 407-411.

[20] Sereebutra, P., Solomons, N., Aliyu, M. H., & Jolly, P. E. (2006). Sociodemographic and environmental predictors of childhood stunting in rural Guatemala. *Nutr Res*, 26, 65-70.

[21] Senshoku, K. S. (2002). Physical Characteristics and Processing method of Chinese Soybean Fiber. *Home Technology Jianghe Tianrongbi. Fiber Co. Ltd. (China).*

[22] Shils, M. E., Olson, J. A., & Moshe, S. (1999). Modern Nutrition in Health and Disease. *Philadelphia Lippincott Williams & Wilkins.*

[23] Tunde-Akintunde, T. Y. (2000). Predictive Models for Evaluating the Effect of Processing Parameters or Yield and Quality of Some Soybean (Glycine max (L) Merri) Products. *Ph.D Thesis- Universityof Ibadan, Nigeria.*

Chapter 24

Food, Nutrition and Health

Eduardo Fuentes, Luis Guzmán, Gilda Carrasco,
Elba Leiva, Rodrigo Moore-Carrasco and
Iván Palomo

Additional information is available at the end of the chapter

http://dx.doi.org/10.5772/52601

1. Introduction

Botanically, soybean belongs to the order *Rosaceae*, family *Leguminosae* or *Papillonaceae* or *Fabaceae*, subfamily *Papilionoidae*, the genus *Glycine* and the cultivar Glycine max. It is an annual plant that measures up to 1.5 m tall, with pubescent leaves and pods; the stems are erect and rigid. In its primary and secondary roots, are located a variable number of nodes. One of the characteristics of the root system development is its sensitivity to variations in the supply and distribution of inorganic nutrients in the soil. The root system has a main root which can reach a meter deep, with an average being between 40 and 50 centimeters [1].

The soybean is a traditional oriental food, a leguminous plant native to eastern Asia, especially in China. Soybean is cultivated worldwide, the United States is the country that grows more than 50% of the world production of this important food, which has been utilized in the diet of humans around the world, due to its high content of essential amino acids (Table 1) and calcium. It is consumed as cooked beans, soy sauce, soymilk and tofu (soybean curd). Also, a vegetable oil is obtained from soybeans, rich in polyunsaturated fatty acids [2].

Soybean is an annual plant, whose seeds are the edible organ. Soybean grains are rich in protein, and also a good source of various phytochemicals such as isoflavones and lignans, molecules with antioxidant and antiplatelet activities, among other effects; also may help fight and prevent various diseases, so constitute a useful source of food. For these reasons, these compounds have been intensively studied at basic and clinical level [3].

Soybean consumption benefits, especially in several chronic diseases, have been related to its important protein content, high levels of essential fatty acids, vitamins and minerals. Consequently, the present chapter aimed at the comprehensive characterization of the anti-

© 2013 Fuentes et al.; licensee InTech. This is a paper distributed under the terms of the Creative Commons Attribution License (http://creativecommons.org/licenses/by/3.0), which permits unrestricted use, distribution, and reproduction in any medium, provided the original work is properly cited.

oxidant and antiplatelet activities of bioactive compounds, of soybean and its derivatives, and the extent to which soybean is a health-promoting food.

Amino acid	g/16 g Nitrogen
Isoleucine	4.54
Leucine	7.78
Lysine	6.38
Methionine	1.26
Cysteine	1.33
Phenylalanine	4.94
Tyrosine	3.14
Threonine	3.86
Tryptophan	1.28
Valine	4.80
Arginine	7.23
Histidine	2.53
Alanine	4.26
Aspartic acid	11.70
Glutamic acid	18.70
Glycine	4.18
Proline	5.49
Serine	5.12

Table 1. Amino acid composition of soybeans seeds. Source: Adapted by authors from FAO (1970) and FAO/WHO (1973).

2. Soybean: foods and bioactive compounds

Soybeans are consumed as cooked beans, which previously should be boiled for at least three hours. With these grains are prepared meals, salads and soups, which in turn, are a source of preparation of other foods. From soybean grains also is possible to obtain soy sauce, which is used especially in oriental foods, such as sushi. The soy sauce is usually made by fermenting soya grains with cracked roasted wheat, which are arranged in blocks and immersed in a cold salt water, the process takes about a year in pots mud, sometimes dried mushrooms are added as mushrooms. In Japan, it is illegal to produce or import artificial soy sauce and therefore all Japanese soy sauces are made by the traditional way [4].

Another food derived from soy is tofu, which is a widely used food in the East as well as vegetarian meals around the world. Required for preparing tofu soybeans are water and a coagulant. Initially, you get the coagulated soymilk, then is pressed and separated the liquid portion from the solid. Tofu has a firm texture similar to cheese milk; the color is cream and served in buckets. Also, from soybeans is possible to obtain vegetable oil, which is characterized by high polyunsaturated fatty acids [5].

Soy is a source not only of proteins, vitamins and minerals, but also of many bioactive compounds, such as isoflavones, protease inhibitors, saponins, and phytates. The great importance of these compounds is based on their biochemical activity, which results in health promotion and disease prevention, by their antioxidant, and antiplatelet activities.

Antioxidant Activity

Soybeans contain a variety of bioactive phytochemicals such as phenolic acids, flavonoids, isoflavones, saponins, phytosterols and sphingolipids; being the phenolic compounds with the highest antioxidant capacity. The key benefits of soy are related to their excellent protein content, its high content of essential fatty acids, numerous vitamins and minerals, their isoflavones and their higher fiber content.

Polyphenolic compounds are a class of secondary metabolites biosynthesized by the vegetal kingdom [6] and involve a wide range of substances that possess one or more aromatic rings with at least one hydroxyl group. Among them, can be mentioned flavonoids, isoflavones, anthraquinones, anthocyanins, xanthones, phenols, hydroxycinnamic acids, lignin and others.

All of them act as scavengers or stabilizers of free radicals, and can produce chelation of metals, those having carboxyl groups at its end. Works have also been reported that its antioxidant action can be attributed to the inhibition of prooxidants enzymes as lipoxygenase [7].

A study by Xu (2008) in 30 samples of soybean from different regions of North Dakota, Minnnesota (USA), found that some cultivars of black soybean had a higher antioxidant capacity measured as ORAC, FRAP and DPPH than yellow soybeans and that the phenolic acid content, isoflavones and antiochians was different, suggesting that some selected cultivars can be used as producers of high quality soy, because it provides a high content of phenolic phytochemicals and antioxidant properties [8].

Isoflavones and equol. Of all plants, soybean contains the highest amount of different isoflavones, a variety of phytoestrogens that have a structure similar to estrogen (figure 1). The interest in soybean isoflavones has gained importance since the 90's to today, there are a lot of evidence that these phytoestrogens possess a powerful and wide range of biological activities. Isoflavones are not a steroid structure, however, has a phenolic ring than is capable of binding to the estrogen receptor (ER) and according to Makela (1995) can act as either an agonist or an antagonist [9].

The discovery of high concentrations of isoflavones in urine of adults who consume soy protein, in addition to the evidence supporting its biological action, elevate the soybean to the category of functional food. The FDA in 1999 gave approval to give foods containing 6.25 g of soy protein the seal of protector of cardiovascular health, increasing significantly the sales of foods fortified with soy and isoflavone constituent.

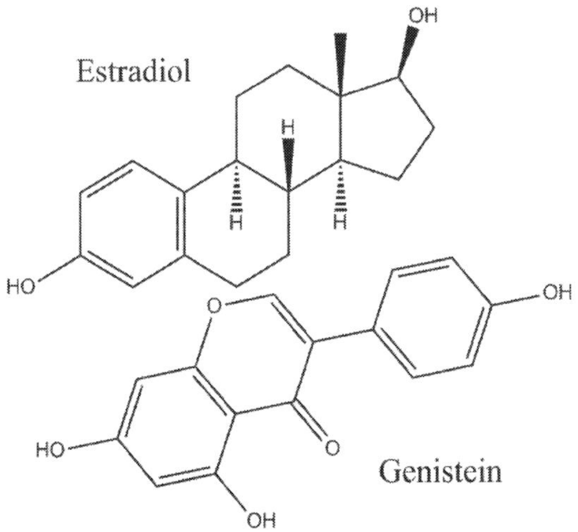

Figure 1. Comparison between the structure of the derivative of Isoflavone (genistein) and estrogen (estradiol), which shows the similarities between the two molecules.

On the other hand, traditional foods in the East, such as extracts and broth of rice and soybeans fermented with microorganisms for 21 days, were reported as antioxidants by Yen (2003) [10] and Yang (2000) [11]. The authors attributed the antioxidant power at the content of polyphenol and the presence of reductons that only occur during the fermentation process. The antioxidant supplements or foods containing antioxidants may be used to reduce the oxidative damage related to age and diseases such as artherosclerosis, diabetes, cancer, cirrhosis, among other [12].

For 10 to 15 years, has been a strong interest in the use of products of botanical origin for the protection, whitening and skin aging. According Baunmann (2009) the mechanism of action of botanical products has been known given the use of advanced technologies applied to research, this is how there are several reports showing that soy components play an important role in the extracellular matrix of the dermis [13].

Moreover, it is accepted that isoflavones act through different mechanisms such as modulation of cell growth and proliferation, extracellular matrix synthesis, inhibition of inflammation and oxidative stress. The isoflavones reduced renal injury by decreasing the concentration of lipoproteins in plasma and acting as an antioxidant reducing the lipid peroxidation [14].

The Equol, whose structure corresponds to 7-hydroxy-3-(4'-hydroxyphenyl)-chroman, is a nonesteroidal estrogen, which was discovered in the early 80's in the urine of adults who consumed soy foods [15]. It has been shown to be a metabolite of daidzein, one of the major isoflavones present in foods containing soy, which is formed after hydrolysis of the isoflavone glycoside [16] at intestinal level and subsequent bacterial biotransformation in the co-lon [17], leaving an intermediary called dihydro-equol [18-20].

Equol does not originate in plants, but is the product of degradation of the isoflavone glyco-side in the intestine [21], situation that was confirmed in infants of 4 months who were fed with formulations containing soy [22, 23]. All mammals can biotransform isoflavone glyco-side permanently, except the man who for reasons still unexplained, only 20-35% of adults produces equol after eating foods made from soybeans or that have been enriched with pure isoflavones [17, 24, 25].

Several studies have suggested that those mammals who are equol producers show a great-er response to isoflavone-enriched diets, leading to the conclusion that equol is a more po-tent isoflavone than genistein and is the only one that has a chiral carbon at position 3 of the furan ring, making two enantiomeric forms, S and R that differ significantly in their confor-mational structure [26].

Gopaul (2012) investigated the effect of equol on gene expression of proteins in the skin, us-ing a cellular model of human dermis and found that equol significantly increased gene ex-pression of collagen, elastin (ELN), and tissue factor inhibitor of metalloproteinases and decreased metalloproteinases (MMPs), causing positive changes in the skin's antioxidant and anti-aging genes. The same occurred in cultured human fibroblasts (hMFC), in which equol significantly increased type I collagen (COL1A1), while 5a-dihydrotestosterone (5a-DHT) significantly decreased cell viability. These findings suggest that equol has great po-tential for topical applications to the skin, for the treatment and prevention of aging of the skin by increasing the extracellular matrix components [27].

Has been found that Equol have affinity for the estrogen receptor beta, which is abundant in keratinocytes of the epidermis and dermal fibroblasts [28-30]. On the other hand, equol is a selective androgen modulator and has the ability to bind to 5a-dihydrotestosterone (5a-DHT) and inhibit its potent action on the skin [31]. In this sense, we can quote the opposite effect that have androgens and estrogens, the former producing an injury to the skin by in-creasing MMPs, while the latter have a positive effect on the aging of the skin by increasing collagen, elastin and decreasing MMPs [13, 32-35].

In turn, Muñoz (2009) studied the inhibitory effect of soy isoflavones and the metabolite equol derived from daidzein, an agonist that has biological effects attributed to an antago-nism of the thromboxane A2 receptor (TXA2R), which helps explain the beneficial effects of dietary isoflavones in the prevention of thrombotic events [36].

Recent works by Ronis (2012) studied the effect of mice fed with extracts of soy protein or isoflavones finding that can reduced the metabolic syndrome in rats via activation of peroxi-some proliferation activated receptor (PPAR), liver X receptor (LXR) and decreased signal-ing protein binding to the sterol regulatory element binding proteins (SREBP) [37].

Anthocyanins. The anthocyanins are known to have antioxidant effects and play an impor-tant role in preventing various degenerative diseases. Structurally, this are a suitable chemi-cal structure to act as antioxidants, since they can donate hydrogens or electrons to free radicals or catch and move them in its aromatic structure. There are about 300 anthocyanins in nature, with different glycosidic substitutions in the basic structure of the ion 2-phenyl-benzopirilio or flavilio [38].

Paik (2012) examined the effect of anthocyanin extracts from the cover of black soybean in an animal model of retinal degeneration (RD), the leading cause of death of the photoreceptor cells that lead to blindness and noted that extracts of anthocyanins may protect retinal neurons from damage induced by degenerative agents such as N-methyl-N-nitrosourea (MNU) at a dose of (50mg/kg), which acts as a methylating agent that causes DNA damage to the photoreceptors [39].

In general, bioactive compounds from soybean are many, but still exists wide variety of information of the beneficial effects and also adverse effects of isoflavones and anthocyanins, so likewise it is necessary to study more thoroughly this compound and its relation to chronic diseases through scientific studies with larger number of patients and longer study periods, in order to clarify all the diffuse concepts, labile or poorly sustained, so as to give the isoflavones, a clear place in the diet therapy.

Antiplatelet Activity

There has been much recent interest in the cardiovascular benefits of dietary soybean on potential anti-thrombogenic and anti-atherogenic effects [40]. Extracts containing isoflavones and soy saponins inhibit the platelet aggregation *ex vivo* induced by ADP and collagen in diabetic rats [41]. Moreover, black soybean extracts inhibited platelet aggregation induced by collagen *in vitro* and *ex vivo*, and attenuates the release of serotonin and P-selectin expression [42].

The effects of soybean products on platelet aggregation were initially described for genistein [43]. In these reports, genistein was able to inhibit platelet activation induced by collagen and thromboxane A2 analog (TXA2), but not by thrombin. Genistein (10 mg/kg) in mouse significantly prolonged the thrombotic occlusion time and significantly inhibited *ex vivo* and *in vitro* (30 µM) platelet aggregation induced by collagen [44]. Genistein is a well-known inhibitor of protein tyrosine kinases, however, on platelet functions *in vitro* genistein inhibits activation of phospholipase C in stimulated platelets, apparently independent of its effects on tyrosine kinases. These results suggest that dietary supplementation of soy may prevent the progression of thrombosis and atherosclerosis [45]. Daidzein, another soy flavonoid that lacks tyrosine kinase inhibitory activity also inhibited the response to collagen and TXA2, suggesting that these flavonoids inhibit platelet aggregation by competition for the thromboxane A2 receptor (TXA2R) rather than through tyrosine kinase inhibition. Genistein and daidzein have effect on platelets, macrophages and endothelial cells: inhibited collagen-induced platelet aggregation in a dose-dependent manner and in macrophage cell line activated with interferon γ, plus lipopolysaccharide inhibit tumoral necrosis factor α (TNF-α) secretion, dose-dependently. Both isoflavones also dose-dependently decreased monocyte chemoattractant protein-1 secretion induced by TNF-α in human umbilical vein endothelial cells [40].

Equol is more active than soy isoflavone itself to compete for binding to TXA2R in human platelets (in the range of micromoles / L), so that inhibits the platelet aggregation and secretion induced by U46619 [36]. Under equilibrium conditions, the following order of the relative affinity in inhibiting [(3)H]-SQ29585 binding was: equol > genistein > daidzein > glycitein > genistin > daidzin > glycitin [36]. Guerrero and colleagues suggested that this competitive binding was due to structural features of these flavonoids such as the presence of a double bond in C2-C3 and a keto group in C4 [36].

From a extraction of soy sauce, two kinds of components with anti-platelet activity were isolated and structurally identified: 1-methyl-1,2,3,4-tetrahydro-β-carboline (MTBC) and 1-methyl-β-carboline (MBC). MTBC shows IC_{50} ranging from 2.3 to 65.8 μg/mL for aggregation response induced by epinephrine, platelet-activating factor (PAF), collagen, ADP and thrombin [46]. Membrane fluidity regulates the platelet function and various membrane-fluidizing agents are known to inhibit platelet aggregation [47]. Certain β-carbolines influence the fluidity of model membranes. The alteration of membrane fluidity may be involved in the antiplatelet effects of MTBC and MBC [48].

Soybean protein inhibits platelet aggregation induced by thrombin, collagen and ADP, and prolongs the clotting time [49]. Also was observed that most fractions obtained of soy protein hydrolysates (gel filtration chromatography, reverse-phase HPLC and cation exchange HPLC) inhibited rat platelet aggregation induced by ADP, which suggests that most peptides have some degree of antiplatelet effect. From the inhibitory fractions, two new peptides were identified, SSGE and DEE, and at concentrations of 480 and 460 μM, respectively, inhibited in 50% platelet aggregation [50].

The diet may be the most important factor influencing the risk of cardiovascular disease. Soybean derivatives can be denominated as functional ingredients as they contain bioactive compounds that inhibit platelet aggregation, which gives a preventive effect on thrombus formation [51].

3. Functional Food

Soybean is a very rich source of essential nutrients and one of the most versatile foodstuffs. It possesses good quality protein and highly digestible (92–100%) and contains all the essential amino acids. Soybean-protein products also contain a high concentration of isoflavones (1 g/kg) [52]. Therefore, consumption of soy-based foods has been associated with multiple health benefits [53, 54]. Among a variety of soybeans, black soybean is known to display diverse biological activities superior to those of yellow and green soybeans, such as antioxidant, anti-inflammatory and anticancer activities [42].

Soy food intake has been shown to have beneficial effects on cardiovascular disease risk factors. Data directly linking soy food intake to clinical outcomes of cardiovascular disease, evidence that soy food consumption may reduce the risk of coronary heart disease in women and may be protective against the development of subarachnoid hemorrhage [54, 55]. Based on that 1% reduction in cholesterol values is associated with an approximate 2-3% reduction in the risk of coronary heart disease, it can be assumed that a daily intake of 20-50 grams of isolated soy protein could result in a 20- 30% reduction in heart disease risk [56, 57].

Several components associated with soy protein have been implicated in lowering cholesterol: trypsin inhibitors, phytic acid, saponins, isoflavins, fiber and proteins [58]. Apparently, there is a synergy among the components of intact soy protein, which provides the maximum hypocholesterolemic benefit. A variety of clinical trials have demonstrated that consuming 25 to 50 g/daily of soy protein is both safe and effective in reducing LDL cholesterol by ≈4% to 8% [58]. Therefore, maturation of SREBP and induction of SRE-regulated genes produce an increase in surface LDL receptor expression that increases the clearance of plasma cholesterol, thus decreasing plasma cholesterol levels [59]. However, other results present direct evidence for the existence of LDL receptor and plasma lipoprotein-independent pathways by which dietary soy protein isolate inhibits atherosclerosis [60, 61].

The addition of soy protein in diet or replacing animal protein in the diet with soy, lowers blood cholesterol. Moreover, defatted soy flour is a widely used in these applications as a partial replacement for nonfat dry milk [62-64]. Soy protein can increase protein content and its used in compounded foods (breads, crackers, doughnuts, and cakes) for their functional properties, including water and fat absorption, emulsification, aeration (whipping), heat setting, and for increasing total protein content and improving the essential amino acids profile [65].

The low breast cancer mortality rates in Asian countries and the putative anti-estrogenic effects of isoflavones have fueled speculation that soyfood intake reduces breast cancer risk [66]. Soy sauce promotes digestion, because the consumption of a cup of clear soup containing soy sauce enhances gastric juice secretion in humans. The feeding of a diet containing 10% soy sauce to male mice for 13 months also reduces the frequency and multiplicity of spontaneous liver tumors [67]. Over the past decades, enormous research efforts have been made to identify bioactive components in soy [68]. The Health effects of soy dietary are variable depending on individuals' metabolism and in particular to their ability to convert daidzein to equol that seems to be restricted to approximately 1/3 of the population. Equol production has been indeed linked to a decreased on arterial stiffness and antiatherosclerotic effects via nitric oxide production [69]. Despite being a biotransformation product of daidzein, the equol at low dosage can prevent skeletal muscle cell damage induced by H_2O_2 [70] and possesses anticancer activity via apoptosis induction in mammary gland tumors of rats [71].

Hydroponic cultivation improved the nutritional quality of soybean seeds with regard to fats and dietary fiber. This suggest that specific cultivars should be selected to obtain the desired nutritional features of the soybean raw material [72]. Irrigation enhanced the isoflavone content of both early- and late-planted soybeans as much as 2.5-fold. Accumulation of individual isoflavones, daidzein and genistein, are also elevated by irrigation [53].

4. Digestion and Absorption

A number of factors including the amount consumed, chemical speciation, interactions with other ingredients, physiological state (e.g., gender, ethnicity, age, health status) and intestinal microflora influence the absorption of dietary isoflavonoids by the gastrointestinal tract [73]. Additionally, the absorption and disposition of isoflavones (daidzein and genistein) appears to be independent of age, menopausal status and probiotic or prebiotic consumption [74, 75].

After absorption, isoflavones are reconjugated predominantly to glucuronic acid and to a lesser extent to sulphuric acid. Only a small portion of the free aglycone has been detected in blood, demonstrating that the rate of conjugation is high [76]. The isoflavone aglycones are absorbed faster and in greater amounts than glucosides in humans, dependent on the initial concentrations. Thus, products rich on aglycone can be more effective in the prevention of chronic diseases [77]. Concentrations of isoflavones and their gut microflora metabolites in the plasma, urine, and feces are significantly higher in the subjects who consume a high-soy diet than in those who consume a low-soy diet [78].

Bioavailability of isoflavone glycosides (daidzein and genistein) as pure compounds or in a soyfood matrix (soymilk) requires initial hydrolysis of the sugar moiety by intestinal β-glucosidases for the uptake to the peripheral circulation [16]. Twenty-four hours after dosing, both plasma and urine isoflavone concentrations were nearly null [79]. The genistein compound is absorbed from the lumen partly unhydrolyzed and transported directly (by an unknown transporter or diffusion) to the vascular side [80]. Conjugates of daidzein are more bioavailable than those of genistein. Thus, after oral administration of soy extract in rats providing 74 micromol of genistein and 77 micromol of daidzein / kg (as Conjugates), were found that plasma concentration of daidzein was maximal at 2 h and it was almost double that of genistein. Since about 50% of the genistein dose is excreted as 4-ethyl phenol (the main end product from genistein) [81]. The end product of the biotransformation of the phytoestrogen daidzein, is the equol, that is not produced in all healthy adults in response to dietary challenge with soy or daidzein [21]. However, plasma genistein concentrations are consistently higher than daidzein when equal amounts of the two isoflavones are administered, and this is accounted for by the more extensive distribution of daidzein (236 L) compared with genistein (161 L). In addition to the conjugated state, the chemical structures of isoflavones play a major role in its pharmacokinetics with marked qualitative and quantitative differences depending on the type of supplement ingested [82-84].

5. Health

Functional food may act as an adjunctive therapy/alternative treatment of different pathologies, and scientific studies are appearing more frequently demonstrating that this hypothesis is, indeed, a reality. Soybean containing isoflavone and protein is considered a functional food item [85].

Epidemiological studies suggest that soybean consumption is associated, at least in part, with lower incidences of a number of chronic diseases. The lower rates of several chronic diseases in Asia, including type 2 diabetes, certain types of cancer and cardiovascular diseases, between others, have been partly attributed to consumption of large quantities of soy foods (figure 2) [86].

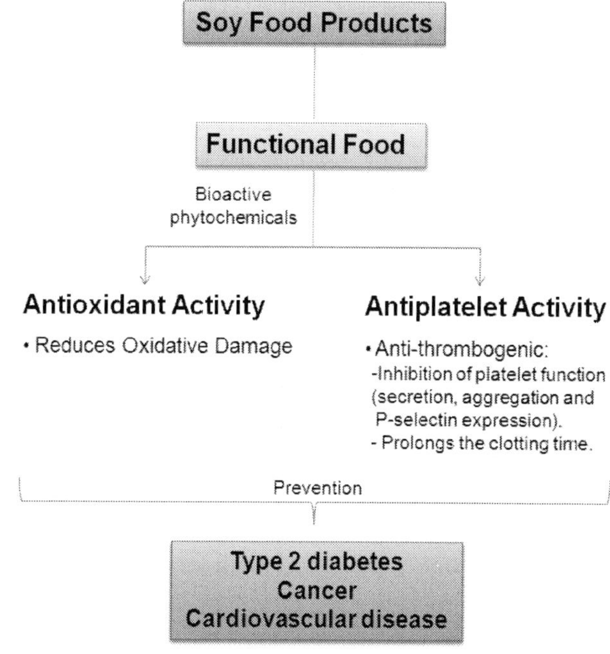

Figure 2. Biological activities from soy food products and its effect on health.

a) Soybean and type 2 diabetes. Type 2 diabetes mellitus is a multifactorial metabolic disorder disease, which results from defects in both insulin secretion and insulin action. Insulin stimulates uptake, utilization and storage of glucose in cells throughout the body by inducing multiple signaling pathways in the tissues that express the transmembrane insulin receptor [87], especially in skeletal muscle that accounts for 75% of whole-body insulin-stimulated glucose uptake. The reduced responsiveness of cells to insulin is due to defective intracellular signaling processes [88, 89]. Millions of people have been diagnosed with type 2 diabetes, and many more are unaware they are at high risk [90]. Obesity is the major risk factor for diabetes and accounts for ≈70% of the variance in the prevalence of this common disease [91]. Dry beans and soyfoods offer benefits in the prevention of diabetes and in the clinical management of established diabetes, soybeans, in particular, have a low glycemic index.

They are rich in phytates, soluble fiber, and tannins, all of which correlate inversely with carbohydrate digestion and glycemic response [62, 92].

Soybean and its natural bioactive products have been studied as an antidiabetic potential. Studies have been conducted to examine the therapeutic effect of different bioactive compounds such as aglycin, phenolic-rich extract and glyceollins, all derivates of soybean. Soybean peptides have been widely used as a natural health food and supplement. It should be good for preventing obesity and diabetes because long-term feeding of soy peptide induced weight loss in obese mice [93]. In healthy and diabetic animal models, soybean peptides decreased blood glucose by increasing insulin sensitivity and improving glucose tolerance [89, 94].

Aglycin, a natural bioactive peptide isolated from soybean, structurally, it has a high stability with six cysteines embedded in three disulfide bonds. It is also resistant to digestion by trypsin, pepsin, Glu-C and bovine rumen fluid and has an antidiabetic potential [95, 96], at this respect, Lu J (2012), studied the effect of aglycin administration on the glucose homeostasis. For this, the diabetes was induced in BALB/c mice fed with a high-fat diet and a single intraperitoneal injection of streptozotocin. With onset of diabetes, the mice were administered daily with aglycin (50 mg/kg/d) for 4 weeks, blood glucose was monitored once a week [89]. The administration of aglycin restored insulin-signaling transduction by maintaining the insulin receptor (IR) and the insulin receptor substrate 1 (IRS1) expression at both the mRNA and protein levels, as well as elevating the expression of p-IR, p-IRS1, p-Akt and membrane GLUT4 protein. The results hence demonstrate that oral administration of aglycin can potentially attenuate or prevent hyperglycemia by increasing insulin receptor signaling pathway in the skeletal muscle of streptozotocin/high-fat-diet-induced diabetic mice [89].

Complex polysaccharides are hydrolyzed by α-amylase to oligosaccharides that are further hydrolyzed to liberate glucose by intestinal α-glucosidase before being absorbed into the intestinal epithelium and entering blood circulation. Therefore, α-amylase and α-glucosidase inhibitors may help to reduce postprandial hyperglycemia by inhibiting the enzymatic hydrolysis of carbohydrates, and hence may delay the absorption of glucose [97]. Therefore, effort has been directed in finding a natural and safer α-amylase and α-glucosidase inhibitors. Phytochemicals such as phenolics with strong antioxidant properties has been reported to be good inhibitors of these enzymes [98]. The phenolic compounds of soybean have also been studied, Ademiluyi (2011) [99] assessed the inhibitory activities of different phenolic-rich extracts from soybean on key-enzyme linked to type 2 diabetes (α-amylase and α-glucosidase) [99]. Their results shown that the different phenolic-rich extracts used inhibited α-amylase, α-glucosidase activities in a dose dependent pattern and the free phenolic extract had higher α-glucosidase inhibitory activity when compared to that of α-amylase; this property confer an advantage on soybean phenolic-rich extracts over commercial antidiabetic drugs with little or no side effect [99].

In turn, the flavonoid family of phytochemicals, particularly those derived from soy, has received attention regarding their estrogenic activity as well as their effects on human health and disease. In addition to these flavonoids other phytochemicals, including phytostilbene, enterolactone, and lignans, possess endocrine activity; the health benefits of soy-based foods may, therefore, be dependent upon the amounts of the various hormonally active phyto-chemicals within these foods, especial attention have recieved the isoflavonoid phytoalexin compounds, glyceollins, in soy plants grown under stressed conditions [100]. A glyceollin-containing fermented soybeans was assessed by Park S (2012), where diabetic mice, induced by intraperitoneal injections of streptozotocin (20 mg/kg bw), were administered a high fat diet with no soybeans (control), 10% unfermented soybeans and 10% fermented soybeans containing glyceollins (FSG), respectively, for 8 weeks; among the diabetic mice, FSG-treated mice exhibited the lowest peak for blood glucose levels with an elevation of serum insulin levels during the first part of oral glucose tolerance testing. FSG also made blood glucose levels drop quickly after the peak and it decreased blood glucose levels more than the con-trol during insulin tolerance testing [101]. The enhancement of glucose homeostasis was comparable to the effect induced by rosiglitazone, a commercial peroxisome proliferator-ac-tivated receptor-γ agonist, but it did not match the level of glucose homeostasis in the non-diabetic mice [101].

In vitro studies suggest that isoflavones have antidiabetic properties such as the inhibition of the intestinal brush border uptake of glucose, α-glucosidase inhibitor actions, and tyrosine kinase inhibitory properties [102]. Animal studies have indicated that soy protein or isofla-vones improve glycemic control, lower insulin requirement, and increase insulin sensitivity [96, 103]. Several observational studies have also suggested that soy intake was associated with improved glycemic control or lowered risk of diabetes [104, 105].

Nevertheless, data from human clinical trials that evaluated the possible beneficial effects of isoflavone-rich soy products on glycemic control and insulin sensitivity have generated mixed findings. Some studies showed that soy products and foods significantly improve glycemic control [106-108], whereas others observed no significant effect [109-111]. This in-consistency could be due to a number of possible reasons. A wide variety of soy products, such as traditional soy foods, isolated soy protein, soy extracts, or purified isoflavones, and a variety of controls have been used [112]. Varied amounts and compositions of protein and isoflavones in soy products and the menopausal status of participants, study duration, base-line health status of participants, intervention adherence, and degree to which dietary intake is controlled may have contributed to variations in studies [113].

b) Soybean and cancer. For a long time has been described the effect of soy on human health, this legume present since ancient times in the Oriental diet is now present all over the world. The eastern population has low rates of various cancers, among these; breast cancer is undoubtedly a classic example.

Liener and Seto in 1955, described an *in vivo* effect in rats inoculated with Walker tumor. Daily injection of a soy extract at 25 mg / kg in these animals results in a decrease in tumor size and weight at the end of the period. This effect was observed only in animals that were treated from the time of inoculation of the tumor and not in those treated after tumor estab-lishment [114]. From Liener investigations have been reported countless works that charac-terize the effect of soy on tumor growth.

Since 1996, it has been reported the use of extracts rich in soy isoflavones in human clinical studies. Has recently been published the results of a clinical study of soy isoflavones used in women at high risk of breast cancer, the results at 6 months of intervention is that soy isoflavones decrease epithelial proliferation in women with risk, modulating the expression of a large number of genes involved in carcinogenesis [115]. One of the markers associated with risk of breast cancer is the IGF-1 (Insulin-like growth factor 1), this growth factor is involved in various processes that stand between growth and cell development. The increase in plasma concentration is associated with up to seven times greater risk of developing this malignancy. However, high intake of soy in American women produces a slight increase in the concentration of IGF-1, these data are extremely complex to interpret. Eating a diet high in soy and algae produce a decrease of up to 40% of the levels of IGF-1 [116].

There is a wide range of cancers in which have been found some degree of association with soy consumption. Recently, a meta-analysis has shown that high-soy intake determined a low risk of lung cancer [117]. In this context, researchers have demonstrated *in vivo*, in nude mice (immune compromised) that were intravenously injected human tumor cell line A549, these cells generate tumor nodules in the lungs of these animals. By analyzing histologically the tissues of mice that received a treatment of 1 mg / day for 30 days orally soybeans, it was determined that the tumor cells were more sensitive to radiation therapy in addition to reducing vascular damage, inflammation and fibrosis caused by radiation on healthy tissue [118].

The relationship between breast cancer and the presence of estrogen receptor in these tumor cells is a fact and is related primarily to the aggressiveness of the tumor and its size. The non-steroidal phytoestrogens chemical compounds present in plants and especially in soy and structurally resemble human estrogen, may play a central role in the effect on the risk of breast cancer in postmenopausal women. This effect, albeit has been reported, is still not entirely clear, while some studies are very conclusive as that conducted by Zaineddin et al in 2012, recently, over 3000 cases and over 5000 controls in a case-control study in German women, the results did show a relationship between the consumption of phytoestrogens including soy and reduced the relative risk of breast cancer in postmenopausal women [119].

Genistein a predominant isoflavonoid in soybeans has long shown a beneficial effect on the prevention and treatment of some cancers. Has been studied this soy isoflavone and the mechanisms that may be associated with its antitumor effect. Among the mechanisms described is the inhibition of nuclear factor NFkappa B a key molecule in tumor cell survival, in 2011 a group of researchers reported that genistein inhibited the proliferation and induced apoptosis in the BGC-823 cell line, a cell line of human gastric cancer, in a dose and time dependent. The mechanism by which this isoflavonoid is capable of producing this effect is by decreasing cyclooxygenase 2 (COX-2), through inhibition of the transcriptional activation of NFkB [120]. Recently another research group reported that genistein is capable of reducing the growth of the cell line of breast cancer MDA-MB-231 inhibit NFkB transcriptional activity through a mechanism dependent signaling pathway Notch-1 [121], These researchers found that genistein negatively regulates the expression of cyclin B1, Bcl-2 and Bcl-xl. Genistein has proven to be a potentiating agent of other compounds with anticarcinogenic effect, a recent example is the reported effect on the cell line A549 lung cancer, where genistein potentiates the apoptotic effect of Trichostatin A, the mechanism involved in this effect would be enhancing the positive regulation of the expression of mRNA encoding the tumor necrosis factor receptor 1(TNFR-1), which play a role of death receptor and therefore may at least partly explain this phenomenon [122]. One line of research to elucidate the mechanisms associated with the anti tumor effect of genistein is the ability of this molecule to inhibit the progression of tumor stem cells, in this context, Zhang et al 2012 have found that genistein has an effect until now not reported to prevent carcinogenesis in a model of early colon cancer using as markers of this process the signaling pathway WNT / beta-cate-nin, some of the genes that are under the control of this pathway are: cyclin D1 and c-myc, being genistein and an extract of soy protein capable of inhibiting various genes involved in the WNT pathway including the mentioned Wnt5a, sFRP1, sFRP2 and Sfrp5 [123]. It has al-so been shown that genistein could also exert its effect by regulating the immune system, is how Iranian researchers showed that genistein is capable of protecting carcinogenesis in a mouse model of cervical cancer by an immune modulatory mechanism [124].

In summary, it appears that soy consumption is a protective agent in carcinogenesis of vari-ous cancers. The mechanisms involved in this process are still a mystery although there are efforts to discover them. One of the molecules characterized in this protective effect is genis-tein, a soy isoflavone, to which today have described multiple benefits and is now the thera-peutic target for the generation of other molecules with structural similarity to her and that enhance its effects *in vivo*.

c) Soybean and cardiovascular disease. Another major concern in today's world is cardiovascu-lar disease, in which the nutritional properties of soybean proteins are well known. Within the past 25 years, numerous studies have reported inverse associations between soy protein intake and plasma cholesterol concentrations; this association is particularly evident in hy-per-cholesterolemic men and women [125-127].

Several studies comparing isoflavone-rich soy diets with isoflavone-free soy diets have been performed in experimental animals and humans. Soy consumption could reduce the cardio-vascular disease risk factors through its beneficial components, including complex carbohy-drates, unsaturated fatty acids, vegetable protein, soluble fiber, oligosaccharides, vitamins, minerals, inositol-derived substances such as lipintol and pinitol, and phytoestrogens, par-ticularly the isoflavones genistein, diadzein, and glycitein [128, 129].

Different studies have been carried out to assess the influence of soy-protein on serum con-centrations of total cholesterol, LDL cholesterol, triacylglycerol, and apoB-100. Beneficial re-sults have also been seen among subjects with different types of diseases [130, 131]. Beneficial effects of soy consumption on blood lipids were the most consistently reported findings by Azadbakht (2007) [108]. In turn, Anderson (1995), showed significant reductions in total cholesterol (9%), LDL cholesterol (13%), and triacylglycerol's (11%) with the con-sumption, on average, of 47 g soy-protein/daily [125].

For its part, Jenkins (2002) studied the effects of high- and low-isoflavone soy-protein foods on both lipid and nonlipid risk factors for coronary artery disease. They found that com-pared with the control diet, however, both soy diets resulted in significantly lower total cholesterol, estimated CAD risk, and ratios of total to HDL cholesterol, LDL to HDL cholesterol, and apolipoprotein B to A-I [132]. No significant sex differences were observed, except for systolic blood pressure, which in men was significantly lower after the soy diets than after the control diet. On the basis of blood lipid and blood pressure changes, the calculated CAD risk was significantly lower with the soy diets, by $10.1 \pm 2.7\%$ [132].

Other group reported that monkeys fed isoflavone-rich soy-protein-isolate diets had significantly better serum lipid values (lower total cholesterol and higher HDL-cholesterol concentrations) than monkeys fed isoflavone-poor soy-protein-isolate diets. Whereas the administration of the antiestrogen tamoxifen is accompanied by an increase in serum triacylglycerol concentrations, soy-protein administration is associated with a decrease in serum triacylglycerol concentrations [133].

Other cardiac benefits of soy intake, independent of cholesterol reduction, have been identified and investigated. Clarkson (1994) [134] used monkeys with experimental atherosclerosis as a model to examine the effects of estrogen administration on vascular dilatation *in vivo*. Using a similar animal model they also showed that an isoflavone-rich soy-protein-isolate diet has a favorable effect on dilatation of coronary arteries similar to that of estrogen administration [62, 135].

The amount of soy protein that should be recommended for use to achieve "therapeutic effects" is unknown. Also, further research is required to determine the safety of isoflavones in pharmaceutical doses. Animal studies suggest that small amounts of isoflavones have favorable effects on lipoprotein oxidation and cholesterol reduction. Much more work is required to determine the minimum amount needed to have a specific beneficial health effect.

6. Conclusion

Epidemiological studies suggest that soybean consumption is associated, at least in part, with lower incidences of a number of chronic diseases, including type 2 diabetes, certain types of cancer and cardiovascular diseases, between others, mainly due antioxidant and antiplatelet activities.

Soybean derivatives can be denominated as functional ingredients as they contain bioactive compounds that inhibit platelet aggregation, which gives a preventive effect on thrombus formation. Also soybeans contain a variety of bioactive phytochemicals such as phenolic acids, flavonoids, isoflavones, saponins, phytosterols and sphingolipids; being the phenolic compounds with the highest antioxidant capacity. The key benefits of soy are related to their excellent protein content, its high content of essential fatty acids, numerous vitamins and minerals, their isoflavones and their higher fiber content. The experimental evidences related soy protein more than soy isoflavones as responsible by effects observed. At the present is not possible to discard another component present in soy as responsible by the effects.

Functional food may act as an adjunctive therapy/alternative treatment of different pathologies, and scientific studies are appearing more frequently demonstrating that this hypothesis is, indeed, a reality.

These evidences were considered by FDA, that published claims that recommended soy protein extract as an alternative to reduce blood cholesterol concentrations and prevent cancer, diabetes and increase protection cardiovascular.

Acknowledgements

The authors acknowledge the support by Centro de Estudios en Alimentos Procesados (CEAP), Conicyt-Regional, Gore Maule, R09I2001, Talca, Chile.

Author details

Eduardo Fuentes[1,3], Luis Guzmán[1], Gilda Carrasco[2,3], Elba Leiva[1],
Rodrigo Moore-Carrasco[1,3] and Iván Palomo[1,3*]

*Address all correspondence to: ipalomo@utalca.cl

1 Departamento de Bioquímica Clínica e Inmunología, Facultad de Ciencias de la Salud, Universidad de Talca, Chile

2 Departamento de Horticultura, Facultad de Ciencias Agrarias, Universidad de Talca, Chile

3 Centro de Estudios en Alimentos Procesados (CEAP), Conicyt-Regional, Gore Maule, R09I2001, Talca, Chile

References

[1] Forde, B., & Lorenzo, H. (2001). The nutritional control of root development. *Plant and Soil*, 232, 51-68.

[2] Liu, X., Jin, H., Wang, G., & Herbert, S. (2008). Soybean yield physiology and development of high-yielding practices in Northeast China. *Field Crops Research*, 105, 157-171.

[3] Setchell, K. D. (1998). Phytoestrogens the biochemistry, physiology, and implications for human health of soy isoflavones. *Am J Clin Nutr*, 68, 1333S -1346S .

[4] Heaney, R. P., Weaver, C. M., & Fitzsimmons, M. L. (1991). Soybean phytate content: effect on calcium absorption. *Am J Clin Nutr*, 53, 745-747.

[5] Liu, Z. S., & Chang, S. K. (2004). Effect of soy milk characteristics and cooking conditions on coagulant requirements for making filled tofu. *J Agric Food Chem*, 52, 3405-3411.

[6] Wood, J. E., Senthilmohan, S. T., & Peskin, A. V. (2002). Antioxidant activity of procyanidin-containing plant extracts at different pHs. *Food Chemistry*, 77, 155-161.

[7] Decker, E. A. (1995). The role of phenolics, conjugated linoleic acid, carnosine, and pyrroloquinoline quinone as nonessential dietary antioxidants. *Nutr Rev*, 53, 49-58.

[8] Xu, B., & Chang, S. K. (2008). Characterization of phenolic substances and antioxidant properties of food soybeans grown in the North Dakota-Minnesota region. *J Agric Food Chem*, 56, 9102-9113.

[9] Makela, S., Poutanen, M., Lehtimaki, J., Kostian, M. L., et al. (1995). Estrogen-specific 17 beta-hydroxysteroid oxidoreductase type 1 (E.C. 1.1.1.62) as a possible target for the action of phytoestrogens. *Proc Soc Exp Biol Med*, 208, 51-59.

[10] Yen, G. C., Chang, Y. C., & Su, S. W. (2003). Antioxidant activity and active compounds of rice koji fermented with Aspergillus candidus. *Food Chemistry*, 83, 49-54.

[11] Yang, J. H., Mau, J. L., Ko, P. T., & Huang, L. C. (2000). Antioxidant properties of fermented soybean broth. *Food Chemistry*, 71, 249-254.

[12] Siddhuraju, P., Mohan, P. S., & Becker, K. (2002). Studies on the antioxidant activity of Indian Laburnum (Cassia fistula L.): a preliminary assessment of crude extracts from stem bark, leaves, flowers and fruit pulp. *Food Chemistry*, 79, 61-67.

[13] Baunmann, L.S. (2009). Cosmetic Dermatology: Principles and Practice, Revised 2nd ed. *McGraw-Hill, Columbus, Ohio.*

[14] Ranich, T., Bhathena, S. J., & Velasquez, M. T. (2001). Protective effects of dietary phytoestrogens in chronic renal disease. *J Ren Nutr*, 11, 183-193.

[15] Axelson, M., Kirk, D. N., Farrant, R. D., Cooley, G., et al. (1982). The identification of the weak oestrogen equol [7-hydroxy-3-(4'-hydroxyphenyl)chroman] in human urine. *Biochem J*, 201, 353-357.

[16] Setchell, K. D., Brown, N. M., Zimmer-Nechemias, L., Brashear, W. T., et al. (2002). Evidence for lack of absorption of soy isoflavone glycosides in humans, supporting the crucial role of intestinal metabolism for bioavailability. *Am J Clin Nutr*, 76, 447-453.

[17] Setchell, K. D., Borriello, S. P., Hulme, P., Kirk, D. N., & Axelson, M. (1984). Nonsteroidal estrogens of dietary origin: possible roles in hormone-dependent disease. *Am J Clin Nutr*, 40, 569-578.

[18] Kelly, G. E., Nelson, C., Waring, M. A., Joannou, G. E., & Reeder, A. Y. (1993). Metabolites of dietary (soya) isoflavones in human urine. *Clin Chim Acta*, 223, 9-22.

[19] Heinonen, S., Wahala, K., & Adlercreutz, H. (1999). Identification of isoflavone metabolites dihydrodaidzein, dihydrogenistein, 6'-OH-O-dma, and cis-4-OH-equol in

human urine by gas chromatography-mass spectroscopy using authentic reference compounds. *Anal Biochem*, 274, 211-219.

[20] Atkinson, C., Berman, S., Humbert, O., & Lampe, J. W. (2004). In vitro incubation of human feces with daidzein and antibiotics suggests interindividual differences in the bacteria responsible for equol production. *J Nutr*, 134, 596-599.

[21] Setchell, K. D., Brown, N. M., & Lydeking-Olsen, E. (2002). The clinical importance of the metabolite equol-a clue to the effectiveness of soy and its isoflavones. *J Nutr*, 132, 3577-3584.

[22] Setchell, K. D., Zimmer-Nechemias, L., Cai, J., & Heubi, J. E. (1998). Isoflavone content of infant formulas and the metabolic fate of these phytoestrogens in early life. *Am J Clin Nutr*, 68, 1453S -1461S .

[23] Setchell, K. D., Zimmer-Nechemias, L., Cai, J., & Heubi, J. E. (1997). Exposure of infants to phyto-oestrogens from soy-based infant formula. *Lancet*, 350, 23-27.

[24] Lampe, J. W., Karr, S. C., Hutchins, A. M., & Slavin, J. L. (1998). Urinary equol excretion with a soy challenge: influence of habitual diet. *Proc Soc Exp Biol Med*, 217, 335-339.

[25] Rowland, I. R., Wiseman, H., Sanders, T. A., Adlercreutz, H., & Bowey, E. A. (2000). Interindividual variation in metabolism of soy isoflavones and lignans: influence of habitual diet on equol production by the gut microflora. *Nutr Cancer*, 36, 27-32.

[26] Setchell, K. D, Clerici, C, Lephart, E. D., Cole, S. J., et al. (2005). S-equol, a potent ligand for estrogen receptor beta, is the exclusive enantiomeric form of the soy isoflavone metabolite produced by human intestinal bacterial flora. *Am J Clin Nutr*, 81, 1072-1079.

[27] Gopaul, R., Knaggs, H. E., & Lephart, E. D. (2012). Biochemical investigation and gene analysis of equol: a plant and soy-derived isoflavonoid with antiaging and antioxidant properties with potential human skin applications. *Biofactors*, 38, 44-52.

[28] Kuiper, G. G., Lemmen, J. G., Carlsson, B., Corton, J. C., et al. (1998). Interaction of estrogenic chemicals and phytoestrogens with estrogen receptor beta. *Endocrinology*, 139, 4252-4263.

[29] Kuiper, G. G., Carlsson, B., Grandien, K., Enmark, E., et al. (1997). Comparison of the ligand binding specificity and transcript tissue distribution of estrogen receptors alpha and beta. *Endocrinology*, 138, 863-870.

[30] Thornton, M. J., Taylor, A. H., Mulligan, K., Al-Azzawi, F., et al. (2003). The distribution of estrogen receptor beta is distinct to that of estrogen receptor alpha and the androgen receptor in human skin and the pilosebaceous unit. J. *Investig Dermatol Symp Proc*, 8, 100-103.

[31] Lund, T. D., Munson, D. J., Haldy, M. E., Setchell, K. D., et al. (2004). Equol is a novel anti-androgen that inhibits prostate growth and hormone feedback. *Biol Reprod*, 70, 1188-1195.

[32] Makrantonaki, E., & Zouboulis, C. C. (2009). Androgens and ageing of the skin. *Curr Opin Endocrinol Diabetes Obes*, 16, 240-245.

[33] Nitsch, S. M., Wittmann, F., Angele, P., Wichmann, M. W., et al. (2004). Physiological levels of 5 alpha-dihydrotestosterone depress wound immune function and impair wound healing following trauma-hemorrhage. *Arch Surg*, 139, 157-163.

[34] Stevenson, S., & Thornton, J. (2007). Effect of estrogens on skin aging and the potential role of SERMs. *Clin Interv Aging*, 2, 283-297.

[35] Verdier-Sevrain, S., Bonte, F., & Gilchrest, B. (2006). Biology of estrogens in skin: implications for skin aging. *Exp Dermatol*, 15, 83-94.

[36] Munoz, Y., Garrido, A., & Valladares, L. (2009). Equol is more active than soy isoflavone itself to compete for binding to thromboxane A(2) receptor in human platelets. *Thromb Res*, 123, 740-744.

[37] Ronis, M. J., Chen, Y., Badeaux, J., & Badger, T. M. (2009). Dietary soy protein isolate attenuates metabolic syndrome in rats via effects on PPAR, LXR, and SREBP signaling. *J Nutr*, 139, 1431-1438.

[38] Harborne, J. B, & Williams, C. A. (2000). Advances in flavonoid research since 1992. *Phytochemistry*, 55, 481-504.

[39] Paik, S. S., Jeong, E., Jung, S. W., Ha, T. J., et al. (2012). Anthocyanins from the seed coat of black soybean reduce retinal degeneration induced by N-methyl-N-nitrosourea. *Exp Eye Res*, 97, 55-62.

[40] Gottstein, N., Ewins, B. A., Eccleston, C., Hubbard, G. P., et al. (2003). Effect of genistein and daidzein on platelet aggregation and monocyte and endothelial function. *Br J Nutr*, 89, 607-616.

[41] Yin, X. Z., Quan, J. S., Takemichi, K., & Makoto, T. (2004). Anti-atherosclerotic effect of soybean isofalvones and soyasaponins in diabetic rats. *Zhonghua Yu Fang Yi Xue Za Zhi*, 38, 26-28.

[42] Kim, K., Lim, K. M., Kim, C. W., Shin, H. J., et al. (2011). Black soybean extract can attenuate thrombosis through inhibition of collagen-induced platelet activation. *J Extract Black*, 22, 964-970.

[43] Mc Nicol, A. (1993). The effects of genistein on platelet function are due to thromboxane receptor antagonism rather than inhibition of tyrosine kinase. *ProstaglandinsLeukot Essent Fatty Acids*, 48, 379-384.

[44] Kondo, K., Suzuki, Y., Ikeda, Y., & Umemura, K. (2002). Genistein, an isoflavone included in soy, inhibits thrombotic vessel occlusion in the mouse femoral artery and in vitro platelet aggregation. *Eur J Pharmacol*, 455, 53-57.

[45] Atsushi, O. (2004). Anti-platelets Effects of Genistein, an Isoflavonoid from Soybean. *Soy Protein Research*, 7, 145-148.

[46] Tsuchiya, H., Sato, M., & Watanabe, I. (1999). Antiplatelet activity of soy sauce as functional seasoning. *J Agric Food Chem*, 47, 4167-4174.

[47] Kitagawa, S., Orinaka, M., & Hirata, H. (1993). Depth-dependent change in membrane fluidity by phenolic compounds in bovine platelets and its relationship with their effects on aggregation and adenylate cyclase activity. *Biochim Biophys Acta*, 1179, 277-282.

[48] Peura, P., Mackenzie, P., Koivusaari, U., & Lang, M. (1982). Increased fluidity of a model membrane caused by tetrahydro-beta-carbolines. *Mol Pharmacol*, 22, 721-724.

[49] Silva, M., Santana, L., Silva-Lucca, R., Lima, A., et al. (2011). Immobilized Cratylia mollis lectin: An affinity matrix to purify a soybean (Glycine max) seed protein with in vitro platelet antiaggregation and anticoagulant activities. *Process Process*, 46, 74-80.

[50] Lee, K., & Kim, S. (2005). SSGE and DEE, new peptides isolated from a soy protein hydrolysate that inhibit platelet aggregation. *Food Chemistry*, 90, 389-393.

[51] Tsuchiya, H., Sato, M., & Watanabe, I. (1999). Antiplatelet activity of soy sauce as functional seasoning. *J Agric Food Chem*, 47, 4167-4174.

[52] Setchell, K. D., & Cassidy, A. (1999). Dietary isoflavones: biological effects and relevance to human health. *J Nutr*, 129, 758S -767S .

[53] Bennett, J. O., Yu, O., Heatherly, L. G., & Krishnan, H. B. (2004). Accumulation of genistein and daidzein, soybean isoflavones implicated in promoting human health, is significantly elevated by irrigation. *J Agric Food Chem*, 52, 7574-7579.

[54] Okamoto, K., & Horisawa, R. (2006). Soy products and risk of an aneurysmal rupture subarachnoid hemorrhage in Japan. *Eur J Cardiovasc Prev Rehabil*, 13, 284-287.

[55] Zhang, X., Shu, X. O., Gao, Y. T., Yang, G., et al. (2003). Soy food consumption is associated with lower risk of coronary heart disease in Chinese women. *J Nutr*, 133, 2874-2878.

[56] Potter, S. M., Bakhit, R. M., Essex-Sorlie, D. L., Weingartner, K. E., et al. (1993). Depression of plasma cholesterol in men by consumption of baked products containing soy protein. *Am J Clin Nutr*, 58, 501-506.

[57] Bakhit, R. M., Klein, B. P., Essex-Sorlie, D., Ham, J. O., et al. (1994). Intake of 25 g of soybean protein with or without soybean fiber alters plasma lipids in men with elevated cholesterol concentrations. *J Nutr*, 124, 213-222.

[58] Erdman, J. W., Jr. (2000). AHA Science Advisory: Soy protein and cardiovascular disease: A statement for healthcare professionals from the Nutrition Committee of the AHA. *Circulation*, 102, 2555-2559.

[59] Mullen, E., Brown, R. M., Osborne, T. F., & Shay, N. F. (2004). Soy isoflavones affect sterol regulatory element binding proteins (SREBPs) and SREBP-regulated genes in HepG2 cells. *J Nutr*, 134, 2942-2947.

[60] Adams, M. R., Golden, D. L., Anthony, M. S., Register, T. C., & Williams, J. K. (2002). The inhibitory effect of soy protein isolate on atherosclerosis in mice does not require the presence of LDL receptors or alteration of plasma lipoproteins. *J Nutr*, 132, 43-49.

[61] Adams, M. R., Register, T. C., Golden, D. L., Anthony, M. S., et al. (2002). The athero-protective effect of dietary soy isoflavones in apolipoprotein E-/- mice requires the presence of estrogen receptor-alpha. *Arterioscler Thromb Vasc Biol*, 22, 1859-1864.

[62] Anderson, J. W., Smith, B. M., & Washnock, C. S. (1999). Cardiovascular and renal benefits of dry bean and soybean intake. *Am J Clin Nutr*, 70, 464S -474S .

[63] Iritani, N., Sugimoto, T., Fukuda, H., Komiya, M., & Ikeda, H. (1997). Dietary soy-bean protein increases insulin receptor gene expression in Wistar fatty rats when di-etary polyunsaturated fatty acid level is low. *J Nutr*, 127, 1077-1083.

[64] Zhan, S., & Ho, S. C. (2005). Meta-analysis of the effects of soy protein containing iso-flavones on the lipid profile. *Am J Clin Nutr*, 81, 397-408.

[65] Lusas, E. W., & Riaz, M. N. (1995). Soy protein products: processing and use. J Nutr ., 125, 573S-580S.

[66] Messina, M. J. (1999). Legumes and soybeans: overview of their nutritional profiles and health effects. *Am J Clin Nutr*, 70, 439S -450S .

[67] Kataoka, S. (2005). Functional effects of Japanese style fermented soy sauce (shoyu) and its components. *J Biosci Bioeng*, 100, 227-234.

[68] Kang, J., Badger, T. M., Ronis, M. J., & Wu, X. Non-isoflavone phytochemicals in soy and their health effects. *J Agric Food Chem*, 58, 8119-8133.

[69] Gil-Izquierdo, A., Penalvo, J. L., Gil, J. I., Medina, S., et al. Soy isoflavones and cardi-ovascular disease epidemiological, clinical and-omics perspectives. *Curr Pharm Bio-technol*, 13, 624-631.

[70] Wei, X., Wu, J., Ni, Y., Lu, L., & Zhao, R. (2011). Antioxidant effect of a phytoestrogen equol on cultured muscle cells of embryonic broilers. *In Vitro Cell Dev Biol Anim*, 47, 735-741.

[71] Choi, E., & Kim, G. (2011). Anticancer mechanism of equol in dimethylbenz(a)anthra-cene-treated animals. 7 12 . *Int J Oncol*, 39, 747-754.

[72] Palermo, M., Paradiso, R., De Pascale, S., & Fogliano, V. Hydroponic cultivation im-proves the nutritional quality of soybean and its products. *J Agric Food Chem*, 60, 250-255.

[73] Hendrich, S., & Fisher, K. (2001). What do we need to know about active ingredients in dietary supplements? Summary of workshop discussion. *J Nutr*, 131, 1387S -8S .

[74] Setchell, K. D., Brown, N. M., Desai, P. B., Zimmer-Nechimias, L., et al. (2003). Bioavailability, disposition, and dose-response effects of soy isoflavones when consumed by healthy women at physiologically typical dietary intakes. *J Nutr*, 133, 1027-1035.

[75] Larkin, T. A., Astheimer, L. B., & Price, W. E. (2009). Dietary combination of soy with a probiotic or prebiotic food significantly reduces total and LDL cholesterol in mildly hypercholesterolaemic subjects. *Eur J Clin Nutr*, 63, 238-245.

[76] Rowland, I., Faughnan, M., Hoey, L., Wahala, K., et al. (2003). Bioavailability of phyto-oestrogens. *Br J Nutr, Suppl 1, S*, 89, 45-58.

[77] Izumi, T., Piskula, M. K., Osawa, S., Obata, A., et al. (2000). Soy isoflavone aglycones are absorbed faster and in higher amounts than their glucosides in humans. *J Nutr*, 130, 1695-1699.

[78] Wiseman, H., Casey, K., Bowey, E. A., Duffy, R., et al. (2004). Influence of 10 wk of soy consumption on plasma concentrations and excretion of isoflavonoids and on gut microflora metabolism in healthy adults. *Am J Clin Nutr*, 80, 692-699.

[79] Xu, X., Wang, H. J., Murphy, P. A., Cook, L., & Hendrich, S. (1994). Daidzein is a more bioavailable soymilk isoflavone than is genistein in adult women. *J Nutr*, 124, 825-832.

[80] Andlauer, W., Kolb, J., & Furst, P. (2000). Isoflavones from tofu are absorbed and metabolized in the isolated rat small intestine. *J Nutr*, 130, 3021-3027.

[81] King, R. A. (1998). Daidzein conjugates are more bioavailable than genistein conjugates in rats. *Am J Clin Nutr*, 68, 1496S -1499S .

[82] Setchell, K. D., Brown, N. M., Desai, P., Zimmer-Nechemias, L., et al. (2001). Bioavailability of pure isoflavones in healthy humans and analysis of commercial soy isoflavone supplements. *J Nutr*, 131, 1362S -75S .

[83] Piazza, C., Privitera, M. G., Melilli, B., Incognito, T., et al. (2007). Influence of inulin on plasma isoflavone concentrations in healthy postmenopausal women. *Am J Clin Nutr*, 86, 775-780.

[84] Benvenuti, C., & Setnikar, I. (2011). Effect of Lactobacillus sporogenes on oral isoflavones bioavailability: single dose pharmacokinetic study in menopausal women. *Arzneimittelforschung*, 61, 605-609.

[85] Miguez, A. C., Francisco, J. C., Barberato, S. H., Simeoni, R., et al. (2012). The functional effect of soybean extract and isolated isoflavone on myocardial infarction and ventricular dysfunction: The soybean extract on myocardial infarction. *J Nutr Biochem*.

[86] Palomo, I., Guzmán, L., Leiva, E., Mujica, V., et al. (2011). Soybean and Health. *Hany El-Shemy: InTech*.

[87] Ruiz-Alcaraz, A. J., Liu, H. K., Cuthbertson, D. J., Mc Manus, E. J., et al. (2005). A novel regulation of IRS1 (insulin receptor substrate-1) expression following short term insulin administration. *Biochem J*, 392, 345-352.

[88] Bjornholm, M., & Zierath, J. R. (2005). Insulin signal transduction in human skeletal muscle: identifying the defects in Type II diabetes. *Biochem Soc Trans*, 33, 354-357.

[89] Lu, J., Zeng, Y., Hou, W., Zhang, S., et al. (2012). The soybean peptide aglycin regulates glucose homeostasis in type 2 diabetic mice via IR/IRS1 pathway. *J Nutr Biochem*.

[90] Mizokami-Stout, K., Cree-Green, M., & Nadeau, K. J. (2012). Insulin resistance in type 2 diabetic youth. *Curr Opin Endocrinol Diabetes Obes*, 19, 255-262.

[91] Everhart, J. E., Pettitt, D. J., Bennett, P. H., & Knowler, W. C. (1992). Duration of obesity increases the incidence of NIDDM. *Diabetes*, 41, 235-240.

[92] Liener, I. E. (1994). Implications of antinutritional components in soybean foods. *Crit Rev Food Sci Nutr*, 34, 31-67.

[93] Ishihara, K., Oyaizu, S., Fukuchi, Y., Mizunoya, W., et al. (2003). A soybean peptide isolate diet promotes postprandial carbohydrate oxidation and energy expenditure in type II diabetic mice. *J Nutr*, 133, 752-757.

[94] Davis, J., Higginbotham, A., O'Connor, T., Moustaid-Moussa, N., et al. (2007). Soy protein and isoflavones influence adiposity and development of metabolic syndrome in the obese male ZDF rat. *Ann Nutr Metab*, 51, 42-52.

[95] Dun, X. P., Wang, J. H., Chen, L., Lu, J., et al. (2007). Activity of the plant peptide aglycin in mammalian systems. *FEBS J*, 274, 751-759.

[96] Lu, M. P., Wang, R., Song, X., Chibbar, R., et al. (2008). Dietary soy isoflavones increase insulin secretion and prevent the development of diabetic cataracts in streptozotocin-induced diabetic rats. *Nutr Res*, 28, 464-471.

[97] Saito, N., Sakai, H., Suzuki, S., Sekihara, H., & Yajima, Y. (1998). Effect of an alpha-glucosidase inhibitor (voglibose), in combination with sulphonylureas, on glycaemic control in type 2 diabetes patients. *J Int Med Res*, 26, 219-232.

[98] Kwon, Y. I., Vattem, D. A., & Shetty, K. (2006). Evaluation of clonal herbs of Lamiaceae species for management of diabetes and hypertension. *Asia Pac J Clin Nutr*, 15, 107-118.

[99] Ademiluyi, A. O., & Oboh, G. (2011). Soybean phenolic-rich extracts inhibit key-enzymes linked to type 2 diabetes (alpha-amylase and alpha-glucosidase) and hypertension (angiotensin I converting enzyme) in vitro. *Exp Toxicol Pathol*.

[100] Burow, M. E., Boue, S. M., Collins-Burow, B. M., Melnik, L. I., et al. (2001). Phytochemical glyceollins, isolated from soy, mediate antihormonal effects through estrogen receptor alpha and beta. *J Clin Endocrinol Metab*, 86, 1750-1758.

[101] Park, S., Kim, da. S., Kim, J. H., Kim, J. S., & Kim, H. J. (2012). Glyceollin-containing fermented soybeans improve glucose homeostasis in diabetic mice. *Nutrition*, 28, 204-211.

[102] Vedavanam, K., Srijayanta, S., O'Reilly, J., Raman, A., & Wiseman, H. (1999). Antioxidant action and potential antidiabetic properties of an isoflavonoid-containing soyabean phytochemical extract (SPE). *Phytother Res*, 13, 601-608.

[103] Ascencio, C., Torres, N., Isoard-Acosta, F., Gomez-Perez, F. J., et al. (2004). Soy protein affects serum insulin and hepatic SREBP-1 mRNA and reduces fatty liver in rats. *J Nutr*, 134, 522-529.

[104] Villegas, R., Gao, Y. T., Yang, G., Li, H. L., et al. (2008). Legume and soy food intake and the incidence of type 2 diabetes in the Shanghai Women's Health Study. *Am J Clin Nutr*, 87, 162-167.

[105] Nanri, A., Mizoue, T., Takahashi, Y., Kirii, K., et al. (2010). Soy product and isoflavone intakes are associated with a lower risk of type 2 diabetes in overweight Japanese women. *J Nutr*, 140, 580-586.

[106] Aubertin-Leheudre, M., Lord, C., Khalil, A., & Dionne, I. J. (2008). Isoflavones and clinical cardiovascular risk factors in obese postmenopausal women: a randomized double-blind placebo-controlled trial. *J Womens Health (Larchmt)*, 17, 1363-1369.

[107] Azadbakht, L., Atabak, S., & Esmaillzadeh, A. (2008). Soy protein intake, cardiorenal indices, and C-reactive protein in type 2 diabetes with nephropathy a longitudinal randomized clinical trial. *Diabetes Care*, 31, 648-654.

[108] Azadbakht, L., Kimiagar, M., Mehrabi, Y., Esmaillzadeh, A., et al. (2007). Soy inclusion in the diet improves features of the metabolic syndrome: a randomized crossover study in postmenopausal women. *Am J Clin Nutr*, 85, 735-741.

[109] Blakesmith, S. J., Lyons-Wall, P. M., George, C., Joannou, G. E., et al. (2003). Effects of supplementation with purified red clover (Trifolium pratense) isoflavones on plasma lipids and insulin resistance in healthy premenopausal women. *Br J Nutr*, 89, 467-474.

[110] Gonzalez, S., Jayagopal, V., Kilpatrick, E. S., Chapman, T., & Atkin, S. L. (2007). Effects of isoflavone dietary supplementation on cardiovascular risk factors in type 2 diabetes. *Diabetes Care*, 30, 1871-1873.

[111] Liao, F. H., Shieh, M. J., Yang, S. C., Lin, S. H., & Chien, Y. W. (2007). Effectiveness of a soy-based compared with a traditional low-calorie diet on weight loss and lipid levels in overweight adults. *Nutrition*, 23, 551-556.

[112] Erdman, J. W., Jr , , Badger, T. M., Lampe, J. W., Setchell, K. D., & Messina, M. (2004). Not all soy products are created equal: caution needed in interpretation of research results. *J Nutr*, 134, 1229S -1233S .

[113] Liu, Z. M., Chen, Y. M., & Ho, S. C. (2011). Effects of soy intake on glycemic control: a meta-analysis of randomized controlled trials. *Am J Clin Nutr*, 93, 1092-1101.

[114] Liener, I. E., & Seto, T. A. (1955). Nonspecific effect of soybean hemagglutinin on tumor growth. *Cancer research*, 15, 407-409.

[115] Khan, S. A., Chatterton, R. T., Michel, N., Bryk, M., et al. (2012). Soy isoflavone supplementation for breast cancer risk reduction: a randomized phase II trial. *Cancer Prev Res (Phila)*, 5, 309-319.

[116] Teas, J., Irhimeh, M. R., Druker, S., Hurley, T. G., et al. (2011). Serum IGF-1 concentrations change with soy and seaweed supplements in healthy postmenopausal American women. *Nutr Cancer*, 63, 743-748.

[117] Yang, W. S., Va, P., Wong, M. Y., Zhang, H. L., & Xiang, Y. B. (2011). Soy intake is associated with lower lung cancer risk: results from a meta-analysis of epidemiologic studies. *Am J Clin Nutr*, 94, 1575-1583.

[118] Hillman, G. G., Singh-Gupta, V., Runyan, L., Yunker, C. K., et al. (2011). Soy isoflavones radiosensitize lung cancer while mitigating normal tissue injury. *Radiotherapy and oncology journal of the European Society for Therapeutic Radiology and Oncology*, 101, 329-336.

[119] Zaineddin, A. K., Buck, K., Vrieling, A., Heinz, J., et al. (2012). The association between dietary lignans, phytoestrogen-rich foods, and fiber intake and postmenopausal breast cancer risk: a german case-control study. *Nutr Cancer*, 64, 652-665.

[120] Li, Y. S., Wu, L. P., Li, K. H., Liu, Y. P., et al. (2011). Involvement of nuclear factor kappaB (NF-kappaB) in the downregulation of cyclooxygenase-2 (COX-2) by genistein in gastric cancer cells. *The Journal of international medical research*, 39, 2141-2150.

[121] Pan, H., Zhou, W., He, W., Liu, X., et al. (2012). Genistein inhibits MDA-MB-231 triple-negative breast cancer cell growth by inhibiting NF-kappaB activity via the Notch-1 pathway. *International journal of molecular medicine*, 30, 337-343.

[122] Wu, T. C., Yang, Y. C., Huang, P. R., Wen, Y. D., & Yeh, S. L. (2012). Genistein enhances the effect of trichostatin A on inhibition of A549 cell growth by increasing expression of TNF receptor-1. *Toxicol Appl Pharmacol*, 262, 247-254.

[123] Zhang, Y., Li, Q., Zhou, D., & Chen, H. (2012). Genistein, a soya isoflavone, prevents azoxymethane-induced up-regulation of WNT/beta-caten in signalling and reduces colon pre-neoplasia in rats. *Br J Nutr*, 1-10.

[124] Ghaemi, A., Soleimanjahi, H., Razeghi, S., Gorji, A., et al. (2012). Genistein induces a protective immunomodulatory effect in a mouse model of cervical cancer. *Iranian journal of immunology : IJI*, 9, 119-127.

[125] Anderson, J. W., Johnstone, B. M., & Cook-Newell, M. E. (1995). Meta-analysis of the effects of soy protein intake on serum lipids. *N Engl J Med*, 333, 276-282.

[126] Carroll, K. K. (1991). Review of clinical studies on cholesterol-lowering response to soy protein. *J Am Diet Assoc*, 91, 820-827.

[127] Wangen, K. E., Duncan, A. M., Xu, X., & Kurzer, M. S. (2001). Soy isoflavones improve plasma lipids in normocholesterolemic and mildly hypercholesterolemic postmenopausal women. *Am J Clin Nutr*, 73, 225-231.

[128] Kim, J. I., Kim, J. C., Kang, M. J., Lee, M. S., et al. (2005). Effects of pinitol isolated from soybeans on glycaemic control and cardiovascular risk factors in Korean patients with type II diabetes mellitus: a randomized controlled study. *Eur J Clin Nutr*, 59, 456-458.

[129] Crisafulli, A., Altavilla, D., Marini, H., Bitto, A., et al. (2005). Effects of the phytoestrogen genistein on cardiovascular risk factors in postmenopausal women. *Menopause*, 12, 186-192.

[130] Hoie, L. H., Graubaum, H. J., Harde, A., Gruenwald, J., & Wernecke, K. D. (2005). Lipid-lowering effect of 2 dosages of a soy protein supplement in hypercholesterolemia. *Adv Ther*, 22, 175-186.

[131] Hermansen, K., Hansen, B., Jacobsen, R., Clausen, P., et al. (2005). Effects of soy supplementation on blood lipids and arterial function in hypercholesterolaemic subjects. *Eur J Clin Nutr*, 59, 843-850.

[132] Jenkins, D. J., Kendall, C. W., Jackson, C. J., Connelly, P. W., et al. (2002). Effects of high- and low-isoflavone soyfoods on blood lipids, oxidized LDL, homocysteine, and blood pressure in hyperlipidemic men and women. *Am J Clin Nutr*, 76, 365-372.

[133] Anthony, M. S., Clarkson, T. B., Hughes, C. L. Jr, Morgan, T. M., & Burke, G. L. (1996). Soybean isoflavones improve cardiovascular risk factors without affecting the reproductive system of peripubertal rhesus monkeys. *J Nutr*, 126, 43-50.

[134] Clarkson, T. B., Anthony, M. S., & Klein, K. P. (1994). Effects of estrogen treatment on arterial wall structure and function. *Drugs*, 2, 42-51.

[135] Williams, J. K., & Clarkson, T. B. (1998). Dietary soy isoflavones inhibit in-vivo constrictor responses of coronary arteries to collagen-induced platelet activation. *Coron Artery Dis*, 9, 759-764.